ADVANCES IN ENTOMOPATHOGENIC NEMATODE TAXONOMY AND PHYLOGENY

Cover illustration (clockwise from top left): *Heterorhabditis georgiana* female head region (after Nguyen *et al.*, 2008); *Steinernema kraussei* female (after Steiner, 1923); DNA alignment (overlay; courtesy Nguyen & Mengistu); *Heterorhabditis safricana* infective juvenile anterior region (after Malan *et al.*, 2008); *Steinernema riobrave* infective juvenile anterior region (after Nguyen & Adams, 2003); DNA chromatogram (overlay; courtesy Nguyen & Mengistu).

ADVANCES IN ENTOMOPATHOGENIC NEMATODE TAXONOMY AND PHYLOGENY

Edited by
David J. Hunt and Khuong B. Nguyen

David J. Hunt and Roland N. Perry (Series Editors)

NEMATOLOGY MONOGRAPHS AND PERSPECTIVES
VOLUME 12

BRILL
LEIDEN-BOSTON
2016

This book is printed on acid-free paper.

Library of Congress Cataloging-in-Publication Data

Names: Hunt, D. J. (David J.), editor. | Nguyen, Khuong B., editor.
Title: Advances in entomopathogenic nematode taxonomy and phylogeny / edited by David J. Hunt and Khuong B. Nguyen.
Description: Boston : Brill, 2016. | Series: Nematology monographs and perspectives ; volume 12 | Includes bibliographical references and index.
Identifiers: LCCN 2016043376 (print) | LCCN 2016044571 (ebook) | ISBN 9789004285330 (hardback : alk. paper) | ISBN 9789004285347 (E-book)
Subjects: LCSH: Insect nematodes–Classification. | Insect nematodes–Phylogeny.
Classification: LCC QL391.N4 A28 2016 (print) | LCC QL391.N4 (ebook) | DDC 592/.57–dc23
LC record available at https://lccn.loc.gov/2016043376

ISBN: 978 90 04 28533 0
E-ISBN: 978 90 04 28534 7

Contents

Contributors

David J. HUNT
CABI Europe-UK
Bakeham Lane
Egham, Surrey, TW20 9TY, UK

Khuong B. NGUYEN
Entomology and Nematology Department, Institute of Food and Agricultural Sciences
University of Florida
Gainesville, FL 32611-0620, USA

Sergei E. SPIRIDONOV
Center of Parasitology
A.N. Severtsov Institute of Ecology and Evolution of the Russian Academy of Sciences
Leninskii prospect 33
Moscow 119071, Russia

Sergei A. SUBBOTIN
Plant Pest Diagnostic Center
California Department of Food and Agriculture
3294 Meadowview Road
Sacramento, CA 95832, USA
and
Center of Parasitology
A.N. Severtsov Institute of Ecology and Evolution of the Russian Academy of Sciences
Leninskii prospect 33
Moscow 119071, Russia

Nematology Monographs & Perspectives, 2016, Vol. 12, xi-xiii

Foreword

A dramatic increase in research on entomopathogenic nematodes occurred in the 1990s and has not diminished. More than 125 publications appeared annually from 1990 to 1998, compared to an average of about 30 per year from 1973 to 1979 (Hominick, 2002). As David Hunt details in Chapter 1, the number of new species described began to increase around 1990 and has continued at a frenetic pace. In 1996, a workshop was convened, with a number of leading taxonomists attempting to establish robust criteria for describing new species (Hominick *et al.*, 1997). Over the next 10 years, the numbers continued to increase and inspired Khuong Nguyen and David Hunt to publish their admirably comprehensive book, documenting 55 valid species of *Steinernema* and 11 of *Heterorhabditis* (Nguyen & Hunt, 2007). However, the torrent of literature continued and the book rapidly became out of date. In late 2014, Hunt and Nguyen wisely and practically decided to compile the present volume to cover just those species described since 2007. It is astonishing that after only a decade, 95 steinernematid and 16 heterorhabditid species are now accepted as valid. Several more are known to be in press (Chapter 1), so more surveys will undoubtedly lead to new discoveries. Nevertheless, we are all indebted to the editors/contributors for applying their extensive knowledge and forensic skills to produce this book and thus provide a solid taxonomic base for future work on these fascinating nematodes. A name is the key to knowledge – if there is no name, there is no information, and if the wrong name is used, wrong information is conveyed. Now, the right name can be applied to the right nematode, and all studies relating to that species can be validated.

This monograph is important not only for documenting new species, but also for assessing all the described species to date and deciding which are no longer valid. There are many more named species, but the authors were able, with their current understanding of specific characters, to reduce them to invalid species. This will certainly help to reduce errors and prevent inappropriate conclusions in future studies.

Molecular methodologies produced a powerful discriminatory tool that has supported the remarkable increase in the proposal of new species, particularly of *Steinernema*. Eighty-five of the 95 valid steinernematid species have been described since 1990 (Chapter 1), and most could not have been sustained by reliance on traditional descriptions. Similarly, since Nguyen & Hunt's book (2007), eight species of *Heterorhabditis* have been proposed, all based on molecular data. However, three of these are now synonymised with existing species (Chapter 2). This illustrates that molecular methodologies are not infallible, and depend on good practice with interpretation/alignment of the molecular sequences, essential for establishing species status. Nevertheless, molecular sequencing now forms a cornerstone for species characterisation within these two families and no description of potentially new taxa should be considered unless supported by good quality molecular data.

This book was not intended to replace the 2007 volume, and should be envisaged as a companion volume. Together, they provide a unique and essential reference for the Steinernematidae and Heterorhabditidae. It is truly admirable that Hunt and Nguyen undertook and completed two books of such complexity, requiring a forensic analysis of masses of data. Few, if any, other workers in the field have the authority, expertise and patience required. Their timely works bring order to a field that was becoming increasingly chaotic and confusing. Thanks to these books, more material will be reliably identified for future comparative studies, and so prevent inappropriate lines of enquiry.

With so many new species now described, our understanding of entomopathogenic nematodes is just beginning. We need to know about the biology of each of these fascinating new nematodes, from their bacterial symbionts to host and habitat preferences, population biology, suitability for mass production methods and potential for biological control. Perhaps some of the new species, adapted to particular habitats, have the potential to perform better than the few species currently used in biological control programmes. Meanwhile, Drs Hunt and Nguyen deserve our gratitude for providing the foundations for all future work.

William M. Hominick
Emeritus Research Fellow and former Director
Biodiversity & Biosystematics, CABI Bioscience, CABI Europe-UK
Bakeham Lane, Egham, Surrey, TW20 9TY, UK

References

HOMINICK, W.M. (2002). Biogeography. In: Gaugler, R. (Ed.). *Entomopathogenic nematodes*. Wallingford, UK, CAB International.

HOMINICK, W.M., BRISCOE, B.R., DEL PINO, F.G., HENG, J., HUNT, D.J., KOZODOY, E., MRÁČEK, Z., NGUYEN, K.B., REID, A.P., SPIRIDONOV, S.E. *ET AL.* (1997). Biosystematics of entomopathogenic nematodes: current status, protocols and definitions. *Journal of Helminthology* 71, 271-298.

NGUYEN, K.B. & HUNT, D.J. (Eds) (2007). *Entomopathogenic nematodes: taxonomy, phylogeny and bacterial symbionts. Nematology Monographs and Perspectives 5* (Series editors: Hunt, D.J. & Perry, R.N.). Leiden, The Netherlands, Brill.

Acknowledgements

We thank all those authors and contributors who made this book possible and the journal editors and publishers for the re-use of illustrations and photographs from published papers. We are also grateful to Vladimir Půža for his suggestions and opinions concerning the status of a number of *Steinernema* species. DJH would also like to thank the Editorial Cat, particularly for proffering advice on mouse-related matters, acting as a dedicated screen saver and for giving frequent paws for thought.

David J. HUNT
Khuong B. NGUYEN
Sergei E. SPIRIDONOV
Sergei A. SUBBOTIN

Nematology Monographs & Perspectives, 2016, Vol. 12, 1-11

Chapter 1

Introduction

David J. HUNT

CABI Europe-UK, Bakeham Lane, Egham, Surrey, TW20 9TY, UK

The previous version of this book was published in 2007 as Volume 5 in the *Nematology Monographs and Perspectives* series (Nguyen & Hunt, 2007a). At the time it was comprehensive in all major respects, but since then has rapidly become out of date, the descriptions of new species, particularly of *Steinernema* Travassos, 1927, continuing at a frenetic pace. In late 2014 we decided to compile an updated volume to cover just those species described since 2007. Due to restrictions on the acceptable length of the book, we decided not to update the symbiotic bacteria chapter so ably written by Heather Koppenhöfer (2007), but to have instead a chapter encompassing phylogeny and phylogeography of *Steinernema* and *Heterorhabditis* Poinar, 1976 (see Chapter 6). We also decided to reduce the length of the species descriptions to avoid much of the repetition of characters that vary little from one species to another, focusing instead on those features carrying more diagnostic weight. The tabular keys to species (see Chapter 3) have been updated and include all species listed in Chapters 4 and 5. For basic techniques the reader is referred to the appropriate sections in the original chapter by Nguyen (2007).

It should be borne in mind that the present book, which is the twelfth in the *Nematology Monographs and Perspectives* series, is not intended to – and indeed does not – replace the previous oeuvre, and should be envisaged as a companion volume, the two together providing a unique (hopefully indispensable!) coverage of entomopathogenic nematodes of the Steinernematidae and Heterorhabditidae.

Perspective

A detailed historical account of entomoparasitic nematology was given by Hunt (2007a) and is not repeated here. With the discovery of *S. kraussei* (Steiner, 1923) Travassos, 1927 in 1917 (although not described until 1923), the study of this group of potentially useful nematodes is fast approaching its 100th birthday. From a slow beginning, the pace of discovery has increased dramatically, progress being given impetus by the increasing world-wide restrictions on chemical biocides, the potential for biological management of insect pests as part of a robust, integrated management strategy, and facilitated by the development of increasingly available molecular techniques for rapid and reliable species diagnostics in the battle against the constraints imposed by a toxic mixture of conserved morphology, variable morphometrics and fewer experienced traditional taxonomists.

The advent of molecular methodologies introduced a powerful discriminatory tool that has facilitated a rapid increase in the proposal of new species, particularly of *Steinernema* where, astonishingly, no fewer than 85 of the 95 valid species have been described since 1990 (Fig. 1.1). Figure 1.2 shows the progress since 1990, with 15 valid species described in the 1990s, 39 from 2000-2009 and 31 from 2010 to the beginning of 2016. Compare this with a mere ten valid species proposed between 1923 and 1989 (and four of those were from the late 1980s).

That *Steinernema* has been shown to be far more speciose than previously suspected would not have been possible by relying on classical taxonomy alone where its discriminatory power would have been unreliable in such a morphologically-conserved genus. To complicate matters, morphometrics, that most traditional of parameters, is highly variable in both the Steinernematidae and Heterorhabditidae, whilst morphology is also subject to much variation, including interpretation problems of three-dimensional structures due, for example, to angle of view, foreshortening effects and the interpretative skills of the observer (many not being professional taxonomists). In addition, the situation is potentially exacerbated by the fact that many descriptions appear to be overly formulaic (*i.e.*, closely following existing descriptive narratives), rather than being based on an objective morphological assessment of the material at hand.

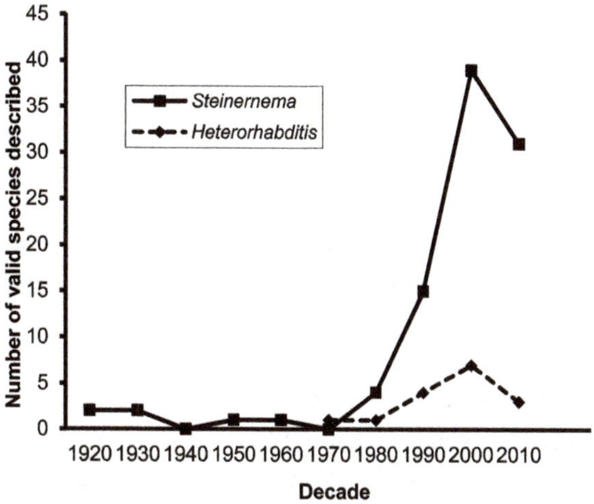

Fig. 1.1. *Number of valid species of* Steinernema *and* Heterorhabditis *described in each decade since 1920.*

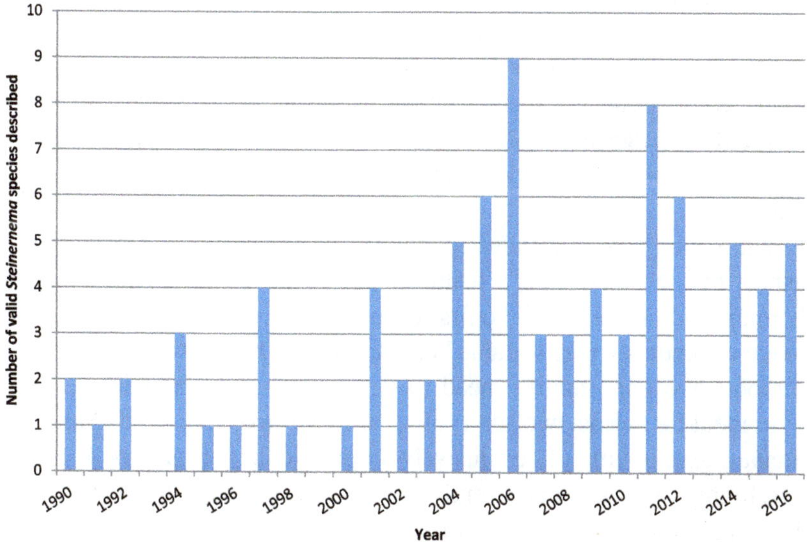

Fig. 1.2. *Number of valid* Steinernema *species described each year since 1990.*

It is not surprising, therefore, that molecular methodologies have been turned to as a potentially more discriminatory tool, although also posing some intriguing challenges to species concept in the two families, particularly concerning problems relating to sequence length, mistakes in sequence reading, alignment errors and, last but not least, interpretation at the taxon level. Although a powerful tool, molecular taxonomy is not free from error or misinterpretation; mistakes range from the simple mis-labelling/confusion of sequences from different species (a surprisingly common error) right through to a lack of understanding of the intrinsic variability of sequence data. To take an example, Půža *et al.* (2015) reported upon intra-individual variability in the ITS region of 49 strains of three species of *Steinernema*: *S. feltiae*, *S. kraussei* and *S. glaseri*. They found a high level of intra-individual variation within the isolates, ranging from 30% in *S. kraussei* to nearly 90% in *S. glaseri*. Within-individual variability ranged from over 1% in *S. kraussei* and *S. feltiae* to as much as 4% in *S. glaseri*. Variation such as this can clearly impact upon taxa where specific status has often been attributed based on a sequence character differing by only a few base pairs from one or more nominal species. Once again, the problems that compromised 'traditional' taxonomy, and which are inherent in any species concept, rear their head.

Despite these issues, however, molecular sequencing forms a pragmatic cornerstone for species characterisation within these two families and no description of potentially new taxa should be entertained unless supported by good quality sequences of both ITS and D2-D3.

The genus *Heterorhabditis* continues to present a far less speciose aspect than *Steinernema*, a feature that is probably a reflection of the hermaphroditic nature of the first generation, itself perhaps an adaption to environments where suitable hosts are few and far between and where reproductive opportunities need to be maximised by avoiding the requirement for at least two adults of opposite sex to be involved. Although *Steinernema* is virtually exclusively amphimictic, hermaphroditism, albeit highly unusual, is apparently not unknown, reportedly occurring in *S. hermaphroditum* Stock, Griffin & Chaerani, 2004, a self-fertile species from Indonesia (see Griffin *et al.*, 2001; Stock *et al.*, 2004).

The genus *Steinernema* Travassos, 1927

In the first edition of the book (Nguyen & Hunt, 2007a), 55 species of *Steinernema* were regarded as valid (Hunt, 2007b; Nguyen *et al.*, 2007). Since then, a further 41 new species have been proposed of which 37 are currently accepted as valid. Many of these new taxa have come from Asia, but there has also been a substantial contribution to the tally from Africa. Analysis of the molecular sequence data indicates that a small number of these new taxa should be regarded as junior synonyms of existing species (Subbotin & Spiridonov, pers. comm.), a course that is followed herein in Chapter 2. Accordingly, the number of valid species in the genus is, at the time of writing, 95, including several known to be in press and listed as an Addendum to Chapter 4.

The genus *Neosteinernema* Nguyen & Smart, 1994

Neosteinernema remains as a monotypic genus (Nguyen & Smart, 1994). The type species lacks molecular data and its phylogenetic position is currently unresolved.

The genus *Heterorhabditis* Poinar, 1976

Since the first edition of the book (Nguyen & Hunt, 2007a), nine species of *Heterorhabditis* have been proposed, all with molecular data. However, of these nine, four are synonymised herein with existing species (see Chapter 2), a study of the molecular data by Subbotin (pers. comm.) indicating that the data in their original descriptions had problems with interpretation/alignment of the molecular sequences that are so essential for establishing species status. The genus, therefore, currently comprises 16 valid species, a substantial increase over the 11 that were accepted in 2007 (Hunt, 2007b; Nguyen & Hunt, 2007b).

Shipshape and Bristol fashion

Although descriptions of new taxa are important to our understanding of the Steinernematidae and Heterorhabditidae, there remains a need to tidy up existing species descriptions. Some of these because they were done many decades ago when criteria were less well established; others because of emerging problems with the standard or application of the molecular methodologies employed. For example, *S. carpocapsae* (Weiser, 1955) Wouts, Mráček, Gerdin & Bedding, 1982 is one of the best known species in the Steinernematidae, yet its detailed characterisation has been less than perfect – hence the recent publication by Mráček *et al.* (2014) where additional data were presented and the species compared in detail with other members of its clade. Work such as this can result in descriptions of later proposed taxa losing credibility, as in the case of Mráček *et al.* (2014), where the authors suggested that *S. anatoliense* Hazir, Stock & Keskin, 2003 and *S. websteri* Cutler & Stock, 2003 appeared on existing data (including the D2-D3, but not the ITS sequence, which is unfortunately unavailable for *S. anatoliense* and *S. websteri*) to be conspecific with *S. carpocapsae*. In Chapter 2 we regard both the above-mentioned species as junior synonyms of *S. carpocapsae*.

Ivanova *et al.* (2013) recently re-described *S. costaricense* Uribe-Lorío, Mora & Stock, 2007, a relatively poorly known, and in some ways confusing, species (Uribe-Lorío *et al.*, 2007). They found a population from the USA, which they attributed to this species on morphological and molecular data. In the course of their work it seems that in the original description the infective juvenile (IJ) length was miscalculated, perhaps being reported at double its actual size. They also concluded that the attribution of the species to the *bicornutum*-group was ill-founded and based on a misinterpretation of IJ morphology, although phylogenetically the actual placement of the species remains unclear. Similarly, it appears that the proposal of *S. bifurcatum* Fayyaz, Yan, Qiu, Han, Gulsher, Khanum & Javed, 2014 by Fayyaz *et al.* (2014) also has problems, the ITS sequence apparently being very similar to that for *S. pakistanense* Shahina, Anis, Reid, Rowe & Maqbool, 2001, whilst the D2-D3 sequence is virtually identical to that reported for *S. abbasi* Elawad, Ahmed & Reid, 1997 (Půža, pers. comm.; Subbotin, pers. comm.). It seems likely, therefore, that the description, or at least the molecular profiling (upon which validity depends), of *S. bifurcatum* was based on a mixture of species and the taxon is probably a junior synonym of either *S. abbasi* or *S. pakistanense*, both of which are known to occur locally. It is also possible, however, that the species is actually valid and that the vitally important molecular sequences were mislabelled/confused. This issue needs to be resolved by further detailed molecular study of the type isolate.

The fate of the above-mentioned taxa has also been shared as a result of studies of the molecular accessions of several other nominal species of *Steinernema*: *S. everestense* Khatri-Chhetri, Waeyenberge, Spiridonov, Manandhar & Moens, 2011 is regarded as a junior synonym of *S. akhursti* Qiu, Hu, Zhou, Mei, Nguyen & Pang, 2005; *S. dharanai* Kulkarni, Rizvi, Kumar, Paunikar & Mishra, 2012 as a junior synonym of *S. hermphroditum* Stock, Griffin & Chaerani, 2004; *S. maqbooli* Fayyaz, Khanum, Gulsher & Javed, 2013 as a junior synonym of *S. pakistanense*; and *S. meghalayense* Ganguly, Rathour & Singh, 2011 as a junior synonym of *S. carpocapsae*.

Similarly, in Heterorhabditidae a number of species fail to justify valid status: *H. gerrardi* Plichta, Joyce, Clarke, Waterfield & Stock, 2009 is a junior synonym of *H. indica* Poinar, Karunakar & David, 1992; *H. sonorensis* Stock, Rivera-Orduño & Flores-Lara, 2009 is a junior synonym of *H. taysearae* Shamseldean, Abou El-Sooud, Abd-Elgawad & Saleh, 1996; *H. somsookae*[1] Maneesakorn, An, Grewal & Chandrapatya, 2015 is a junior synonym of *H. baujardi* Phan, Subbotin, Nguyen & Moens, 2003 (see Chapter 2 for details); and *H. pakistanensis*[2] Shahina, Tabassum, Salma, Mehreen & Knoetze, 2016 is a junior synonym of *H. indica*.

Doubtless other taxa will suffer the same fate, yet their demise will surely enhance the robustness of our understanding of these fascinating nematodes.

References

CUTLER, G.C. & STOCK, S.P. (2003). *Steinernema websteri* sp. n. (Rhabditida: Steinernematidae), a new entomopathogenic nematode from China. *Nematologia Mediterranea* 31, 215-224.

ELAWAD, S., AHMED, W. & REID, A.P. (1997). *Steinernema abbasi* sp. n. (Nematoda: Steinernematidae) from the Sultanate of Oman. *Fundamental and Applied Nematology* 20, 435-442.

FAYYAZ, S., KHANUM, T.A., GULSHER, M. & JAVED, S. (2013). *Steinernema maqbooli* sp. n. (Nematoda: Steinernematidae), a species of the '*bicornutum*' group from Pakistan. *International Journal of Nematology* 23, 59-72.

[1] Emended ending from the original, and incorrect, *somsookii*.

[2] Emended to *pakistanensis* from the incorrectly formed '*pakistanense*' in the original paper.

FAYYAZ, S., YAN, X., QIU, L., HAN, R., GULSHER, M., KHANUM, T.A. & JAVED, S. (2014). A new entomopathogenic nematode, *Steinernema bifurcatum* n. sp. (Rhabditida: Steinernematidae) from Punjab, Pakistan. *Nematology* 16, 821-836.

GANGULY, S., RATHOUR, K.S. & SINGH, M. (2011). *Steinernema meghalayensis* sp. n. (Rhabditida: Steinernematidae) from northeastern hilly region of India. *Indian Journal of Nematology* 41, 83-97.

GRIFFIN, C.T., O'CALLAGHAN, K.M. & DIX, I. (2001). A self-fertile species of *Steinernema* from Indonesia: further evidence of convergent evolution amongst entomopathogenic nematodes? *Parasitology* 122, 181-186.

HAZIR, S., STOCK, S.P. & KESKIN, N. (2003). A new entomopathogenic nematode, *Steinernema anatoliense* n. sp. (Rhabditida: Steinernematidae), from Turkey. *Systematic Parasitology* 55, 211-220.

HUNT, D.J. (2007a). Introduction. In: Nguyen, K.B. & Hunt, D.J. (Eds). *Entomopathogenic nematodes: taxonomy, phylogeny and bacterial symbionts. Nematology Monographs and Perspectives 5* (Series editors: Hunt, D.J. & Perry, R.N.). Leiden, The Netherlands, Brill, pp. 1-26.

HUNT, D.J. (2007b). Overview of taxonomy and systematics. In: Nguyen, K.B. & Hunt, D.J. (Eds). *Entomopathogenic nematodes: taxonomy, phylogeny and bacterial symbionts. Nematology Monographs and Perspectives 5* (Series editors: Hunt, D.J. & Perry, R.N.). Leiden, The Netherlands, Brill, pp. 27-57.

IVANOVA, E.S., SHEPELEVA, N.S. & SPIRIDONOV, S.E. (2013). Morphological and molecular characterisation of *Steinernema costaricense* Uribe-Lorío, Mora & Stock, 2007 (Panagrolaimorpha: Steinernematidae) isolate from Bush Augusta State Park, Missouri, USA. *Systematic Parasitology* 85, 219-234.

KHATRI-CHHETRI, H.B., WAEYENBERGE, L., SPIRIDONOV, S., MANANDHAR, H.K. & MOENS, M. (2011). *Steinernema everestense* n. sp. (Rhabditida: Steinernematidae), a new species of entomopathogenic nematode from Pakhribas, Dhankuta, Nepal. *Nematology* 13, 443-462.

KOPPENHÖFER, H.S. (2007). Bacterial symbionts of *Steinernema* and *Heterorhabditis*. In: Nguyen, K.B. & Hunt, D.J. (Eds). *Entomopathogenic nematodes: taxonomy, phylogeny and bacterial symbionts. Nematology Monographs and Perspectives 5* (Series editors: Hunt, D.J. & Perry, R.N.). Leiden, The Netherlands, Brill, pp. 735-808.

KULKARNI, N., RIZVI, A.N., KUMAR, V., PAUNIKAR, S. & MISHRA, V.K. (2012). Morphological and molecular characterization of *Steinernema dharanaii* sp. n.: a new entomopathogenic nematode from India. *Indian Journal of Tropical Biodiversity* 20, 107-116.

MANEESAKORN, P., AN, R., GREWAL, P.S. & CHANDRAPATYA, A. (2015). *Heterorhabditis somsookii* sp. nov. (Rhabditida: Heterorhabditidae): a new entomopathogenic nematode from Thailand. *International Journal of Nematology* 11, 1-17.

MRÁČEK, Z., PŮŽA, V. & NERMUT', J. (2014). Addendum to the description of *Steinernema carpocapsae* (Weiser, 1955) Wouts, Mráček, Gerdin & Bedding, 1982. *Russian Journal of Nematology* 22, 109-120.

NGUYEN, K.B. (2007). Methodology, morphology and identification. In: Nguyen, K.B. & Hunt, D.J. (Eds). *Entomopathogenic nematodes: taxonomy, phylogeny and bacterial symbionts. Nematology Monographs and Perspectives 5* (Series editors: Hunt, D.J. & Perry, R.N.). Leiden, The Netherlands, Brill, pp. 59-119.

NGUYEN, K.B. & HUNT, D.J. (Eds) (2007a). *Entomopathogenic nematodes: taxonomy, phylogeny and bacterial symbionts. Nematology Monographs and Perspectives 5* (Series editors: Hunt, D.J. & Perry, R.N.). Leiden, The Netherlands, Brill.

NGUYEN, K.B. & HUNT, D.J. (2007b). Heterorhabditidae: species descriptions. In: Nguyen, K.B. & Hunt, D.J. (Eds). *Entomopathogenic nematodes: taxonomy, phylogeny and bacterial symbionts. Nematology Monographs and Perspectives 5* (Series editors: Hunt, D.J. & Perry, R.N.). Leiden, The Netherlands, Brill, pp. 611-692.

NGUYEN, K.B. & SMART JR, G.C. (1994). *Neosteinernema longicurvicauda* n. gen., n. sp. (Rhabditida: Steinernematidae), a parasite of the termite *Reticulitermes flavipes* (Koller). *Journal of Nematology* 26, 162-174.

NGUYEN, K.B., HUNT, D.J. & MRÁČEK, Z. (2007). Steinernematidae: species descriptions. In: Nguyen, K.B. & Hunt, D.J. (Eds). *Entomopathogenic nematodes: taxonomy, phylogeny and bacterial symbionts. Nematology Monographs and Perspectives 5* (Series editors: Hunt, D.J. & Perry, R.N.). Leiden, The Netherlands, Brill, pp. 121-609.

PHAN, K.L., SUBBOTIN, S.A., NGUYEN, N.C. & MOENS, M. (2003). *Heterorhabditis baujardi* sp. n. (Rhabditida: Heterorhabditidae) from Vietnam and morphometric data for *H. indica* populations. *Nematology* 5, 367-382.

PLICHTA, K.L., JOYCE, S.A., CLARKE, D., WATERFIELD, N. & STOCK, S.P. (2009). *Heterorhabditis gerrardi* n. sp. (Nematoda: Heterorhabditidae): the hidden host of *Photorhabdus asymbiotica* (Enterobacteriaceae: γ-Proteobacteria). *Journal of Helminthology* 83, 309-320.

POINAR JR, G.O. (1976). Description and biology of a new insect parasitic rhabditoid, *Heterorhabditis bacteriophora* n. gen. n. sp. (Rhabditida; Heterorhabditidae n. fam.). *Nematologica* 21, 463-470.

POINAR JR, G.O., KARUNAKAR, G.K. & DAVID, H. (1992). *Heterorhabditis indicus* n. sp. (Rhabditida, Nematoda) from India: separation of *Heterorhabditis* spp. by infective juveniles. *Fundamental and Applied Nematology* 15, 467-472.

PŮŽA, V., CHUNELOVÁ, D., NERMUT', J., ŽUROVCOVÁ, M. & MRÁČEK, Z. (2015). Intra-individual variability of ITS regions in entomopathogenic nematodes (Steinernematidae: Nematoda): implications for their taxonomy. *BioControl* 60, 547-554. DOI:10.1007/s10526-015-9664-5.

QIU, L., HU, X., ZHOU, Y., MEI, S., NGUYEN, K.B. & PANG, Y. (2005). *Steinernema akhursti* n. sp. (Nematoda: Steinernematidae) from Yunnan, China. *Journal of Invertebrate Pathology* 90, 151-160.

SHAHINA, F., ANIS, M., REID, A.P., ROWE, J. & MAQBOOL, M.A. (2001). *Steinernema pakistanense* sp. n. (Rhabditida: Steinernematidae) from Pakistan. *International Journal of Nematology* 11, 124-133.

SHAHINA, F., TABASSUM, K.A., SALMA, J., MEHREEN, G. & KNOETZE, R. (2016). *Heterorhabditis pakistanense* n. sp. (Nematoda: Heterorhabditidae) a new entomopathogenic nematode from Pakistan. *Journal of Helminthology*. DOI:10.1017/S0022149X16000158.

SHAMSELDEAN, M.M., ABOU EL-SOOUD, A.B., ABD-ELGAWAD, M.M. & SALEH, M.M. (1996). Identification of a new heterorhabditid species from Egypt, *Heterorhabditis taysearae* n. sp. (Rhabditida: Heterorhabditidae). *Egyptian Journal of Biological Pest Control* 6, 129-138.

STEINER, G. (1923). *Aplectana kraussei* n. sp., eine in der Blattwespe *Lyda* sp. parasitierende Nematodenform, nebst Bemerkungen über das Seitenorgan der parasitischen Nematoden. *Zentralblatt für Bakteriologie, Parasitenkunde, Infektions krankheiten und Hygiene, Abteilung II* 59, 14-18.

STOCK, S.P., GRIFFIN, C.T. & CHAERANI, R. (2004). Morphological and molecular characterisation of *Steinernema hermaphroditum* n. sp. (Nematoda: Steinernematidae), an entomopathogenic nematode from Indonesia, and its phylogenetic relationships with other members of the genus. *Nematology* 6, 401-412.

STOCK, S.P., RIVERA-ORDUÑO, B. & FLORES-LARA, Y. (2009). *Heterorhabditis sonorensis* n. sp. (Nematoda: Heterorhabditidae), a natural pathogen of the seasonal cicada *Diceroprocta ornea* (Walker) (Homoptera: Cicadidae) in the Sonoran desert. *Journal of Invertebrate Pathology* 100, 175-184.

TRAVASSOS, L. (1927). Uma nova *Capillaria* parasita de peixes be agua doce: *Capillaria sentinosa* n. sp. *Boletim Biologico, São Paulo* 10, 215-217.

URIBE-LORÍO, L., MORA, M. & STOCK, S.P. (2007). *Steinernema costaricense* n. sp. and *S. puntauvense* n. sp. (Rhabditida: Steinernematidae), two new entomopathogenic nematodes from Costa Rica. *Systematic Parasitology* 68, 167-182.

WEISER, J. (1955). *Neoaplectana carpocapsae* n. sp. (Anguillulata, Steinernematinae), novy cizopasník housenek obaleče jablečného, *Carpocapsa pomonella* L. *Věstník Československé Zoologické Společnosti* 19, 44-52.

WOUTS, W.M., MRÁČEK, Z., GERDIN, S. & BEDDING, R.A. (1982). *Neoaplectana* Steiner, 1929 a junior synonym of *Steinernema* Travassos, 1927 (Nematoda: Rhabditida). *Systematic Parasitology* 4, 147-154.

Nematology Monographs & Perspectives, 2016, Vol. 12, 13-58

Chapter 2

Taxonomy and systematics

David J. HUNT [1] and Sergei A. SUBBOTIN [2,3]

[1] *CABI Europe-UK, Bakeham Lane, Egham, Surrey, TW20 9TY, UK*
[2] *Plant Pest Diagnostic Center, California Department of Food and Agriculture, 3294 Meadowview Road, Sacramento, CA 95832, USA*
[3] *Center of Parasitology, A.N. Severtsov Institute of Ecology and Evolution of the Russian Academy of Sciences, Leninskii prospect 33, Moscow 119071, Russia*

Introduction

The following account is a revised and updated version of the chapter by Hunt (2007). Following the description of *S. unicornum* (see Edgington *et al.*, 2009a), a minor change to the morphological diagnosis of the infective juvenile (IJ) of *Steinernema* has been made to reflect the labial tooth-like structure found in the IJ of that species and previously regarded as being restricted to *Heterorhabditis*. The species lists for *Steinernema* and *Heterorhabditis* have been updated and a number of new synonymies proposed, mostly on a re-appraisal of their molecular sequence data (Table 2.1). Several previously overlooked or more recent invalid names found in the literature are appended and/or discussed. The chapter includes species described from January 2016, although in Chapter 4 full diagnostic accounts are restricted to those taxa described up to the end of December 2015. Species described in 2016 are not included in the tabular keys of Chapter 3 and lack full descriptive narratives, although they are represented by their modified abstracts in an Addendum to Chapter 4.

Table 2.1. *List of synonyms proposed on the basis of sequence analysis.*

Binomen	Junior synonym of:
Steinernema anatoliense	*Steinernema carpocapsae*
Steinernema dharanai	*Steinernema hermaphroditum*
Steinernema everestense	*Steinernema akhursti*
Steinernema maqbooli	*Steinernema pakistanense*
Steinernema meghalayense	*Steinernema carpocapsae*
Steinernema tbilisiense	*Steinernema thesami*
Steinernema websteri	*Steinernema carpocapsae*
Heterorhabditis gerrardi	*Heterorhabditis indica*
Heterorhabditis pakistanensis	*Heterorhabditis indica*
Heterorhabditis somsookae	*Heterorhabditis baujardi*
Heterorhabditis sonorensis	*Heterorhabditis taysearae*

Family Steinernematidae

History

The first species of Steinernematidae to be described was *Aplectana kraussei* Steiner, 1923. This was collected by Dr Krausse in Germany in 1917 and came from *Lyda potrophica* Hartig (= *Cephalcia abietis* L.), a hymenopterous sawfly (Krausse, 1917). The nematode was subsequently described by Steiner (1923). Steiner tentatively placed his species in the oxyurid genus *Aplectana* Railliet & Henry, 1916, a group of nematodes otherwise known from amphibians, and perhaps a rather unlikely repository for an insect nematode. It did not remain long in that taxon as a few years later, Travassos (1927a), working in Brazil, transferred the species to his new genus *Steineria* Travassos, 1927, nominating the former *A. kraussei* as the type species. Travassos named *Steineria* in honour of Steiner, but unfortunately overlooked the fact that this name was already occupied by *Steineria* Micoletzky, 1922 and was therefore an objective junior homonym. According to the rules of zoological nomenclature, a new name was therefore required. On realizing his error, Travassos (1927b) immediately proposed *Steinernema* Travassos, 1927 as the replacement epithet, thereby neatly solving the problem yet at the same time maintaining the patronymic attribution to Steiner. For many years, *Steinernema kraussei*, the type species of the genus, remained obscure, the only record being the original description from 1923. Skrjabin *et al.* (1951) even transferred the species to the genus

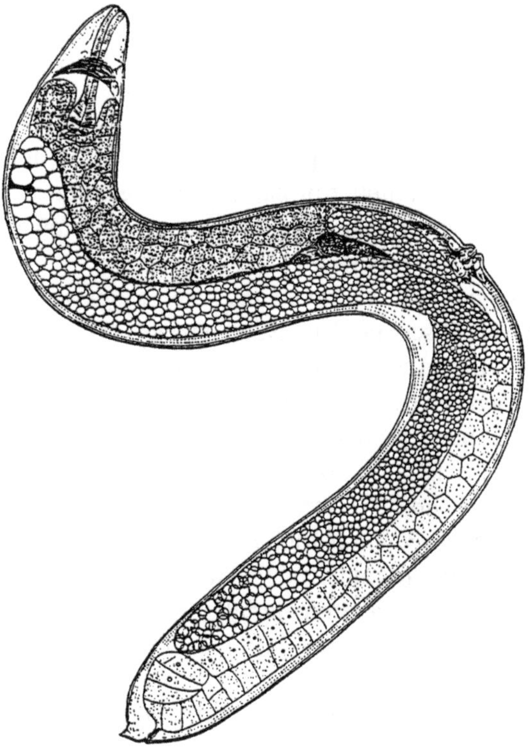

Fig. 2.1. Aplectana kraussei. *After original illustration by Steiner (1923).*

Oxysomatium, a move ignored by other authors and subsequently repudiated by Veremchuk (1964).

The original description of *S. kraussei* was based on five males and 18 females. The morphometrics were given as the usual Cobb formula although Steiner also included certain of the de Man indices, namely α, β and γ (nowadays universally referred to as a, b and c). The detailed description was illustrated by a rather good drawing of the mature female (Fig. 2.1) and a series of drawings depicting the anterior end, tail region and male copulatory organs. An *en face* view depicted the lateral sensory organs (the amphids) plus four submedian papillae, two of which were subdorsal and the other two subventral in location. No labial papillae were illustrated or mentioned in the description.

Steiner (1929) erected the genus *Neoaplectana* Steiner, 1929 and differentiated it from *Steinernema* on the basis: *i*) that the former had two circles of six cephalic papillae whereas *Steinernema* was described as having but a single ring of four; and *ii*) that *Neoaplectana* had more male copulatory papillae in a different arrangement. In fact, Steiner misinterpreted the sense organs; *Neoaplectana* actually has an outer circle of four cephalic papillae plus two large amphidial apertures and an inner circle of six labial papillae. The supposed difference between the two genera in the number of papillae on the anterior extremity was not corrected for over 50 years and resulted in confusion and uncertainty about the status of the genera.

The type species of *Neoaplectana* was *N. glaseri* Steiner, 1929. This species was discovered in New Jersey, USA, where it attacked white grubs, the larvae of the scarabaeid beetle, *Popillia japonica*. *Neoaplectana glaseri* (now known as *Steinernema glaseri*) was the first nematode to be mass-produced and used for biological control of an insect pest under field conditions.

Neoaplectana menozzii Travassos, 1932, the next species to be described (Travassos, 1932), was found inside *Conorrhynchus* (= *Cleonus*) *mendicus* Gyll., a curculionid beetle, in Italy by C. Menozzi. This species was described from preserved material only, is not well known and has not been re-identified since the original discovery. It is possible that *N. menozzii* is conspecific with *Neoaplectana feltiae* Filipjev, 1934, a species originally discovered in 1926. Other suggestions include the possibility that *menozzii* is conspecific with either *affine* or *carpocapsae*. All things considered, it is probably best to regard the poorly known *menozzii* as a *species inquirenda*.

The description of *N. feltiae* by Filipjev (1934) was based on a nematode first isolated in Russia in 1926. *Neoaplectana feltiae* came from the cutworm, *Feltia segetum* (Schiff.) Herbst (Insecta: Lepidoptera), the specific epithet being derived from the genus of its insect host. In the same paper, Filipjev proposed the subfamily Steinernematinae Filipjev, 1934 (actually he called it Steinerneminae, but this name is improperly formed and must be corrected), the type genus being *Steinernema* Travassos, 1927. The subfamily was later recognised at family rank by Chitwood & Chitwood (1937), although the family authority should still be cited as Steinernematidae Filipjev, 1934 (rather than the oft-seen Steinernematidae (Filipjev, 1934) Chitwood & Chitwood, 1937) as the Principle of Coordination states that ranks created at one level in the family group (*i.e.*, subfamily, family and superfamily) are simultaneously created at the other ranks and carry the same author and date (see International Commission on Zoological

Nomenclature, 1999). Filipjev (1934) also placed *Neoaplectana* in the Steinernematinae and indicated that it could well be congeneric with *Steinernema*.

The next species to be described were *N. bibionis* Bovien, 1937 and *N. affinis* Bovien, 1937, both recorded from bibionid flies (Diptera: Bibionidae) from several localities in Denmark (Bovien, 1937). Prosper Bovien, in a delightfully crafted work that is still well worth reading, commented that one of his species may have been identical to the inadequately described *N. menozzii*. The quality of Bovien's illustrations comfortably exceed that of many modern descriptions; his care and attention to the disposition of the genital papillae, for example, serve as an example that later authors would have done well to follow (Fig. 2.2). Of the two species described by Bovien, *N. bibionis* was synonymised under *S. feltiae* whilst Poinar (1988) redescribed *N. affinis*.

Little progress was made in the next 15 years or so until Sobolev (1953) proposed the family Neoaplectanidae Sobolev, 1953. This family subsequently became a junior synonym of Steinernematidae when *Neoaplectana*, its type genus, was eventually recognised as being a junior synonym of *Steinernema*.

Neoaplectana carpocapsae Weiser, 1955, perhaps the most famous of all steinernematids, was described by Weiser (1955) from the larvae of the codling moth in Germany. The nematode was more or less simultaneously discovered by Dutky & Hough (1955), also from codling moth, but in the USA. This species (particularly the Dutky and Hough isolate which was known as DD-136) was widely used as a potential biocontrol agent and a number of different isolates were named and maintained under laboratory conditions. Weiser & Köhler (1955) also described *N. janickii* Weiser & Köhler, 1955, although this species is now regarded as a *species inquirenda*.

In the 1950s, 1960s and 1970s, a number of species of *Neoaplectana* were described from the then USSR (Kirjanova & Putschkova, 1955; Kakuliya & Veremchuk, 1965; Veremchuk, 1969; Veremchuk & Litvinchuk, 1971; Rubtsov & Polevik, 1979), and others were named but not formally described (Veremchuk, 1966, 1970). As these latter names have no formal descriptions they are regarded as *nomina nuda*, or naked names and have no official standing. Many of these 'species' would appear to be synonymous with either *S. feltiae* or *S. carpocapsae*, although it is impossible to be sure and the names are best left in peace. Veremchuk (1964) regarded Steinernematidae as invalid and replaced it

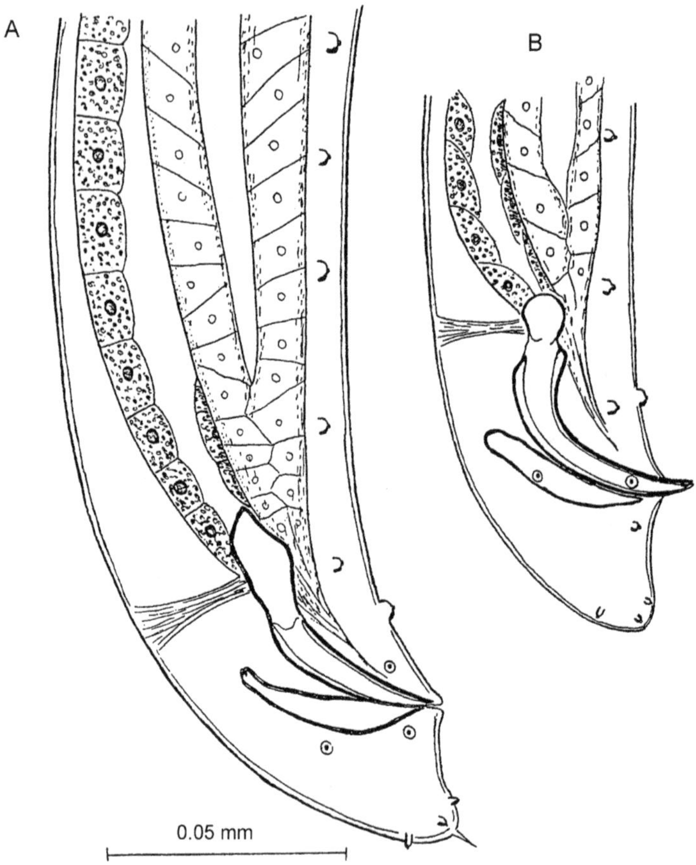

Fig. 2.2. Neoaplectana bibionis *(A) and* N. affinis *(B). Male tail, lateral view. After original illustration by Bovien (1937).*

with the later Neoaplectanidae. *Steinernema kraussei* was also removed from *Oxysomatium*, where it had been placed by Skrjabin *et al.* (1951), and returned to *Neoaplectana*.

Mráček (1977), in a determined attempt to re-establish *S. kraussei*, claimed to have rediscovered the type species in *Cephalcia abietis*, a sawfly found in coniferous forests in southern Bohemia. Although the type host of the original *S. kraussei* was the same as Mráček's material, the type locality differed, with Mráček's material coming from Bohemia (Czechoslovakia) and Steiner's from Westphalia, Germany. His claim was met with some resistance by other workers, including Poinar (1986), as it was disputed whether Mráček had actually rediscovered *S. kraussei* or had merely found a species of *Neoaplectana* infecting the same host. Mráček & Weiser (1979) and Mráček *et al.* (1981) continued to study the cephalic sensory organs of Steinernematidae, the latter paper using the SEM to elucidate papillal arrangement.

Eventually Wouts *et al.* (1982) published what was to become a landmark paper in which they proposed that *Steinernema* and *Neoaplectana* be synonymised on the basis that both nominal genera had the same arrangement of cephalic and labial papillae. Wouts *et al.* (1982) studied the population of *S. kraussei* from the former Czechoslovakia (= Czech Republic) plus surviving type material of the species that had been discovered in the USDA collection. They clearly demonstrated that *S. kraussei* had, in addition to the four cephalic papillae originally mentioned by Steiner, an inner circle of six labial papillae, *i.e.*, precisely the same arrangement that had been used to define *Neoaplectana*. This discovery meant that *Neoaplectana*, which was largely based on the presence of an outer ring of six (actually there are only four, the other two corresponding to the rather prominent amphidial apertures) cephalic papillae and an inner ring of six labial papillae, *vs* only a single ring of four cephalic papillae in *Steinernema*, became a junior synonym of *Steinernema*, the older and therefore senior genus.

Although this move gained an almost immediate general acceptance, it was not universal: Poinar, a determined champion of *Neoaplectana*, refused to accept that the true *S. kraussei* had been rediscovered in Czechoslovakia and opined that this material did not correspond to *S. kraussei* and was, in fact, a *Neoaplectana* – hence the undeniable SEM demonstration of the occurrence of four cephalic and six labial papillae, the diagnostic character of the latter genus. Poinar (1986) seemed to overlook or dismiss the fact that the type specimen of *S. kraussei* also showed the presence of labial papillae (the material was poorly preserved, thereby making morphological observations difficult and any conclusions contentious). This dispute rumbled on for the next decade and a half with both generic combinations circulating in the literature to the confusion of one and all. Eventually, Poinar (1990), without further explanation, renounced *Neoaplectana* in favour of *Steinernema*, although he suggested that *S. kraussei* was not validly designated as type species of the genus and proposed that *S. glaseri*, the second species to be described, be used in its stead. Thankfully, this proposal was not followed, and the taxonomy of the genus finally settled down according to the proposal by Wouts *et al.* (1982).

Subsequently, Mráček *et al.* (1991) successfully re-identified the type locality of *S. kraussei* near Neuenherrsee, Westphalia, Germany, whilst Mráček (1992) provided a full description of topotype material of the type species, thus stabilising the systematics of the group beyond all reasonable contention. There now only remained to be resolved some minor skirmishing over the appropriate endings of various specific epithets (the International Commission on Zoological Nomenclature, 1999 states that the gender of the species name must match that of the genus name, *Neoaplectana* being feminine and *Steinernema* neuter in gender).

The proposal to adjust the endings of the species names to conform to the Rules was made by Hominick *et al.* (1997) and reiterated by Stock *et al.* (1997). Minor tussles over the decision to change the ending of certain names were soon resolved, the situation being summarised, apparently to everyone's satisfaction, by Adams & Nguyen (2002).

Species descriptions in the genus *Steinernema* have risen at an impressive rate in the last decade, partly as a result of increased interest in the group as potential biocontrol agents, particularly in previously unresearched geographic areas such as Asia and Africa, but also because of the utility of molecular techniques in reliably characterising species in a group that is increasingly difficult to access using classical methodologies. Molecular characterisation, at first with RFLP profiles of the ITS region but later with sequences of the ITS region and D2-D3 expansion region, have provided an excellent tool to probe beyond the conserved morphology of this fascinating group of entomopathogenic nematodes, revealing a hitherto unsuspected complexity in their taxonomy – 95 species are now accepted as valid, most of which could not have been sustained by reliance on traditional descriptions. The rate at which new species have been described has increased enormously since 1990. Previous to that date, only two currently valid species were described in the 1920s, two more in the 1930s, one in the 1950s, one in the 1960s, and four in the 1980s – a total of ten valid taxa. Since 1990, however, another 85 valid species have been proposed, 15 of these in the 1990s, another 39 in the 2000s and 31 from 2010-2016. One may question if *Steinernema* really is that diverse and whether there is a need to recalibrate species criteria based on differences in molecular sequences. Indeed, in this update, a number of species, despite their molecular characterisation, have been synonymised because of apparent errors in sequence reading, alignment, or other uncertainty, whilst doubts exist about the true status of several other recently published taxa (see the Notes section under *Steinernema*).

TAXONOMY AND SYSTEMATICS

Family Steinernematidae Filipjev, 1934
= Neoaplectanidae Sobolev, 1953

DIAGNOSIS (after Hunt, 2007)

Panagrolaimomorpha: Strongyloidoidea. Parasites of insects and other arthropods. Infective juveniles non-feeding, carrying symbiotic entomopathogenic bacteria. First and second generation adults usually reproducing by amphimixis, but hermaphroditic in *S. hermaphroditum*. Oviparous to ovoviviparous.

Adults

Medium- to large-sized, fairly plump, nematodes, Labial region with six labial and four cephalic papillae. Lateral fields indistinct or absent. Cheilorhabdions forming sclerotised ring, posterior stoma funnel-shaped. Pharynx rhabditoid, comprising cylindroid procorpus with slight metacorpal swelling, short isthmus and rounded basal bulb with small valve plates. Nerve ring located in isthmus region, excretory pore distinct, usually located anterior to nerve ring. Male gonad monorchic, reflexed. Spicules paired, robust, capitulum distinct. Gubernaculum large, bifurcate in ventral view. Bursa absent. Copulatory papillae typically comprising single midventral, precloacal papilla plus 10-14 paired papillae (occasionally one of a given pair may be missing). Male tail short, terminus rounded, may bear terminal mucron of variable length. Female vulva in form of median transverse slit, genital system amphididelphic, reflexed. Female tail either short, conoid (*Steinernema*), or considerably longer than anal body diam. (*Neosteinernema*).

Infective juvenile

Third-stage infective juvenile slender, vermiform. Labial region lacking a dorsal tooth-like process except in *S. unicornum*. Cuticle finely annulated. Mouth and anus closed, stoma collapsed, pharynx non-functional, reduced; symbiotic bacteria usually visible in intestinal vesicle located in lumen just posterior to basal bulb. Lateral field prominent with up to eight ridges (*i.e.*, nine incisures). Excretory pore located anterior to nerve ring. Tail medium, conoid, tapering to pointed terminus (*Steinernema*), or longer, tapering to a filiform terminus (*Neosteinernema*).

TYPE GENUS

Steinernema Travassos, 1927b
= *Steineria* Travassos, 1927a *nec Steineria* Micoletzky, 1922
= *Neoaplectana* Steiner, 1929
= *Patanodontus* de Villalobos & Camino, 1997

OTHER GENUS

Neosteinernema Nguyen & Smart, 1994

Genus *Steinernema* Travassos, 1927b

= *Steineria* Travassos, 1927a *nec Steineria* Micoletzky, 1922
= *Neoaplectana* Steiner, 1929
= *Patanodontus* de Villalobos & Camino, 1997

DIAGNOSIS (modified after Adams & Nguyen, 2002; Hunt, 2007)

Male

Medium-sized, fairly plump nematodes, heat-relaxed habitus usually J- or C-shaped. Labial region with six labial and four cephalic papillae. Perioral disc usually present. Deirids in cervical region reported for some species. Pharynx comprising cylindroid procorpus with slight metacorpal swelling, short isthmus and rounded basal bulb with small valve plates. Nerve ring located in isthmus region, excretory pore distinct, usually located anterior to nerve ring. Testis monorchic, reflexed. Spicules paired, robust, often yellowish to brownish in colour, capitulum distinct. Gubernaculum large, bifurcate in ventral view. Bursa absent. Copulatory papillae typically comprising single midventral, precloacal papilla plus 10-14 (usually 10-11) paired papillae. Precloacal subventral series numbering seven to ten pairs. Postcloacal papillae may be subventral, lateral or subdorsal in position. Tail short, terminus rounded, mucron present or absent.

Female

Large, rather plump nematodes. Heat-relaxed habitus often C-shaped or spiralled. Labial region with six partially or completely fused lips, each bearing a papilla; four cephalic papillae; and two small amphidial apertures. Cuticle appearing smooth or annulated, lateral fields absent. Cheilorhabdions sclerotised, forming a ring, posterior stoma funnel-shaped. Pharynx similar to male. Vulva in form of median transverse slit, epiptygmata present or absent. Genital system amphididelphic, reflexed, oviparous or ovoviviparous. Tail short, conoid, *ca* one or less anal body diam. long in first generation but longer in second generation. Projections present on tail may have taxonomic significance (see Mráček *et al.*, 2016). Phasmids may be inconspicuous.

Infective juvenile

Third-stage infective juvenile slender, vermiform, non-feeding. Heat-relaxed habitus usually straight or slightly ventrally arcuate. Cuticle finely annulated, retained sheath from second-stage moult may be present. Dorsal tooth-like process on labial region almost invariably absent (but present in *S. unicornum*). Mouth and anus closed, stoma collapsed, pharynx non-functional, reduced, basal bulb usually directed dorsad; symbiotic bacteria usually visible in intestinal vesicle just posterior to basal bulb. Lateral field well developed with up to eight ridges (*i.e.*, nine incisures). Excretory pore prominent, located anterior to nerve ring. Tail medium conoid, tapering to pointed terminus. Phasmids located about mid-point of tail.

TYPE SPECIES

Steinernema kraussei (Steiner, 1923) Travassos, 1927b
= *Aplectana kraussei* Steiner, 1923
= *Steineria kraussei* (Steiner, 1923) Travassos, 1927a
= *Oxysomatium kraussei* (Steiner, 1923) Skrjabin, Shikhobalova & Mozgovoi, 1951

OTHER SPECIES

S. abbasi Elawad, Ahmad & Reid, 1997
= *Steinernema thermophilum* Sudershan & Singh, 2000
S. aciari Qiu, Yan, Zhou, Nguyen & Pang, 2005a
S. affine (Bovien, 1937) Wouts, Mráček, Gerdin & Bedding, 1982

= *Neoaplectana affinis* Bovien, 1937

S. akhursti Qiu, Hu, Zhou, Mei, Nguyen & Pang, 2005b

= *Steinernema everestense* Khatri-Chhetri, Waeyenberge, Spiridonov, Manandhar & Moens, 2011a **n. syn.**

S. apuliae Triggiani, Mráček & Reid, 2004

S. arasbaranense Nikdel, Niknam & Ye, 2011

S. arenarium (Artyukhovsky, 1967) Wouts, Mráček, Gerdin & Bedding, 1982

= *Neoaplectana arenaria* Artyukhovsky, 1967

= *Neoaplectana anomali* Kozodoi, 1984

= *Steinernema anomalae* (Kozodoi, 1984) Curran, 1989

S. ashiuense Phan, Takemoto & Futai, 2006

S. asiaticum Anis, Shahina, Reid & Rowe, 2002

S. australe Edgington, Buddie, Tymo, Hunt, Nguyen, France, Merino & Moore, 2009b

S. backanense Phan, Spiridonov, Subbotin & Moens, 2006

S. balochiense Fayyaz, Khanum, Ali, Solangi, Gulsher & Javed, 2015

S. beddingi Qiu, Hu, Zhou, Pang & Nguyen, 2005c

S. beitlechemi Çimen, Půža, Nermut', Hatting, Ramakuwela, Faktorová & Hazir, 2016

S. bicornutum Tallosi, Peters & Ehlers, 1995

S. bifurcatum Fayyaz, Yan, Qiu, Han, Gulsher, Khanum & Javed, 2014

S. boemarei Lee, Sicard, Skeie & Stock, 2009

S. brazilense Nguyen, Ginarte, Leite, Santos & Harakava, 2010

S. cameroonense Kanga, Trinh, Waeyenberge, Spiridonov, Hauser & Moens, 2012

S. carpocapsae (Weiser, 1955) Wouts, Mráček, Gerdin & Bedding, 1982

= *Neoaplectana carpocapsae* Weiser, 1955

= *Neoaplectana feltiae sensu* Stanuszek, 1974a *nec* Filipjev, 1934

= *Neoaplectana feltiae pieridarum* Stanuszek, 1974b

= *Neoaplectana pieridarum* Stanuszek, 1974b

= *Steinernema feltiae pieridarum* (Stanuszek, 1974b) Wouts, Mráček, Gerdin & Bedding, 1982

= *Steinernema pieridarum* (Stanuszek, 1974b) Wouts, Mráček, Gerdin & Bedding, 1982

= *Neoaplectana carpocapsae pieridarum* Stanuszek, 1974b

= *Neoaplectana dutkyi* Turco, Thames & Hopkins, 1971

= *Steinernema dutkyi* (Turco, Thames & Hopkins, 1971) Wouts, Mráček, Gerdin & Bedding, 1982

= *Steinernema anatoliense* Hazir, Stock & Keskin, 2003 **n. syn.**
= *Steinernema websteri* Cutler & Stock, 2003 **n. syn.**
= *Steinernema meghalayense*[4] Ganguly, Rathour & Singh, 2011 **n. syn.**
S. caudatum Xu, Wang & Li, 1991
S. ceratophorum Jian, Reid & Hunt, 1997
S. changbaiense Ma, Chen, De Clercq, Han & Moens, 2012
S. cholashanense Nguyen, Půža & Mráček, 2008
S. citrae Stokwe, Malan, Nguyen, Knoetze & Tiedt, 2011
S. colombiense López-Núñez, Plichta, Góngora-Botero & Stock, 2008
S. costaricense Uribe-Lorío, Mora & Stock, 2007
S. cubanum Mráček, Hernandez & Boemare, 1994
S. cumgarense Phan, Spiridonov, Subbotin & Moens, 2006
S. diaprepesi Nguyen & Duncan, 2002
S. eapokense Phan, Spiridonov, Subbotin & Moens, 2006
S. ethiopiense Tamiru, Waeyenberge, Hailu, Ehlers, Půža & Mráček, 2012
S. fabii Abate, Malan, Tiedt, Wingfield, Slippers & Hurley, 2016
S. feltiae (Filipjev, 1934) Wouts, Mráček, Gerdin & Bedding, 1982
= *Neoaplectana feltiae* Filipjev, 1934
= *Neoaplectana bibionis* Bovien, 1937
= *Steinernema bibionis* (Bovien, 1937) Wouts, Mráček, Gerdin & Bedding, 1982
= *Neoaplectana leucaniae* Hoy, 1954
= *Steinernema leucaniae* (Hoy, 1954) Wouts, Mráček, Gerdin & Bedding, 1982
S. glaseri (Steiner, 1929) Wouts, Mráček, Gerdin & Bedding, 1982
= *Neoaplectana glaseri* Steiner, 1929
S. goweni San-Blas, Morales-Montero, Portillo, Nermut' & Půža, 2016
S. guangdongense Qiu, Fang, Zhou, Pang & Nguyen, 2004
S. hebeiense Chen, Li, Yan, Spiridonov & Moens, 2006
S. hermaphroditum Stock, Griffin & Chaerani, 2004
= *Steinernema dharanai*[5] Kulkarni, Rizvi, Kumar, Paunikar & Mishra, 2012 **n. syn.**
S. huense Phan, Mráček, Půža, Nermut' & Jarošová, 2014
S. ichnusae Tarasco, Mráček, Nguyen & Triggiani, 2008
S. innovationi Çimen, Lee, Hatting, Hazir & Stock, 2015

[4] Emended from the wrongly formed *-ensis* as used in the original description.
[5] Emended spelling from the original *dharanaii* (incorrectly formed ending).

S. intermedium (Poinar, 1986) Mamiya, 1988
= *Neoaplectana intermedia* Poinar, 1986
S. jeffreyense Malan, Knoetze & Tiedt, 2015
S. jollieti Spiridonov, Krasomil-Osterfeld & Moens, 2004
S. karii Waturu, Hunt & Reid, 1997
S. khoisanae Nguyen, Malan & Gozel, 2006
S. kushidai Mamiya, 1988
S. lamjungense Khatri-Chhetri, Waeyenberge, Spiridonov, Manandhar & Moens, 2011b
S. leizhouense Nguyen, Qiu, Zhou & Pang, 2006
S. litorale Yoshida, 2005
S. loci Phan, Nguyen & Moens, 2001a
S. longicaudum Shen & Wang, 1992
= *Steinernema serratum* Liu, 1992 [= *nomen nudum*]
S. minutum Maneesakorn, Grewal & Chandrapatya, 2010
S. monticolum Stock, Choo & Kaya, 1997
S. nguyeni Malan, Knoetze & Tiedt, 2016
S. neocurtillae Nguyen & Smart, 1992
S. nepalense Khatri-Chhetri, Waeyenberge, Spiridonov, Manandhar & Moens, 2011c
S. nyetense Kanga, Trinh, Waeyenberge, Spiridonov, Hauser & Moens, 2012
S. oregonense Liu & Berry, 1996a
S. pakistanense Shahina, Anis, Reid, Rowe & Maqbool, 2001
= *Steinernema maqbooli* Fayyaz, Khanum, Gulsher & Javed, 2013 **n. syn.**
S. papillatum San-Blas, Portillo, Nermut', Půža & Morales-Montero, 2015
S. phyllophagae Nguyen & Buss, 2011
S. poinari Mráček, Půža & Nermut', 2014a
S. puertoricense Román & Figueroa, 1994
S. pui Qiu, Zhao, Wu, Lv & Pang, 2011
S. puntauvense Uribe-Lorío, Mora & Stock, 2007
S. pwaniense[6] Půža, Nermut', Mráček, Gengler & Haukeland, 2016
S. rarum (de Doucet, 1986) Mamiya, 1988
= *Neoaplectana rara* de Doucet, 1986
S. riobrave Cabanillas, Poinar & Raulston, 1994

[6] Emended from the wrongly formed *-ensis* as used in the original description.

S. ritteri de Doucet & Doucet, 1990
S. robustispiculum Phan, Subbotin, Waeyenberge & Moens, 2005
S. sacchari Nthenga, Knoetze, Berry, Tiedt & Malan, 2014
S. sangi Phan, Nguyen & Moens, 2001
S. sasonense Phan, Spiridonov, Subbotin & Moens, 2006
S. scapterisci Nguyen & Smart, 1990
S. scarabaei Stock & Koppenhöfer, 2003
S. schliemanni Spiridonov, Waeyenberge & Moens, 2010
S. siamkayai Stock, Somsook & Reid, 1998
S. sichuanense Mráček, Nguyen, Tailliez, Boemare & Chen, 2006
S. silvaticum Sturhan, Spiridonov & Mráček, 2005
S. surkhetense Khatri-Chhetri, Waeyenberge, Spiridonov, Manandhar & Moens, 2011c
S. tami Luc, Nguyen, Reid & Spiridonov, 2000
S. texanum Nguyen, Stuart, Andalo, Gozel & Rogers, 2007
S. thanhi Phan, Nguyen & Moens, 2001a
S. thesami (Gorgadze, 1988) Gorgadze & Lortkipanidze,[7] 2004[8]
= *Neoaplectana thesami* Gorgadze, 1988
= *Steinernema tbilisiense*[9] Gorgadze, Lortkhipanidze, Ogier, Tailliez & Burjanadze, 2015 **n. syn.**
S. tielingense Ma, Chen, Li, Han, Khatri-Chhetri, De Clercq & Moens, 2012
S. tophus Çimen, Lee, Hatting, Hazir & Stock, 2014
S. unicornum Edgington, Buddie, Tymo, France, Merino & Hunt, 2009a
S. vulcanicum Clausi, Longo, Rappazzo, Tarasco & Vinciguerra, 2011
S. weiseri Mráček, Sturhan & Reid, 2003
S. xinbinense Ma, Chen, De Clercq, Waeyenberge, Han & Moens, 2012
S. xueshanense Mráček, Liu & Nguyen, 2009
S. yirgalemense Nguyen, Tesfamariam, Gozel, Gaugler & Adams, 2004

[7] The spelling of this surname varies between publications. She refers to herself in her CV as 'Lortkipanidze' but publications also have 'Lordkipanidze' and 'Lortkhipanidze'. In this chapter, we follow the spelling used in the title of the original article.

[8] The combining authority is assumed to be the earliest published combination of the species name with the generic name *Steinernema* that we have been able to find.

[9] Emended from the wrongly formed *-ensis* as used in the original description.

Species inquirendae

Steinernema acaudatum (de Villalobos & Camino, 1997) Sudhaus & Fürst von Lieven, 2003
= *Patanodontus acaudatum* de Villalobos & Camino, 1997
Steinernema agriotos (Veremchuk, 1969) Wouts, Mráček, Gerdin & Bedding, 1982
= *Neoaplectana agriotos* Veremchuk, 1969
Steinernema belorussica (Veremchuk, 1969) Wouts, Mráček, Gerdin & Bedding, 1982
= *Neoaplectana belorussica* Veremchuk, 1969
Steinernema bothynoderi (Kirjanova & Putschkova, 1955) Wouts, Mráček, Gerdin & Bedding, 1982
= *Neoaplectana bothynoderi* Kirjanova & Putschkova, 1955
Steinernema disparica Gorgadze, 2001
= *Steinernema* (*Neoaplectana*) *disparica* Gorgadze, 2001
= *Neoaplectana disparica* Gorgadze, 2001
Steinernema georgica (Kakuliya & Veremchuk, 1965) Wouts, Mráček, Gerdin & Bedding, 1982
= *Neoaplectana georgica* Kakuliya & Veremchuk, 1965
Steinernema gurgistanica Gorgadze & Lordkipanidze, 2006
Steinernema janickii (Weiser & Köhler, 1955) Wouts, Mráček, Gerdin & Bedding, 1982
= *Neoaplectana janickii* Weiser & Köhler, 1955
Steinernema kirjanovae (Veremchuk, 1969) Wouts, Mráček, Gerdin & Bedding, 1982
= *Neoaplectana kirjanovae* Veremchuk, 1969
Steinernema masoodi Ali, Shaheen, Pervez & Hussain, 2005
Steinernema melolonthae (Weiser, 1958) Wouts, Mráček, Gerdin & Bedding, 1982
= *Neoaplectana melolonthae* Weiser, 1958
Steinernema menozzii (Travassos, 1932) Wouts, Mráček, Gerdin & Bedding, 1982
= *Neoaplectana menozzii* Travassos, 1932
Steinernema octodontum (Camino & de Villalobos, 1996) Sudhaus & Fürst von Lieven, 2003
= *Diplogaster octodontus* Camino & de Villalobos, 1996
Steinernema qazii Ali, Shaheen, Asif & Akhtar, 2009
Steinernema sayeedae Ali & Azra, 2011

Steinernema seemae Ali, Shaheen, Pervez & Hussain, 2005
Steinernema semiothisae (Veremchuk & Litvinchuk, 1971) Wouts, Mráček, Gerdin & Bedding, 1982
= *Neoaplectana semiothisae* Veremchuk & Litvinchuk, 1971
Steinernema tabanivorum (Rubstov & Polevik, 1979) Wouts, Mráček, Gerdin & Bedding, 1982
= *Neoaplectana tabanivora* Rubstov & Polevik, 1979

Nomina nuda

Neoaplectana brevilarvalis pieridarum Sandner & Stanuszek, 1972
Neoaplectana chresima Steiner *in* Glaser, McCoy & Girth, 1942
Neoaplectana dutkii Welch, 1963
= *Neoaplectana dutkyi* Turco, Thames & Hopkins, 1971 (= a junior synonym of *S. carpocapsae*)
Neoaplectana elateridicola Veremchuk, 1970
Neoaplectana titovi Veremchuk, 1966
Steinernema serratum Liu, 1992
Steinernema solanense Walia & Bajaj, 2005

Notes

1) Poinar (1990) has been regarded as being the combining authority for a number of *Steinernema* species formally placed in *Neoaplectana* (*S. affine*, *S. anomalae*, *S. carpocapsae*, *S. intermedium* and *S. rarum*), but this is not correct. When Wouts *et al.* (1982) formally proposed *Neoaplectana* as a junior synonym of *Steinernema*, it is clear that they regarded all nominal species formerly in *Neoaplectana*, whether valid or not, as belonging to *Steinernema*. Although they did not formally propose the combination '*Steinernema carpocapsae*', for example, the action nevertheless occurred, notwithstanding the possibility that the authors themselves did not realise that such was the case. This applies to all the species placed in *Neoaplectana* (but not previously combined under *Steinernema*) prior to the paper by Wouts *et al.* (1982). Combinations made subsequent to that date take the combining authority as normal. A similar situation prevailed with combinations under the plant-parasitic genus *Globodera* and discussion with The Secretariat of the International Commission on Zoological Nomenclature (see Loof & Bakker, 1992) confirmed this point.

2) Many of the species listed as *species inquirenda* have been proposed as synonyms of several other species, notably the widespread

and common *S. carpocapsae* and *S. feltiae*. Although this is a convenient way of disposing of the names and of tidying up the nomenclature, the fact remains that these species are too poorly characterised to allow such conclusions, particularly now that many more species have been and are being proposed and the importance of molecular data is being realised. Accordingly, it is preferable to follow a more objective course whilst realising and accepting that the ultimate fate of these binomina may never be unambiguously resolved.

Of the *species inquirendae* or *nomina nuda* listed above, Mráček (pers. comm., 2006) indicates from his research that: *i*) *S. agriotos*, *S. belorussica*, *S. elateridicola* and *S. semiothisae* are likely to be junior synonyms of *S. carpocapsae*; *ii*) *S. kirjanovae* is probably a junior synonym of *S. kraussei*; and *iii*) *S. bothynoderi* and *S. tabanivora* are probably junior synonyms of *S. feltiae*.

3) The species *menozzii* may be a senior synonym of *affine* (see Wouts, 1980), *carpocapsae* (Wouts, unpublished letter) or *feltiae*.

4) In 1963 Welch, in an unpublished thesis, used the binomen *Neoaplectana dutkii* for the nematode then known as DD-136 and subsequently shown to be *N. carpocapsae*. The name was also employed by Jackson (1965), although it has no validity under the Code. Turco *et al.* (1971), cited the species as '*N. dutkyi* Jackson, 1965' in a key – despite the fact that Jackson had not validly characterised the species. In differentiating the species in a key, however, Turco *et al.* (1971) can be regarded as validly establishing the binomen and hence the correct citation is *N. dutkyi* Turco, Thames & Hopkins, 1971.

5) *Steinernema thezami* (Gorgadze, 1988) appears as an alternative spelling to *thesami* and may be either a *lapsus* or different transliteration. The species seems to have been named after 'Tezami', apparently a river, but maybe also a place, about 30 km north of Tbilisi. The more widespread '*S. thesami*' is assumed herein to be the correct spelling.

6) According to Sudhaus & Fürst von Lieven (2003), two species described from Argentina as diplogastrids actually belong to *Steinernema*, namely *Diplogaster octodontus* Camino & de Villalobos, 1996 and *Patanodontus acaudatum* de Villalobos & Camino, 1997. Apparently, both species have all the features of *Steinernema* and were therefore combined under that genus by Sudhaus & Fürst von Lieven (2003). These authors also synonymised under *Steinernema* the genus *Patanodontus* de Villalobos & Camino, 1997, which was proposed with *P. acaudatum* as type species (de Villalobos & Camino, 1997).

7) *Steinernema ritteri* de Doucet & Doucet, 1990 and *S. caudatum* Xu, Wang & Li, 1991 lack molecular data, but are nevertheless regarded as valid species herein.

8) *Steinernema thermophilum* Sudershan & Singh, 2000 is regarded as a junior synonym of *S. abbasi* because the sequences of the ITS region of the two nematodes are identical (Nguyen, pers. comm., 2006). Interestingly, an extensive and detailed study (Tailliez *et al.*, 2006) on the bacterial associates of *Steinernema* spp. found that both *S. abbasi* and *S. thermophilum* shared the same symbiont, namely *Xenorhabdus indica* Somvanshi, Lang, Ganguly, Swiderski, Saxena & Stackebrandt, 2006, a species that had been described from material identified as *S. thermophilum*. The authority for *S. thermophilum* is usually cited as 'Sudershan & Singh, 2000' yet also occurs as 'Ganguly & Singh, 2000'; the former is correct.

9) *Steinernema anatoliense* Hazir, Stock & Keskin, 2003 and *S. websteri* Cutler & Stock, 2003 are both regarded herein as junior synonyms of *S. carpocapsae*. Neither species has an ITS sequence available and the D2-D3 sequences are virtually identical to that of *S. carpocapsae*. Mráček *et al.* (2014b) remarked upon the doubtful status of *S. anatoliense* and *S. websteri* and comparison of the D2-D3 of 28S rRNA gene sequence of *S. websteri*, deposited in GenBank by Nadler *et al.* (2006), with those of *S. carpocapsae* did not reveal significant differences. Recently, Gökçe *et al.* (2015) reported that analysis of the 28S rDNA region of a Turkish isolate of *Steinernema*, which they attributed to *S. websteri*, shared a 100% similarity with the rDNA sequences of three *S. websteri* isolates in GenBank (accession nos EF217326, AY841762, and GU569058). Sequence comparisons also indicated a 99% similarity with *S. carpocapsae* (HM140688) and a 98% similarity with *S. anatoliense* (GU569043).

10) The original proposals of *S. masoodi* Ali, Shaheen, Pervez & Hussain, 2005 and *S. seemae* Ali, Shaheen, Pervez & Hussain, 2005 (Ali *et al.*, 2005) were compromised by exceptionally poor and inaccurate descriptions and both taxa were regarded as *species inquirendae* by Hunt (2007). The original descriptions were not supported by molecular data, although Ali *et al.* (2010) attempted to rectify this lack. Our analysis of sequences of *S. masoodi* and *S. seemae* published by Ali *et al.* (2010) reveals that these sequences do not belong to entomoparasitic nematodes. Thus, *S. masoodi* and *S. seemae* are still not molecularly characterised and remain as *species inquirendae*.

11) *Steinernema mushtaqi* Pervez, Ali & Asif, 2009 appears to be a *nomen nudum*.

12) The validity of *Steinernema qazii* Ali, Shaheen, Asif & Akhtar, 2009 is currently uncertain and the taxon is regarded as *species inquirenda* herein, the original description lacking molecular data. The species was isolated from Kanpur, Uttar Pradesh, India, and was said to be similar to *S. abbasi* and *S. masoodi*. In a reference cited by Kulkarni *et al.* (2012) this species appears as '*S. qaqiii*', the article title being the same as cited herein for Ali *et al.* (2009) although the journal volume and pagination are different.

13) *Steinernema arasbaranense* Nikdel, Niknam & Ye, 2011 may be a chimeric taxon, the sequence belonging to one, possibly new, species in the *affine*-group, whereas the morphological description belongs to another species in the *kraussei*-group (Půža, pers. comm.). Pending further information, we suspend judgement on this taxon, although it may better be regarded as a *species inquirenda*.

14) The validity of *Steinernema sayeedae* Ali & Azra, 2011 is currently uncertain and the taxon is regarded as *species inquirenda* herein, the original description lacking molecular data. The authors state that it has a very small IJ (*ca* 400 μm) and is most similar to *S. masoodi* and *S. diaprepesi*. Being named after a woman, the specific epithet is correctly formed. The nematode was isolated from soil in a banana plantation in Uttar Pradesh, India.

15) *Steinernema meghalayense* Ganguly, Rathour & Singh, 2011 is, based on molecular data, herein regarded as a junior synonym of *S. carpocapsae*. *Steinernema meghalayense* was described from Meghalaya state, India, by Ganguly *et al.* (2011). In their article, the authors provided the ITS rRNA gene sequence alignment of this species with the related species *S. tami*, *S. siamkayai*, *S. carpocapsae* and *S. scapterisci*. Visual analysis of the alignment showed that significant nucleotide differences of this species from all other species occurred in the first 100 and last 80 nucleotides, indicating mistakes in sequence reading for these positions. Excluding these positions from the alignment revealed a 100% sequence similarity of *S. meghalayense* with *S. carpocapsae*.

16) *Steinernema everestense* Khatri-Chhetri, Waeyenberge, Spiridonov, Manandhar & Moens, 2011a is, based on molecular data, herein regarded as a junior synonym of *S. akhursti*. *Steinernema akhursti* was found from soil samples collected from Yunnan Province, the People's

Republic of China, and described by Qiu *et al.* (2005b). The accession numbers of the partial 28S rRNA and the ITS rRNA gene sequences indicated in the original publication were mistaken and belong to another species. After searching in GenBank on the isolate names as indicated in the original article, two partial 28S rRNA gene sequences of this species were revealed, which were deposited for this species under numbers AY177194 and AY177188 (both in December 2003). The ITS rRNA gene sequence for this species was only released in July 2010 under the number AY603397.

Steinernema everestense was recovered from soil samples collected from Pakhribas, Dhankuta district, Nepal, and described by Khatri-Chhetri *et al.* (2011a). The authors obtained the partial 28S rRNA and the ITS rRNA gene sequences for this species, but did not compare them with those of *S. akhursti*. Our comparison of the ITS rRNA gene sequences of both species showed that they differed by four nucleotides and one four nucleotide indel (735 bp and 739 bp) from each other, whereas the partial 28S rRNA gene sequences (459 bp and 460 bp) of these species differed by one nucleotide only. These differences are minimal and are considered as intraspecific variation of *S. akhursti*.

17) *Steinernema dharanai*[10] Kulkarni, Rizvi, Kumar, Paunikar & Mishra, 2012 is, based on sequence analysis, accepted as a junior synonym of *S. hermaphroditum*. The original description of *S. hermaphroditum* from Indonesia given by Stock *et al.* (2004) contains only the D2-D3 of 28S rRNA gene sequence of this species. The ITS rRNA gene sequence of this species was released in GenBank by Ogier *et al.* (unpubl.) in 2012 with comments that this species from Indonesia was identified by 'Stock S.P.'. Our sequence analysis of the ITS rRNA gene sequences of *S. hermaphroditum* (JQ687355) and *S. dharanai* (KC252604) revealed a difference in one nucleotide only. Thus, based on molecular evidence we consider *S. dharanai* as a junior synonym of *S. hermaphroditum*, a conclusion also supported by Půža (pers. comm.).

18) *Steinernema maqbooli* Fayyaz, Khanum, Gulsher & Javed, 2013 is, based on molecular data, herein regarded as a junior synonym of *S. pakistanense*. *Steinernema maqbooli* was isolated and described from soil along Clifton beach, Karachi, Pakistan, by Fayyaz *et al.* (2013). The authors stated that the ITS rRNA sequence of *S. maqbooli* significantly differed from that of *S. pakistanense*. Our analysis of the

[10] Emended spelling from the original *dharanaii* (incorrectly formed ending).

alignment of ITS rRNA sequences of *S. maqbooli*, *S. pakistanense*, *S. bicornutum*, *S. riobrave*, *S. abbasi* and other species, however, revealed many nucleotide changes for *S. maqbooli* in conserved regions of the alignment, an indication of mistakes in sequence reading for *S. maqbooli*. Excluding these positions from the alignment resulted in the clustering of *S. maqbooli* with *S. pakistanense* and showed that differences between these taxa are within the range of intraspecific variation for *S. pakistanense*. The D2-D3 of 28S rRNA gene sequence of *S. maqbooli* also showed a very high similarity with that of *S. pakistanense*.

19) *Steinernema balochiense* Fayyaz, Khanum, Ali, Solangi, Gulsher & Javed, 2015 may, based on sequence data, be a junior synonym of *S. nepalense*. Sequence comparison of the ITS rRNA and the D2-D3 of 28S rRNA gene sequence of *S. balochiense* and *S. nepalense*, after excluding, because of possible reading mistakes, the 100 nucleotides at the 5′ end from the ITS alignment and the 20 nucleotides at the 3′ end from the 28S alignment, revealed minimal differences between these species: five nucleotides (0.7%) and four nucleotides (0.7%), respectively. Thus, based on molecular evidence we consider *S. balochiense* as a junior synonym of *S. nepalense*, a conclusion also supported by Půža (pers. comm.).

20) The endings of some of the specific epithets combined under *Steinernema* have been corrected to reflect either the fact that *-nema* is neuter in gender, or that they were originally wrongly formed by their author(s).

21) The '*Steinernema litura*' *in litt.* is apparently a *lapsus calami* for *Spodoptera litura*.

Genus *Neosteinernema* Nguyen & Smart, 1994

DIAGNOSIS (after Adams & Nguyen, 2002)

Male

Similar to *Steinernema* in general respects. Smaller than female. Spicule foot-shaped with dorsal hump or swelling, gubernaculum almost as long as spicule. Genital papillae arranged as single midventral papilla located just anterior to cloacal aperture and 13-14 pairs of which eight pairs precloacal. Tail conoid with digitate terminus, phasmids conspicuous.

Female

Similar to *Steinernema*. Labial region with four cephalic and six labial papillae. Ovoviviparous. Tail longer than anal body diam. ($c' = 1.1$-1.68), phasmids prominent, located on protuberance in posterior half of tail.

Infective juvenile

Similar in general respects to *Steinernema*. Labial region slightly swollen, tail long ($c =$ *ca* 5.5) elongate to filiform, as long as pharynx, usually curved near extremity, phasmid conspicuous.

TYPE AND ONLY SPECIES

N. longicurvicauda Nguyen & Smart, 1994

Family Heterorhabditidae

HISTORY

The first species of heterorhabditid to be described may have been *Rhabditis hambletoni* Pereira, 1937 (Pereira, 1937), although this is currently regarded as a *species inquirenda*. The first definitive description, however, came relatively late in the history of entomopathogenic nematology when Poinar (1976) proposed *Heterorhabditis bacteriophora* Poinar, 1976. The description of this species occurred in the same year as the proposal of the new genus and new species *Chromonema heliothidis* Khan, Brooks & Hirschmann, 1976 by Khan *et al.* (1976). The latter binomen was apparently based on the same original isolate as Poinar's species but, because Poinar published first, *Heterorhabditis bacteriophora* has priority and the latter becomes a junior synonym.

The family Heterorhabditidae Poinar, 1976 was proposed by Poinar (1976) at the same time as he described *Heterorhabditis*, the type genus. Although the family was initially regarded as belonging to the Rhabditida (together with the biologically similar Steinernematidae), recent molecular evidence has supported the contention of Poinar (1993) that the two families have evolved separately. Similarities in life cycle, and to some extent morphology, having therefore evolved in parallel. Current molecular evidence places the Heterorhabditidae with the strongylids (Blaxter *et al.*, 1998; De Ley & Blaxter, 2002, 2004).

The number of nominal *Heterorhabditis* species (currently only 16) is rather small when compared to the 95 valid species of *Steinernema*. This may partly be a consequence of previous under-reporting, a large proportion of the 16 valid species having been described in the last 15 years, molecular techniques playing a major role in diagnostics. The overall pattern of species descriptions is similar to that for *Steinernema*, only one species being described in the 1970s, one in the 1980s, four in the 1990s, nine in the 2000s (of which two are synonymised herein) and five since 2010, two of which we regard as junior synonyms. The pace of new descriptions therefore appears to be stable post-2000 although the genus is clearly not as speciose as *Steinernema*, probably due to the fact that the first generation is hermaphroditic rather than amphimictic, the latter strategy opening up the opportunity for greater genetic diversity in the ensuing progeny.

TAXONOMY AND SYSTEMATICS

Family Heterorhabditidae Poinar, 1976

DIAGNOSIS

Rhabditomorpha: Strongyloidea. Obligate parasites of insects and other arthropods. Infective juveniles non-feeding, carrying symbiotic entomopathogenic bacteria. First generation adults hermaphroditic, second generation adults amphimictic. Oviparous to ovoviviparous.

Adults

Labial region bearing six, separate, conoid lips, each with terminal papilla. Amphidial apertures small. Stoma wide, shallow, cheilostom forming a ring, other parts fused. Pharyngeal collar present. Pharynx short, with cylindroid procorpus, no metacorpal swelling, isthmus weak, basal bulb spheroid, with small valve plates. Nerve ring located in isthmus region, excretory pore usually opening posterior to basal bulb. Vulva median, transverse. Hermaphrodite with two opposed, reflexed, ovotestes; amphimictic female with amphididelphic, reflexed, ovaries. Tail short conoid. Male spicules paired, separate, slender, more or less straight with small, offset, capitulum. Gubernaculum linear in lateral view, plate-like in ventral view. Bursa present. Usually nine pairs of genital papillae or rays, sometimes fewer. Tail medium conoid, tapering to pointed terminus.

Infective juvenile

Third-stage infective juvenile usually enclosed in longitudinally ridged sheath of second-stage. Sheath tessellate anteriorly. Infective juvenile vermiform, non-feeding, bearing apical dorsal tooth-like structure on labial region. Lateral field with two ridges, although central incisure not easily visible with LM so that area between bands appearing smooth. Mouth and anus closed, stoma and pharynx reduced. Excretory pore located posterior to nerve ring. Symbiotic bacteria retained in intestine. Tail medium conoid, tapering to pointed terminus.

TYPE AND ONLY GENUS

Heterorhabditis Poinar, 1976
= *Chromonema* Khan, Brooks & Hirschmann, 1976

Genus *Heterorhabditis* Poinar, 1976
= *Chromonema* Khan, Brooks & Hirschmann, 1976

DIAGNOSIS (modified after Adams & Nguyen, 2002)

Hermaphroditic female

Large, plump nematodes. Labial region truncate to slightly rounded, lips six, well developed, separate, each with terminal papilla. One or two small raised structures may be present at base of each lip. Amphidial aperture small. Stoma wide, shallow, cheilorhabdions forming a cuticularised refractive ring, other stoma parts fused to form collapsed posterior portion. Posterior part of stoma surrounded by pharyngeal collar. Pharynx lacking metacorpus, isthmus slender, basal bulb swollen with reduced valve plates. Nerve ring located at middle of isthmus, excretory pore usually located posterior to pharynx. Vulva median, slit-like, surrounded by elliptical rings. Ovotestes two, opposed, reflexed. Oviparous at first, ovoviviparous later. Tail conoid, longer than anal body diam., postanal swelling usually present.

Amphimictic female

Similar to, but usually smaller than, hermaphrodite. Labial papillae prominent. Reproductive tract amphididelphic reflexed. Vulva not functional for egg laying (eclosion in body) but functional for mating.

Male

Genital tract monorchic, reflexed. Spicules paired, separate, slightly curved ventrally. Spicule capitulum short, offset from lamina by constriction. Gubernaculum usually *ca* half of spicule length. Bursa well-developed, peloderan, usually nine pairs of ray-like genital papillae (occasionally fewer) present.

Infective juvenile

Third-stage infective juvenile usually enclosed in sheath formed from cuticle of second-stage. Sheath with longitudinal ridges and tessellated anteriorly. Cuticle of infective juvenile striated, lateral field with two ridges although central incisure not easily visible with LM so that area between bands appears smooth. Labial region with prominent dorsal tooth-like process. Mouth and anus closed, non-functional. Stoma collapsed with parallel walls. Pharynx and intestine reduced. Excretory pore located posterior to nerve ring. Symbiotic bacteria retained within intestine. Tail conoid, tapering to pointed terminus.

TYPE SPECIES

Heterorhabditis bacteriophora Poinar, 1976
= *Chromonema heliothidis* Khan, Brooks & Hirschmann, 1976
= *Heterorhabditis heliothidis* (Khan, Brooks & Hirschmann, 1976) Poinar, Thomas & Hess, 1977
= *H. argentinensis* Stock, 1993

OTHER SPECIES

H. amazonensis Andaló, Nguyen & Moino, 2007
H. atacamensis Edgington, Buddie, Moore, France, Merino & Hunt, 2011
H. baujardi Phan, Subbotin, Nguyen & Moens, 2003
= *Heterorhabditis somsookae*[11] Maneesakorn, An, Grewal & Chandrapatya, 2015 **n. syn.**
H. beicherriana Li, Liu, Nermut', Půža & Mráček, 2012
H. downesi Stock, Griffin & Burnell, 2002
H. floridensis Nguyen, Gozel, Koppenhöfer & Adams, 2006

[11] Termination of specific epithet emended, in accordance with ICZN Rules, from '*somsookii*' to *somsookae*, the species being stated as being named after a woman, Mrs Vacharee Somsook.

H. georgiana Nguyen, Shapiro-Ilan & Mbata, 2008
H. indica Poinar, Karunakar & David, 1992
= *Heterorhabditis brevicaudis* Liu, 1994
= *Heterorhabditis hawaiiensis* Gardner, Stock & Kaya, 1994
= *Heterorhabditis gerrardi* Plichta, Joyce, Clarke, Waterfield & Stock, 2009 **n. syn.**
= *Heterorhabditis pakistanensis*[12] Shahina, Tabassum, Salma, Mehreen & Knoetze, 2016 **n. syn.**
H. marelatus Liu & Berry, 1996b
= *Heterorhabditis hepialius* Stock, Strong & Gardner, 1996
H. megidis Poinar, Jackson & Klein, 1987
H. mexicana Nguyen, Shapiro-Ilan, Stuart, McCoy, James & Adams, 2004
H. noenieputensis Malan, Knoetze & Tiedt, 2014
H. safricana Malan, Nguyen, De Waal & Tiedt, 2008
H. taysearae Shamseldean, Abou El-Sooud, Abd-Elgawad & Saleh, 1996
= *Heterorhabditis sonorensis* Stock, Rivera-Orduño & Flores-Lara, 2009 **n. syn.**
H. zealandica Poinar, 1990
= *Heterorhabditis heliothidis apud* Wouts, 1979 *nec* Khan, Brooks & Hirschmann, 1976

Species inquirendae

H. egyptii Abd-Elgawad & Ameen, 2005
H. hambletoni (Pereira, 1937) Poinar, 1976
= *Rhabditis hambletoni* Pereira, 1937
H. hoptha (Turco, 1970) Poinar, 1979
= *Neoaplectana hoptha* Turco, 1970
H. poinari Kakuliya & Mikaia, 1997

Nomina nuda

Heterorhabditis downesi Hass, Downes & Griffin, 2001 *nec H. downesi* Stock, Griffin & Burnell, 2002
Heterorhabditis minutus Prabhuraj, Viraktamath & Kumar, 2003

[12] Emended to *pakistanensis* from the incorrectly formed '*pakistanense*' in the original paper.

Notes

1) Hass *et al.* (2001) published the combination *Heterorhabditis downesi* before the species was formally described by Stock *et al.* (2002), the former paper being published on 27 December 2001 and the latter in February 2002. Hass *et al.* (2001) did not provide sufficient information to diagnose the species and hence their regrettable use of the name prior to its official publication must be regarded as a *nomen nudum*. In view of this situation, the subsequent proposal of the binomen *Heterorhabditis downesi* by Stock *et al.* (2002) must be regarded as valid.

2) *Heterorhabditis minutus* Prabhuraj, Viraktamath & Kumar, 2003 is regarded as a *nomen nudum*, it being referred to as a new species in a book chapter, apparently without formal description.

3) '*Heterorhabditis riobravus*' is a *lapsus* for *Heterorhabditis baujardi* (see Costa *et al.*, 2007 and their correction in Costa *et al.*, 2008).

4) *Heterorhabditis gerrardi* Plichta, Joyce, Clarke, Waterfield & Stock, 2009 is, on the basis of molecular data, herein regarded as a junior synonym of *H. indica*. *Heterorhabditis gerrardi* was found in sandy soils from Kingscliff beach, Australia, and described as a new species by Plichta *et al.* (2009). The authors stated that the ITS rRNA gene sequence of this species differed by eight nucleotides from that of *H. indica* (AY321483, India). However, our sequence comparison of *H. gerrardi* (FJ152545) with *H. indica*, after excluding mistakes in sequence reading in the ITS2 for *H. gerrardi* (AY321483) and even after adding several other published sequences of *H. indica* samples (India, Nepal), did not reveal any differences in the ITS sequences between *H. gerrardi* and *H. indica*.

5) *Heterorhabditis sonorensis* Stock, Rivera-Orduño & Flores-Lara, 2009 was isolated and described by Stock *et al.* (2009) from nymphal stages of the seasonal cicada, *Diceroprocta ornea* (Walker), in an asparagus field in the state of Sonora, Mexico. Comparison of the ITS rRNA gene sequences with other *Heterorhabditis* revealed a high similarity with those identified as *H. taysearae* from Palestine (Darissa *et al.*, unpubl.; Iraki & Dar Issa, unpubl.; Regeai & Burnell, unpubl.), small sequence differences being considered as being within the range of intraspecific variation. Thus, based on molecular characterisation *H. sonorensis* is considered herein as a junior synonym of *H. taysearae*.

6) *Heterorhabditis somsookae* Maneesakorn, An, Grewal & Chandrapatya, 2015 is, on the basis of molecular data, herein regarded as a junior synonym of *H. baujardi*. *Heterorhabditis somsookae* was discovered and described from soil samples collected in Kanchanaburi Province, Western Thailand by Maneesakorn *et al.* (2015). The authors provided the ITS rRNA gene sequence for this species and compared it with other *Heterorhabditis* spp., including a sequence of an isolate under accession number EU363039, which was originally identified as *H. baujardi* by Dolinski *et al.* (2008), but might indeed be *H. amazonensis* or an undescribed species. Comparison of the ITS rRNA gene sequence of *H. somsookae* with that of the type population of *H. baujardi* (AF548768) revealed only four nucleotide differences between these species, *i.e.*, within the range of intraspecific variation for *H. baujardi*.

7) *Heterorhabditis pakistanensis*[13] Shahina, Tabassum, Salma, Mehreen & Knoetze, 2016 is, on the basis of molecular sequence data, regarded as a junior synonym of *H. indica*. Although the authors claimed that their taxon differed from *H. indica* by 9 bp and from *H. gerrardi* (herein also regarded as a junior synonym of *H. indica*) by 7 bp, this is well within the range of intraspecific variation for *H. indica*.

References

Abate, B.A., Malan, A.P., Tiedt, L.R., Wingfield, M.J., Slippers, B. & Hurley, B.P. (2016). *Steinernema fabii* n. sp. (Rhabditida: Steinernematidae), a new entomopathogenic nematode from South Africa. *Nematology* 18, 235-255.

Abd-Elgawad, M.M.M. & Ameen, H.H. (2005). *Heterorhabditis egyptii* n. sp. (Rhabditida: Heterorhabditidae) from Egypt. *Egyptian Journal of Agricultural Research-National Research Center* 2, 855-870.

Adams, B.J. & Nguyen, K.B. (2002). Taxonomy and systematics. In: Gaugler, R. (Ed.). *Entomopathogenic nematology*. Wallingford, UK, CABI Publishing, pp. 1-33.

Ali, S.S. & Azra, S. (2011). *Steinernema sayeedae* sp. n. a heat tolerant EPN from banana rhizosphere of Koshambhi district, U.P. India. *Trends in Biosciences* 4, 123-125.

Ali, S.S., Shaheen, A., Pervez, R. & Hussain, M.A. (2005). *Steinernema masoodi* sp. n. and *S. seemae* sp. n. (Nematoda: Rhabditidae [*sic*]:

[13] Emended to *pakistanensis* from the incorrectly formed '*pakistanense*' in the original paper.

Steinernematidae) from India. *International Journal of Nematology* 15, 89-99.

ALI, S.S., SHAHEEN, A., ASIF, M. & AKHTAR, M.H. (2009). *Steinernema qazii* sp. n. (Nematoda: Steinernematidae) from Kanpur, India. *Trends in Biosciences* 2, 59-64.

ALI, S.S., RIYAZ, A., VERMA, V., AZRA, S., RASHID, P. & SOBIA, A. (2010). Molecular characterization of *Steinernema masoodi, S. seemae* and other Indian isolates of *Steinernema* spp. *Trends in Biosciences* 3, 112-116.

ANDALÓ, V., NGUYEN, K.B. & MOINO, A. (2007). *Heterorhabditis amazonensis* n. sp. (Rhabditida: Heterorhabditidae) from Amazonas, Brazil. *Nematology* 8(2006), 853-867.

ANIS, M., SHAHINA, F., REID, A.P. & ROWE, J. (2002). *Steinernema asiaticum* sp. n. (Rhabditida: Steinernematidae) from Pakistan. *International Journal of Nematology* 12, 220-231.

ARTYUKHOVSKY, A.K. (1967). [*Neoaplectana arenaria* nov. sp. (Steinernematidae, Nematoda) inducing nematode disease in chafers of the Voronezh region.] *Trudy Voroneszhskogo Gosudarstvennogo Zapovednika* 15, 94-100.

BLAXTER, M.L., DE LEY, P., GAREY, J.R., LIU, L.X., SCHELDMAN, P., VIERSTRAETE, A., VANFLETEREN, J.R., MACKEY, L.Y., DORRIS, M., FRISSE, L.M. *ET AL.* (1998). A molecular evolutionary framework for the phylum Nematoda. *Nature* 392, 71-75.

BOVIEN, P. (1937). Some types of association between nematodes and insects. *Videnskabelige Meddelelser fra Dansk Naturhistorisk Forening, København* 101, 1-114.

CABANILLAS, H.E., POINAR JR, G.O. & RAULSTON, J.R. (1994). *Steinernema riobravis* sp. nov. (Rhabditida: Steinernematidae) from Texas. *Fundamental and Applied Nematology* 17, 123-131.

CAMINO, N.B. & DE VILLALOBOS, L.C. (1996). *Diplogaster octodontus* n. sp. (Rhabditida, Diplogasteridae) parasito de ninfas de *Gryllodes laplatae* (Orthoptera, Gryllidae) en la Argentina. *Iheringia Serie Zoologia* 89, 183-186.

CHEN, S., LI, X., YAN, A., SPIRIDONOV, S.E. & MOENS, M. (2006). A new entomopathogenic nematode, *Steinernema hebeiense* sp. n. (Rhabditida: Steinernematidae) from North China. *Nematology* 8, 563-574.

CHITWOOD, B.G. & CHITWOOD, M.B. (1937). *An introduction to nematology. Section 1, anatomy*. Baltimore, MD, USA, Monumental Printing Co.

ÇIMEN, H., LEE, M.-M., HATTING, J., HAZIR, S. & STOCK, S.P. (2014). *Steinernema tophus* sp. n. (Nematoda: Steinernematidae), a new entomopathogenic nematode from South Africa. *Zootaxa* 3821, 337-353.

ÇIMEN, H., LEE, M.M., HATTING, J., HAZIR, S. & STOCK, S.P. (2015). *Steinernema innovationi* n. sp. (Panagrolaimomorpha: Steinernematidae), a new entomopathogenic nematode species from South Africa. *Journal of Helminthology* 89, 415-427. DOI:10.1017/S0022149X14000182.

Çimen, H., Půža, V., Nermut', J., Hatting, J., Ramakuwela, T., Faktorová, L. & Hazir, S. (2016). *Steinernema beitlechemi* n. sp., a new entomopathogenic nematode (Nematoda: Steinernematidae) from South Africa. *Nematology* 18, 439-453.

Clausi, M., Longo, A., Rappazzo, G., Tarasco, E. & Vinciguerra, M.T. (2011). *Steinernema vulcanicum* n. sp. (Rhabditida: Steinernematidae), a new entomopathogenic nematode species from Sicily (Italy). *Nematology* 13, 409-423.

Costa, J.C., Dias, R.J. & Morenz, M.J. (2007). Determining the adaptation potential of entomopathogenic nematode multiplication of *Heterorhabditis riobravus* and *Steinernema carpocapsae* (Rhabditida: Heterorhabditidae, Steinernematidae) in larvae of *Alphitobius diaperinus* (Coleoptera: Tenebrionidae) and *Galleria mellonella* (Lepidoptera: Pyralidae). *Parasitological Research* 102, 139-144.

Costa, J.C., Dias, R.J. & Morenz, M.J. (2008). Erratum: Determining the adaptation potential of entomopathogenic nematode multiplication of *Heterorhabditis baujardi* and *Steinernema carpocapsae* (Rhabditida: Heterorhabditidae, Steinernematidae) in larvae of *Alphitobius diaperinus* (Coleoptera: Tenebrionidae) and *Galleria mellonella* (Lepidoptera: Pyralidae). *Parasitological Research* 102, 823.

Curran, J. (1989). Chromosome numbers of *Steinernema* and *Heterorhabditis* species. *Revue de Nématologie* 12, 145-148.

Cutler, G.C. & Stock, S.P. (2003). *Steinernema websteri* sp. n. (Rhabditida: Steinernematidae), a new entomopathogenic nematode from China. *Nematologia Mediterranea* 31, 215-224.

de Doucet, M.M.A. (1986). A new species of *Neoaplectana* Steiner, 1929 (Nematoda: Steinernematidae) from Cordoba, Argentina. *Revue de Nématologie* 9, 317-323.

de Doucet, M.M.A. & Doucet, M.E. (1990). *Steinernema ritteri* n. sp. (Nematoda: Steinernematidae) with a key to the species of the genus. *Nematologica* 36, 257-265.

De Ley, P. & Blaxter, M.L. (2002). Systematic position and phylogeny. In: Lee, D.L. (Ed.). *The biology of nematodes*. London, UK, Taylor & Francis, pp. 1-30.

De Ley, P. & Blaxter, M.L. (2004). A new system for Nematoda: combining morphological characters with molecular trees, and translating clades into ranks and taxa. In: Cook, R. & Hunt, D.J. (Eds). *Proceedings of the fourth international congress of nematology, Tenerife, Spain, 8-13 June 2002. Nematology Monographs and Perspectives 2.* Leiden, The Netherlands, Brill, pp. 633-653.

DE VILLALOBOS, C. & CAMINO, N.B. (1997). *Patanodontus acaudatum* gén. n. et sp. n. (Nematoda-Diplogasteridae) un parasito de larves de *Diloboderus abderus* Sturm. (Coleoptera-Scarabaeidae) en Argentina. *Estudo de Biologia (Curitiba)* 42, 33-39.

DOLINSKI, C., KAMITANI, F.L., MACHADO, I.R. & WINTER, C.E. (2008). Molecular and morphological characterization of heterorhabditid entomopathogenic nematodes from the tropical rainforest in Brazil. *Memórias do Instituto Oswaldo Cruz* 103, 150-159.

DUTKY, S.R. & HOUGH, W.S. (1955). Note on a parasitic nematode from codling moth larvae, *Carpocapsa pomonella* (Lepidoptera, Olethreutidae). *Proceedings of the Entomological Society of Washington* 57, 244.

EDGINGTON, S., BUDDIE, A.G., TYMO, L., FRANCE, A., MERINO, L. & HUNT, D.J. (2009a). *Steinernema unicornum* sp. n. (Panagrolaimomorpha: Steinernematidae), a new entomopathogenic nematode from Chile. *Journal of Nematode Morphology and Systematics* 12, 113-131.

EDGINGTON, S., BUDDIE, A.G., TYMO, L., HUNT, D.J., NGUYEN, K.B., FRANCE, A.I., MERINO, L.M. & MOORE, D. (2009b). *Steinernema australe* n. sp. (Panagrolaimomorpha: Steinernematidae), a new entomopathogenic nematode from Isla Magdalena, Chile. *Nematology* 11, 699-717.

EDGINGTON, S., BUDDIE, A.G., MOORE, D., FRANCE, A., MERINO, L. & HUNT, D.J. (2011). *Heterorhabditis atacamensis* n. sp. (Nematoda: Heterorhabditidae), a new entomopathogenic nematode from the Atacama Desert, Chile. *Journal of Helminthology* 85, 381-394.

ELAWAD, S., AHMED, W. & REID, A.P. (1997). *Steinernema abbasi* sp. n. (Nematoda: Steinernematidae) from the Sultanate of Oman. *Fundamental and Applied Nematology* 20, 435-442.

FAYYAZ, S., KHANUM, T.A., GULSHER, M. & JAVED, S. (2013). *Steinernema maqbooli* sp. n. (Nematoda: Steinernematidae), a species of the '*bicornutum*' group from Pakistan. *International Journal of Nematology* 23, 59-72.

FAYYAZ, S., YAN, X., QIU, L., HAN, R., GULSHER, M., KHANUM, T.A. & JAVED, S. (2014). A new entomopathogenic nematode, *Steinernema bifurcatum* n. sp. (Rhabditida: Steinernematidae) from Punjab, Pakistan. *Nematology* 16, 821-836.

FAYYAZ, S., KHANUM, T.A., ALI, S., SOLANGI, G.S., GULSHER, M. & JAVED, S. (2015). *Steinernema balochiense* n. sp. (Rhabditida: Steinernematidae) a new entomopathogenic nematode from Pakistan. *Zootaxa* 3904, 387-402.

FILIPJEV, I.N. (1934). Miscellanea Nematologica 1. Eine neue Art der Gattung *Neoaplectana* Steiner nebst Bermerkungen über die systematische Stellung der letzteren. *Parasitologicheskii Sbornik* 4, 229-240.

GANGULY, S., RATHOUR, K.S. & SINGH, M. (2011). *Steinernema meghalayensis* [*sic*] sp. n. (Rhabditida: Steinernematidae) from northeastern hilly region of India. *Indian Journal of Nematology* 41, 83-97.

GARDNER, S.L., STOCK, S.P. & KAYA, H.K. (1994). A new species of *Heterorhabditis* from the Hawaiian Islands. *Journal of Parasitology* 80, 100-106.

GLASER, R.W., MCCOY, E.E. & GIRTH, H.B. (1942). The biology and culture of *Neoaplectana chresima*, a new nematode parasitic in insects. *Journal of Parasitology* 28, 122-136.

GÖKÇE, C., ERBAŞ, Z., YILMAZ, H., DEMIRBAĞ, Z. & DEMIR, I. (2015). A new entomopathogenic nematode species from Turkey, *Steinernema websteri* (Rhabditida: Steinernematidae), and its virulence. *Turkish Journal of Biology* 39, 167-174.

GORGADZE, O. (1988). [A new species of nematode *Neoaplectana thezami* sp. n. found in *Operophthera brumata* L.] *Bulletin of the Academy of Science, Georgian SSR* 130, 405-408.

GORGADZE, O. (2001). The entomopathogenic nematode new species *Steinernema* (*Neoaplectana*) *disparica* sp. n. (Rhabditidae, Steinernematidae) from Georgia. *Bulletin of the Georgian Academy of Sciences* 164, 161-163.

GORGADZE, O. & LORTKIPANIDZE, M. (2004). The invasion of colorado beetle (*Leptinotarsa decemlineata* Say) by using the nematode *Steinernema thesami* (Steinernematidae). *Bulletin of the Georgian Academy of Sciences* 169, 592-594.

GORGADZE, O. & LORDKIPANIDZE, M. (2006). Nematoda *Steinernema gurgistana* sp. n. (Rhabditida: Steinernematidae) from *Agriotes gurgistana* F. *Proceedings of the Georgian Academy of Science, Biological Series, B* 4, 117-122.

GORGADZE, O., LORTKHIPANIDZE, M., OGIER, J.-C., TAILLIEZ, P. & BURJANADZE, M. (2015). *Steinernema tbilisiensis* sp. n. (Nematoda: Steinernematidae) – a new species of entomopathogenic nematode from Georgia. *Journal of Agricultural Science and Technology A* 5, 264-276.

HASS, B., DOWNES, M.J. & GRIFFIN, C.T. (2001). Correlation between survival in water and persistence of infectivity in soil of *Heterorhabditis* spp. isolates. *Nematology* 3, 573-579.

HAZIR, S., STOCK, S.P. & KESKIN, N. (2003). A new entomopathogenic nematode, *Steinernema anatoliense* n. sp. (Rhabditida: Steinernematidae), from Turkey. *Systematic Parasitology* 55, 211-220.

HOMINICK, W.M., BRISCOE, B., DEL PINO, F.G., HENG, J., HUNT, D.J., KOZODOY, E., MRÁČEK, Z., NGUYEN, K.B., REID, A.P., SPIRIDONOV, S. *ET AL.* (1997). Biosystematics of entomopathogenic nematodes: current status, protocols and definitions. *Journal of Helminthology* 71, 271-298.

HOY, J.M. (1954). The biology and host range of *Neoaplectana leucaniae*, a new species of insect-parasitic nematode. *Parasitology* 44, 392-399.

HUNT, D.J. (2007). Overview of taxonomy and systematics. In: Nguyen, K.B. & Hunt, D.J. (Eds). *Entomopathogenic nematodes: taxonomy, phylogeny and bacterial symbionts. Nematology Monographs and Perspectives 5* (Series editors: Hunt, D.J. & Perry, R.N.). Leiden, The Netherlands, Brill, pp. 27-57.

INTERNATIONAL COMMISSION ON ZOOLOGICAL NOMENCLATURE (1999). *International Code of Zoological Nomenclature*, 4th edition. London, UK, International Trust for Zoological Nomenclature.

JACKSON, G.J. (1965). Differentiation of three species of *Neoaplectana* (Nematoda: Rhabditida), grown axenically. *Parasitology* 55, 571-578.

JIAN, H., REID, A.P. & HUNT, D.J. (1997). *Steinernema ceratophorum* n. sp. (Nematoda: Steinernematidae), a new entomopathogenic nematode from north-east China. *Systematic Parasitology* 37, 115-125.

KAKULIYA, G. & MIKAIA, N. (1997). [New species of the nematode *Heterorhabditis poinari* sp. nov. (Rhabditida; Heterorhabditidae).] *Soobshcheniya Akademii Nauk Gruzii SSR* 155, 457-459.

KAKULIYA, G.A. & VEREMCHUK, G.V. (1965). [*Neoaplectana georgica* n. sp. (Nematoda: Steinernematidae) in *Amphimallon solstitialis*.] *Soobshcheniya Akademii Nauk Gruzii SSR* 40, 713-718.

KANGA, F.N., TRINH, P.Q., WAEYENBERGE, L., SPIRIDONOV, S.E., HAUSER, S. & MOENS, M. (2012). Two new species of *Steinernema* Travassos, 1927 from the humid forest of southern Cameroon. *Russian Journal of Nematology* 20, 15-36.

KHAN, A., BROOKS, W.M. & HIRSCHMANN, H. (1976). *Chromonema heliothidis* n. gen., n. sp. (Steinernematidae, Nematoda), a parasite of *Heliothis zeae* (Noctuidae, Lepidoptera), and other insects. *Journal of Nematology* 8, 159-168.

KHATRI-CHHETRI, H.B., WAEYENBERGE, L., SPIRIDONOV, S., MANANDHAR, H.K. & MOENS, M. (2011a). *Steinernema everestense* n. sp. (Rhabditida: Steinernematidae), a new species of entomopathogenic nematode from Pakhribas, Dhankuta, Nepal. *Nematology* 13, 443-462.

KHATRI-CHHETRI, H.B., WAEYENBERGE, L., SPIRIDONOV, S., MANANDHAR, H.K. & MOENS, M. (2011b). *Steinernema lamjungense* n. sp. (Rhabditida: Steinernematidae), a new species of entomopathogenic nematode from Lamjung district, Nepal. *Nematology* 13, 589-605.

KHATRI-CHHETRI, H.B., WAEYENBERGE, L., SPIRIDONOV, S.E., MANADHAR, H.M. & MOENS, M. (2011c). Two new species of *Steinernema* Travassos, 1927 with short IJ from Nepal. *Russian Journal of Nematology* 19, 53-74.

KIRJANOVA, E.S. & PUTSCHKOVA, L.V. (1955). [A new parasite of the beet weevil – *Neoaplectana bothynoderi* Kirjanova et Putschkova n. sp. (Nematoda).] *Trudi Zoologicheskogo Instituta Akademiya Nauk SSR* 18, 53-62.

KOZODOI, E.M. (1984). [A new entomophagous nematode *Neoaplectana anomali* sp. n. (Rhabditida: Steinernematidae) and its biology.] *Zoologichesky Zhurnal* 63, 1605-1909.

KRAUSSE, A. (1917). Forstentomologische Exkursionen ins Eggegebirge zum Studium der Massenvermehrung der *Cephaleia abietis* L. *Archiv für Naturgeschichte Berlin* A6, 46-49.

KULKARNI, N., RIZVI, A.N., KUMAR, V., PAUNIKAR, S. & MISHRA, V.K. (2012). Morphological and molecular characterization of *Steinernema dharanaii* sp. n.: a new entomopathogenic nematode from India. *Indian Journal of Tropical Biodiversity* 20, 107-116.

LEE, M.M., SICARD, M., SKEIE, M. & STOCK, S.P. (2009). *Steinernema boemarei* n. sp. (Nematoda: Steinernematidae), a new entomopathogenic nematode from southern France. *Systematic Parasitology* 72, 127-141.

LI, X.Y., LIU, Q.Z., NERMUT', J., PŮŽA, V. & MRÁČEK, Z. (2012). *Heterorhabditis beicherriana* n. sp. (Nematoda: Heterorhabditidae), a new entomopathogenic nematode from the Shunyi district of Beijing, China. *Zootaxa* 3569, 25-40.

LIU, J. (1992). Taxonomic study of the genus: *Steinernema* Travassos and *Heterorhabditis* Poinar. In: *Proceedings of the XIX international congress of entomology, 28 June-4 July 1992, Beijing, China*, p. 318. [Abstr.]

LIU, J. (1994). [A new species of the genus *Heterorhabditis* from China (Rhabditida: Heterorhabditidae).] *Acta Zootaxonomica Sinica* 19, 268-272.

LIU, J. & BERRY, R.E. (1996a). *Steinernema oregonensis* n. sp. (Rhabditida: Steinernematidae) from Oregon, U.S.A. *Fundamental and Applied Nematology* 19, 375-380.

LIU, J. & BERRY, R.E. (1996b). *Heterorhabditis marelatus* n. sp. (Rhabditida: Heterorhabditidae) from Oregon. *Journal of Invertebrate Pathology* 67, 48-54.

LOOF, P.A.A. & BAKKER, J. (1992). Authorities of specific names in, and transfers to, the nominal genus *Globodera* Skarbilovich, 1959. *Nematologica* 38, 385-386.

LÓPEZ-NÚÑEZ, J.C., PLICHTA, K., GÓNGORA-BOTERO, C.E. & STOCK, S.P. (2008). A new entomopathogenic nematode, *Steinernema colombiense* n. sp. (Nematoda: Steinernematidae) from Colombia. *Nematology* 10, 561-574.

LUC, P.V., NGUYEN, K.B., REID, A.P. & SPIRIDONOV, S.E. (2000). *Steinernema tami* sp. n. (Rhabditida: Steinernematidae) from Cat Tien Forest, Vietnam. *Russian Journal of Nematology* 8, 33-43.

MA, J., CHEN, S., DE CLERCQ, P., HAN, R. & MOENS, M. (2012). *Steinernema changbaiense* sp. n. (Rhabditida: Steinernematidae), a new species of entomopathogenic nematode from Northeast China. *Russian Journal of Nematology* 20, 97-112.

MA, J., CHEN, S., DE CLERCQ, P., WAEYENBERGE, L., HAN, R. & MOENS, M. (2012). A new entomopathogenic nematode, *Steinernema xinbinense* n. sp. (Nematoda: Steinernematidae), from North China. *Nematology* 14, 723-739.

MA, J., CHEN, S., LI, X., HAN, R., KHATRI-CHHETRI, H.B., DE CLERCQ, P. & MOENS, M. (2012). A new entomopathogenic nematode, *Steinernema tielingense* n. sp. (Rhabditida: Steinernematidae), from north China. *Nematology* 14, 321-338.

MALAN, A.P., NGUYEN, K.B., DE WAAL, J.Y. & TIEDT, L. (2008). *Heterorhabditis safricana* n. sp. (Nematoda: Heterorhabditidae), a new entomopathogenic nematode from South Africa. *Nematology* 10, 381-396.

MALAN, A.P., KNOETZE, R. & TIEDT, L. (2014). *Heterorhabditis noenieputensis* n. sp. (Rhabditida: Heterorhabditidae), a new entomopathogenic nematode from South Africa. *Journal of Helminthology* 88, 139-151.

MALAN, A.P., KNOETZE, R. & TIEDT, L.R. (2015). *Steinernema jeffreyense* n. sp. (Rhabditida: Steinernematidae), a new entomopathogenic nematode from South Africa. *Journal of Helminthology* 90, 262-278.

MALAN, A.P., KNOETZE, R. & TIEDT, L.R. (2016). *Steinernema nguyeni* n. sp. (Rhabditida: Steinernematidae), a new entomopathogenic nematode from South Africa. *Nematology* 18, 571-590.

MAMIYA, Y. (1988). *Steinernema kushidai* n. sp. (Nematoda: Steinernematidae) associated with scarabaeid beetle larvae from Shizuoka, Japan. *Applied Entomology and Zoology* 23, 313-320.

MANEESAKORN, P., GREWAL, P.S. & CHANDRAPATYA, A. (2010). *Steinernema minutum* sp. nov. (Rhabditida: Steinernema): a new entomopathogenic from Thailand. *International Journal of Nematology* 20, 27-42.

MANEESAKORN, P., AN, R., GREWAL, P.S. & CHANDRAPATYA, A. (2015). *Heterorhabditis somsookii* sp. nov. (Rhabditida: Heterorhabditidae): a new entomopathogenic nematode from Thailand. *International Journal of Nematology* 11, 1-17.

MICOLETZKY, H. (1922). Die freilebenden Erd-Nematoden mit besonderer Berücksichtigung der Steiermark und der Bukowina, zuglich mit einer Revision sämtlicher nicht mariner, freilebender Nematoden in Form von Genus – Beschreibungen und Bestimmungsschlüsseln. *Archiv für Naturgeschichte Berlin* 87, 1-320 (p. 119).

MRÁČEK, Z. (1977). *Steinernema kraussei*, a parasite of the body cavity of the sawfly, *Cephaleia abietis* in Czechoslovakia. *Journal of Invertebrate Pathology* 30, 87-94.

MRÁČEK, Z. (1992). *Steinernema kraussei* (Steiner, 1923) (Nematoda: Rhabditida: Steinernematidae): redescription of its topotype from Westphalia. *Folia Parasitologica* 41, 59-64.

MRÁČEK, Z. & WEISER, J. (1979). The head papillae of Steinernematidae. *Journal of Invertebrate Pathology* 34, 310-311.

MRÁČEK, Z., WEISER, J. & GERDIN, S. (1981). Head and cuticular structures of some species in the family Steinernematidae (Nematoda). *Nematologica* 27, 443-448.

MRÁČEK, Z., WEISER, J., BURES, M. & KAHOUNOVA, L. (1991). *Steinernema kraussei* (Steiner, 1923) (Nematoda: Rhabditida): rediscovery of its type locality in Germany. *Folia Parasitologica* 39, 181-182.

MRÁČEK, Z., HERNANDEZ, E.A. & BOEMARE, N.E. (1994). *Steinernema cubana* sp. n. (Nematoda: Rhabditida: Steinernematidae) and the preliminary characterisation of its associated bacterium. *Journal of Invertebrate Pathology* 64, 123-129.

MRÁČEK, Z., STURHAN, D. & REID, A. (2003). *Steinernema weiseri* n. sp. (Rhabditida, Steinernematidae), a new entomopathogenic nematode from Europe. *Systematic Parasitology* 56, 37-47.

MRÁČEK, Z., NGUYEN, K.B., TAILLIEZ, P., BOEMARE, N. & CHEN, S. (2006). *Steinernema sichuanense* n. sp. (Rhabditida, Steinernematidae) a new species of entomopathogenic nematode from the province of Sichuan, east Tibetan Mts., China. *Journal of Invertebrate Pathology* 93, 157-169.

MRÁČEK, Z., LIU, Q.-Z. & NGUYEN, K.B. (2009). *Steinernema xueshanense* n. sp. (Rhabditida, Steinernematidae), a new species of entomopathogenic nematode from the province of Yunnan, southeast Tibetan Mts., China. *Journal of Invertebrate Pathology* 102, 69-78.

MRÁČEK, Z., PŮŽA, V. & NERMUT', J. (2014a). *Steinernema poinari* sp. n. (Nematoda: Steinernematidae) a new entomopathogenic nematode from the Czech Republic. *Zootaxa* 3760, 336-350.

MRÁČEK, Z., PŮŽA, V. & NERMUT', J. (2014b). Addendum to the description of *Steinernema carpocapsae* (Weiser, 1955) Wouts, Mráček, Gerdin & Bedding, 1982. *Russian Journal of Nematology* 22, 109-120.

MRÁČEK, Z., PŮŽA, V. & NERMUT', J. (2016). The significance of the tail projections in female entomopathogenic nematodes within the family Steinernematidae. *Journal of Helminthology* 90, 249-251.

NADLER, S.A., BOLOTIN, E. & STOCK, S.P. (2006). Phylogenetic relationships of *Steinernema* Travassos, 1927 (Nematoda: Cephalobina: Steinernematidae) based on nuclear, mitochondrial and morphological data. *Systematic Parasitology* 63, 161-181.

NGUYEN, K.B. & BUSS, E.A. (2011). *Steinernema phyllophagae* n. sp. (Rhabditida: Steinernematidae), a new entomopathogenic nematode from Florida, USA. *Nematology* 13, 425-442.

NGUYEN, K.B. & DUNCAN, L.W. (2002). *Steinernema diaprepesi* n. sp. (Rhabditida: Steinernematidae), a parasite of the citrus root weevil *Diaprepes abbreviatus* (L.) (Coleoptera: Curculionidae). *Journal of Nematology* 34, 159-170.

NGUYEN, K.B. & HUNT, D.J. (Eds) (2007). *Entomopathogenic nematodes: taxonomy, phylogeny and bacterial symbionts. Nematology Monographs and Perspectives 5* (Series editors: Hunt, D.J. & Perry, R.N.). Leiden, The Netherlands, Brill.

NGUYEN, K.B. & SMART JR, G.C. (1990). *Steinernema scapterisci* n. sp. (Rhabditida: Steinernematidae). *Journal of Nematology* 22, 187-199.

NGUYEN, K.B. & SMART JR, G.C. (1992). *Steinernema neocurtillis* n. sp. (Rhabditida: Steinernematidae) and a key to species of the genus *Steinernema*. *Journal of Nematology* 24, 463-477.

NGUYEN, K.B. & SMART JR, G.C. (1994). *Neosteinernema longicurvicauda* n. gen., n. sp. (Rhabditia: Steinernematidae), a parasite of the termite *Reticulitermes flavipes* (Koller). *Journal of Nematology* 26, 162-174.

NGUYEN, K.B., SHAPIRO-ILAN, D.I., STUART, R.J., MCCOY, C.W., JAMES, R.R. & ADAMS, B.J. (2004). *Heterorhabditis mexicana* n. sp. (Rhabditida: Heterorhabditidae) from Tamaulipas, Mexico, and morphological studies of the bursa of *Heterorhabditis* spp. *Nematology* 6, 231-244.

NGUYEN, K.B., TESFAMARIAM, M., GOZEL, U., GAUGLER, R. & ADAMS, B.J. (2004). *Steinernema yirgalemense* n. sp. (Rhabditida: Steinernematidae) from Ethiopia. *Nematology* 6, 839-856.

NGUYEN, K.B., GOZEL, U., KOPPENHÖFER, H.S. & ADAMS, B.J. (2006). *Heterorhabditis floridensis* n. sp. (Rhabditida: Heterorhabditidae) from Florida. *Zootaxa* 1177, 1-19.

NGUYEN, K.B., MALAN, A.P. & GOZEL, U. (2006). *Steinernema khoisanae* n. sp. (Rhabditida: Steinernematidae), a new entomopathogenic nematode from South Africa. *Nematology* 8, 157-175.

NGUYEN, K.B., QIU, L., ZHOU, Y. & PANG, Y. (2006). *Steinernema leizhouense* sp. n. (Nematoda: Steinernematidae), a new entomopathogenic nematode from southern China. *Russian Journal of Nematology* 14, 101-118.

NGUYEN, K.B., STUART, R.J., ANDALO, V., GOZEL, U. & ROGERS, M.E. (2007). *Steinernema texanum* n. sp. (Rhabditida: Steinernematidae), a new entomopathogenic nematode from Texas, USA. *Nematology* 9, 379-396.

NGUYEN, K.B., PUŽA, V. & MRÁČEK, Z. (2008a). *Steinernema cholashanense* n. sp. (Rhabditida, Steinernematidae) a new species of entomopathogenic nematode from the province of Sichuan, Chola Shan Mountains, China. *Journal of Invertebrate Pathology* 97, 251-264.

NGUYEN, K.B., SHAPIRO-ILAN, D.I. & MBATA, G.N. (2008b). *Heterorhabditis georgiana* n. sp. (Rhabditida: Heterorhabditidae) from Georgia, USA. *Nematology* 10, 433-448.

NGUYEN, K.B., GINARTE, C.M.A., LEITE, L.G., SANTOS, J.M. DOS & HARAKAVA, R. (2010). *Steinernema brazilense* n. sp. (Rhabditida: Steinernematidae), a new entomopathogenic nematode from Mato Grosso, Brazil. *Journal of Invertebrate Pathology* 103, 8-20.

NIKDEL, M., NIKNAM, G. & YE, W. (2011). *Steinernema arasbaranense* n. sp. (Nematoda: Steinernematidae), a new entomopathogenic nematode from Arasbaran forests, Iran. *Nematologia Mediterranea* 39, 17-28.

NTHENGA, I., KNOETZE, R., BERRY, S., TIEDT, L.R. & MALAN, A.P. (2014). *Steinernema sacchari* n. sp. (Rhabditida: Steinernematidae), a new entomopathogenic nematode from South Africa. *Nematology* 16, 475-494.

PEREIRA, C. (1937). *Rhabditis hambletoni* n. sp., nema apparentemente semiparasito da 'bróca do algodoeiro' (*Gasterocercodes brasiliensis*). *Archivo do Instituto Biologico* 8, 215-230.

PERVEZ, R., ALI, S.S. & ASIF, M. (2009). A new species of entomopathogenic nematodes *Steinernema mushtaqi* sp. n. (Nematoda: Rhabditida: Steinernematidae) from chickpea rhizosphere. In: *International Conference on Legumes (ICGL), Indian Institute of Pulses Research, Kanpur, February, 2009*, pp. 14-16.

PHAN, K.L., NGUYEN, N.C. & MOENS, M. (2001a). *Steinernema loci* sp. n. and *Steinernema thanhi* n. sp. (Rhabditida: Steinernematidae) from Vietnam. *Nematology* 3, 503-514.

PHAN, K.L., NGUYEN, N.C. & MOENS, M. (2001b). *Steinernema sangi* sp. n. (Rhabditida: Steinernematidae) from Vietnam. *Russian Journal of Nematology* 9, 1-7.

PHAN, K.L., SUBBOTIN, S.A., NGUYEN, N.C. & MOENS, M. (2003). *Heterorhabditis baujardi* sp. n. (Rhabditida: Heterorhabditidae) from Vietnam and morphometric data for *H. indica* populations. *Nematology* 5, 367-382.

PHAN, K.L., SUBBOTIN, S.A., WAEYENBERGE, L. & MOENS, M. (2005). A new entomopathogenic nematode, *Steinernema robustispiculum* n. sp. (Rhabditida: Steinernematidae), from Chumomray National Park in Vietnam. *Systematic Parasitology* 60, 23-32.

PHAN, K.L., SPIRIDONOV, S.E., SUBBOTIN, S.A. & MOENS, M. (2006). Four new species of *Steinernema* Travassos, 1928 with short infective juveniles from Vietnam. *Russian Journal of Nematology* 14, 11-29.

PHAN, K.L., TAKEMOTO, S. & FUTAI, K. (2006). *Steinernema ashiuense* sp. n. (Nematoda: Steinernematidae), a new entomopathogenic nematode from Japan. *Nematology* 8, 681-690.

PHAN, K.L., MRÁČEK, Z., PŮŽA, V., NERMUT', J. & JAROŠOVÁ, A. (2014). *Steinernema huense* sp. n., a new entomopathogenic nematode (Nematoda: Steinernematidae) from Vietnam. *Nematology* 16, 761-775.

PLICHTA, K.L., JOYCE, S.A., CLARKE, D., WATERFIELD, N. & STOCK, S.P. (2009). *Heterorhabditis gerrardi* n. sp. (Nematoda: Heterorhabditidae): the hidden host of *Photorhabdus asymbiotica* (Enterobacteriaceae: γ-Proteobacteria). *Journal of Helminthology* 83, 309-320.

POINAR JR, G.O. (1976). Description and biology of a new insect parasitic rhabditoid, *Heterorhabditis bacteriophora* n. gen. n. sp. (Rhabditida; Heterorhabditidae n. fam.). *Nematologica* 21, 463-470.

POINAR JR, G.O. (1979). *Nematodes for biological control of insects.* Boca Raton, FL, USA, CRC Press.

POINAR JR, G.O. (1986). *Neoaplectana intermedia* n. sp. (Steinernematidae: Nematoda) from South Carolina. *Revue de Nématologie* 8(1985), 321-327.

POINAR JR, G.O. (1988). Redescription of *Neoaplectana affinis* Bovien (Rhabditida: Steinernematidae). *Revue de Nématologie* 11, 143-147.

POINAR JR, G.O. (1990). Taxonomy and biology of Steinernematidae and Heterorhabditidae. In: Gaugler, R. & Kaya, H.K. (Eds). *Entomopathogenic nematodes in biological control.* Boca Raton, FL, USA, CRC Press, pp. 23-61.

POINAR JR, G.O. (1993). Origins and phylogenetic relationships of the entomophilic rhabditids, *Heterorhabditis* and *Steinernema*. *Fundamental and Applied Nematology* 16, 333-338.

POINAR JR, G.O., THOMAS, G.M. & HESS, R. (1977). Characteristics of the specific bacterium associated with *Heterorhabditis bacteriophora* (Heterorhabditidae: Rhabditida). *Nematologica* 23, 97.

POINAR JR, G.O., JACKSON, T. & KLEIN, M. (1987). *Heterorhabditis megidis* sp. n. (Heterorhabditidae: Rhabditida), parasitic in the Japanese beetle, *Popillia japonica* (Scarabaeidae: Coleoptera), in Ohio. *Proceedings of the Helminthological Society of Washington* 54, 53-59.

POINAR JR, G.O., KARUNAKAR, G.K. & DAVID, H. (1992). *Heterorhabditis indicus* n. sp. (Rhabditida, Nematoda) from India: separation of *Heterorhabditis* spp. by infective juveniles. *Fundamental and Applied Nematology* 15, 467-472.

PRABHURAJ, A., VIRAKTAMATH, C.A. & KUMAR, A.R.V. (2003). Pathogenicity of two entomopathogenic nematodes against *Holotrichia serrata* (Fabricius) (Coleoptera: Scarabaeidae) and *Spodoptera litura* (Fabricius) (Lepidoptera: Noctuidae). In: Tandon, P.L., Ballal, C.R., Jalali, S.K. & Rabindra, R.J. (Eds). *Biological control of lepidopteran pests. Proceedings of the symposium of biological control of lepidopteran pests, July 17-18, 2002, Bangalore, India*, pp. 205-209.

PŮŽA, V., NERMUT', J., MRÁČEK, Z., GENGLER, S. & HAUKELAND, S. (2016). *Steinernema pwaniensis* n. sp., a new entomopathogenic nematode (Nematoda: Steinernematidae) from Tanzania. *Journal of Helminthology*. DOI: 10.1017/S0022149X15001157.

QIU, L., FANG, Y., ZHOU, Y., PANG, Y. & NGUYEN, K.B. (2004). *Steinernema guangdongense* sp. n. (Nematoda: Steinernematidae), a new entomopathogenic nematode from southern China with a note on *S. serratum* (*nomen nudum*). *Zootaxa* 704, 1-20.

QIU, L., YAN, X., ZHOU, Y., NGUYEN, K.B. & PANG, Y. (2005a). *Steinernema aciari* sp. n. (Nematoda: Steinernematidae), a new entomopathogenic nematode from Guangdong, China. *Journal of Invertebrate Pathology* 88, 58-69.

QIU, L., HU, X., ZHOU, Y., MEI, S., NGUYEN, K.B. & PANG, Y. (2005b). *Steinernema akhursti* n. sp. (Nematoda: Steinernematidae) from Yunnan, China. *Journal of Invertebrate Pathology* 90, 151-160.

QIU, L., HU, X., ZHOU, Y., PANG, Y. & NGUYEN, K.B. (2005c). *Steinernema beddingi* n. sp. (Nematoda: Steinernematidae), a new entomopathogenic nematode from Yunnan, China. *Nematology* 7, 737-749.

QIU, L., ZHAO, J., WU, Z., LV, Z. & PANG, Y. (2011). *Steinernema pui* sp. n. (Rhabditida, Steinernematidae), a new entomopathogenic nematode from Yunnan, China. *Zootaxa* 2767, 1-13.

RAILLIET, A. & HENRY, A. (1916). Sur les Oxyuridés. *Compte rendu des séances de la Société de Biologie, Paris* 79, 113-115.

ROMÁN, J. & FIGUEROA, W. (1994). *Steinernema puertoricensis* n. sp. (Rhabditida: Steinernematidae), a new entomopathogenic nematode from Puerto Rico. *Journal of Agriculture of the University of Puerto Rico* 78, 167-175.

RUBSTOV, I.A. & POLEVIK, N.L. (1979). [First occurrence of *Neoaplectana tabanivora* (Nematoda: Steinernematidae) parasitic in gad flies.] *Vestnik Zoologii* 3, 16-19.

SAN-BLAS, E., PORTILLO, E., NERMUT', J., PŮŽA, V. & MORALES-MONTERO, P. (2015). *Steinernema papillatum* n. sp. (Rhabditida: Steinernematidae), a new entomopathogenic nematode from Venezuela. *Nematology* 17, 1081-1097.

SAN-BLAS, E., MORALES-MONTERO, P., PORTILLO, E., NERMUT', J. & PŮŽA, V. (2016). *Steinernema goweni* n. sp. (Rhabditida: Steinernematidae), a new entomopathogenic nematode from Zulia State, Venezuela. *Zootaxa* 4067, 200-214.

SANDNER, H. & STANUSZEK, S. (1972). Differentiation of nematode populations of the genus *Neoaplectana* Steiner, 1929 (Nematoda; Rhabditoidea; Steinernematidae) into generations of 'giant' and small forms. In: *Interna-*

tional symposium of nematology (11th), European Society of Nematologists, Reading, UK, 3-8 September, 1972, p. 60. [Abstr.]

SHAHINA, F., ANIS, M., REID, A.P., ROWE, J. & MAQBOOL, M.A. (2001). *Steinernema pakistanense* sp. n. (Rhabditida: Steinernematidae) from Pakistan. *International Journal of Nematology* 11, 124-133.

SHAHINA, F., TABASSUM, K.A., SALMA, J., MEHREEN, G. & KNOETZE, R. (2016). *Heterorhabditis pakistanense* n. sp. (Nematoda: Heterorhabditidae) a new entomopathogenic nematode from Pakistan. *Journal of Helminthology*. DOI:10.1017/S0022149X16000158.

SHAMSELDEAN, M.M., ABOU EL-SOOUD, A.B., ABD-ELGAWAD, M.M. & SALEH, M.M. (1996). Identification of a new heterorhabditid species from Egypt, *Heterorhabditis taysearae*, n. sp. (Rhabditida: Heterorhabditidae). *Egyptian Journal of Biological Pest Control* 6, 129-138.

SHEN, C.P. & WANG, G.H. (1992). [Description of an entomopathogenic nematode, *Steinernema longicaudum* sp. nov. and its application.] In: *Proceedings of the XIX international congress of entomology, Beijing, China*, pp. 220-231.

SKRJABIN, K.I., SHIKHOBALOVA, N.P. & MOZGOVOI, A.A. (1951). [*Descriptive catalogue of parasitic nematodes. Volume 2. Oxyurata and Ascaridata.*] Moscow, USSR, Izdatelstvo Akademii Nauk SSR.

SOBOLEV, A.A. (1953). [To define more accurately the information on Rhabditata (superfamilies Rhabditoidea and Aphelenchoidea), parasitic in insects.] In: [*Papers on helminthology presented to Academician K. I. Skryabin on his 75th birthday.*] Moscow, USSR, Izdatelstvo Akademii Nauk SSR, pp. 676-684.

SOMVANSHI, V.S., LANG, E., GANGULY, S., SWIDERSKI, J., SAXENA, A.K. & STACKEBRANDT, E. (2006). A novel species of *Xenorhabdus*, family Enterobacteriaceae: *Xenorhabdus indica* sp. nov., symbiotically associated with entomopathogenic nematode *Steinernema thermophilum* Ganguly and Singh, 2000. *Systematic and Applied Microbiology* 29, 519-525.

SPIRIDONOV, S.E., KRASOMIL-OSTERFELD, K. & MOENS, M. (2004). *Steinernema jollieti* sp. n. (Rhabditida: Steinernematidae), a new entomopathogenic nematode from the American Midwest. *Russian Journal of Nematology* 12, 85-95.

SPIRIDONOV, S.E., WAEYENBERGE, L. & MOENS, M. (2010). *Steinernema schliemanni* sp. n. (Steinernematidae; Rhabditida) – a new species of steinernematids of the '*monticolum*' group from Europe. *Russian Journal of Nematology* 12, 175-190.

STANUSZEK, S. (1974a). [*Neoaplectana feltiae* complex (Nematoda: Rhabditoidea, Steinernematidae) its taxonomic position within the genus

Neoaplectana and intraspecific structure.] *Zeszyty Problemowe Postepow Nauk Rolniczych* 154, 331-360.

STANUSZEK, S. (1974b). [*Neoaplectana feltiae pieridarum*, n. ecotype (Nematoda: Rhabditoidea, Steinernematidae) – a parasite of *Pieris brassicae* L. and *Mamestra brassicae* L. in Poland, morphology and biology.] *Zeszyty Problemowe Postepow Nauk Rolniczych* 154, 361-393.

STEINER, G. (1923). *Aplectana kraussei* n. sp., eine in der Blattwespe *Lyda* sp. parasitierende Nematodenform, nebst Bemerkungen über das Seitenorgan der parasitischen Nematoden. *Zentralblatt für Bakteriologie, Parasitenkunde, Infektions krankheiten und Hygiene, Abteilung II* 59, 14-18.

STEINER, G. (1929). *Neoaplectana glaseri* n. g. n. sp. (Oxyuridae), a new nemic parasite of the Japanese beetle (*Popillia japonica* Newm.). *Journal of the Washington Academy of Science* 19, 436-440.

STOCK, S.P. (1993). A new species of the genus *Heterorhabditis* Poinar, 1976 (Nematoda: Heterorhabditidae) parasitizing *Graphognathus* sp. larvae (Coleoptera: Curculionidae) from Argentina. *Research and Reviews in Parasitology* 53, 103-107.

STOCK, S.P. & KOPPENHÖFER, A.M. (2003). *Steinernema scarabaei* n. sp. (Rhabditida: Steinernematidae), a natural pathogen of scarab beetle larvae (Coleoptera: Scarabaeidae) from New Jersey, USA. *Nematology* 5, 191-204.

STOCK, S.P., STRONG, D. & GARDNER, S.L. (1996). Identification of *Heterorhabditis* (Nematoda: Heterorhabditidae) from California with a new species isolated from larvae of the ghost moth *Hepialus californicus* (Lepidoptera: Hepialidae) from the Bodega Bay Natural Reserve. *Fundamental and Applied Nematology* 19, 585-592.

STOCK, S.P., CHOO, H.Y. & KAYA, H.K. (1997). An entomopathogenic nematode, *Steinernema monticolum* sp. n. (Rhabditida: Steinernematidae) from Korea with a key to other species. *Nematologica* 43, 15-29.

STOCK, S.P., SOMSOOK, V. & REID, A.P. (1998). *Steinernema siamkayai* n. sp. (Rhabditida: Steinernematidae), an entomopathogenic nematode from Thailand. *Systematic Parasitology* 41, 105-113.

STOCK, S.P., GRIFFIN, C.T. & BURNELL, A.M. (2002). Morphological characterisation of three isolates of *Heterorhabditis* Poinar, 1976 from the 'Irish group' (Nematoda: Rhabditida: Heterorhabditidae) and additional evidence supporting their recognition as a distinct species, *H. downesi* n. sp. *Systematic Parasitology* 51, 95-106.

STOCK, S.P., GRIFFIN, C.T. & CHAERANI, R. (2004). Morphological and molecular characterisation of *Steinernema hermaphroditum* n. sp. (Nematoda: Steinernematidae), an entomopathogenic nematode from Indonesia, and its phylogenetic relationships with other members of the genus. *Nematology* 6, 401-412.

STOCK, S.P., RIVERA-ORDUÑO, B. & FLORES-LARA, Y. (2009). *Heterorhabditis sonorensis* n. sp. (Nematoda: Heterorhabditidae), a natural pathogen of the seasonal cicada *Diceroprocta ornea* (Walker) (Homoptera: Cicadidae) in the Sonoran desert. *Journal of Invertebrate Pathology* 100, 175-184.

STOKWE, N.F., MALAN, A.P., NGUYEN, K.B., KNOETZE, R. & TIEDT, L. (2011). *Steinernema citrae* n. sp. (Rhabditida: Steinernematidae), a new entomopathogenic nematode from South Africa. *Nematology* 13, 567-587.

STURHAN, D., SPIRIDONOV, S.E. & MRÁČEK, Z. (2005). *Steinernema silvaticum* sp. n. (Rhabditida: Steinernematidae), a new entomopathogenic nematode from Europe. *Nematology* 7, 227-241.

SUDERSHAN, G. & SINGH, L.K. (2000). *Steinernema thermophilum* sp. n. (Rhabditida: Steinernematidae) from India. *International Journal of Nematology* 10, 183-191.

SUDHAUS, W. & FÜRST VON LIEVEN, A. (2003). A phylogenetic classification and catalogue of the Diplogastridae (Secernentea, Nematoda). *Journal of Nematode Morphology and Systematics* 6, 43-90.

TAILLIEZ, P., PAGÈS, S., GINIBRE, N. & BOEMARE, N. (2006). New insight into diversity in the genus *Xenorhabdus*, including the description of ten novel species. *International Journal of Systematic and Evolutionary Microbiology* 56, 2805-2818.

TALLOSI, B., PETERS, A. & EHLERS, R.-U. (1995). *Steinernema bicornutum* sp. n. (Rhabditida: Nematoda) from Vojvodina, Yugoslavia. *Russian Journal of Nematology* 3, 71-80.

TAMIRU, T., WAEYENBERGE, L., HAILU, T., EHLERS, R.-U., PŮŽA, V. & MRÁČEK, Z. (2012). *Steinernema ethiopiense* sp. n. (Rhabditida: Steinernematidae), a new entomopathogenic nematode from Ethiopia. *Nematology* 14, 741-757.

TARASCO, E., MRÁČEK, Z., NGUYEN, K.B. & TRIGGIANI, O. (2008). *Steinernema ichnusae* sp. n. (Nematoda: Steinernematidae) a new entomopathogenic nematode from Sardinia Island (Italy). *Journal of Invertebrate Pathology* 99, 173-185.

TRAVASSOS, L. (1927a). Sobre o genero *Oxysomatium*. *Boletim Biologico, São Paulo* 5, 20-21.

TRAVASSOS, L. (1927b). Uma nova *Capillaria* parasita de peixes be agua doce: *Capillaria sentinosa* n. sp. *Boletim Biologico, São Paulo* 10, 215-217.

TRAVASSOS, L. (1932). Uma nova especie do genero *Neoaplectana* Steiner, 1929 (Nematoda). *Boletim Biologico, São Paulo* 19, 150-154.

TRIGGIANI, O., MRÁČEK, Z. & REID, A. (2004). *Steinernema apuliae* sp. n. (Rhabditida: Steinernematidae): a new entomopathogenic nematode from southern Italy. *Zootaxa* 460, 1-12.

TURCO, C.P. (1970). *Neoaplectana hoptha* sp. n. (Neoaplectanidae: Nematoda), a parasite of the Japanese beetle, *Popillia japonica* Newm. *Proceedings of the Helminthological Society of Washington* 37, 119-121.

TURCO, C.P., THAMES, W.H. & HOPKINS, S.H. (1971). On the taxonomic status and comparative morphology of species of the genus *Neoaplectana* Steiner (Neoaplectanidae: Nematoda). *Proceedings of the Helminthological Society of Washington* 38, 68-79.

URIBE-LORÍO, L., MORA, M. & STOCK, S.P. (2007). *Steinernema costaricense* n. sp. and *Steinernema puntauvense* n. sp. (Rhabditida: Steinernematidae), two new entomopathogenic nematodes from Costa Rica. *Systematic Parasitology* 68, 167-182.

VEREMCHUK, G.V. (1964). [On the systematic position of Steinernematidae Chitwood & Chitwood, 1937 (Nematoda: Rhabditida).] *Materialy Nauchnoi Konferentsii Vsesoiuznogo Obshchestva Gel'mintologov* 1, 55-58. [Abstr.]

VEREMCHUK, G.V. (1966). [Entomopathogenic nematodes of the genus *Neoaplectana*, family Steinernematidae Chitwood & Chitwood, 1937.] In: *Tez. Doklady IX Mezdunarodnogo Kongressa po Mikrobiologii* [*Abstr. IX international congress on microbiology*], *Moscow*, pp. 326-327.

VEREMCHUK, G.V. (1969). [New species of *Neoaplectana* (Rhabditida: Steinernematidae), nematodes pathogenic to Elateridae.] *Materialy Nauchnoi Konferentsii Vsesoiuznogo Obshchestva Gel'mintologov* 1, 44-50.

VEREMCHUK, G.V. (1970). [A new species of *Neoaplectana* (Rhabditida: Steinernematidae) pathogenic to insects.] *Parazitologiya* 3, 249-252.

VEREMCHUK, G.V. & LITVINCHUK, L.N. (1971). [Nematodes of the genus *Neoaplectana* (Rhabditida: Steinernematidae) parasites of the larch looper moth, *Semiothisa pumila* Kusn. (Lepidoptera; Geometridae).] *Novye I maloizvestnye vidy fauny Sibiri* 4, 92-96.

WALIA, K.K. & BAJAJ, H.K. (2005). *Steinernema solanense* – a new entomopathogenic nematode species from Himachal Pradesh, India. In: *Sustainable plant protection strategies: health and environmental concerns. National symposium held at BSKKV, Dapoli, Maharashtra, India, 15-17 October, 2005*, p. 108.

WATURU, C.N., HUNT, D.J. & REID, A.P. (1997). *Steinernema karii* sp. n. (Nematoda: Steinernematidae), a new entomopathogenic nematode from Kenya. *International Journal of Nematology* 7, 68-75.

WEISER, J. (1955). *Neoaplectana carpocapsae* n. sp. (Anguillulata, Steinernematinae), novy cizopasník housenek obaleče jablečného, *Carpocapsa pomonella* L. *Věstník Československé Zoologické Společnosti* 19, 44-52.

WEISER, J. (1958). Ein neuer Nematode als Parasit der Engerlinge des Maikäfers, *Melolontha melolontha* in der Tschechoslowakei. In: *Transactions of the 1st International Conference of Insect Pathology and Biological Control, Prague, Czech Republic*, pp. 331-336.

WEISER, J. & KOHLER, W. (1955). *Neoaplectana janickii* n. sp. novy cizopasnik larvy ploskohrbetky *Acantholyda nemoralis* Thoms. v Polsku. *Roczniki Nauk Lesnych* 11, 93-110.

WELCH, H.E. (1963). Unpublished thesis.

WOUTS, W.M. (1979). The biology and life cycle of a New Zealand population of *Heterorhabditis heliothidis* (Heterorhabditidae). *Nematologica* 25, 191-202.

WOUTS, W.M. (1980). Biology, life cycle and redescription of *Neoaplectana bibionis* Bovien, 1937 (Nematoda: Steinernematidae). *Journal of Nematology* 12, 62-72.

WOUTS, W.M., MRÁČEK, Z., GERDIN, S. & BEDDING, R.A. (1982). *Neoaplectana* Steiner, 1929 a junior synonym of *Steinernema* Travassos, 1927 (Nematoda: Rhabditida). *Systematic Parasitology* 4, 147-154.

XU, Z., WANG, G. & LI, X. (1991). [A new species of the genus *Steinernema* (Rhabditida: Steinernematidae).] *Zoological Research* 12, 17-20.

YOSHIDA, M. (2005). *Steinernema litorale* n. sp. (Rhabditida: Steinernematidae) a new entomopathogenic nematode from Japan. *Nematology* 6(2004), 819-838.

ZHANG, C., LIU, J., XU, M., SUN, J., YANG, S., AN, X., GAO, G., LIN, M., LAI, R., HE, Z. ET AL. (2008). *Heterorhabditidoides chongmingensis* gen. nov., sp. nov. (Rhabditida: Rhabditidae), a novel member of the entomopathogenic nematodes. *Journal of Invertebrate Pathology* 98, 153-168.

Nematology Monographs & Perspectives, 2016, Vol. 12, 59-109

Chapter 3

Tabular keys to species of *Steinernema* and *Heterorhabditis*

Khuong B. NGUYEN

Entomology and Nematology Department, Institute of Food and Agricultural Sciences, University of Florida, Gainesville, FL 32611-0620, USA

As of the end of 2015, 95 valid species of *Steinernema* and 16 species of *Heterorhabditis* have been described or are in press. As the number of species increases, the identification of individual species becomes progressively more difficult. The identification of species based on one stage may not be accurate because there are few differentiating morphological characteristics between species, and morphometric ranges of several species overlap. Characteristics of males, females and infective juveniles (IJ) must be used for accurate identification. Diagnostic characteristics of various stages of the different species are presented in this part so as to help in the identification of species of entomopathogenic nematodes. Morphological and morphometrical characteristics can be used to identify species of *Steinernema*, but they are less reliable for species differentiation in *Heterorhabditis*. Molecular characterisation is necessary for final confirmation of identity.

Morphology of the families Steinernematidae and Heterorhabditidae

The morphology of this group of nematodes is presented in detail by Nguyen (2007).

IDENTIFICATION OF *STEINERNEMA* AND *HETERORHABDITIS* SPECIES

The diagnostic characters for the identification of species of *Steinernema* are presented in Table 3.1 and those for *Heterorhabditis* in

Table 3.2. The nematode morphometrics are arranged in descending order of body length of IJ and characters considered important are in boldface.

Ratios and abbreviations used in the keys

L = body length,
MBD = max. body diam.,
EP = distance from anterior end to excretory pore,
NR = distance from anterior end to nerve ring,
ES = pharynx length,
T = tail length,
a = L/MBD,
b = L/ES,
c = L/T,
D% = EP/ES × 100,
E% = EP/T × 100,
ABD = anal body diam.,
SL = spicule length,
GL = gubernaculum length,
IJ = infective juvenile,
SW% = spicule length/ABD × 100,
GS% = gubernaculum/spicule × 100,
MUC = mucron,
A = absent,
P = present,
na = not available.

How to use the tables for identification

To identify a nematode, the following steps should be followed:

- The nematode should be reared in *Galleria mellonella*.
- Make measurements of ten IJ and ten males of the target nematode.
- Match the IJ body length of the nematode with nematodes that have a similar body length in the key.
- Compare the collected data with the measurements in **boldface** letters and **specific characters** of selected nematodes.
- Decide which nematodes in the key are most similar to the one being identified.
- Compare with the original descriptions to confirm the identity.

Tabular key to *Steinernema* species (Table 3.1)

Table 3.1. *Tabular key for the identification of species of* Steinernema *(arranged in descending order of body length). All measurements are in μm. Abbreviations as given on page 60.*

Species	Morphometric characters (mean, range)										
	L	MBD	EP	NR	ES	T	a	b	c	D%	E%
S. australe											
Infective juvenile	**1316**	38	**110**	126	169	103	35	7.8	12.8	65	107
	1162-1484	34-41	95-125	102-143	140-194	92-114	31-38	6.6-9.2	11-14.9	57-78	94-122

Male: **SL = 72** (55-78); GL = 45 (36-51); MBD = 75 (61-97); D% = 71 (59-87); **SW% = 172** (118-196); GS% = 62 (45-71); MUC = A

Spicule and gubernaculum as in the micrographs:

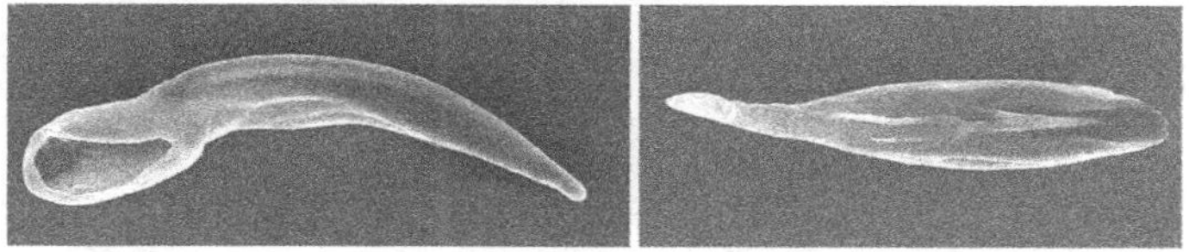

Specific characters: Female with well developed epiptygmata; cuneus V-shaped

Table 3.1. *(Continued).*

Species	Morphometric characters (mean, range)										
	L	MBD	EP	NR	ES	T	a	b	c	D%	E%
S. phyllophagae											
Infective juvenile	**1289**	41	**99**	112	146	**89**	31	8.9	14.6	70	**110**
	1133-1395	37-43	84-120	101-135	120-178	77-109	27.6-35.5	7.6-11.0	12.4-15.9	58-100	70-139
Male	**SL** = **72** (65-77); GL = 51 (46-56); MBD = 65 (58-79); **D%** = **75** (68-80); SW% = 1.87; GS% = 0.71; **MUC** = **A**										
	Spicule and gubernaculum shapes as in the micrographs:										
	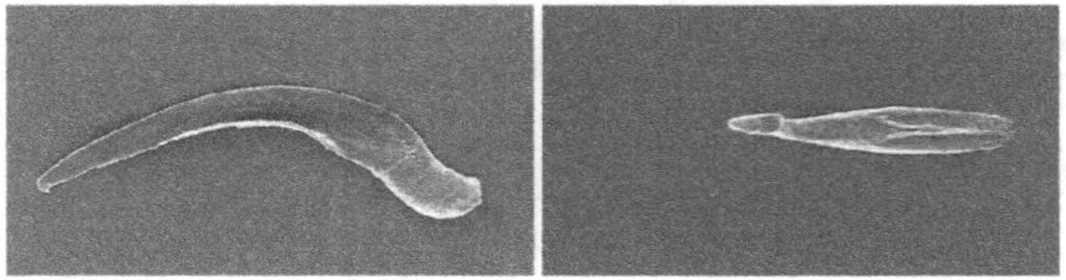										
	Specific characters: Female with low epiptygmata; several papilla-like structures near tail tip										
S. cubanum											
Infective juvenile	**1283**	37	**106**	116	148	**67**	35	8.6	19.2	70	**160**
	1149-1508	33-46	101-114	106-130	135-159	61-77	na	na	na	na	na
Male	**SL** = **58** (50-67); GL = 39 (37-42); MBD = 97 (77-117); D% = 70; SW% = 141; GS% = 70; MUC = A										
	Spicule and gubernaculum shapes as in the micrographs:										
	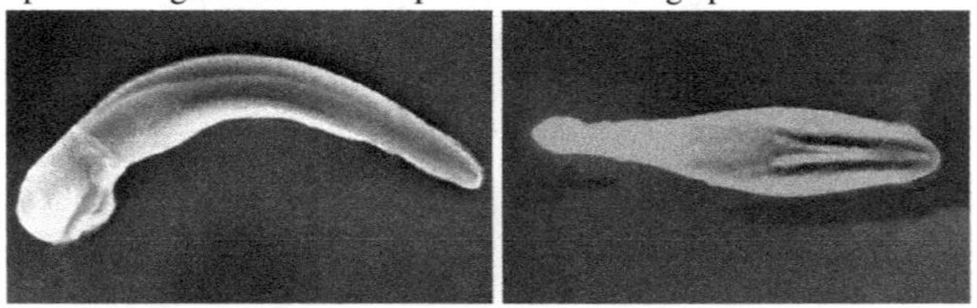										
	Specific character: Spicule without notch at tip										

Table 3.1. *(Continued).*

Species	Morphometric characters (mean, range)										
	L	MBD	EP	NR	ES	T	a	b	c	D%	E%
S. vulcanicum											
Infective juvenile	**1204**	39	93	117	157	**86**	31.5	7.7	14.1	60	**109**
	1081-1323	29-55	77-107	98-136	134-173	74-100	21.6-39	6.8-9.2	11.3-16.8	47-68	83-135
Male	**SL** = **71** (46-88); GL = 50 (32-65); MBD = 110 (48-159); **D%** = **70** (42-90); **SW%** = **140** (120-170); GS% = 70 (60-90); MUC = A										
	Spicule and gubernaculum shapes as in the micrographs: 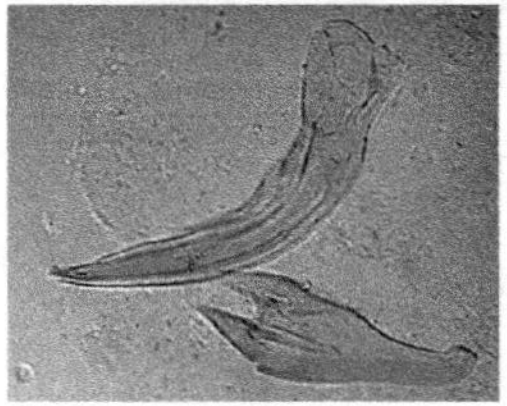										
	Specific character: Female with double-flapped epiptygmata										
S. puertoricense											
Infective juvenile	**1171**	51	**95**	117	143	**94**	23	8.2	12.4	66	**101**
	1057-1238	47-54	90-102	111-121	138-147	88-107	20-24	7.4-8.6	11.6-13.6	62-74	88-108
Male	**SL** = **78** (71-88); GL = 40 (36-45); MBD = 101 (67-148); **D%** = **77**; SW% = 152; GS% = 50; **MUC** = **A**										
	Spicule and gubernaculum shapes as in the micrographs: 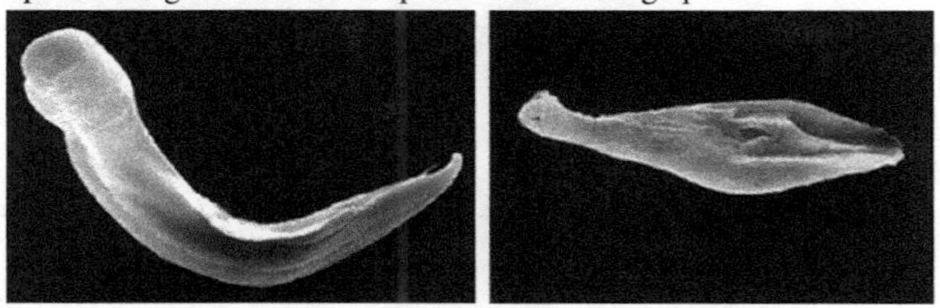										
	Specific character: Female with small double-flapped epiptygmata										

Table 3.1. *(Continued).*

Species	Morphometric characters (mean, range)										
	L	MBD	EP	NR	ES	T	a	b	c	D%	E%
S. brazilense											
Infective juvenile	**1157**	42	94	114	148	**88**	28	7.8	13.0	63	106
	1023-1284	39-49	87-102	108-121	139-154	80-104	26-31	7.2-8.9	12.0-14.5	58-70	95-118
Male	**SL** = **83** (75-89); GL = 47 (41-56); MBD = 94 (78-99); D% = 68 (57-80); **SW%** = **192** (158-208); GS% = 57 (48-65); MUC = A										
	Spicule and gubernaculum shapes as in the micrographs: 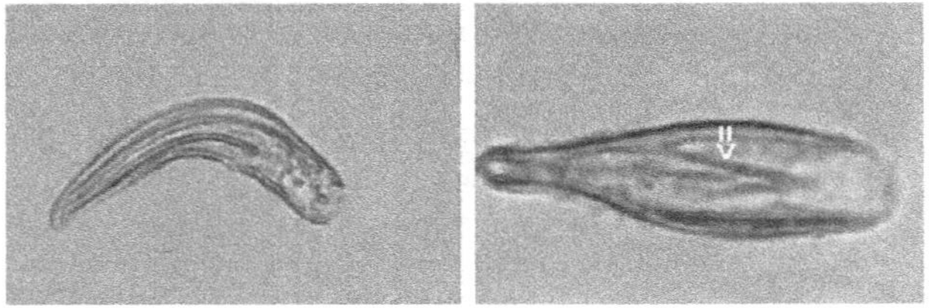										
	Specific characters: Male lateral field with 1 ridge; IJ lateral field formula: 2, 4, 6, 8, 6, 2										
S. glaseri											
Infective juvenile	**1130**	43	**102**	120	162	**78**	29	7.3	14.7	65	**131**
	864-1448	31-50	87-110	112-126	158-168	62-87	26-35	6.3-7.8	13.6-15.7	58-71	122-138
Male	**SL** = **77** (64-90); GL = 55 (44-59); MBD = 72 (54-92); **D%** = **70** (60-78); **SW%** = **210** (164-243); GS% = 70 (60-85); **MUC** = **A**										
	Spicule and gubernaculum shapes as in the micrographs: 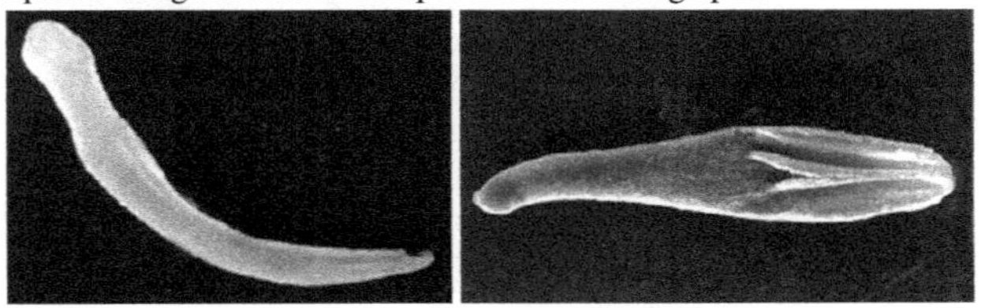										
	Specific characters: Spicule tip with a notch; lateral field formula: 2, 5, 6, 8, 2										

Table 3.1. *(Continued).*

Species	Morphometric characters (mean, range)										
	L	MBD	EP	NR	ES	T	a	b	c	D%	E%
S. aciari											
Infective juvenile	**1113**	47	**95**	114	146	**78**	24	7.6	14.4	65	**123**
	975-1250	42-53	87-100	106-125	135-157	68-88	20-27	7.0-8.3	14-16	60-70	113-134
Male	**SL** = **86** (75-95); GL = 56 (48-65); MBD = 104 (88-125); **D%** = **76** (69-88); **SW%** = **204** (180-240); GS% = 65 (57-77); **MUC** = **A**										
	Spicule and gubernaculum shapes as in the micrographs: 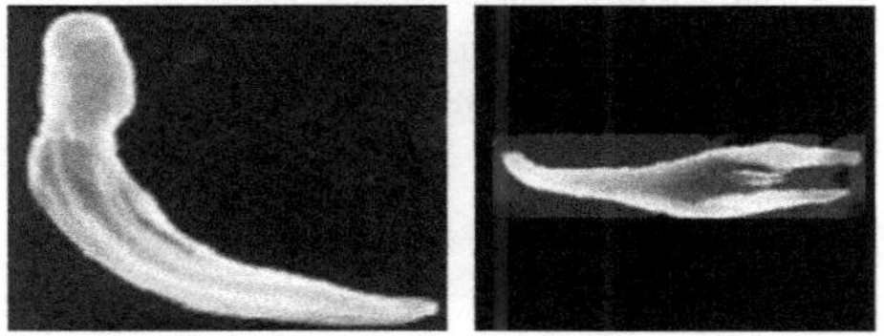										
	Specific character: Female with small double-flapped epiptygmata										
S. caudatum											
Infective juvenile	**1106**	36	**82**	109	156	**88**	31	7.1	12.6	52	**94**
	933-1269	34-41	76-89	104-126	149-175	80-100	na	na	na	na	87-100
Male	**SL** = **75**; GL = 52; MBD = 91; **D%** = **71**; SW% = 222; GS% = 70; **MUC** = **A**										
	Spicule and gubernaculum shapes as in the illustration (redrawn from the original description): 										
	Specific characters: EP and T short; spicule tip not hook-like in shape										

Table 3.1. *(Continued).*

Species	Morphometric characters (mean, range)										
	L	MBD	EP	NR	ES	T	a	b	c	D%	E%
S. boemarei											
Infective juvenile	**1103**	39	91	112	143	86	28	8.0	13.0	63	**106**
		34-46	82-111	100-142	131-182	77-109	24-33	7.0-8.4	12-14	56-68	94-122
Male	**SL** = **79** (54-96); GL = 52 (43-65); MBD = 96 (70-122); **D%** = **85** (68-99); **SW%** = **170** (120-240); GS% = 70 (50-90); MUC = A										
	Spicule and gubernaculum shapes as in the drawing: 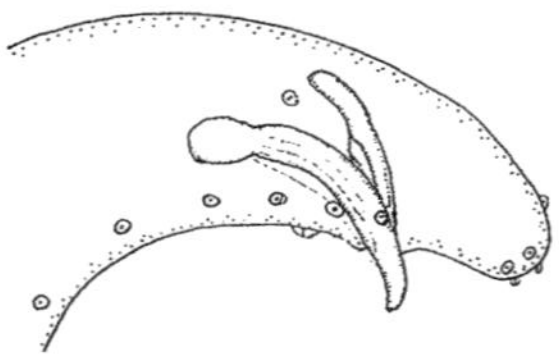										
	Specific character: First generation male with 23 genital papillae										
S. apuliae											
Infective juvenile	**1064**	37	**95**	111	145	**71**	29	7.3	15.0	**66**	**130**
	945-1212	34-41	86-102	103-124	137-159	63-80	26-31	6.8-7.8	14-17	63-70	120-150
Male	**SL** = **72** (64-80); GL = 50 (46-54); MBD = 104 (86-124); **D%** = **81** (75-92); **SW%** = **160** (150-180); GS% = 70 (70-80); **MUC** = **A**										
	Spicule and gubernaculum shapes as in the micrograph: 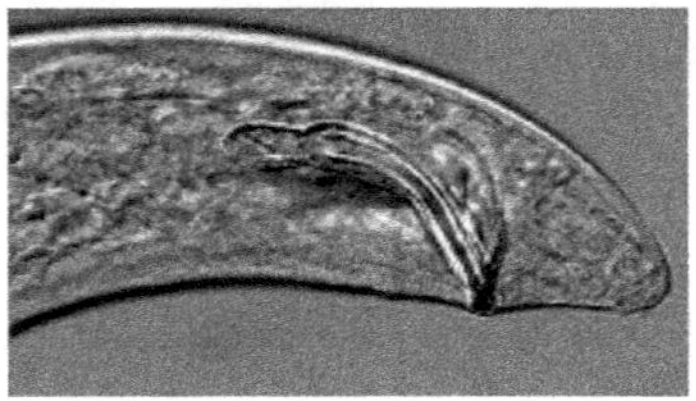										
	Specific characters: H% = 42 (38-47); V > 50 (50-58)										

Table 3.1. *(Continued).*

Species	Morphometric characters (mean, range)										
	L	MBD	EP	NR	ES	T	a	b	c	D%	E%
S. longicaudum											
Infective juvenile	**1063**	40	**81**	107	145	**95**	27	7.3	11.2	56	**85**
	(No ranges available)										
Male	**SL = 77**; GL = 48; MBD = 155; **D% = 62**; SW% = 160; GS% = 62; **MUC = A**										
	Spicule and gubernaculum shapes as in the micrographs:										
	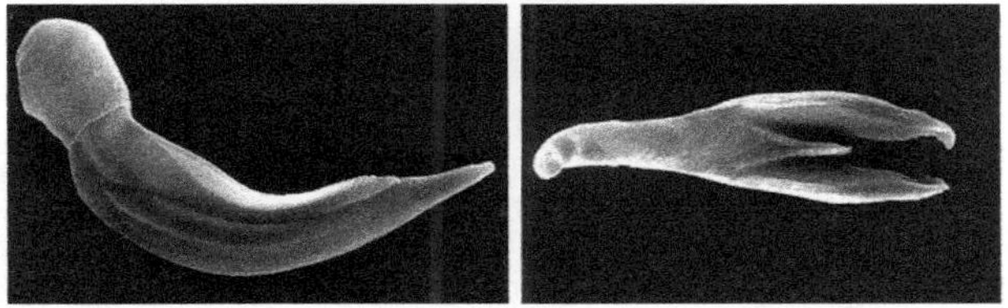										
	Specific characters: IJ tail long; spicule reduced in width suddenly to form a tip with flattened terminus										
S. khoisanae											
Infective juvenile	**1062**	31	**95**	109	140	**85**	**35**	7.6	**12.5**	**68**	**112**
	994-1159	27-34	87-101	101-118	130-155	76-97	32-37	6.6-8.4	10.6-13.8	60-73	95-128
Males	**SL = 85** (70-88); GL = 56 (50-63); MBD = 108 (90-132); **D% = 85** (71-99); **SW% = 203** (167-227); GS% = 70 (60-80); **MUC = A**										
	Spicule and gubernaculum shapes as in the micrograph:										
	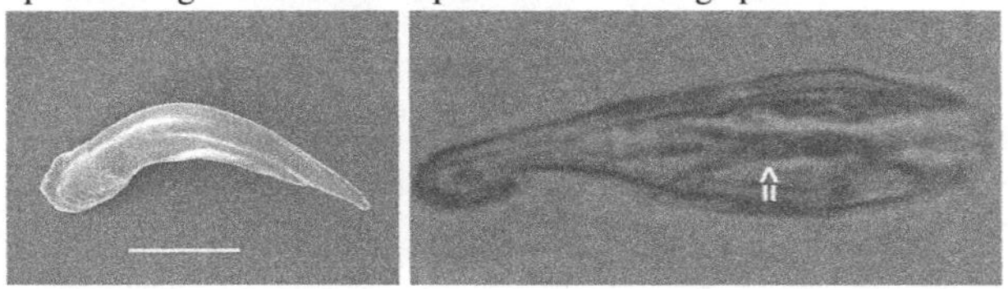										
	Specific characters: Male excretory pore near basal bulb; lateral field formula: 2, 7, 8, 6, 4, 2										

Table 3.1. *(Continued).*

Species	Morphometric characters (mean, range)										
	L	MBD	EP	NR	ES	T	a	b	c	D%	E%
S. guangdongense											
Infective juvenile	**1055**	42	**80**	102	134	**91**	25	7.9	11.6	**59**	**88**
	987-1145	30-48	71-85	88-111	123-144	82-103	22-35	7.3-8.5	10.2-12.9	54-65	74-100
Male	**SL** = **86** (80-94); **GL** = **64** (47-73); MBD = 112 (90-135); **D%** = **70** (67-78); **SW%** = **175** (152-216); GS% = 75 (59-82); MUC = A										
	Spicule and gubernaculum shapes as in the micrographs:										
	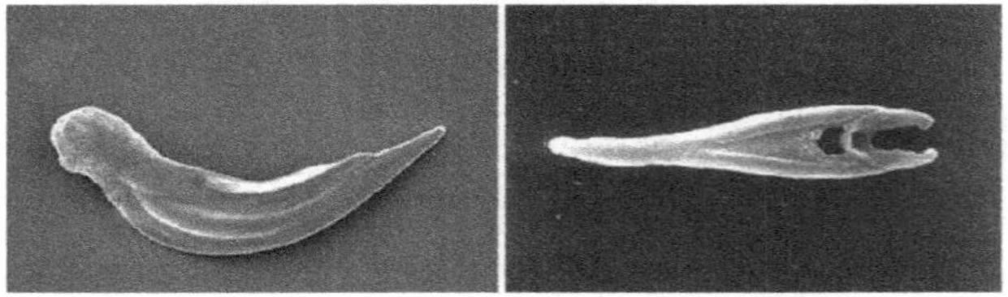										
	Specific characters: Closer to *S. longicaudum* but spicule longer; lateral field formula: 2, 7, 8, 7, 4, 2										
S. innovationi											
Infective juvenile	**1054**	37	88	108	150	**76**	29	7.0	–	**58**	**115**
	1000-1103	31-50	82-91	97-117	142-160	66-80	21-34	6.5-7.6		54-63	104-137
Male	**SL** = **81** (74-91); GL = 58 (54-63); MBD = 139 (100-171); **D%** = **82** (67-95); **SW%** = **140** (110-190); GS% = 70 (60-80); **MUC** = **A**										
	Spicule shape as in the micrograph:										
	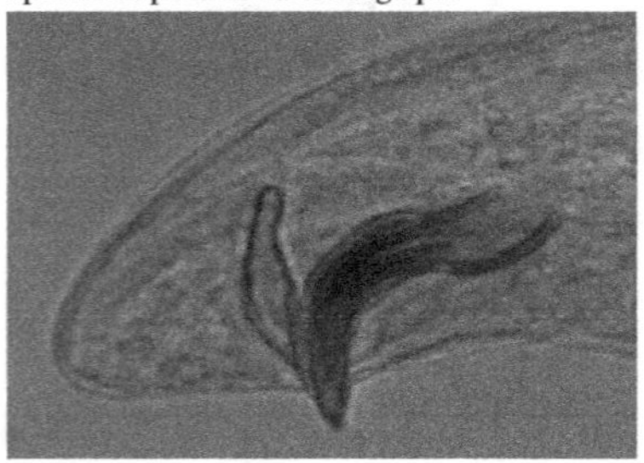										

Table 3.1. *(Continued).*

Species	Morphometric characters (mean, range)										
	L	MBD	EP	NR	ES	T	a	b	c	D%	E%
S. tophus											
Infective juvenile	**1046**	38	**92**	105	146	**70**	28	7.2	–	**63**	**132**
	966-1126	30-50	86-97	98-111	134-156	58-80	21-34	6.5-8.1		58-71	120-156
Male	**SL = 79** (70-91); GL = 59 (47-65); MBD = 162 (112-185); **D% = 92** (71-103); **SW% = 120** (100-160); GS% = 70 (60-90); **MUC = A**										
	Spicule and gubernaculum shapes as in the micrographs: 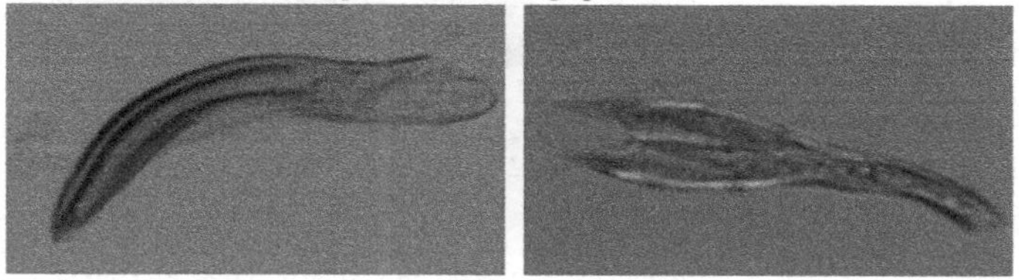										
S. arenarium											
Infective juvenile	**1034**	46	**83**	109	138	**75**	26	7.6	13.8	55	**119**
	724-1408	28-77	76-86	100-120	123-160	64-84	17-34	5.9-10.8	9.4-16.9	52-59	106-130
Male	**SL = 84** (81-91); GL = 55 (49-60); MBD = 188 (184-219); **D% = 93** (88-102); SW% = 210; GS% = 65 (60-66); **MUC = A**										
	Spicule and gubernaculum shapes as in the micrographs: 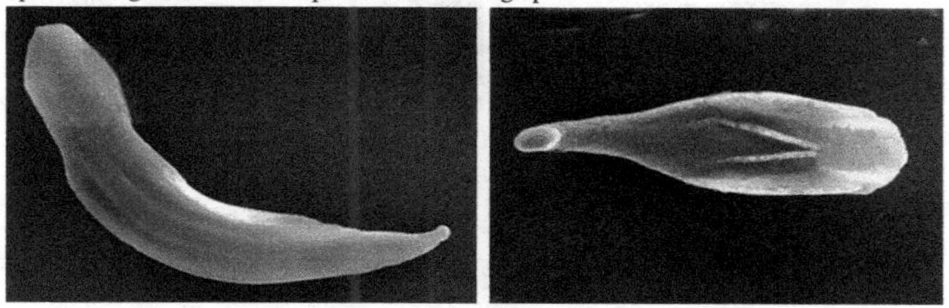										
	Specific characters: Spicule tip ball-shaped; cuneus V-shaped										

Table 3.1. *(Continued).*

Species	Morphometric characters (mean, range)										
	L	MBD	EP	NR	ES	T	a	b	c	D%	E%
S. pui											
Infective juvenile	**1004**	36	**85**	117	144	69	28	7.0	15.0	59.5	**125**
	900-1120	33-40	80-95	100-125	130-150	60-80	24-33	6-8.2	13-16	55-66	109-142
Male	**SL** = **84** (78-88); GL = 62 (58-65); MBD = 137 (118-180); **D%** = **77** (70-91); **SW%** = **152** (140-184); GS% = 74 (69-81); MUC = A										
	Spicule and gubernaculum shapes as in the micrographs: 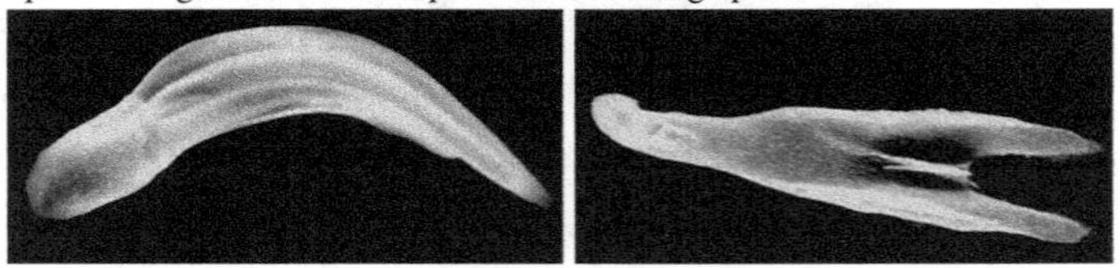										
	Specific character: An aperture present on spicule tip										
S. diaprepesi											
Infective juvenile	**1002**	34	**74**	102	138	**83**	30	7.3	12.1	**54**	**90**
	880-1133	30-42	66-83	74-109	111-152	65-91	23-35	6.5-8.3	10.4-13.2	30-70	78-114
Male	**SL** = **79** (71-90); GL = 54 (45-61); **D%** = **80** (68-86); **SW%** = **180** (150-200); GS% = 69 (59-79); **MUC** = **A**										
	Spicule gubernaculum shapes as in the micrographs: 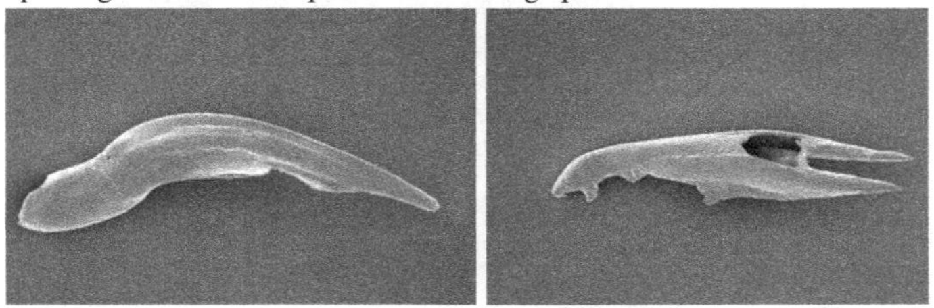										
	Specific characters: Epiptygma present; female tail terminus usually with 5 papillae-like structures; lateral field formula: 2, 6, 7, 8, 4, 2										

Table 3.1. *(Continued).*

Species	Morphometric characters (mean, range)										
	L	MBD	EP	NR	ES	T	a	b	c	D%	E%
S. loci											
Infective juvenile	**986**	37	**80**	102	141	**75**	27	7.0	13.0	**57**	**107**
	896-1072	30-45	71-86	89-111	126-155	66-83	24-30	6-8	11-15	52-63	94-120
Male	**SL** = **71** (60-80); GL = 46 (40-52); **D%** = **73** (61-80); SW% = 190 (170-210); GS% = 70 (50-80); **MUC** = **A**										
	Spicule, gubernaculum and lateral field as in the micrographs:										
	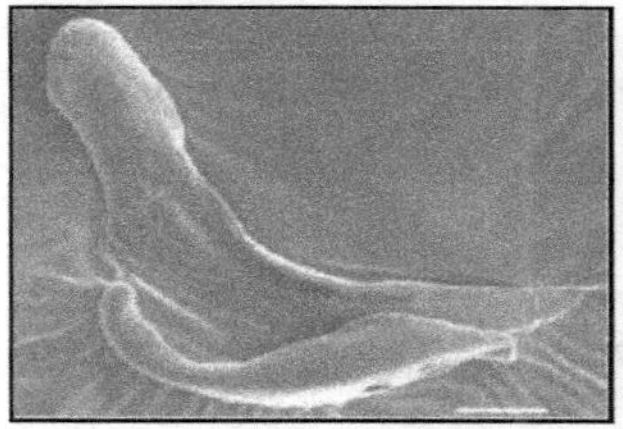 										
	Specific character: Lateral field with 2 ridges anteriorly then changing to 6, 7										
S. leizhouense											
Infective juvenile	**983**	38	**83**	101	133	**66**	28	7.5	**13.9**	**59**	**120**
	900-1050	35-43	80-88	95-108	125-143	58-75	23-38	6.8-8.9	11.0-16.8	54-63	105-140
Male	**SL** = **82** (75-93); GL = 55 (50-60); MBD = 104 (75-143); **D%** = **75** (67-83); **SW%** = **213** (177-251); GS% = 67 (60-74); **MUC** = **A**										
	Spicule and gubernaculum shapes as in the micrographs:										
	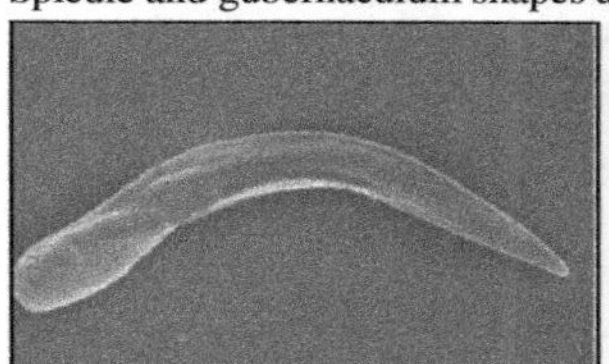 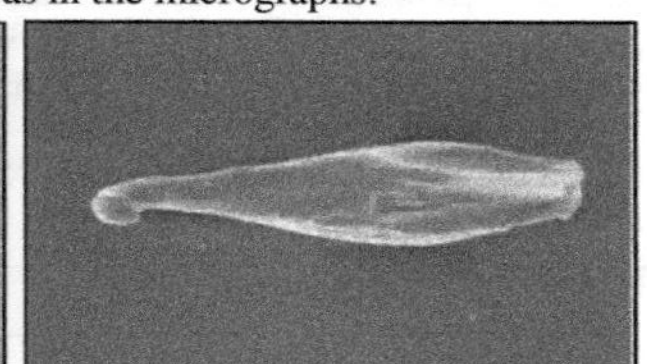										
	Specific characters: Hyaline portion of IJ 38%; lateral field formula: 2, 6, 8, 4, 2										

Table 3.1. *(Continued).*

Species	Morphometric characters (mean, range)										
	L	MBD	EP	NR	ES	T	a	b	c	D%	E%
S. oregonense											
Infective juvenile	**980**	34	**66**	na	132	**70**	30	7.6	14.0	50	**100**
	820-1110	28-38	60-72	na	116-148	64-78	24-37	6.0-8.0	12.0-16.0	40-60	90-110
Male	**SL** = **71** (65-73); GL = 56 (52-59); MBD = 138 (105-161); **D%** = **73** (64-75); SW% = 151; GS% = 79; **MUC** = **A**										
	Spicule and gubernaculum shapes as in the micrographs: 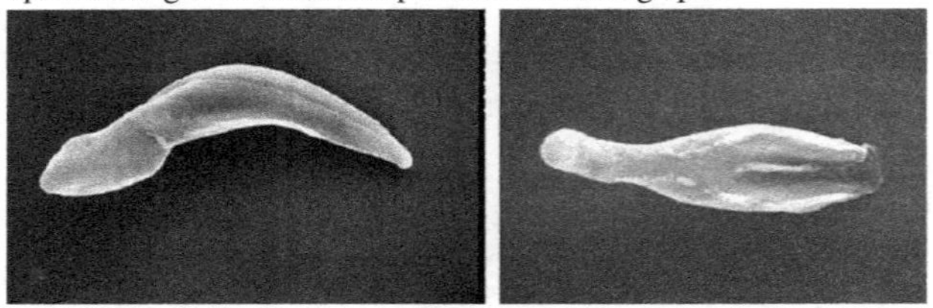										
	Specific characters: This nematode is very similar to *Steinernema feltiae* except male without a mucron and IJ longer										
S. kraussei											
Infective juvenile	**951**	33	**63**	105	134	**79**	29	7.1	12.1	47	**80**
	797-1102	30-36	50-66	99-111	119-145	63-86	na	na	na	na	na
Male	**SL** = **55** (52-57); GL = 33 (23-38); MBD = 128 (110-144); **D%** = **53**; SW% = 110; GS% = 71; **MUC** = **P**										
	Spicule and gubernaculum shapes as in the micrographs: 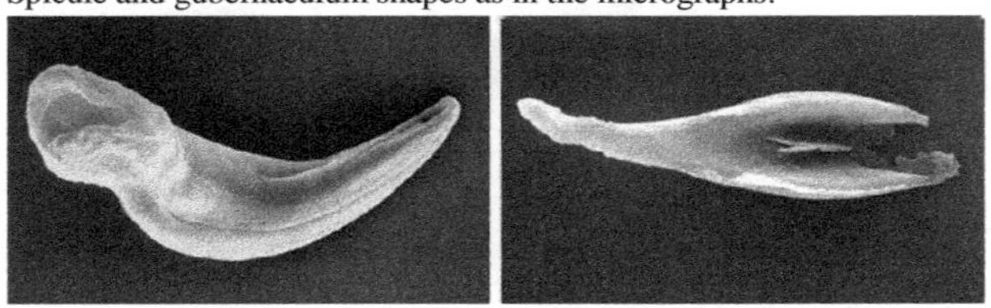										
	Specific characters: Spicule short; D% low										

Table 3.1. *(Continued).*

Species	Morphometric characters (mean, range)										
	L	MBD	EP	NR	ES	T	a	b	c	D%	E%
S. schliemanni											
Infective juvenile	**934**	35	72	–	148	**88**	26	6.3	10.6	47.6	–
	842-1008	30-38	61-80		127-162	76-95	23-30	5.6-6.7	9.6-12.2	42-55	
Male	SL = 72 (61-81); GL = 53 (43-64); BMD = 87 (76-120); D% = 54 (50-58); MUC = P										
	Spicule and gubernaculum shapes as in the micrographs:										
	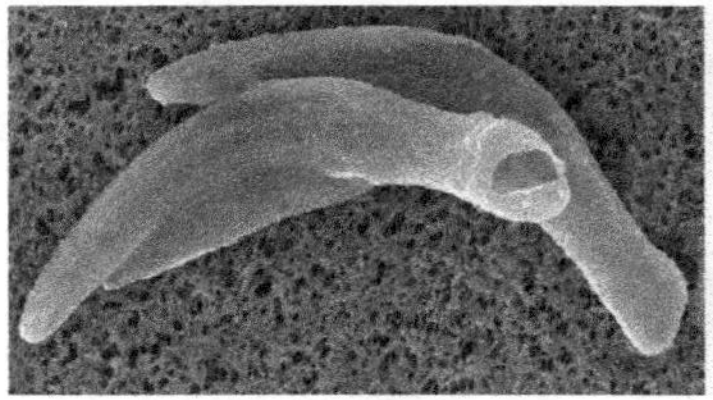 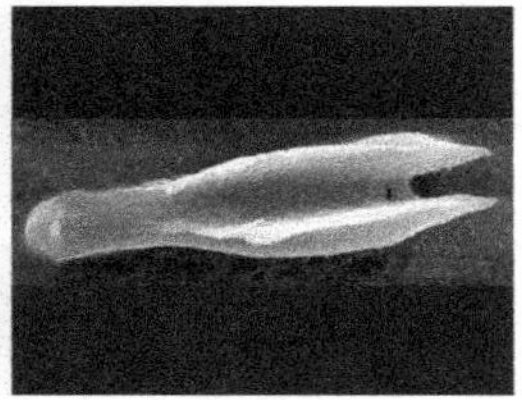										
S. karii											
Infective juvenile	**932**	33	**74**	105	134	**74**	29	6.8	12.6	57	**96**
	876-982	31-35	68-80	97-112	122-147	64-80	na	na	na	na	na
Male	**SL** = **83** (73-91); GL = 57 (42-64); MBD = 136 (107-166); **D%** = **66** (57-78); **MUC** = **A**										
	Spicule and gubernaculum shapes as in the drawing:										
	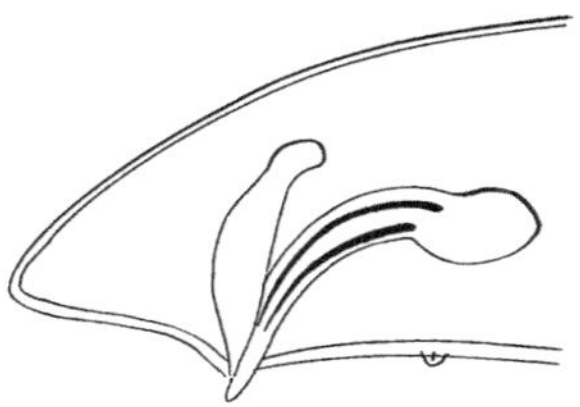										
	Specific character: Spicule long										

Table 3.1. *(Continued).*

Species	Morphometric characters (mean, range)										
	L	MBD	EP	NR	ES	T	a	b	c	D%	E%
S. hermaphroditum											
Infective juvenile	**928**	–	**65**	102	–	**77**	29	6.2	10.6	**50**	**85**
	700-1100		63-68	95-113		65-83	23-35	5.6-7.1	9.3-11.7	47-55	76-100
Male	**SL** = **68** (65-70); **GL** = **48** (47-50); **D%** = **48** (44-50); SW% = 150 (140-160); GS% = 71 (70-75); **MUC** = **A**										
	Spicule and gubernaculum shapes as in the micrographs: 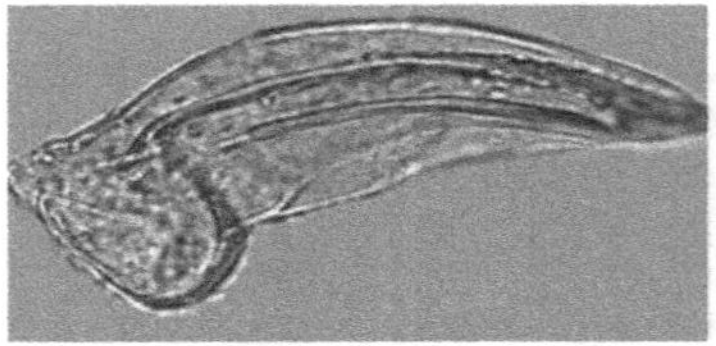 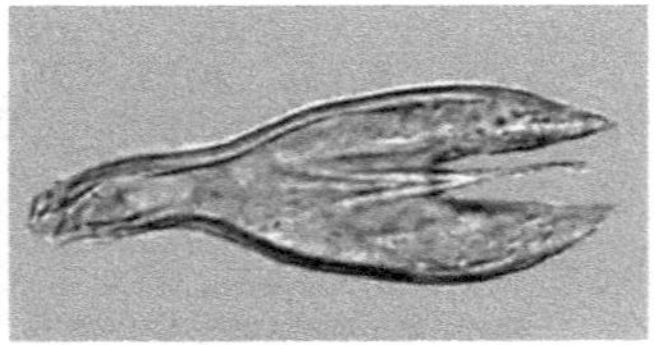										
	Specific character: Hermaphrodites present										
S. jeffreyense											
Infective juvenile	**926**	35	**87**	100	133	**81**	26	7.0	12.0	**66**	**109**
	784-1043	23-43	78-107	85-114	116-149	50-96	20-35	6.1-7.9	9.7-19.0	57-85	86-169
Male	**SL** = **88** (79-95); GL = 57 (51-61); MBD = 139 (94-167); **D%** = **47** (34-68); **SW%** = **215** (171-295); **GS%** = **65** (61-71); MUC = A										
	Spicule and gubernaculum shapes as in the micrographs: 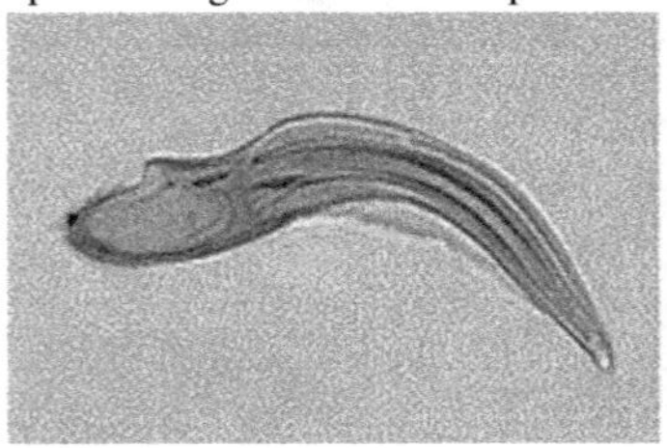 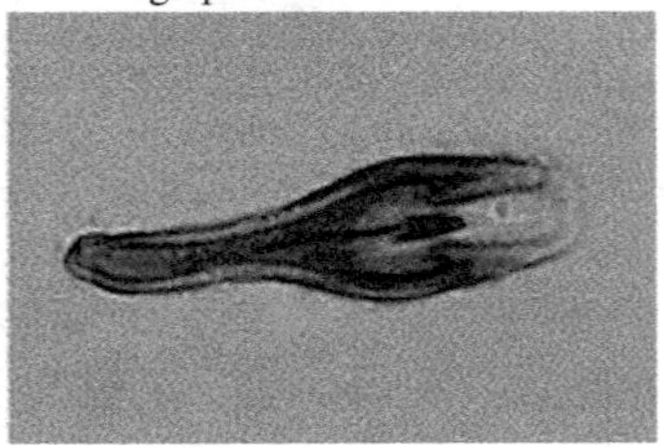										

Table 3.1. *(Continued).*

Species	Morphometric characters (mean, range)										
	L	MBD	EP	NR	ES	T	a	b	c	D%	E%
S. scarabaei											
Infective juvenile	**918**	31	**77**	120	156	**76**	31	6.6	12.0	**60**	**100**
	890-959	25-37	72-82	116-127	145-171	71-80	26-36	5.4-7.6	11.0-13.0	50-75	90-110
Male	**SL** = **75** (67-83); **GL** = **44** (36-50); MBD = 96 (55-120); **D%** = **66** (53-77); **SW%** = **170** (150-200); GS% = 60 (50-65); MUC = P										
	Spicule and gubernaculum as in the micrographs: 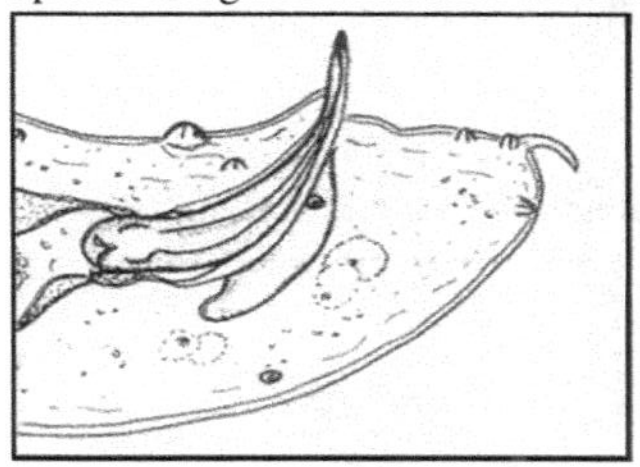 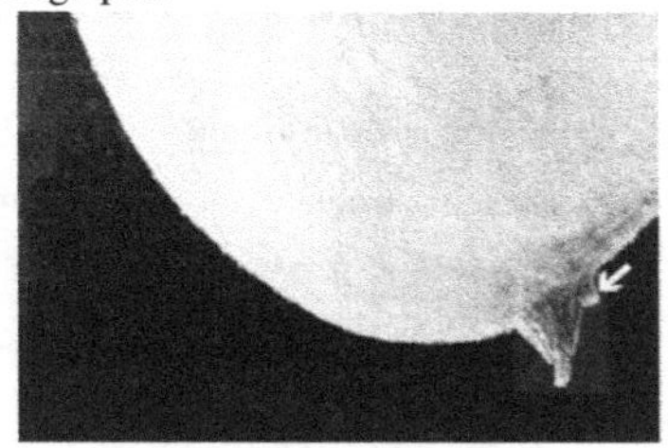										
	Specific character: First generation female tail with bifurcated mucron										
S. tielingense											
Infective juvenile	**915**	35	69	98	128	**81**	26	7.0	11.0	55	**88**
	824-979	32-38	64-73	90-105	120-135	74-85	23-28	6.0-8.0	9.0-13.0	47-61	85-94
Male	**SL** = **88** (79-98); GL = 61 (49-70); MBD = 129 (111-159); **D%** = **71** (64-78); **SW%** = **191** (176-212); GS% = 73 (59-82); MUC = A										
	Spicule and gubernaculum shapes as in the micrographs:  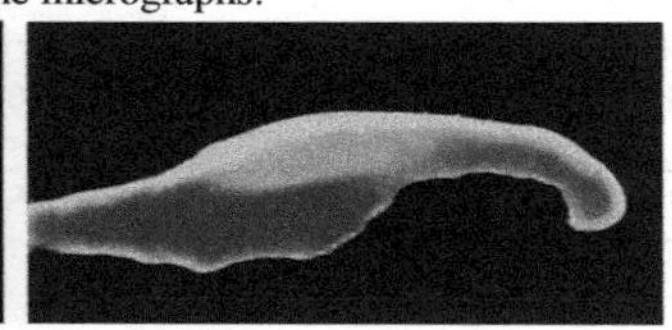										
	Specific character: Lateral field formula: 2, 6, 7, 8, 2										

Table 3.1. *(Continued).*

Species	Morphometric characters (mean, range)										
	L	MBD	EP	NR	ES	T	a	b	c	D%	E%
S. litorale											
Infective juvenile	**909**	31	**61**	96	125	**83**	30	**7.3**	11.0	**49**	**73**
	834-988	28-33	54-69	89-104	114-133	72-91	27-31	6.7-7.9	9.7-11.9	44-56	68-84
Male	**SL** = **75** (67-89); GL = 53 (44-64); MBD = 96 (82-111); **D%** = **65** (55-72); **SW%** = **174** (154-200); GS% = 71 (62-81); **MUC** = **P** Spicule and gubernaculum shapes as in the micrographs: 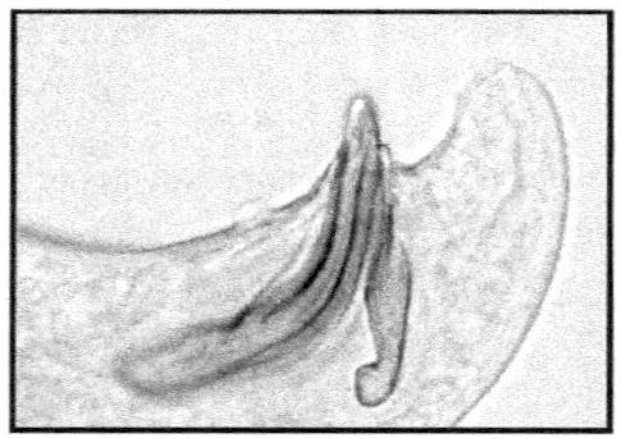 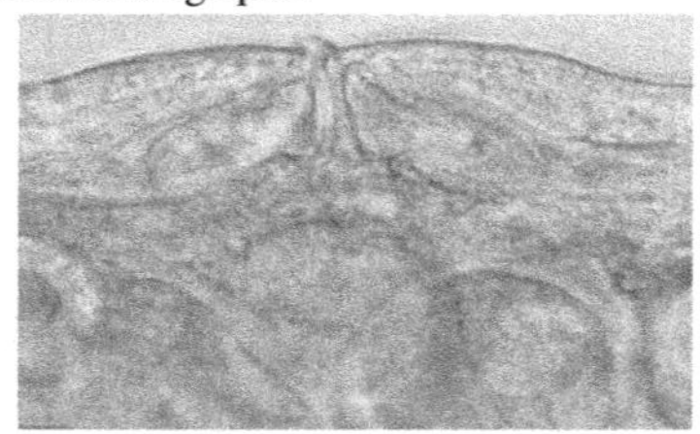 **Specific characters**: First generation female tail with spicule head twice as long as wide; vulva with epiptygmata										
S. ethiopiense											
Infective juvenile	**898**	34	**78**	104	140	**73**	27	6.4	12.0	**56**	**107**
	769-1010	32-43	65-84	83-113	127-153	64-80	20-30	5.7-7.1	11-14	51-58	91-116
Male	**SL** = **73** (69-77); GL = 49 (46-57); MBD = 92 (81-99); **D%** = **57** (54-61); **SW%** = **164** (154-175); GS% = 67 (63-70); MUC = A Spicule and gubernaculum shapes as in the drawings: 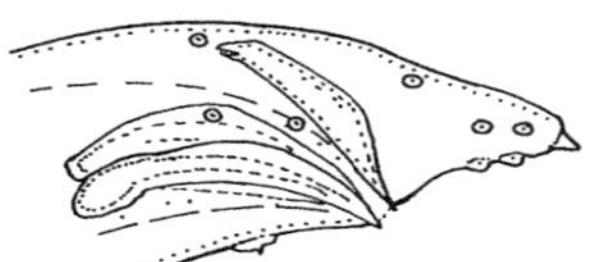 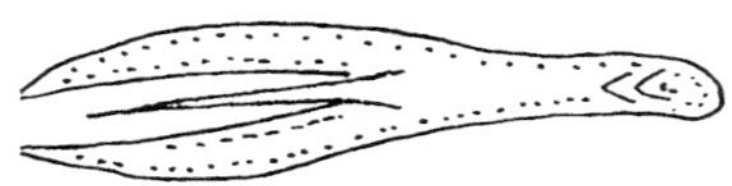										

Table 3.1. *(Continued).*

Species	Morphometric characters (mean, range)										
	L	MBD	EP	NR	ES	T	a	b	c	D%	E%
S. neocurtillae											
Infective juvenile	**885**	34	**18**	107	144	**80**	26	6.1	11.2	**12**	**23**
	741-988	28-42	14-22	100-119	130-159	64-97	22-29	5.4-6.7	9.1-14.2	10.0-15.0	18-30
Male	**SL** = **58** (52-64); GL = 52 (44-59); MBD = 111 (77-144); **D%** = **19** (13-26); SW% = 143 (118-164); **GS%** = **89** (82-93); **MUC** = **P**										
	Spicule and gubernaculum shapes as in the micrographs: 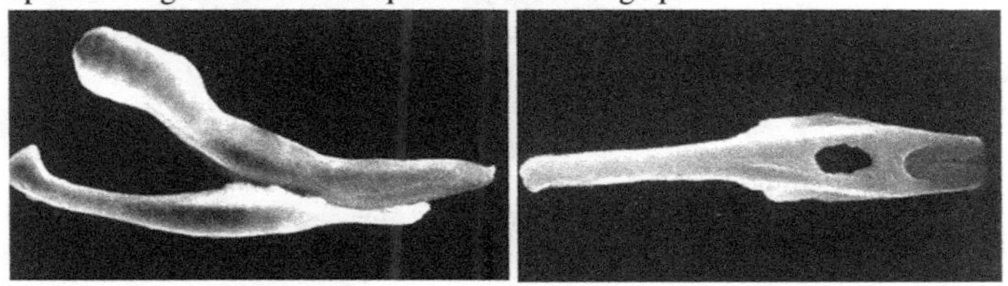										
	Specific characters: EP extremely short; GS very high; cannot be produced in *Galleria*										
S. ichnusae											
Infective juvenile	**866**	31	63	102	138	**81**	28	6.3	11.0	46	**77**
	767-969	27-35	59-68	94-108	119-148	76-89	24-32	5.6-6.9	8.8-12.0	42-49	68-83
Male	**SL** = **66** (64-67); GL = 44 (43-46); MBD = 137 (73-204); **D%** = **62** (59-65); **SW%** = **139** (120-162); GS% = 67 (64-69); MUC = A										
	Spicule and gubernaculum shapes as in the micrographs: 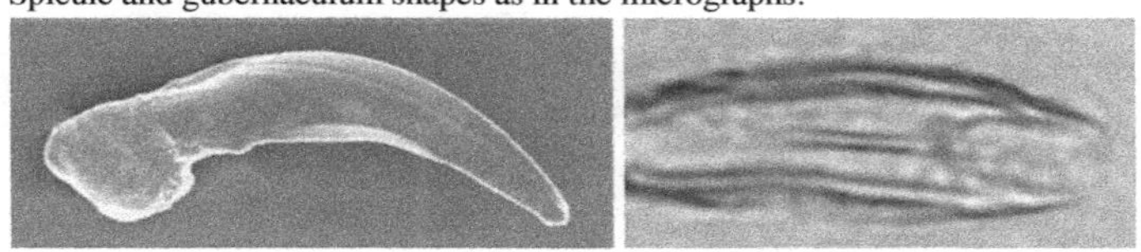										
	Specific character: Lateral field formula: 2, 8, 6, 4, 2										

Table 3.1. *(Continued).*

Species	Morphometric characters (mean, range)										
	L	MBD	EP	NR	ES	T	a	b	c	D%	E%
S. silvaticum											
Infective juvenile	**860**	30	**62**	96	121	**75**	29	7.1	11.4	**50**	**83**
	670-975	26-35	51-73	75-109	100-141	63-86	23-33	6.3-7.7	9.9-13.1	46-56	
Male	**SL** = **51** (42-64); GL = 37 (30-43); MBD = 65 (52-78); D% = 60 (45-63); SW% = 155; GS% = 73; **MUC** = **P**										
	Spicule and gubernaculum as in the micrograph: 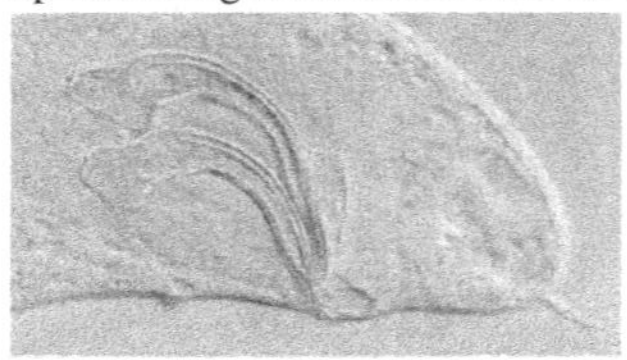										
	Specific characters: Male tail with prominent mucron; spicule head wide										
S. thanhi											
Infective juvenile	**851**	31	**75**	96	129	**63**	27	7.0	**14.0**	**58**	**119**
	720-960	27-39	68-84	83-112	115-146	52-72	23-33	6.0-8.0	12.0-16.0	52-67	101-138
Male	**SL** = **72** (67-78); GL = 49 (40-56); MBD = 97 (76-124); **D%** = **73** (64-82); **SW%** = **180** (150-210); GS% = 70 (50-80); **MUC** = **A**										
	Spicule and lateral field as in the micrographs: 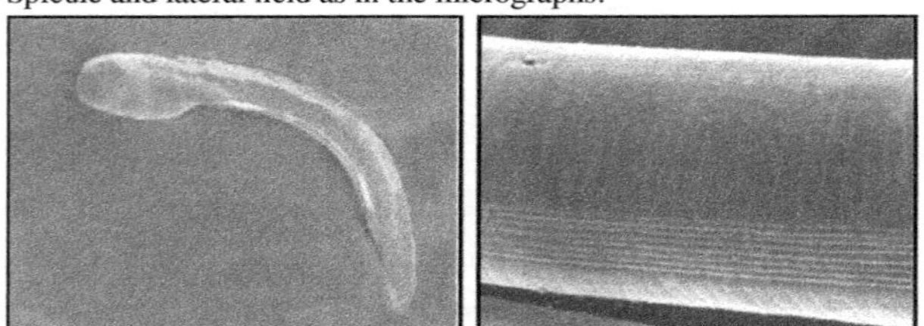										
	Specific character: The presence of 13 pairs of genital papillae on first generation male										

Table 3.1. *(Continued).*

Species	Morphometric characters (mean, range)										
	L	MBD	EP	NR	ES	T	a	b	c	D%	E%
S. feltiae											
Infective juvenile	**849**	26	**62**	99	136	**81**	31	6.0	10.4	45	**78**
	736-950	22-29	53-67	88-112	115-150	70-92	29-33	5.3-6.4	9.2-12.6	42-51	69-86
Male	**SL** = **70** (65-77); GL = 41 (34-47); MBD = 75 (60-90); **D%** = **60**; SW% = 113 (99-130); GS% = 59 (52-61); **MUC** = **P**										
	Spicule and gubernaculum shapes as in the micrographs: 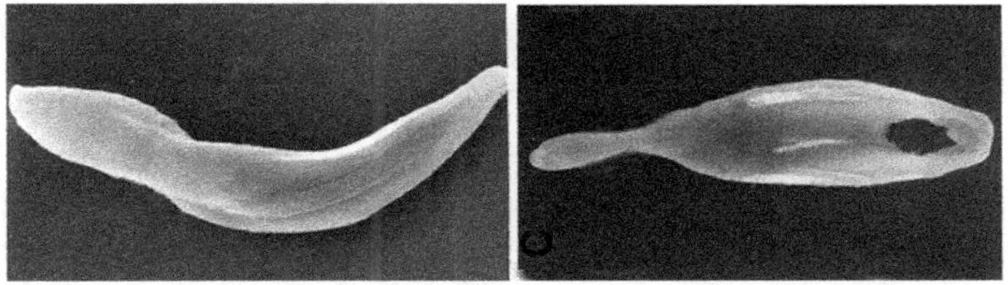										
	Specific characters: Spicule head very long; male with long mucron; lateral field formula: 2, 7, 8, 7, 4, 2										
S. lamjungense											
Infective juvenile	**832**	27	68	93	127	88	30	6.6	9.6	**54**	79
	690-950	23-31	61-73	82-100	111-142	67-103	23-40	5.8-7.2	7.7-10.8	48-63	69-97
Male	**SL** = **87** (81-94); GL = 57 (50-66); MBD = 154 (116-205); **D%** = **61** (44-91); **SW%** = **173** (151-193); GS% = 65 (56-77); **MUC** = **A**										
	Spicule and gubernaculum shapes as in the micrographs: 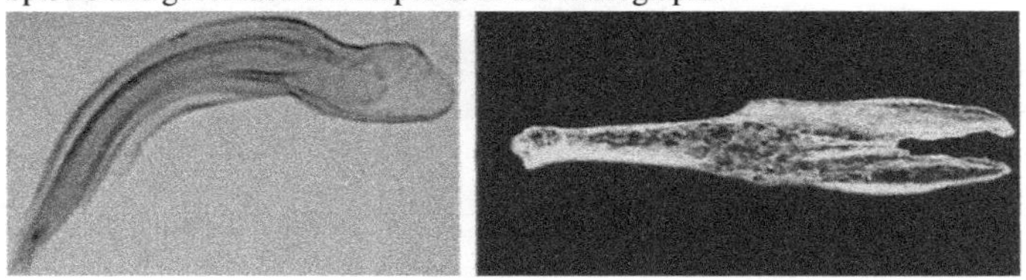										
	Specific character: Lateral field formula: 2, 6, 7, 8, 3, 2										

Table 3.1. *(Continued).*

Species	Morphometric characters (mean, range)										
	L	MBD	EP	NR	ES	T	a	b	c	D%	E%
S. akhursti											
Infective juvenile	**812**	**33**	**59**	90	119	**73**	**24**	6.8	11.0	47	**77**
	770-835	33-35	55-60	83-95	115-123	68-75	23-26	6.6-7.2	10.0-12.0	45-50	73-86
Male	**SL** = **90** (85-100); GL = 64 (58-68); MBD = 131 (115-150); D% = 56 (52-61); **SW%** = **180** (140-200); GS% = 71 (65-77); **MUC** = **P**										
	Spicule and gubernaculum shapes as in the micrographs: 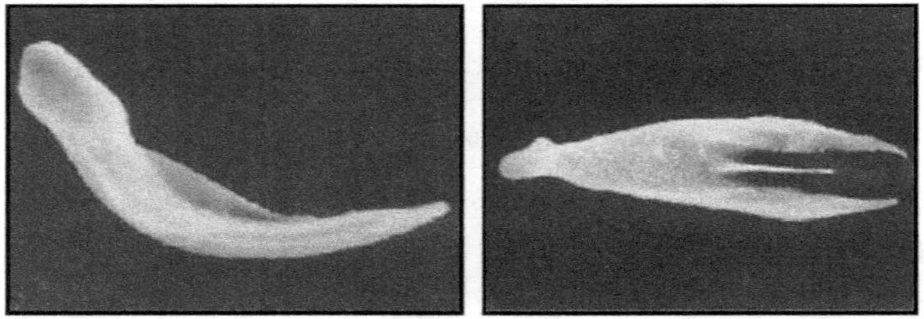										
	Specific characters: Lateral field formula: 2, 6, 2; male with the longest spicule in *S. feltiae* group										
S. everestense											
Infective juvenile	**775**	29	55	92	119	71	27	6.5	11.0	**47**	**78**
	705-838	24-31	45-61	84-102	110-130	64-78	25-31	6.0-7.0	10.0-12.0	36-51	60-89
Male	**SL** = **79** (65-87); GL = 50 (39-57); MBD = 92 (76-116); **D%** = **55** (44-66); **SW%** = **226** (197-273); **GS%** = **64** (52-74); MUC = P										
Syn. *S. akhursti*	Spicule and gubernaculum shapes as in the micrographs: 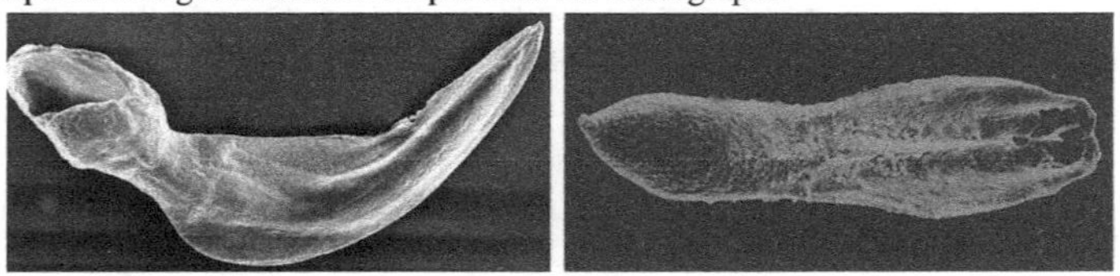										
	Specific character: Lateral field formula: 2, 8, 6, 4, 3										

Table 3.1. *(Continued).*

Species	Morphometric characters (mean, range)										
	L	MBD	EP	NR	ES	T	a	b	c	D%	E%
S. bicornutum											
Infective juvenile	**769**	29	**61**	92	124	**72**	27	6.2	10.7	50	**84**
	648-873	25-33	53-65	88-100	113-135	63-78	23-29	5.6-6.9	9.7-12.0	40-60	80-100
Male	**SL** = **65** (53-70); GL = 48 (38-50); MBD = 109 (80-127); **D%** = **52** (50-60); **SW%** = **222** (218-226); **GS%** = **72**; **MUC** = **A**										
	Spicule and gubernaculum as in the micrographs:										
	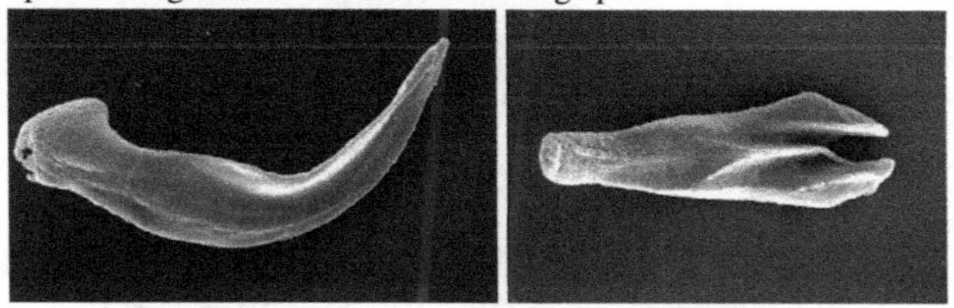										
	Specific character: 2 horn-like structures on lip region of IJ										
S. ashiuense											
Infective juvenile	**768**	30	**55**	86	119	**71**	25	6.0	11.0	**46**	**78**
	720-800	28-33	51-59	77-91	113-128	66-76	24-27	6.0-7.0	10.0-12.0	43-50	70-85
Male	**SL** = **59** (50-65); GL = 37 (25-43); MBD = 106 (80-125); **D%** = **50** (44-56); **SW%** = **149** (128-167); GS% = 63 (43-73); **MUC** = **P**										
	Spicule and gubernaculum as in the micrographs:										
	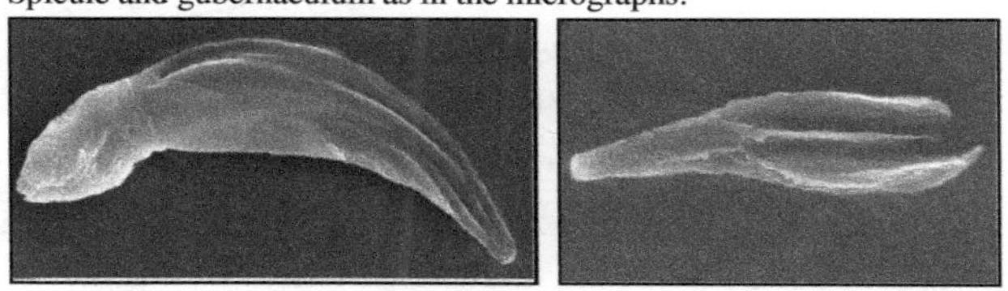										
	Specific characters: Lateral field with 5 ridges; hyaline portion = 1/3 of tail										

Table 3.1. *(Continued).*

Species	Morphometric characters (mean, range)										
	L	MBD	EP	NR	ES	T	a	b	c	D%	E%
S. poinari											
Infective juvenile	**758** 687-848	32 29-37	**64** 56-70	–	141 119-151	77 68-87	24 21-28	5.5 4.8-6.3	10.0 8.7-12.0	46 40-55	**84** 76-95
Male	**SL** = **70** (62-78); GL = 49 (45-57); MBD = 158 (118-219); D% = 54 (42-64); **SW%** = **109** (98-123); **GS%** = **70** (58-87); **MUC** = **A**										

Spicule and gubernaculum as in the micrographs:

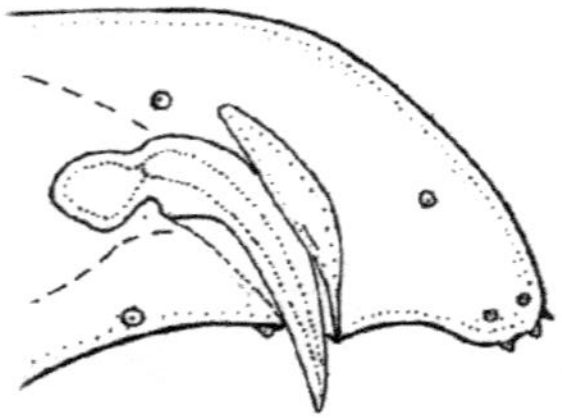

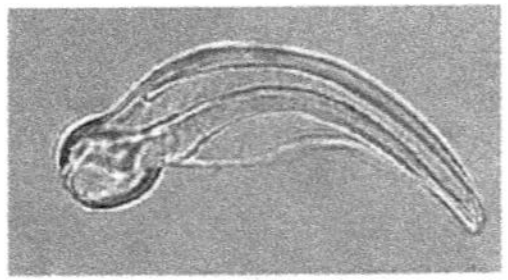

Specific character: Lateral field formula: 2, 5, 6, 5, 2, 1

Species	L	MBD	EP	NR	ES	T	a	b	c	D%	E%
S. sangi											
Infective juvenile	**753** 704-784	35 30-40	**51** 46-54	91 78-97	127 120-138	**81** 76-89	**22** 19-25	5.9 5.6-6.3	9.3 8.7-10.2	**40** 36-44	**62** 56-70
Male	**SL** = **63** (58-80); GL = 40 (34-46); MBD = 159 (120-225); D% = 49 (42-63); **SW%** = **150** (120-160); GS% = 60 (50-70); **MUC** = **P**										

Spicule shape as in the micrograph:

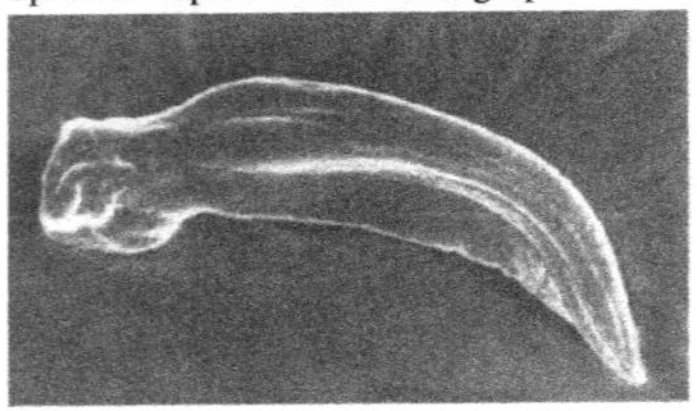

Specific characters: This species is closer to *S. kraussei* but IJ body is shorter and SW% is larger

Table 3.1. *(Continued).*

Species	Morphometric characters (mean, range)										
	L	MBD	EP	NR	ES	T	a	b	c	D%	E%
S. beddingi											
Infective juvenile	**743** 700-790	–	**70** 64-75	96 80-113	125 113-130	**77** 72-83	**26.7** 24-30	5.8 5.0-6.8	10.1 9.6-10.8	53 44-64	91 80-108
Male	**SL = 67** (63-73); **GL = 40** (35-45); **D% = 59** (53-62); SW% = 122 (110-131); GS% = 60 (56-66); **MUC = A** Spicules and gubernaculum shapes as in the micrographs: 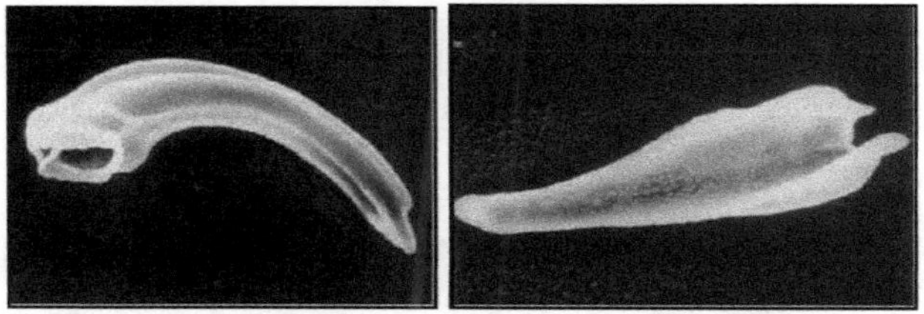 **Specific characters**: Spicule shape is unique; gubernaculum without cuneus; lateral fields of IJ has 4 ridges										
S. weiseri											
Infective juvenile	**740** 586-828	25 24-29	**57** 43-65	84 72-92	113 95-119	**60** 49-68	**29** 25-33	6.6 5.7-7.2	12.0 10.0-14.0	51 44-55	**95**
Male	**SL = 68** (62-72); GL = 53 (46-57); MBD = 112 (84-138); **D% = 49** (39-60); **SW% = 180** (150-240); GS% = 80 (70-85); **MUC = A** Spicule and gubernaculum as in the micrograph: 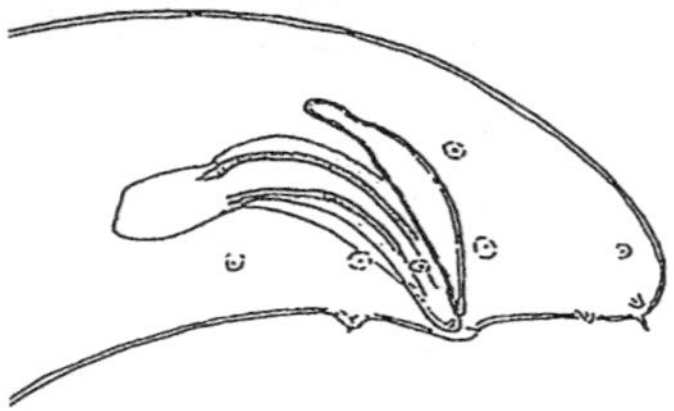 **Specific character**: IJ has very short hyaline portion, 1/3 of tail length										

Table 3.1. *(Continued).*

Species	Morphometric characters (mean, range)										
	L	MBD	EP	NR	ES	T	a	b	c	D%	E%
S. unicornum											
Infective juvenile	**732**	29	55	88	124	**60**	25	5.9	12.2	**45**	**91**
	654-786	25-33	48-62	77-99	106-143	50-71	22-29	5.3-7.3	10.7-14.5	38-55	77-116
Male	**SL** = **73** (60-82); GL = 48 (44-52); MBD = 130 (94-151); **D%** = **36** (24-49); **SW%** = **198** (184-217); **GS%** = **68** (60-74); MUC = P (10%)										
	Labial region of IJ and vulva of first generation as in the micrographs:										

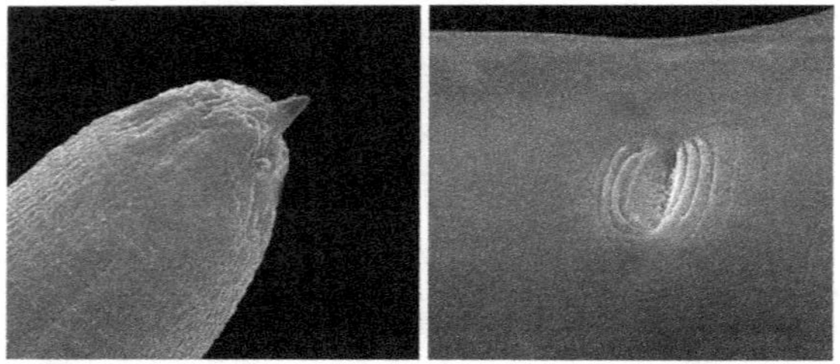

Species	L	MBD	EP	NR	ES	T	a	b	c	D%	E%
S. robustispiculum											
Infective juvenile	**712**	28	56	84	120	**75**	25	6.0	10.0	46	**75**
	642-778	26-35	50-68	80-100	115-152	68-92	18-29	3.6-6.3	6.0-11.0	43-59	67-87
Male	**SL** = **58** (51-65); GL = 41 (36-44); MBD = 127 (105-150); **D%** = **56** (50-63); **SW%** = **129** (111-150); GS% = 70 (64-79); **MUC** = **P**										
	Spicule large as in the drawing:										

Specific character: Spicule length/spicule width = 4.5 (3.8-5.6)

Table 3.1. *(Continued).*

Species	Morphometric characters (mean, range)										
	L	MBD	EP	NR	ES	T	a	b	c	D%	E%
S. jollieti											
Infective juvenile	**711**	23	**60**	–	123	**68**	31	5.7	11.0	**48**	88
	625-820	20-28	53-65		115-135	60-73	25-34	4.9-6.4	9.0-12.0	46-50	

Male: **SL** = **64** (55-70); GL = 54 (45-60); MBD = 115 (98-135); **D%** = **64** (53-83); SW% = 145; GS% = 84; **MUC** = **A**

Spicule and gubernaculum as in the drawing:

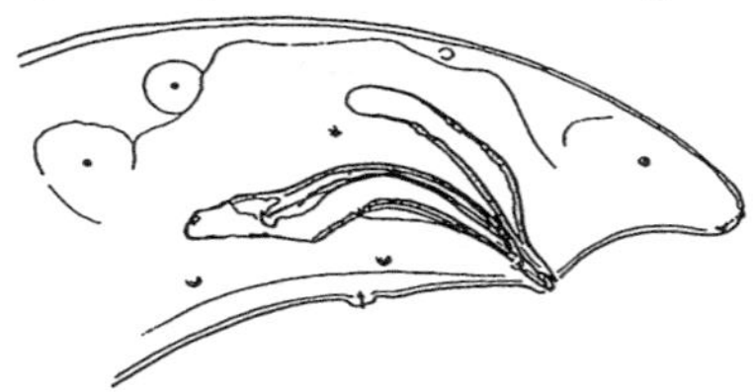

Specific character: Lateral field with 6 ridges

Species	L	MBD	EP	NR	ES	T	a	b	c	D%	E%
S. sichuanense											
Infective juvenile	**710**	28	**64**	99	131	**72**	25	5.4	10.0	**49**	**89**
	606-787	24-32	57-68	91-108	121-142	64-76	23-28	4.9-6.0	8.9-11.0	46-53	

Male: **SL** = **88** (83-101); GL = 47 (40-51); MBD = 122 (91-141); D% = 51 (45-56); SW% = 160 (140-180); **GS%** = **80** (60-80); **MUC** = **A**

Spicule and gubernaculum are in the micrographs:

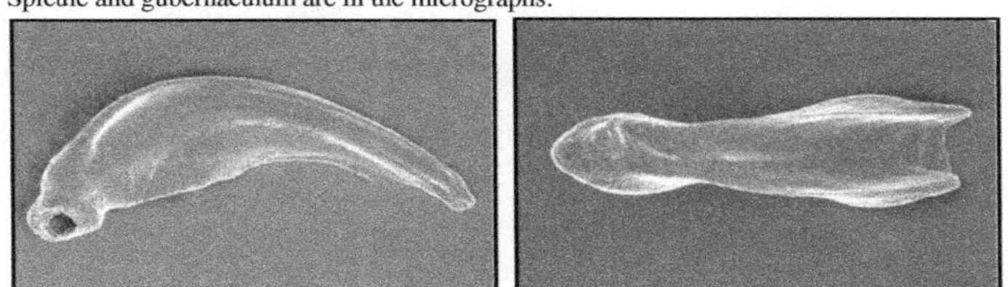

Specific characters: Lateral field with 6 ridges; lateral field formula: 2, 5, 6, 4, 2

Table 3.1. *(Continued).*

Species	Morphometric characters (mean, range)										
	L	MBD	EP	NR	ES	T	a	b	c	D%	E%
S. monticolum											
Infective juvenile	**706**	37	**58**	88	124	**77**	19	5.7	9.3	47	**76**
	612-821	32-46	54-62	81-93	120-131	71-95	14-22	5.0-6.4	7.6-11.1	44-50	63-86
Male	**SL = 70** (61-80); GL = 45 (35-54); MBD = 160 (117-206); **D% = 55** (49-61); SW% = 140 (120-150); GS% = 60 (50-70); **MUC = P** Spicule and gubernaculum shapes as in the micrographs: 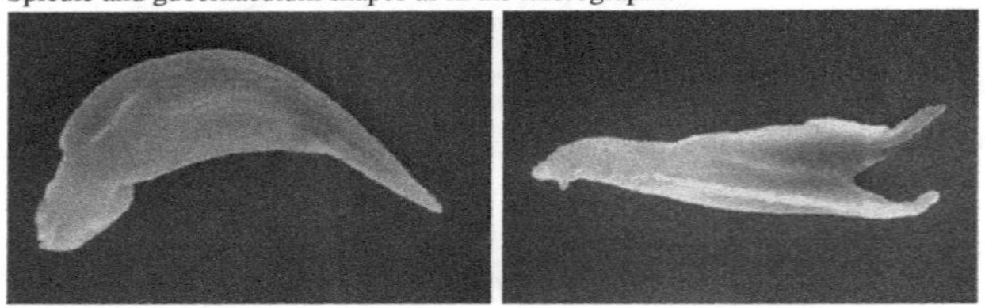 **Specific characters**: IJ tail long; E% low										
S. ceratophorum											
Infective juvenile	**706**	27	**55**	92	123	**66**	26	na	10.6	45	**84**
	591-800	23-34	47-70	79-103	108-144	56-74	24-28	na	8.8-12.9	40-56	74-96
Male	**SL = 71** (54-90); GL = 40 (25-45); MBD = 146 (104-185); **D% = 51** (33-65); **SW% = 140** (100-200); GS% = 6 (40-80); **MUC = A** Spicule and gubernaculum shapes as in the micrographs: 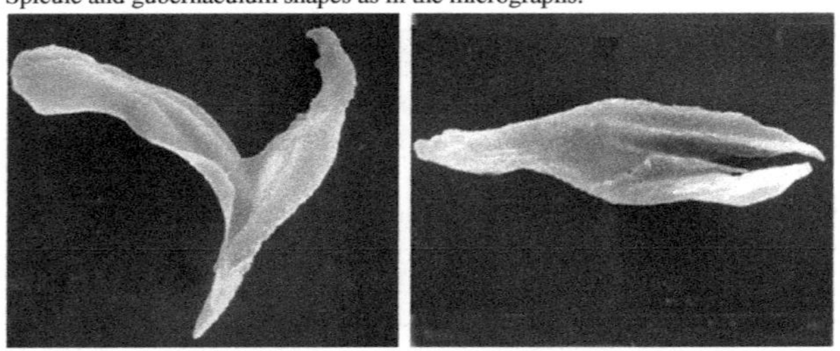 **Specific character**: 2 horn-like structures on lip region of IJ										

Table 3.1. *(Continued).*

Species	Morphometric characters (mean, range)										
	L	MBD	EP	NR	ES	T	a	b	c	D%	E%
S. affine											
Infective juvenile	**693** 608-880	30 28-34	**62** 51-69	95 88-104	126 115-134	**66** 64-74	23 21-28	5.5 5.1-6.0	10.5 9.5-11.5	49 43-53	**94** 74-108
Male	**SL = 70** (67-86); GL = 46 (37-56); MBD = 118 (95-164); **D% = 61**; SW% = 117; GS% = 66; **MUC = P** Spicule and gubernaculum shapes as in the micrographs: 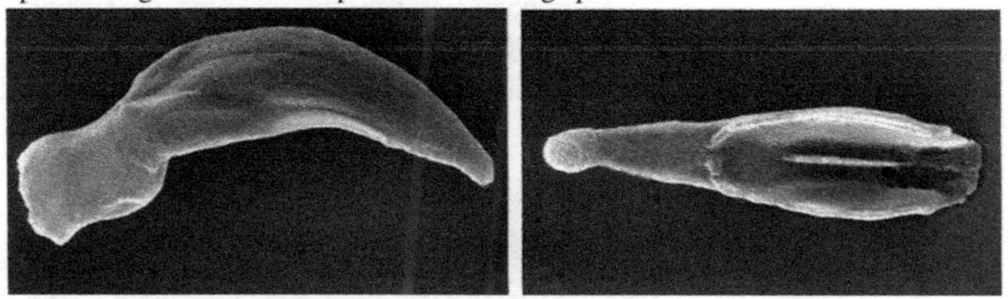 **Specific character**: Spine-like structure in IJ tail tip										
S. pakistanense											
Infective juvenile	**683** 649-716	27 24-29	**54** 49-58	80 76-83	113 108-122	58 53-62	24 21-27	6.0 5.0-6.0	11.0 10.0-12.0	**47** 42-53	**91** 87-102
Male	**SL = 68** (62-73); GL = 41 (36-45); MBD = 102 (80-128); D% = 60 (50-60); SW% = 180 (100-220) GS% = 60 (50-60); MUC = A Spicule and gubernaculum shapes as in the drawings: 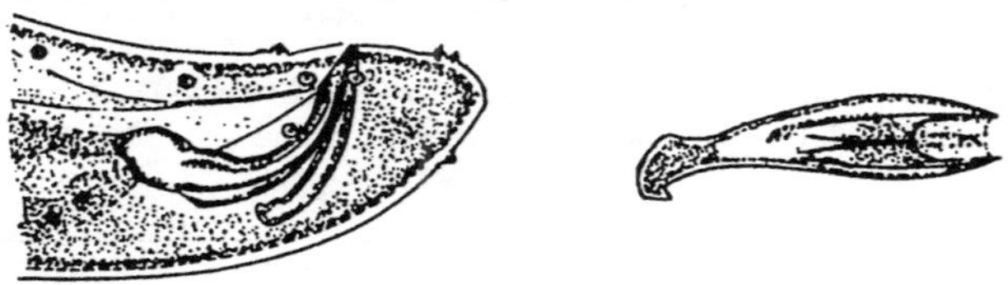 **Specific character**: IJ with horn-like structure on labial region										

Table 3.1. *(Continued).*

Species	Morphometric characters (mean, range)										
	L	MBD	EP	NR	ES	T	a	b	c	D%	E%
S. intermedium											
Infective juvenile	**671**	29	**65**	93	123	**66**	23	5.3	10.0	51	**96**
	608-800	25-32	59-69	85-99	110-133	53-74	20-26	5.0-6.0	9.3-10.8	48-58	89-108
Male	**SL** = **91** (84-100); GL = 64 (56-75); MBD = 168 (113-207); **D%** = **67** (58-76); SW% = 124 (103-139); GS% = 69 (62-77); MUC = A										

Spicule and gubernaculum shapes as the micrographs:

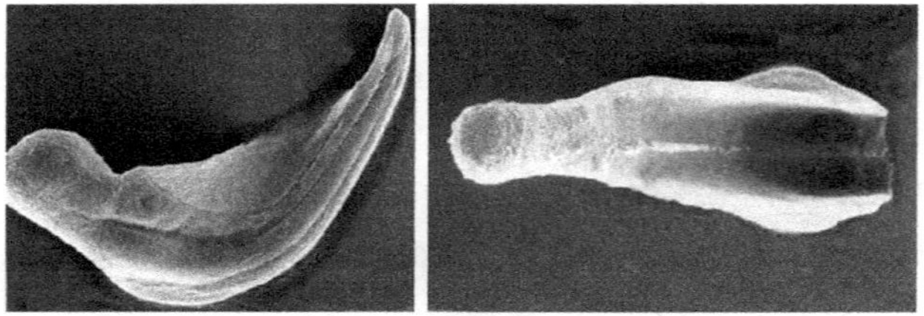

Specific character: Spicule is the largest in *Steinernema* spp.

Species	L	MBD	EP	NR	ES	T	a	b	c	D%	E%
S. hebeiense											
Infective juvenile	**658**	26	**48**	78	107	**66**	26	6.2	10.0	**45**	**72**
	610-710	23-28	43-51	73-83	100-111	63-71	24-28	5.7-6.7	9.4-11.0	40-50	65-80
Male	**SL** = **57** (51-63); GL = 46 (38-50); MBD = 86 (74-98); **D%** = **51** (48-59); SW% = 140 (120-170); GS% = 80 (60-90); **MUC** = **A**										

Spicule and gubernaculum as in the micrographs:

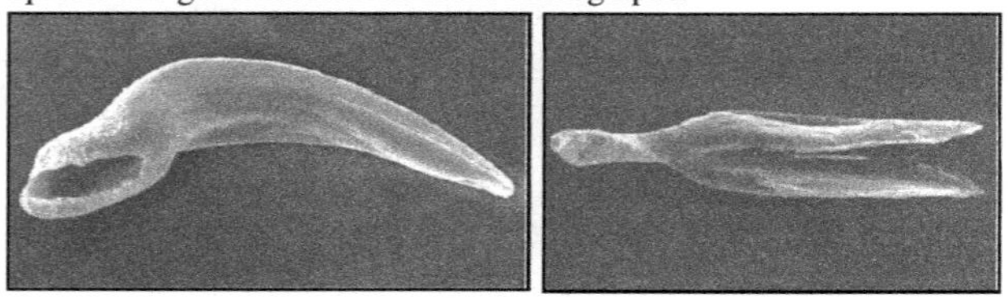

Specific character: Gubernaculum *ca* 80% of spicule

Table 3.1. *(Continued).*

Species	Morphometric characters (mean, range)										
	L	MBD	EP	NR	ES	T	a	b	c	D%	E%
S. papillatum											
Infective juvenile	**652**	24	50	88	110	**54**	27	5.9	12.1	**46**	**93**
	572-720	21-31	44-58	81-96	103-121	40-78	22-30	5.0-6.4	8.3-15.0	40-52	66-121
Male	**SL** = **58** (47-66); GL = 31 (23-36); **MBD** = **69** (54-87); **D%** = **54** (43-65); **SW%** = **156** (125-194); **GS%** = **59** (48-70); **MUC** = **A** Spicule shape as in the micrograph: 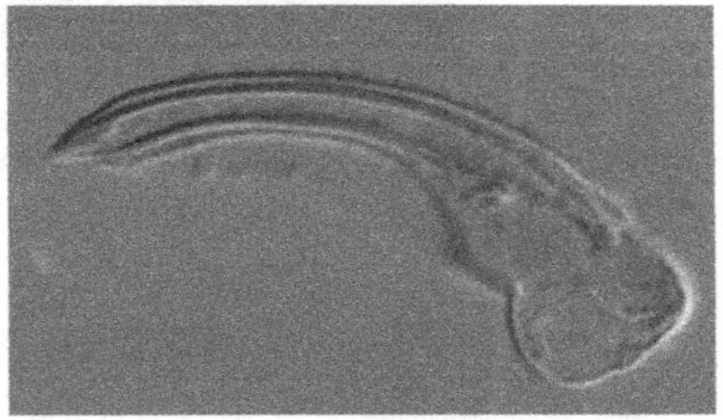 **Specific character**: IJ with horn-like structure on labial region										
S. nyetense											
Infective juvenile	**648**	32	**52**	85	114	**82**	21	6.0	8.0	46	**66**
	565-708	25-37	46-57	72-102	104-128	54-113	19-26	5.0-6.0	6.0-11.0	37-50	44-89
Male	**SL** = **80** (67-98); GL = 53 (40-62); MBD = 106 (62-159); **D%** = **55** (40-70); **SW%** = **199** (125-283); **GS%** = **66** (51-77); **MUC** = **P** Spicule and gubernaculum as in the micrographs: 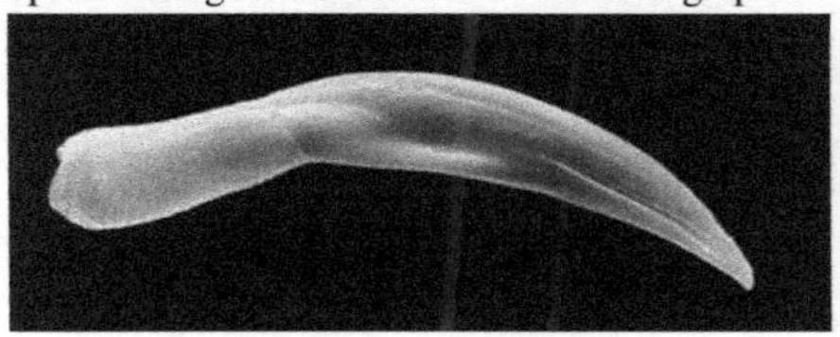 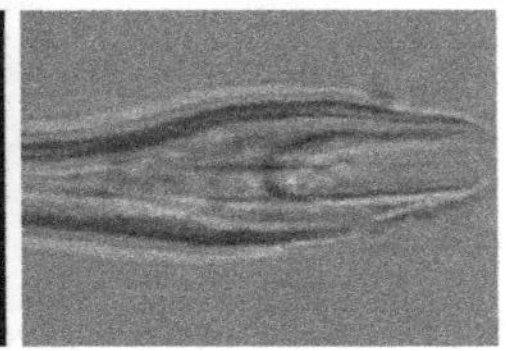 **Specific character**: Lateral field formula: 2, 4, 5, 4, 3, 2										

Table 3.1. *(Continued).*

Species	Morphometric characters (mean, range)										
	L	MBD	EP	NR	ES	T	a	b	c	D%	E%
S. yirgalemense											
Infective juvenile	**635**	29	**51**	88	121	**62**	21	5.2	10.3	**42**	83
	548-693	24-33	45-59	82-93	115-128	57-67	20-25	4.8-5.9	9.2-11.2	38-48	67-98
Male	**SL** = **64** (51-77); GL = 48 (42-54); MBD = 112 (97-138); D% = 58 (50-66); **SW%** = **171** (121-213); GS% = 74 (65-85); **MUC** = **A**										
	Spicule and gubernaculum shapes as in the micrographs: 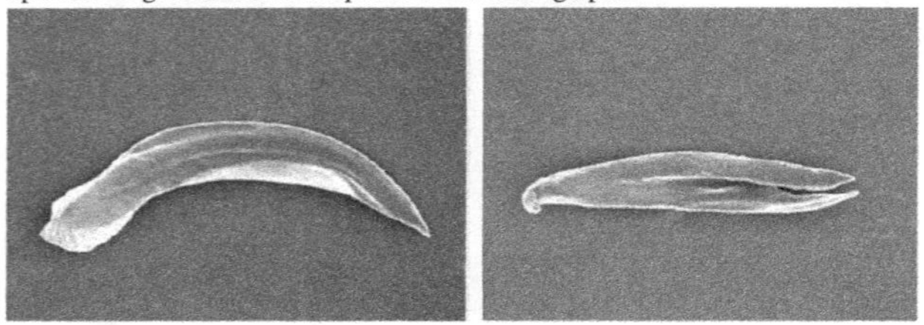										
	Specific characters: 2 horn-like structure present; the presence of 12 pairs of genital papillae in male										
S. cameroonense											
Infective juvenile	**622**	30	54	85	113	**76**	21	6.0	9.0	**48**	**75**
	490-694	24-35	45-64	69-100	105-125	52-107	17-25	5.0-6.0	6.0-12.0	42-56	48-116
Male	**SL** = **69** (51-85); GL = 45 (37-57); MBD = 90 (65-124); **D%** = **64** (48-76); SW% = 170 (131-201); GS% = 64 (47-76); **MUC** = **P**										
	Spicule shape as in the micrograph: 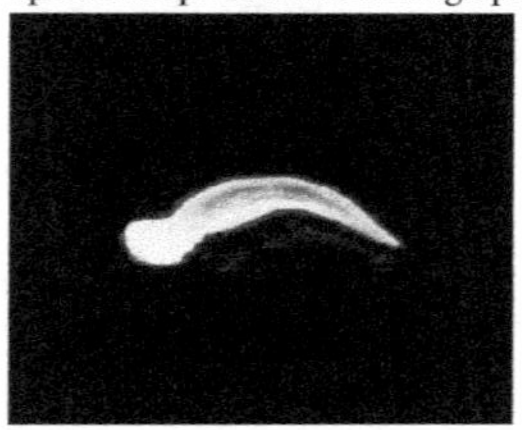										
	Specific character: Lateral field formula: 2, 4, 5, 6, 3, 2										

Table 3.1. *(Continued).*

Species	Morphometric characters (mean, range)										
	L	MBD	EP	NR	ES	T	a	b	c	D%	E%
S. riobrave											
Infective juvenile	**622**	28	**56**	87	114	**54**	23	5.4	11.6	49	**105**
	561-701	26-30	51-64	84-89	109-116	46-59	20-24	4.9-6.0	10.1-12.4	45-55	93-111
Male	**SL** = **67** (62.5-75.0); GL = 51 (47.5-56.2); MBD = 133 (116-159); **D%** = **71** (60-80); SW% = 114; GS% = 76; **MUC** = **A**										
	Spicule and gubernaculum shapes as in the micrographs: 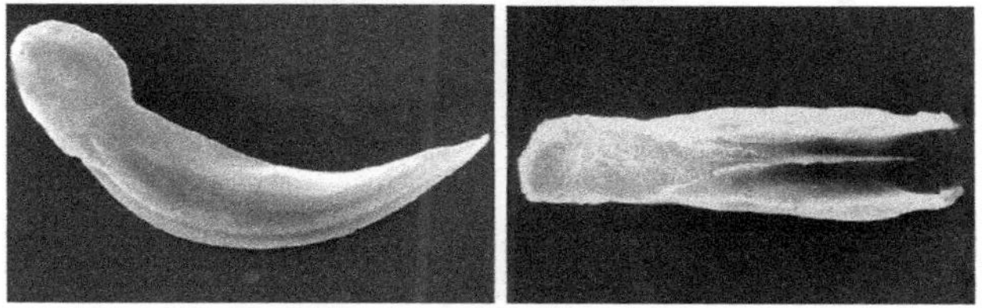										
	Specific character: Width of gubernaculum neck as wide as posterior corpus										
S. kushidai											
Infective juvenile	**589**	26	**46**	76	111	**50**	22.5	5.3	11.7	41	**92**
	524-662	22-31	42-50	70-84	106-120	44-59	19-25	4.9-5.9	9.9-12.9	38-44	84-95
Male	**SL** = **63**; GL = 44; MBD = 97; **D%** = **51**; SW% = 150; GS% = 70; **MUC** = **A**										
	Spicule and gubernaculum shapes as in the micrographs: 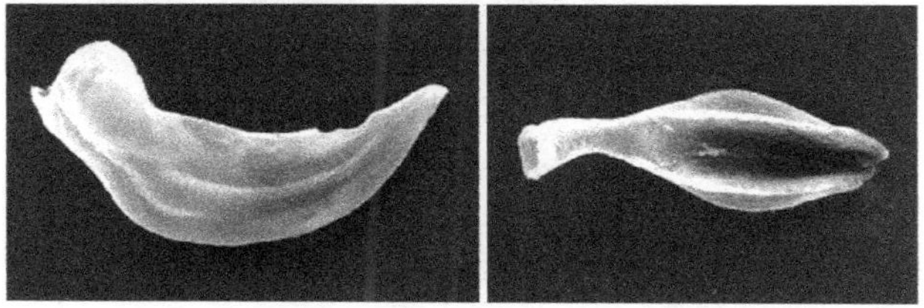										
	Specific characters: SL/spicule width about 3.9; cannot reproduce in *Galleria*										

Table 3.1. *(Continued).*

Species	Morphometric characters (mean, range)										
	L	MBD	EP	NR	ES	T	a	b	c	D%	E%
S. scapterisci											
Infective juvenile	**572**	24	**39**	97	127	**54**	24	4.5	10.7	31	**73**
	517-609	18-30	36-48	83-106	113-134	48-60	20-31	4.0-4.6	9.2-11.7	27-40	60-80

Male: **SL** = **83** (72-92); GL = 65 (59-75); MBD = 156 (97-213); **D%** = **38** (32-44); **SW%** = **252** (204-280); GS% = 78; **MUC** = **P**

Spicule and gubernaculum shapes as in the micrographs:

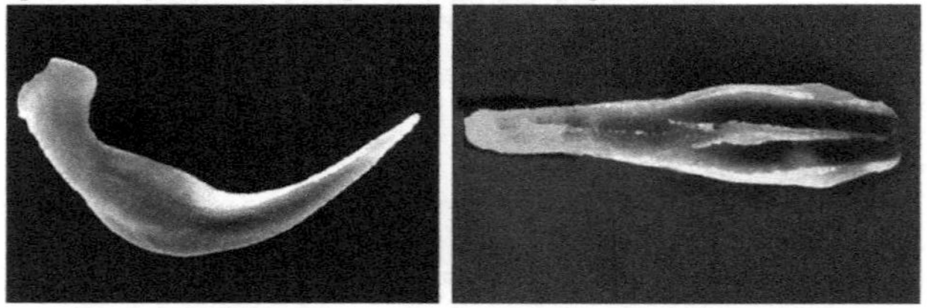

Specific characters: Female: Epiptygmata very large; excretory duct with ellipse-shaped structure; male with long spicule.

Note: Develops poorly in *Galleria*

Species	L	MBD	EP	NR	ES	T	a	b	c	D%	E%
S. carpocapsae											
Infective juvenile	**558**	25	**38**	85	120	**53**	21	4.4	10.0	26	**60**
	438-650	20-30	30-60	76-99	103-190	46-61	19-24	4.0-4.8	9.1-11.2	23-28	54-66

Male: **SL** = **66** (58-77); GL = 47 (39-55); MBD = 101 (77-130); **D%** = **41** (27-55); **SW%** = **172** (140-200); GS% = 71 (59-88); **MUC** = **P**

Spicule and gubernaculum shapes as in the micrographs:

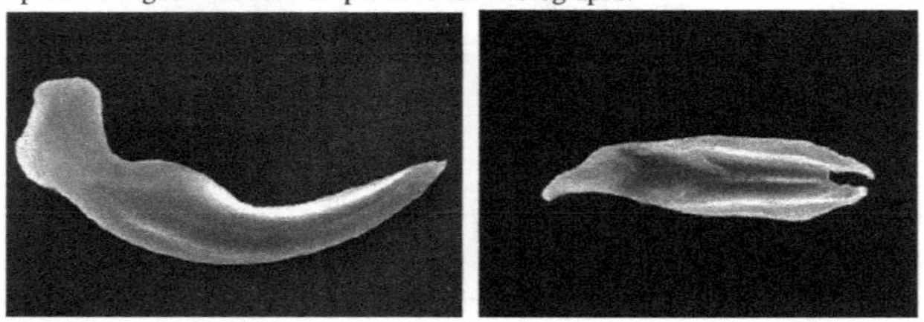

Specific characters: Spicule head wider than long; gubernaculum with short anterior part

Table 3.1. *(Continued).*

Species	Morphometric characters (mean, range)										
	L	MBD	EP	NR	ES	T	a	b	c	D%	E%
S. websteri											
Infective juvenile	**584**	21	**36**	88	115	**47**	28	5.1	12.6	31	**77**
	553-631	17-25	29-40	83-95	107-122	37-56	24-35	4.8-5.6	11.0-15.5	24-34	62-102
Male	**SL = 68** (64-72); GL = 49 (42-56); MBD = 147 (119-175); **D% = 40** (30-50); **SW% = 180** (160-210); GS% = 70 (60-80); **MUC = P**										
	Spicule and gubernaculum shapes as in the micrographs: 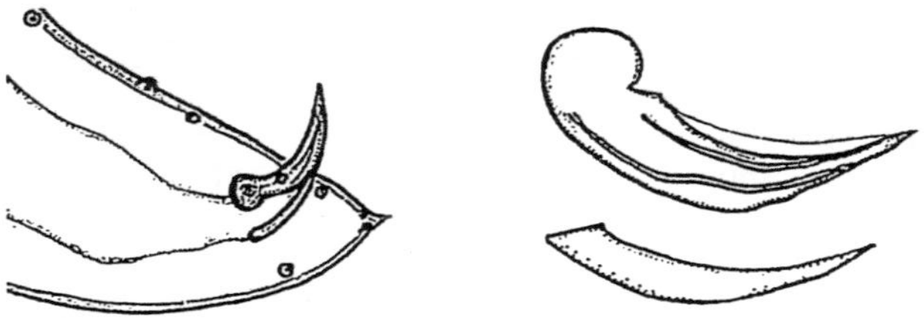										
	Specific character: Spicule shape is unique										
S. anatoliense											
Infective juvenile	**545**	24.5	37	76	107	**52**	22	5.1	10.5	35	**72**
	507-580	21-28	36-39	71-82	97-124	46-58	19-27	4.6-5.7	9.4-11.8	32-39	64-82
Male	**SL = 74** (68-84); GL = 47 (42-59); MBD = 109 (87-150); D% = 49 (47-55); SW% = 175 (160-190); GS% = 55 (40-65); **MUC = P**										
	Spicule and gubernaculum shapes as in the drawing: 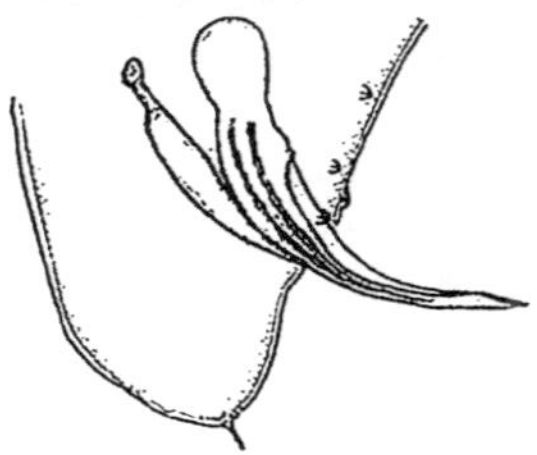										
	Specific characters: Spicule longer than that of *S. carpocapsae*; spicule head elongate										

Table 3.1. *(Continued).*

Species	Morphometric characters (mean, range)										
	L	MBD	EP	NR	ES	T	a	b	c	D%	E%
S. abbasi											
Infective juvenile	**541**	29	**48**	68	89	**56**	18	6.0	9.8	53	**86**
	496-579	27-30	46-51	64-72	85-92	52-61	17-20	5.5-6.6	8.1-10.8	51-58	79-94
Male	**SL** = **65** (57-74); GL = 45 (33-50); MBD = 87 (82-98); **D%** = **60** (51-68); SW% = 156 (107-187); GS% = 70 (58-85); **MUC** = **A**										
	Spicule and gubernaculum shapes as in the drawing: 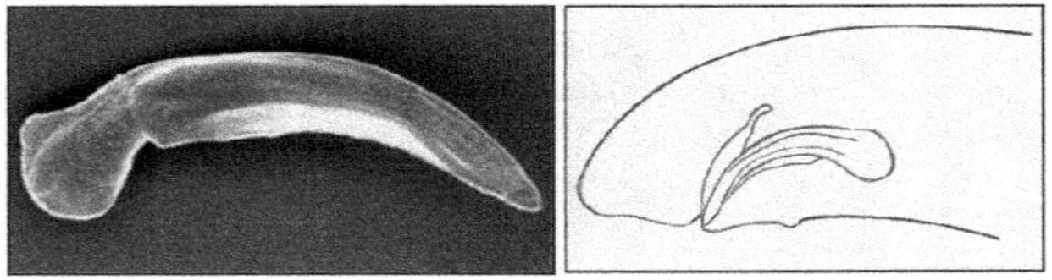										
	Specific character: Mucron in male absent										
S. tami											
Infective juvenile	**530**	23	**36**	na	117	**50**	23	5.0	10.0	31	**73**
	400-600	19-29	34-41	na	110-123	42-57	19-28	3.7-5.1	9.0-11.0	28-34	67-86
Male	**SL** = **77** (71-84); GL = 48 (38-55); MBD = 129 (98-161); **D%** = **44** (30-60); **SW%** = **200** (140-300); GS% = 160; **MUC** = **P**										
	Spicule and gubernaculum shapes as in the micrographs: 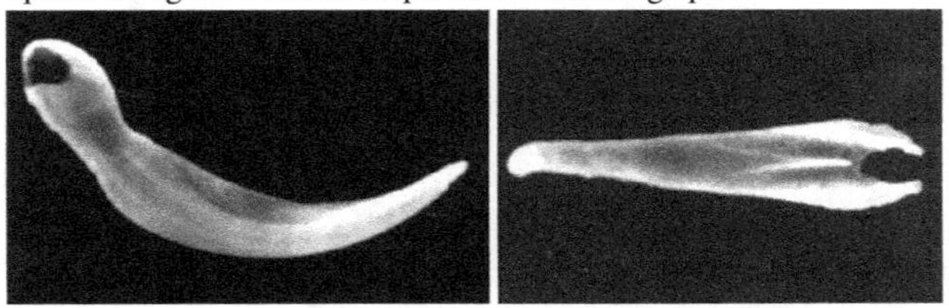										
	Specific characters: Short nematode with long spicule; shaft prominent; SW greater than that of other short nematodes										

Table 3.1. *(Continued).*

Species	Morphometric characters (mean, range)										
	L	MBD	EP	NR	ES	T	a	b	c	D%	E%
S. rarum											
Infective juvenile	**511**	23	**38**	70	102	**51**	23	4.7	9.8	35	**72**
	443-573	18-26	32-40	60-88	89-120	44-56	20-26	4.1-5.6	8.7-11.0	30-39	63-80
Male	**SL** = **47** (42-52); GL = 34 (23-38); MBD = 123 (100-142); **D%** = **50** (44-51); SW% = 94 (91-105); GS% = 71 (55-73); **MUC** = **P**										
	Spicule and gubernaculum shapes as in the micrographs: 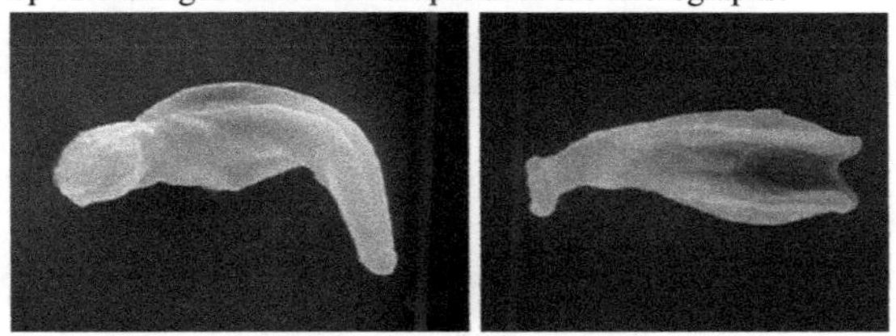										
	Specific characters: Short nematode; spicule short, strongly curved posteriorly; male has 2 openings posterior to edge of cloaca										
S. ritteri											
Infective juvenile	**510**	22	**43**	73	92	**49**	24	5.5	10.6	46	**88**
	470-590	19-24	40-46	68-85	85-95	44-54	19-31	4.9-6.3	9.2-13.1	44-50	79-97
Male	**SL** = **69** (58-75); GL = 44 (33-50); MBD = 130 (110-176); **D%** = **47** (44-50); SW% = 156 (144-157); GS% = 64 (57-67); **MUC** = **A**										
	Spicule as in the drawing: 										
	Specific characters: Short nematode; mucron absent in first generation male										

Table 3.1. *(Continued).*

Species	Morphometric characters (mean, range)										
	L	MBD	EP	NR	ES	T	a	b	c	D%	E%
S. sasonense											
Infective juvenile	**486**	22	**32**	69	100	**47**	22	4.8	10.0	**32**	**69**
	456-516	19-23	30-34	65-72	96-104	42-52	21-24	4.6-5.1	9.4-11.0	30-34	61-76
Male	**SL** = **71** (66-78); GL = 53 (50-56); MBD = 113 (105-120); D% = 40 (37-45); **SW%** = **250** (219-306); GS% = 75 (69-82); **MUC** = **P**										
	Spicule and gubernaculum as in the drawing: 										
	Specific character: IJ body very short										
S. nepalense											
Infective juvenile	**479**	24	**38**	72	100	**53**	20	4.8	9.1	**38**	**71**
	424-513	21-27	31-46	66-80	91-115	48-61	17-23	4.4-5.4	7.0-10.0	31-51	58-87
Male	**SL** = **71** (56-78); GL = 56 (47-60); MBD = 105 (76-135); **D%** = **46** (34-54); **SW%** = **218** (165-260); GS% = 79 (70-93); **MUC** = **P**										
	Spicule and gubernaculum as in the micrographs: 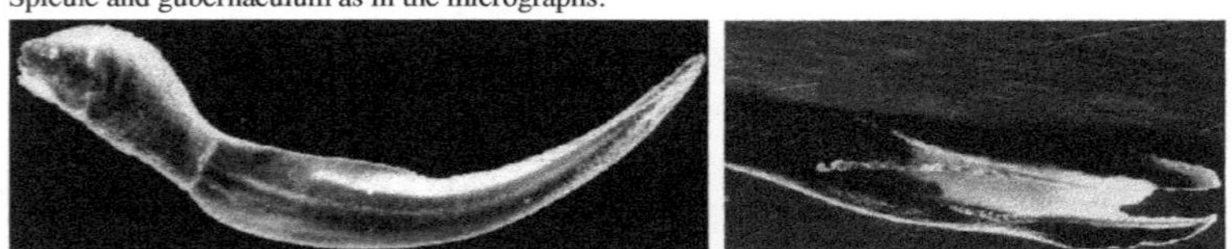										
	Specific characters: IJ body short; female with double-flapped epiptygmata										

Table 3.1. *(Continued).*

Species	Morphometric characters (mean, range)										
	L	MBD	EP	NR	ES	T	a	b	c	D%	E%
S. balochiense											
Infective juvenile	**455**	22.8	**35**	65	90.6	**44.3**	18.5	4.9	10.2	**39**	**80**
	415-528	22-25	32-38	60-72	85-98	40-51	16-21	4.4-5.8	9.4-11.6	36-44	70-92
Male	**SL** = **70** (60-76); GL = 44 (40-47); MBD = 120 (95-135); **D%** = **44** (40-51); **SW%** = **200** (173-217); **GS%** = **64** (53-75); **MUC** = **P** (50%)										
	Spicule and gubernaculum as in the micrographs:										

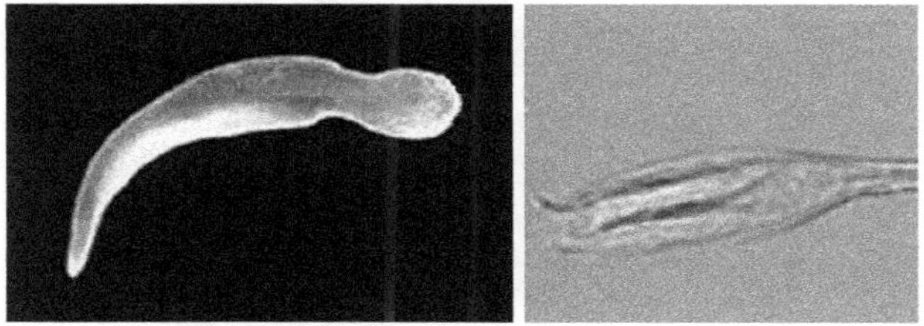

Specific characters: IJ with short body; GS% small

Species	L	MBD	EP	NR	ES	T	a	b	c	D%	E%
S. siamkayai											
Infective juvenile	**446**	21	**35**	72	95	**36**	21	4.7	11.3	**37**	**96**
	398-495	18-24	29-38	68-80	80-107	31-41	19-23	4.0-6.1	10.3-14.8	31-43	95-112
Male	**SL** = **77.5** (75-80); GL = 54 (47-65); MBD = 140 (107-159); D% = 42 (35-49); SW% = 170 (140-220); GS% = 70 (60-80); **MUC** = **P**										
	Spicule and gubernaculum as in the drawing:										

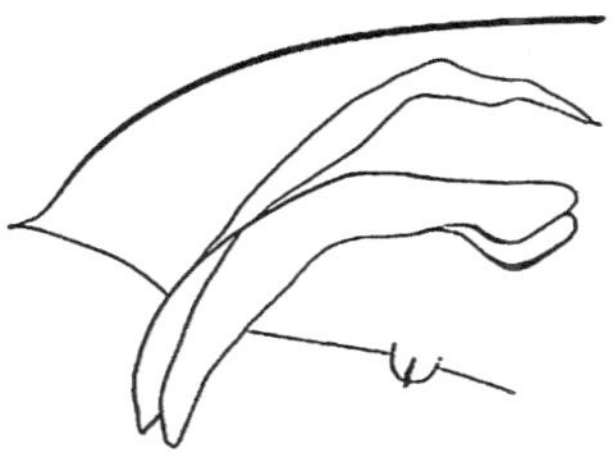

Specific characters: Very short nematode; spicule long for a short nematode; epiptygmata present in females

Table 3.1. *(Continued).*

Species	Morphometric characters (mean, range)										
	L	MBD	EP	NR	ES	T	a	b	c	D%	E%
S. backanense											
Infective juvenile	**438**	20	31	62	91	**42**	22	4.8	10.0	34	**73**
	391-473	18-22	28-33	58-65	87-97	39-46	20-27	4.5-5.2	9.8-11.0	30-37	64-83
Male	**SL** = **58** (53-62); **GL** = **43** (36-48); MBD = 101 (75-135); D% = 41 (30-48); **SW%** = **219** (184-256); GS% = 74 (66-79); **MUC** = **P**										
	Spicule and gubernaculum as in the drawing: 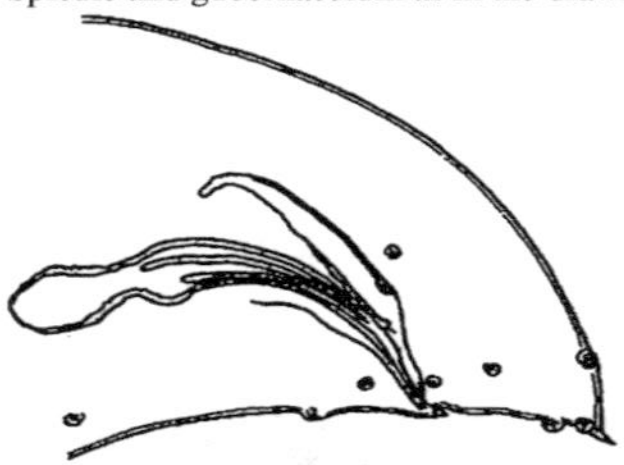										
	Specific character: IJ body very short										
S. asiaticum											
Infective juvenile	**425**	23	32	56	96	**55**	32	5.2	16.1	**32**	**78**
	360-450	20-25	28-34	49-61	82-104	52-57	25-42	4.7-6.1	13.9-20.0	30-36	60-90
Male	SL = 68 (61-74); GL = 53 (46-62); MBD = 104 (86-130); SW% = 200; GS% = 76										
	Spicule and gubernaculum as in the drawings: 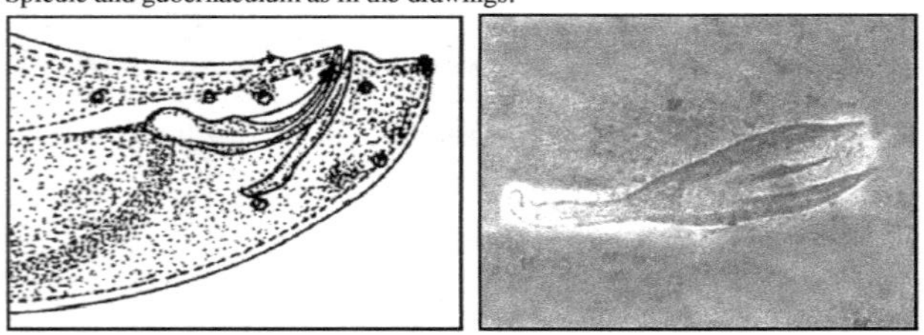										
	Specific character: IJ body very short										

Table 3.1. *(Continued).*

Species	Morphometric characters (mean, range)										
	L	MBD	EP	NR	ES	T	a	b	c	D%	E%
S. surkhetense											
Infective juvenile	**415**	21	32	63	92	45	19	5.0	9.0	**35**	**72**
	393-450	18-25	28-34	57-70	84-101	38-53	16-24	4.3-5.8	8.3-10	31-40	54-84
Male	**SL = 70** (58-78); GL = 52 (42-63); MBD = 105 (75-132); **D% = 48** (37-64); **SW% = 230** (168-330); GS% = 75 (66-84); **MUC = P**										
	Spicule and gubernaculum as in the micrographs: 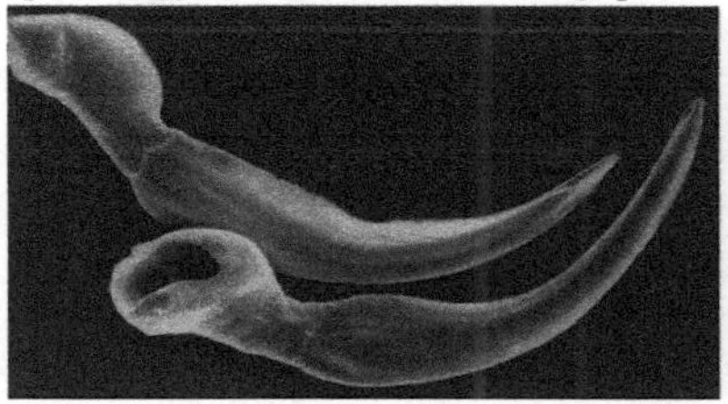 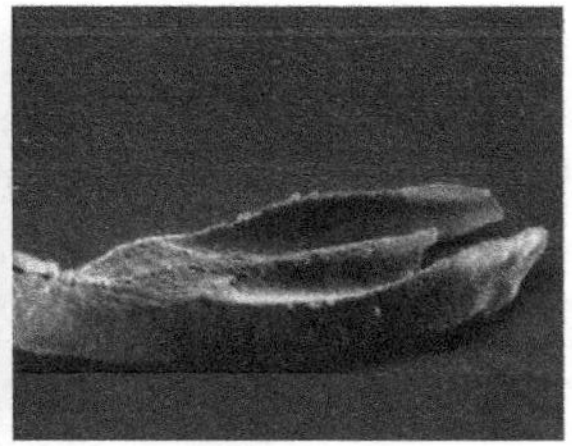										
	Specific characters: IJ body short; female with double-flapped epiptygmata										
S. minutum											
Infective juvenile	**403**	21	34	66	94	**41**	19	4.3	9.9	**36**	**83**
	364-455	19.6-22.7	31-36	61-73	85-99	37-47	17-22	3.9-4.9	8.5-11.2	33-39	76-95
Male	**SL = 75** (68-84); GL = 58 (50-66); MBD = 124 (104-150); **D% = 51** (47-61); **SW% = 115** (84-132); **GS% = 77** (66-87); MUC = A										
	IJ body very short; spicules and gubernaculum as in drawings:  										

Table 3.1. *(Continued).*

Species	L	MBD	EP	NR	ES	T	a	b	c	D%	E%
	Morphometric characters (mean, range)										
S. eapokense											
Infective juvenile	**402**	19	29	58	84	**39**	21	4.8	10.2	34	73
	370-434	18-21	26-31	54-62	79-88	35-45	18-23	4.3-5.3	9.6-11.4	32-37	66-80

Male: **SL** = **69** (63-74); GL = 50 (47-56); MBD = 87 (75-90); D% = 43 (37-48); **SW%** = **246** (221-282); GS% = 73 (69-80); **MUC** = **P**

Spicule and gubernaculum as in the drawing:

Specific character: IJ body extremely short

Species	L	MBD	EP	NR	ES	T	a	b	c	D%	E%
S. cumgarense											
Infective juvenile	**402**	22	**35**	58	86	36	18	4.7	11.0	**45**	**99**
	384-432	21-25	33-38	54-60	80-91	34-38	17-19	4.4-4.9	10.5-12.1	35-52	87-103

Male: **SL** = **76** (68-81); GL = 58 (48-66); MBD = 96 (90-120); D% = 45 (35-52); **SW%** = **277** (250-313); GS% = 76 (71-83); MUC = P

Spicule and gubernaculum as in the drawing:

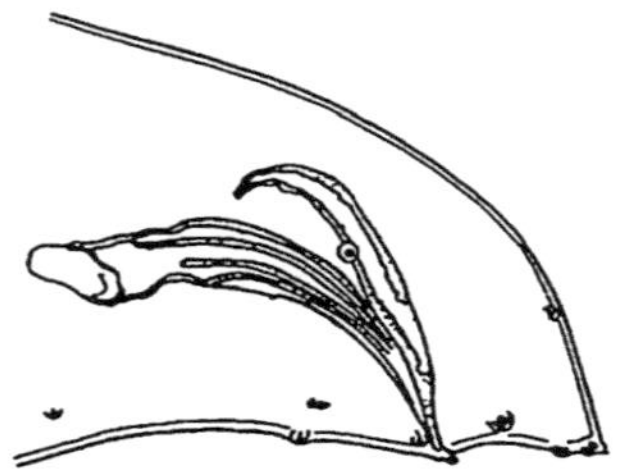

Specific character: IJ body extremely short

Tabular key to *Heterorhabditis* species (Table 3.2)

Table 3.2. *Tabular key for the identification of valid species of* Heterorhabditis *(in descending order of body length). All measurements are in μm. Abbreviations as defined on page 60.*

Species	Morphometric characters (mean, range)										
	L	MBD	EP	NR	ES	T	a	b	c	D%	E%
H. megidis											
Infective juvenile	**768**	29	**131**	109	155	**119**	26	5.0	**6.5**	85	**110**
	736-800	27-32	123-142	104-115	147-160	112-128	23-28	4.6-5.9	6.1-6.9	81-91	103-120
Male	**SL = 49** (46-54); GL = 21 (17-24); MBD = 47 (44-50); **D% = 122**; TR = 128; **SW% = 135** (130-140); GS% = 43										

Dorsal tooth with membranous ring, spicule, gubernaculum and bursa as in micrographs:

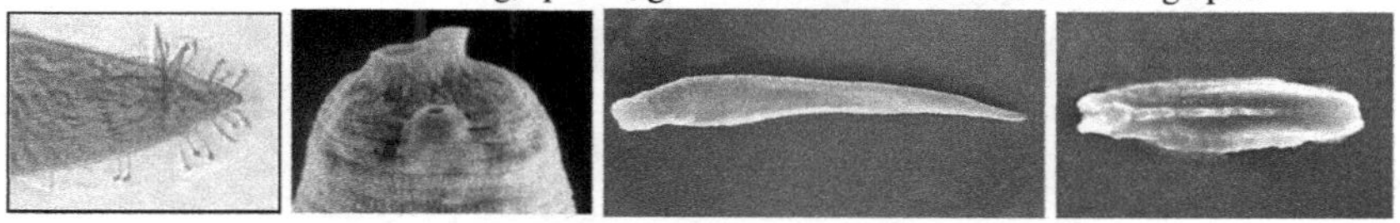

Specific character: Long IJ with membranous ring around oral aperture

Species	L	MBD	EP	NR	ES	T	a	b	c	D%	E%
H. zealandica											
Infective juvenile	**685**	27	**112**	100	140	**102**	25	4.9	**6.6**	80	**108**
	570-740	22-30	94-123	90-107	135-147	87-119	24-26	4.2-5.0	6.2-6.7	70-84	103-109
Male	SL = 51 (48-55); GL = 22 (19-25); MBD = 41 (36-45); D% = 118; TR = 132 (88-173); SW% = 181 (160-209); GS% = 43										

Specific characters: Vulva pattern, spicule and bursa as in the micrographs:

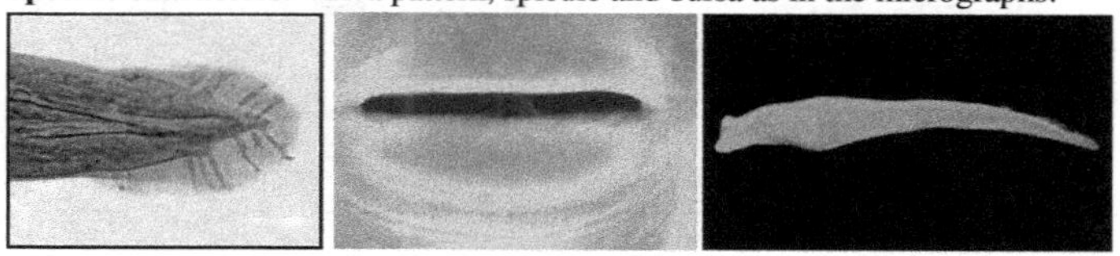

Table 3.2. *(Continued).*

Species	Morphometric characters (mean, range)										
	L	MBD	EP	NR	ES	T	a	b	c	D%	E%
H. marelatus											
Infective juvenile	**654**	28	**102**	99	133	**107**	24	4.9	**6.1**	77	**96**
	588-700	24-32	81-113	83-113	121-139	99-117	21-29	4.7-5.4	5.5-6.6	60-86	89-110
Male	**SL** = **45** (42-50); GL = 19 (18-22); MBD = 51 (48-56); **D%** = **113**; TR = 91 (67-136); **SW%** = **175** (173-178); GS% = 41										
Syn. *H. hepialius*	**Specific characters**: Bursa, spicule as in the micrographs: 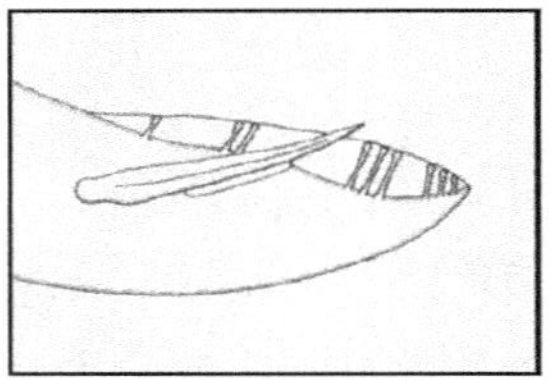 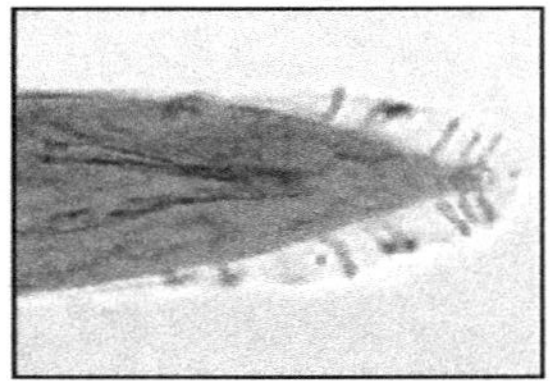										
H. beicherriana											
Infective juvenile	**639**	24	113	95	137	**101**	27	4.7	6.4	**83**	**112**
	566-687	21-25	100-122	85-106	118-146	86-111	24-29	4.2-4.9	5.9-6.8	80-93	103-121
Male	**SL** = **44** (40-49); **GL** = **24** (22-27); MBD = 58 (51-73); D% = 112 (102-120); TR = 107 (90-117); SW% = 180 (153-208); GS% = 54 (48-59) 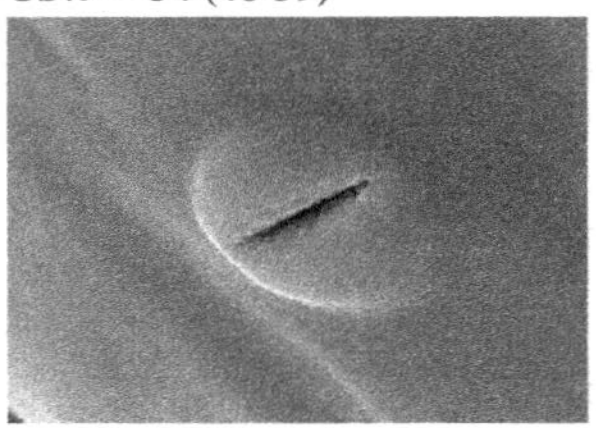 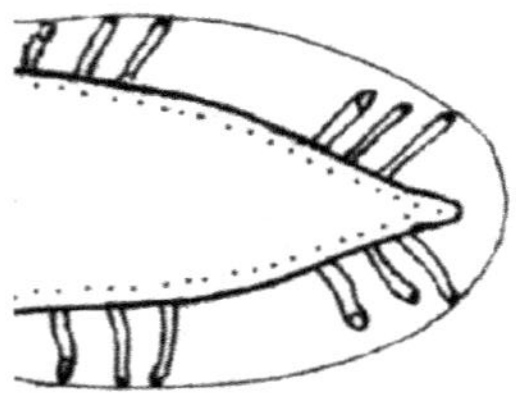 **Specific characters**: Vulvar pattern smooth; 3 pairs of papillae in terminal group										

Table 3.2. *(Continued).*

Species	Morphometric characters (mean, range)										
	L	MBD	EP	NR	ES	T	a	b	c	D%	E%
H. downesi											
Infective juvenile	**637**	18	**115**	101	135	**68**	35	4.7	9.5	**85**	**170**
	588-692	15-22	96-128	96-105	126-141	62-74	29-42	4.4-5.3	8.5-10.5	76-98	160-180
Male	**SL** = **43** (41-47); GL = 18 (17-20); MBD = 36 (33-40); **D%** = **88**; TR = 150 (108-181); **SW%** = **180** (170-220); GS% = 43 (36-47)										

Specific characters: Hermaphrodite and male tail as in the drawings:

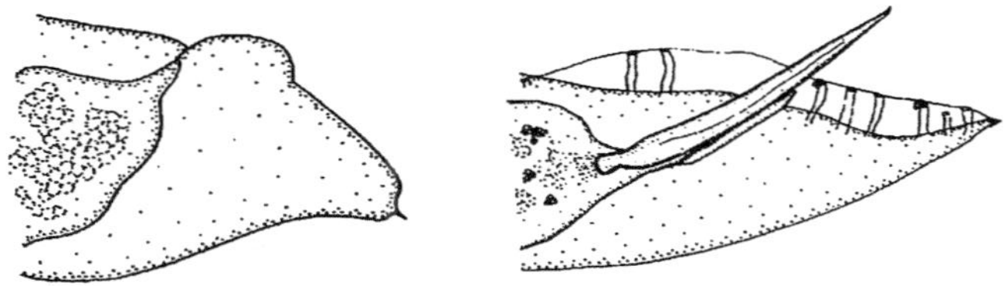

Species	L	MBD	EP	NR	ES	T	a	b	c	D%	E%
H. atacamensis											
Infective juvenile	**611**	22	115	92	135	**101**	28	5.3	6.1	85	**165**
	578-666	19-26	101-126	79-101	124-144	94-107	24.6-30.7	4.8-5.7	5.7-7.1	79-94	149-182
Male	**SL** = **45** (40-49); GL = 20 (17-22); MBD = 48 (42-55); **D%** = **117** (108-126); TR = 120 (109-135); **SW%** = **205** (179-249); GS% = 44 (38-51)										

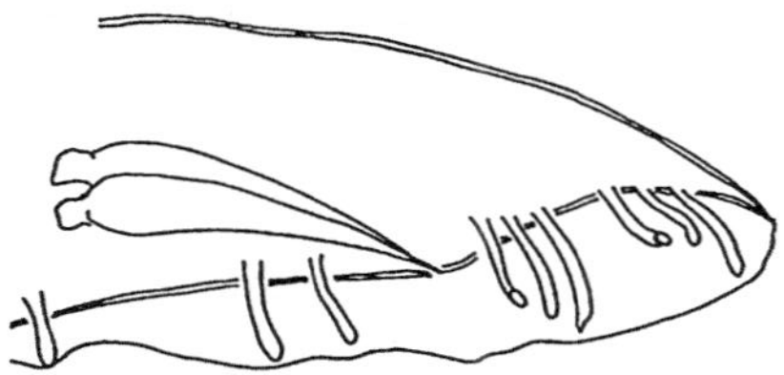

Specific character: Cuticular fold immediately anterior and posterior to vulva

Table 3.2. *(Continued).*

Species	Morphometric characters (mean, range)										
	L	MBD	EP	NR	ES	T	a	b	c	D%	E%
H. safricana											
Infective juvenile	**600**	21	**110**	93	131	93	28.6	4.5	6.4	84	**119**
	550-676	19-23	103-122	86-101	125-141	86-108	24.8-31.8	3.9-4.9	5.4-7.5	80-90	99-133
Male	**SL** = **45** (35-54); GL = 24 (19-27); MBD = 49 (40-58); **D%** = **117** (92-133); TR = 93 (52-129); SW% = 196 (130-259); GS% = 54 (43-62)										
	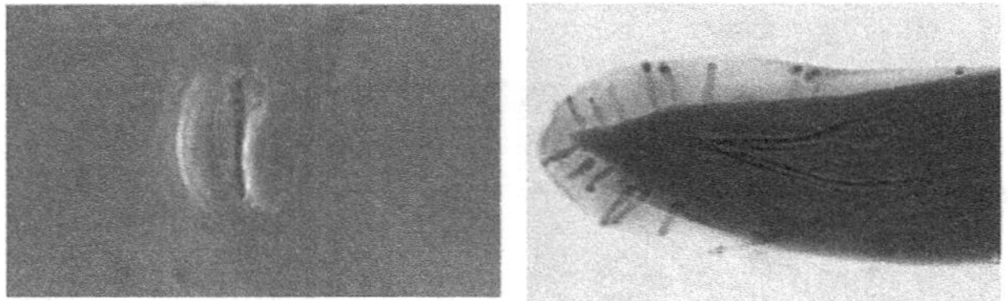										
	Specific characters: Vulvar pattern; 3 pairs of papillae in terminal group										
H. georgiana											
Infective juvenile	**598**	22	**104**	85	127	98	27	4.7	6.1	82	**106**
	547-651	17-26	97-113	74-94	110-139	86-108	23-34	4.1-5.3	5.5-6.9	70-93	
Male	SL = 44 (41-49); GL = 25 (20-28); MBD = 49 (43-55); **D%** = **110** (100-122); TR = 101 (75-120); **SW%** = **172** (150-200); GS% = 56 (51-64)										
	Specific characters: Spicule, bursa and vulva pattern, as in the micrographs:										
	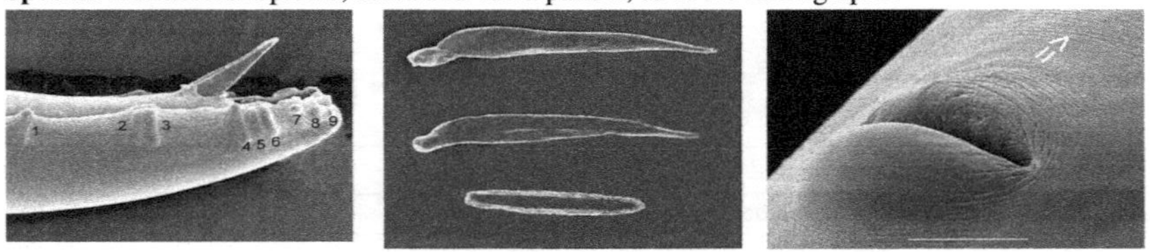										

Table 3.2. *(Continued).*

Species	Morphometric characters (mean, range)										
	L	MBD	EP	NR	ES	T	a	b	c	D%	E%
H. amazonensis											
Infective juvenile	**589**	23	**107**	85	121	**107**	26	4.9	5.5	**88**	**100**
	567-612	20-24	89-115	76-93	107-132	98-115	24-29	4.4-5.5	5.1-6.1	83-92	89-109
Male	**SL** = **41** (35-45); GL = 21 (19-23); MBD = 41 (36-43); **D%** = **103** (95-109); **TR** = **59** (42-85); **SW%** = **152** (120-187); GS% = 51 (44-56)										
	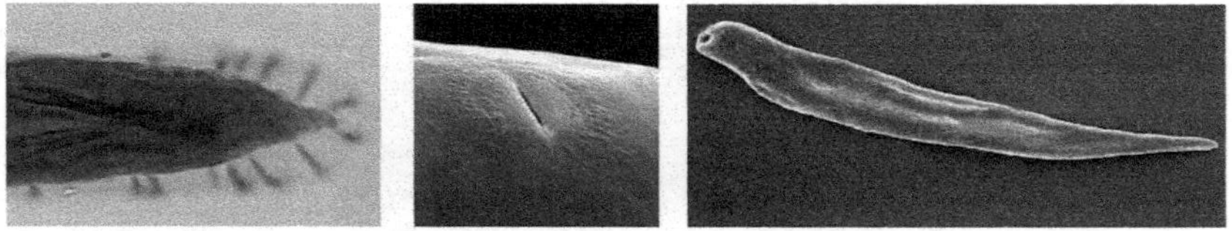										
	Specific characters: Bursa with 2 pairs of ribs or fewer in terminal group; vulva pattern smooth, elliptical										
H. bacteriophora											
Infective juvenile	**588**	23	**103**	85	125	98	25	4.5	**6.2**	84	**112**
	512-671	18-31	87-110	72-93	100-139	83-112	17-30	4.0-5.1	5.5-7.0	76-92	103-130
Male	SL = 40 (36-44); GL = 20 (18-25); MBD = 43 (38-46); D% = 117; TR = 76 (61-89); SW% = 128 (120-134); GS% = 50										
Syn. *H. argentinensis*	**Specific characters**: Bursa, vulva patterns and spicule shape as in the micrographs:										
	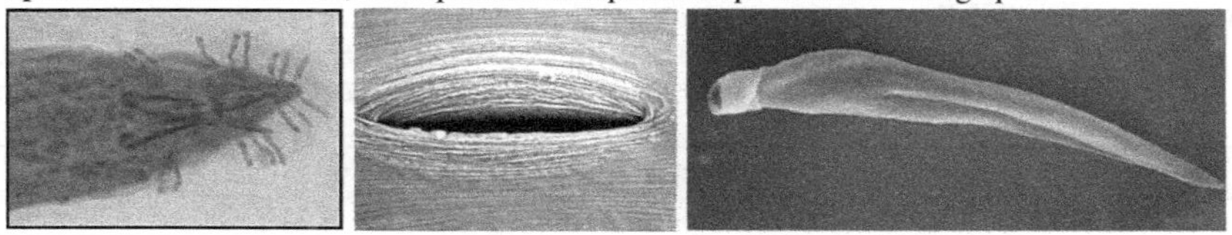										

Table 3.2. *(Continued).*

Species	Morphometric characters (mean, range)										
	L	MBD	EP	NR	ES	T	a	b	c	D%	E%
H. mexicana											
Infective juvenile	**578**	23	**102**	81	122	**99**	25	4.6	5.9	**81**	104
	530-620	20-24	83-109	74-88	104-142	91-106	24-28	4.2-5.1	5.5-6.3	72-86	87-111
Male	**SL** = **41** (30-47); GL = 23 (18-32); MBD = 42 (38-47); D% = 129 (114-149); TR = 96 (65-130); SW% = 167 (130-196); GS% = 56 (43-70)										
	Specific characters: Bursa, vulva patterns and spicule as in the micrographs: 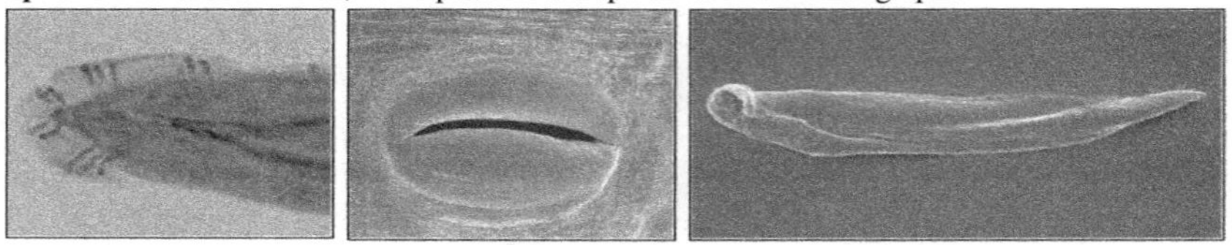										
H. floridensis											
Infective juvenile	**562**	21	**109**	86	**135**	**103**	28	4.3	5.6	**81**	**105**
	554-609	19-23	101-122	68-107	123-142	91-113	25-32	3.9-4.9	5.3-6.6	71-90	95-134
Male	**SL** = **42** (36-46); GL = 23 (17-30); MBD = 48 (43-50); **D%** = **112** (105-119); TR = 93 (78-116); SW% = 157 (133-209); GS% = 54 (47-65)										
	Specific characters: Bursa and vulva patterns as in the micrographs: 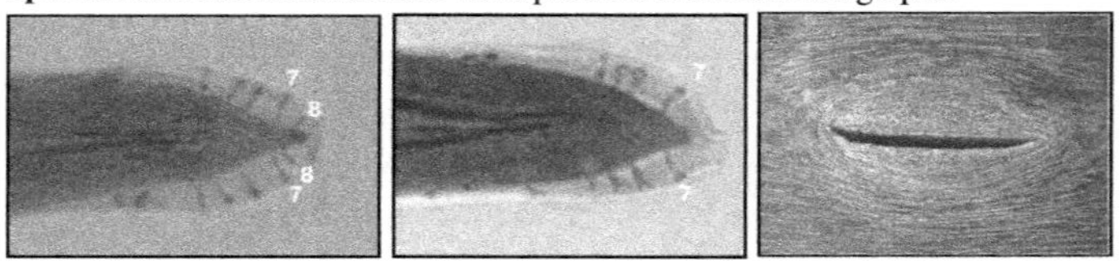										

Table 3.2. *(Continued).*

Species	Morphometric characters (mean, range)										
	L	MBD	EP	NR	ES	T	a	b	c	D%	E%
H. baujardi											
Infective juvenile	**551**	20	**97**	81	115	**90**	28	4.8	6.0	**84**	**108**
	497-595	18-22	91-103	75-86	107-120	83-97	26-30	4.5-5.1	6.0-6.7	78-88	98-114
Male	**SL** = **40** (33-45); GL = 20 (18-22); MBD = 49 (43-53); **D%** = **79**; TR = 91 (63-106); SW% = 182 (138-208); GS% = 50 (44-61)										
	Specific characters: Bursa, spicule and gubernaculum as in the micrographs: 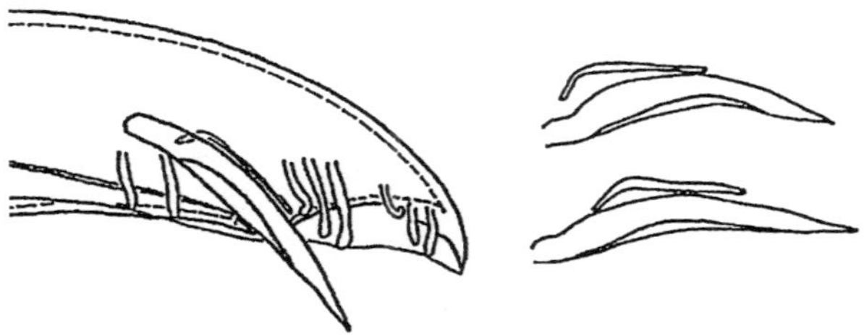										
H. noenieputensis											
Infective juvenile	**536**	23	97	81	106	**86**	24	4.9	6.2	**89**	**113**
	484-578	21-25	88-105	69-96	79-115	78-95	21-27	4.3-5.2	5.5-6.8	81-95	99-125
Male	**SL** = **43** (37-49); GL = 20 (17-24); MBD = 41 (34-46); **D%** = **90** (81-108); TR = 82 (67-104); **SW%** = **231** (202-301); GS% = 47										
	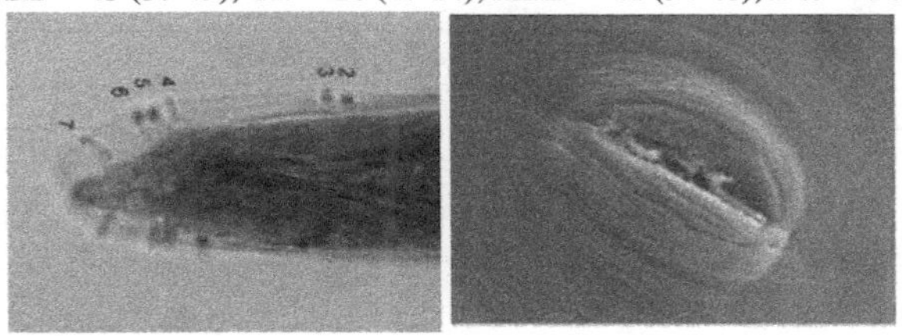 **Specific characters:** Vulva pattern; 1 pair of papillae in terminal group										

Table 3.2. *(Continued).*

Species	Morphometric characters (mean, range)										
	L	MBD	EP	NR	ES	T	a	b	c	D%	E%
H. indica											
Infective juvenile	**528**	20	**98**	82	117	**101**	26	4.5	**5.3**	84	**94**
	479-573	19-22	88-107	72-85	109-123	93-109	25-27	4.3-4.8	4.5-5.6	79-90	83-103
Male	**SL** = **43** (35-48); GL = 21 (18-23); MBD = 42 (35-46); **D%** = **122**; **TR** = **106** (78-132); **SW%** = **190** (180-200); GS% = 49										
Syn. *H. hawaiiensis*	**Specific characters**: Vulva, spicule and gubernaculum shapes as in the micrographs:										
Syn. *H. brevicaudis* Syn. *H. gerrardi*	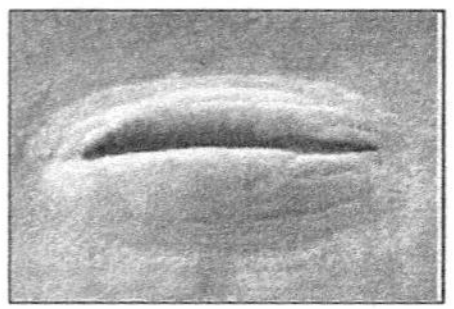 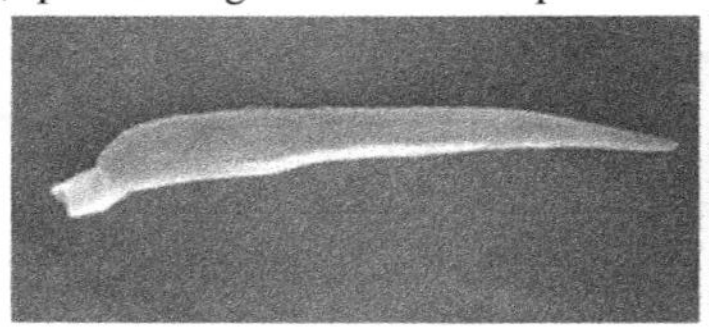 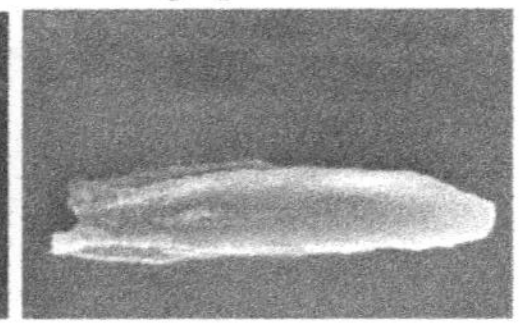										
H. taysearae											
Infective juvenile	**418**	20	**90**	–	110	**55**	21	3.8	7.7	**82**	**180**
	332-499	17-23	74-113		96-130	44-70	18-27	2.4-4.2	6.5-8.7	71-96	110-230
Male	**SL** = **39** (30-42); GL = 18 (14-21); MBD = 44 (38-48); TR = 122 (100-146); SW% = 127; GS% = 46										
Syn. *H. sonorensis*	**Note**: The tail looks short. The authors may have measured tail length without the sheath making E% very high										

Reference

NGUYEN, K.B. (2007). Methodology, morphology and identification. In: Nguyen, K.B. & Hunt, D.J. (Eds). *Entomopathogenic nematodes: systematics, phylogeny and bacterial symbionts. Nematology Monographs and Perspectives 5* (Series editors: Hunt, D.J. & Perry, R.N.). Leiden, The Netherlands, Brill, pp. 59-119.

Nematology Monographs & Perspectives, 2016, Vol. 12, 111-375

Chapter 4

Steinernematidae: species descriptions

David J. HUNT [1], Khuong B. NGUYEN [2] and
Sergei E. SPIRIDONOV [3]

[1] *CABI Europe-UK, Bakeham Lane, Egham, Surrey, TW20 9TY, UK*
[2] *Entomology and Nematology Department, Institute of Food and Agricultural Sciences, University of Florida, Gainesville, FL 32611-0620, USA*
[3] *Center of Parasitology, A.N. Severtsov Institute of Ecology and Evolution of the Russian Academy of Sciences, Leninskii prospect 33, Moscow 119071, Russia*

This chapter is an update of the chapter by Nguyen *et al.* (2007a) in Nguyen & Hunt (2007). All *Steinernema* species described since then, and regarded by us as being valid (see Chapter 2 for a complete list), are presented here in a slightly condensed format that we hope will enable the reader to focus on the more important diagnostic characters. The tabular key in Chapter 3 includes all valid *Steinernema* species and should be used in conjunction with this chapter and the sister chapter in Nguyen & Hunt (2007). Everything related to the order of description, the collection of morphological, morphometric data and illustrations was stated in that chapter and is not repeated here.

Notes concerning the species descriptions

The morphological characteristics of the anterior region of the male, female and infective juvenile (IJ) stages of *Steinernema* species are very similar and apparently have very little diagnostic value. We believe that it is uninformative and unnecessary to re-state such information in every individual species description unless such characters are deemed to have diagnostic utility. However, for completeness sake, the general morphology of steinernematids is given below.

MALE (FIRST GENERATION)

Body strongly curved posteriorly or J-shaped when heat-killed. Cuticle smooth under light microscope but with striation under SEM. Lateral field not observed or with one ridge. Labial region slightly swollen, flattened anteriorly in some specimens. Face view with six labial papillae, two amphidial apertures and four cephalic papillae (Fig. 4.1A, B). Cephalic papillae larger than labial papillae. Stoma shallow and narrow, usually with small cheilorhabdions, posterior part funnel-shaped, slightly cuticularised (Fig. 4.2B). Excretory pore anterior (or posterior) to nerve ring, anterior part of excretory duct well cuticularised, excretory cells not observed. Pharynx with cylindrical procorpus, metacorpus slightly swollen, isthmus present. Nerve ring surrounding isthmus, basal bulb distinct. Pharyngo-intestinal valve present (Fig. 4.2B).

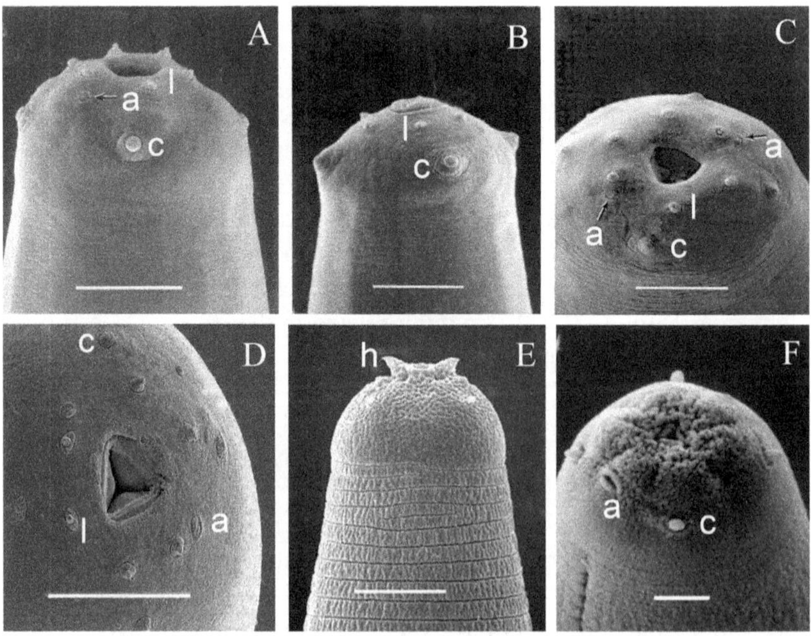

Fig. 4.1. *Anterior region of* Steinernema *male (A, B), female (C, D) and infective juvenile (E, F) showing mouth, labial papillae (l), cephalic papillae (c), amphids (a) and horn-like structures (h). (Scale bars: A = 8.8 μm; B-D = 10 μm; E = 4.3 μm; F = 6 μm.) Compiled by Khuong B. Nguyen.*

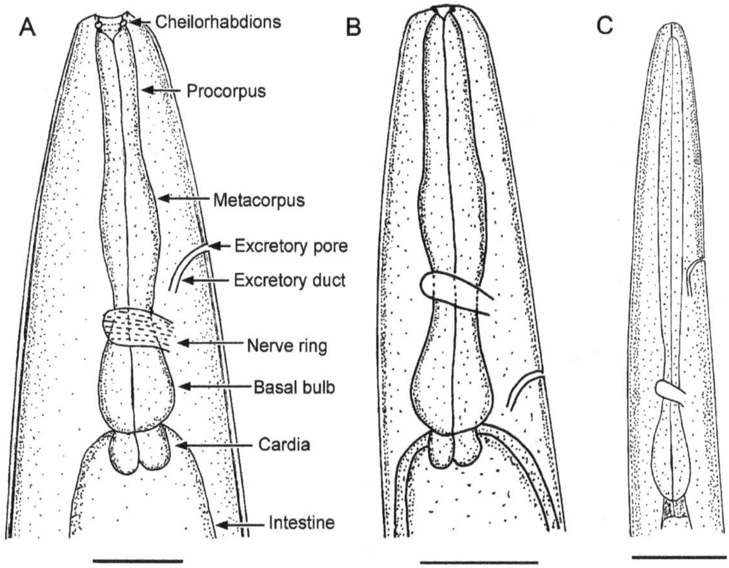

Fig. 4.2. *Diagrammatic anterior region of* Steinernema. *A: Female; B: Male; C: Infective juvenile. (Scale bars: A, B = 50 μm; C = 25 μm.) Compiled by Khuong B. Nguyen.*

FEMALE (FIRST GENERATION)

Body C-shaped or strongly spiral when heat-killed. Labial region slightly swollen in young female but rounded and continuous with body when mature. Face view with six labial papillae, two amphidial apertures and four cephalic papillae (Fig. 4.1C, D). Cephalic papillae larger than labial papillae. Cuticle smooth or with faint annules. Lateral fields not observed. Phasmids inconspicuous. Cheilorhabdions small or large, well sclerotised (Fig. 4.2A). Another smaller sclerotised structure posterior to cheilorhabdions present, posterior part of stoma funnel-shaped. Dorsal wall of stoma well sclerotised, ventral wall shorter, less sclerotised. Pharynx with procorpus cylindrical, muscular, metacorpus slightly swollen, isthmus distinct, basal bulb enlarged, valvate. Nerve ring surrounding isthmus just anterior to basal bulb. Pharyngo-intestinal valve present (Fig. 4.2A).

INFECTIVE JUVENILE

Body elongate, almost straight or slightly curved. Sheath (second-stage cuticle) present immediately after harvesting, but many IJ losing sheath in storage. Exsheathed juvenile with four cephalic papillae. Labial region smooth, continuous with body. Amphidial apertures prominent (Fig. 4.1F). Cuticle marked with prominent transverse striations. Pharynx with narrow corpus, slightly swollen metacorpus, distinct isthmus and basal bulb elongate with visible valve. Excretory pore located anterior to nerve ring (Fig. 4.2C). Nerve ring usually located at mid-point of isthmus. Cardia present (Fig. 4.2C).

Abbreviations

In addition to the standard de Man ratios, the following abbreviations are used throughout the text and tables:

D% = labial region to excretory pore/pharynx length × 100,
E% = labial region to excretory pore/tail length × 100,
SW% = spicule length/cloacal body diam. × 100,
GS% = gubernaculum length/spicule length × 100.

Descriptions of the species

Steinernema arasbaranense[4,5] Nikdel, Niknam & Ye, 2011
(Figs 4.3, 4.4)

Measurements

See Table 4.1.

Diagnostic description

Male (first generation)

Anterior region typical for *Steinernema* male. Excretory pore located at mid-metacorpus. Testis monorchic, reflexed. Spicules paired, symmetrical, slightly brownish in colour, somewhat curved, length/width ratio = 6.5, labial region (manubrium) moderately elongate, shaft (calomus) separating manubrium from lamina, lamina (blade) slightly curved, supported by two internal ribs and almost straight posteriorly, velum

[4] The new species is named after the *Arasbaran* forests, a mountainous area in northwest Iran where the nematode was collected.

[5] Description after Nikdel *et al.* (2011). Revised by Khuong B. Nguyen.

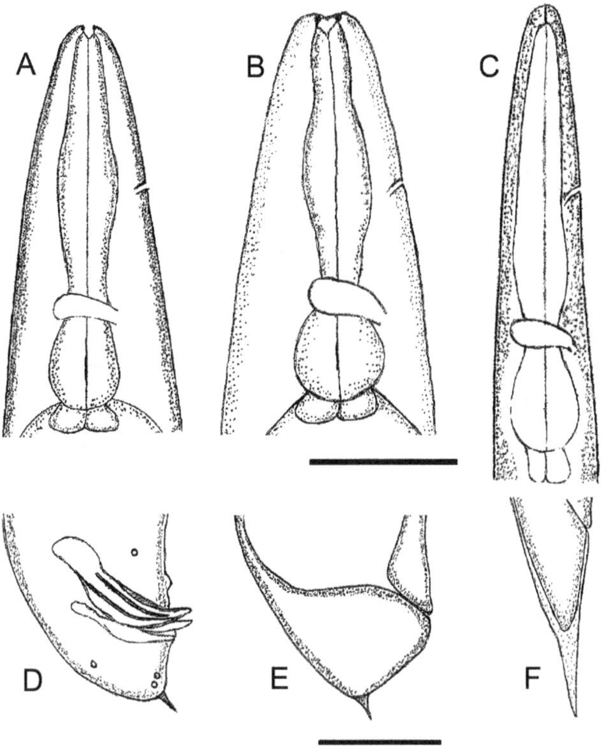

Fig. 4.3. Steinernema arasbaranense. *A-C: Anterior regions of male, female and infective juvenile, respectively; D-F: Posterior regions of male, female and infective juvenile, respectively. (Scale bars: A = 65 μm; B = 72 μm; C, E = 45 μm; D = 54 μm; F = 50 μm.) Modified after Nikdel* et al. *(2011), with acknowledgement to* Nematologia Mediterranea.

present, extending from base of rostrum or beginning of lamina to distal part of lamina, not covering spicule tip. Gubernaculum *ca* 63% of spicule length, boat-shaped in lateral view, swollen at middle, with slightly ventrally curved knob at proximal end. Genital papillae totalling 23, including a single precloacal papilla and 11 pairs arranged as follows: six pairs ventral precloacal (three pairs sublateral precloacal, three pairs subventral precloacal), two pairs lateral adcloacal, one pair dorsal postcloacal and two pairs subterminal postcloacal. Tail conoid, terminating in a 8.5 μm long mucron, mucron always present. Phasmids inconspicuous.

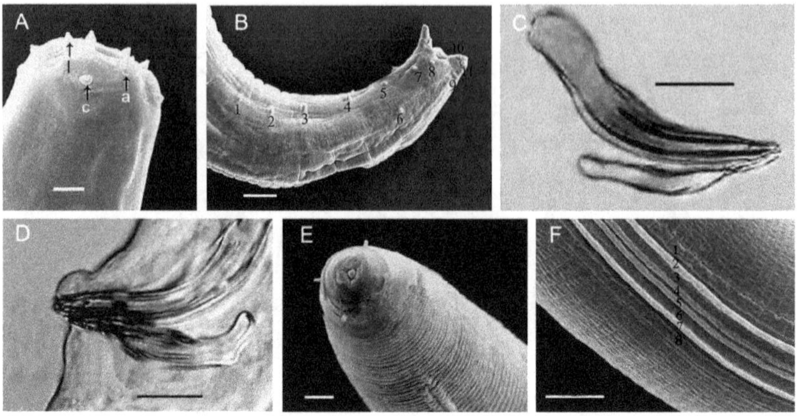

Fig. 4.4. Steinernema arasbaranense. *A-D: Male. A: Anterior region showing labial papillae (l), cephalic papillae (c) and amphid (a); B: Posterior region showing 11 pairs of genital papillae; C: Spicule and gubernaculum; D: Cuticle at cloacal region and posterior ends of spicule and gubernaculum. E, F: Infective juvenile. E: Anterior region showing mouth, four cephalic papillae and anterior end of lateral field; F: Lateral field at mid-body showing eight ridges. (Scale bars: A = 13 μm; B = 28 μm; C = 21 μm; D = 14 μm; E = 8 μm; F = 10 μm.) Modified after Nikdel* et al. *(2011), with acknowledgement to* Nematologia Mediterranea.

Male (second generation)

Similar to first generation male except for shorter body and smaller body diam. Testis flexure extending more posteriorly, mucron more distinct, *ca* 12 μm long.

Female (first generation)

Anterior region typical for *Steinernema* female. Excretory pore usually located at mid-pharynx level and excretory duct cuticularised. Cardia prominent, protruding into intestinal lumen. Genital system didelphic and amphidelphic reflexed, filled with eggs. Vulva posterior to mid-body, in form of a symmetrical and transverse slit, protruding from body, with double-flapped epiptygmata. Vagina short, oblique with muscular walls. Tail dome-shaped, shorter than anal body diam. and with a terminal peg.

Female (second generation)

Similar to first generation in general morphology, but smaller. Tail conoid with mucron at its end. Post-anal swelling not developed.

Table 4.1. *Morphometrics of* Steinernema arasbaranense. *All measurements are in μm and in the form: mean ± s.d. (range).*

Character	First generation			Second generation		Infective juvenile
	Holotype (male)	Male	Female	Male	Female	
n	–	22	15	20	15	20
L	1240	1265 ± 175 (1057-1645)	3720 ± 1053 (2320-5345)	863 ± 37 (824-921)	2332 ± 249 (2063-2553)	731 ± 30 (693-795)
a	8.5	10 ± 1 (8-12)	15 ± 2 (12-18)	13 ± 1 (11-15)	13 ± 2 (12-16)	16.4 ± 1 (14-18)
b	9.2	9 ± 1 (8-11)	19 ± 5 (13-26)	7 ± 0.5 (6-7)	14 ± 2 (13-17)	6.5 ± 0.4 (6-7)
c	35	37 ± 6 (29-47)	112 ± 7 (89-128)	27 ± 1 (26-29)	51 ± 5 (48-57)	8.7 ± 0.7 (7-10)
c′	0.9	1 ± 0.1 (0.7-1)	1 ± 0 (1-1)	1 ± 0 (1-1)	1 ± 0.3 (0.7-1.3)	3.6 ± 0.2 (3-5)
V	–	–	56 ± 3 (52-60)	–	59 ± 2 (56-62)	–
Body diam.	133	125 ± 13 (98-143)	242 ± 37 (186-314)	64 ± 6 (56-73)	170 ± 19 (151-192)	44 ± 2 (41-50)
Excretory pore	73	65 ± 13 (51-87)	90 ± 24 (88-149)	61 ± 5 (55-70)	67 ± 6 (60-76)	56 ± 2.5 (53-60)
Nerve ring	107	106 ± 11 (98-122)	125 ± 11 (118-138)	–	–	84 ± 5 (79-88)
Pharynx	136	133 ± 5 (126-143)	192 ± 16 (154-215)	125 ± 6 (116-132)	152 ± 5 (147-161)	112 ± 3 (109-118)
Tail length	34	32 ± 3 (31-36)	38 ± 7 (29-46)	32 ± 2 (29-34)	45 ± 9 (36-57)	84 ± 6 (79-101)
Anal body diam.	41	40 ± 3 (33-44)	50 ± 4 (45-57)	34 ± 3 (30-38)	39 ± 1 (37-42)	23 ± 2 (20-27)
Mucron length	–	8.5 ± 0.2 (7.5-9.0)	–	12 ± 2 (9-16)	–	–
Spicule length	78	75 ± 3 (70-82)	–	59 ± 5 (51-67)	–	–
Spicule width	12	11 ± 1 (11-13)	–	9 ± 2 (7-11)	–	–
Gubernaculum length	49	48 ± 4 (42-53)	–	40 ± 3 (35-44)	–	–
D%	52	50 ± 7 (43-59)	47 ± 8 (38-57)	49 ± 3 (43-53)	42 ± 5 (36-47)	50 ± 3 (45-54)
E%	189	187 ± 27 (157-223)	239 ± 33 (205-284)	197 ± 16 (168-215)	149 ± 19 (129-171)	67 ± 6 (53-75)
SW%	184	187 ± 12 (174-213)	–	179 ± 23 (149-210)	–	–
GS%	66	63 ± 5 (52-70)	–	61 ± 6 (49-72)	–	–

Infective juvenile

Anterior region typical for *Steinernema* IJ. Lip region with four projecting cephalic papillae. Excretory pore located near middle of corpus. Hemizonid distinct, positioned anterior to base of basal bulb. Bacterial vesicle relatively elongate. Lateral fields pattern is 2, 4, 5, 8, 4, 2 with eight equally developed ridges (*i.e.*, nine lines or incisures) at mid-body, ridge numbers 2, 4, 5 and 7 prominent. Deirids not observed. Tail conical with dorsal depression, no spine-like structure seen on tail tip, hyaline region distinct, *ca* 48% of tail length. Phasmids indistinct.

TYPE HABITAT AND LOCALITY

The type host of this species in nature is unknown as the nematode was recovered from soil by baiting with *Galleria mellonella*. The soil sample was collected from an oak (*Quercus macrantera* L.) forest habitat near Kerengan village, Kaleibar town, East Azarbaijan Province, Iran.

TYPE MATERIAL

Holotype male, paratype males, females and IJ deposited in the Nematode Collection, Herbarium and Insect Museum of Natural Resources Research Institute, Tabriz, Iran. Three male, three female and five IJ paratypes on separate microscopic slides are maintained in the Nematode Collection, Faculty of Agriculture, University of Tabriz, Tabriz, Iran. Some paratype specimens also deposited in the University of California Davis Nematode Collection, Davis, CA, USA.

RELATIONSHIPS

Steinernema arasbaranense is characterised by features of the male and IJ. For IJ, average body length of 731 μm, lateral field pattern 2, 4, 5, 8, 4, 2 with 2, 4, 5 and 7 being prominent, E% = 67 (53-75), four projecting cephalic papillae and tail length of 84 (79-101) μm with a dorsal depression. A velum and mucron are present in both male generations, the mucron being longer in the second generation male, and the SW ratio is high. More information on relationships is available in Nikdel *et al.* (2011).

Molecular characterisation

The sequence of the ITS region (accession no. FJ860039) of *S. arasbaranense*, including primers TW81 and AB28, can be recognised by its length of 869 bp. More information about molecular characterisations is available in Nikdel *et al.* (2011).

Symbiotic bacterium

No information.

Bionomics

No information.

Distribution

Steinernema arasbaranense was collected near Kerengan village, Kaleibar town, East Azarbaijan Province, Iran.

Steinernema australe[6,7] Edgington, Buddie, Tymo, Hunt, Nguyen, France, Merino & Moore, 2009

(Figs 4.5-4.8)

Measurements

See Table 4.2.

Diagnostic description

Male (first generation)

Anterior region typical for *Steinernema* male. Body J-shaped when heat-relaxed. Cuticle smooth under light microscope. Labial region relatively low compared to its diam., rounded anteriorly, usually continuous with body contour, but demarcated by slight depression in some specimens. Excretory pore located more or less opposite nerve ring, anterior portion of excretory duct well cuticularised, posterior section slightly

[6] Specific epithet derived from the *Austral* region of Chile, from where the species was collected.

[7] Description after Edgington *et al.* (2009a). Revised by David J. Hunt.

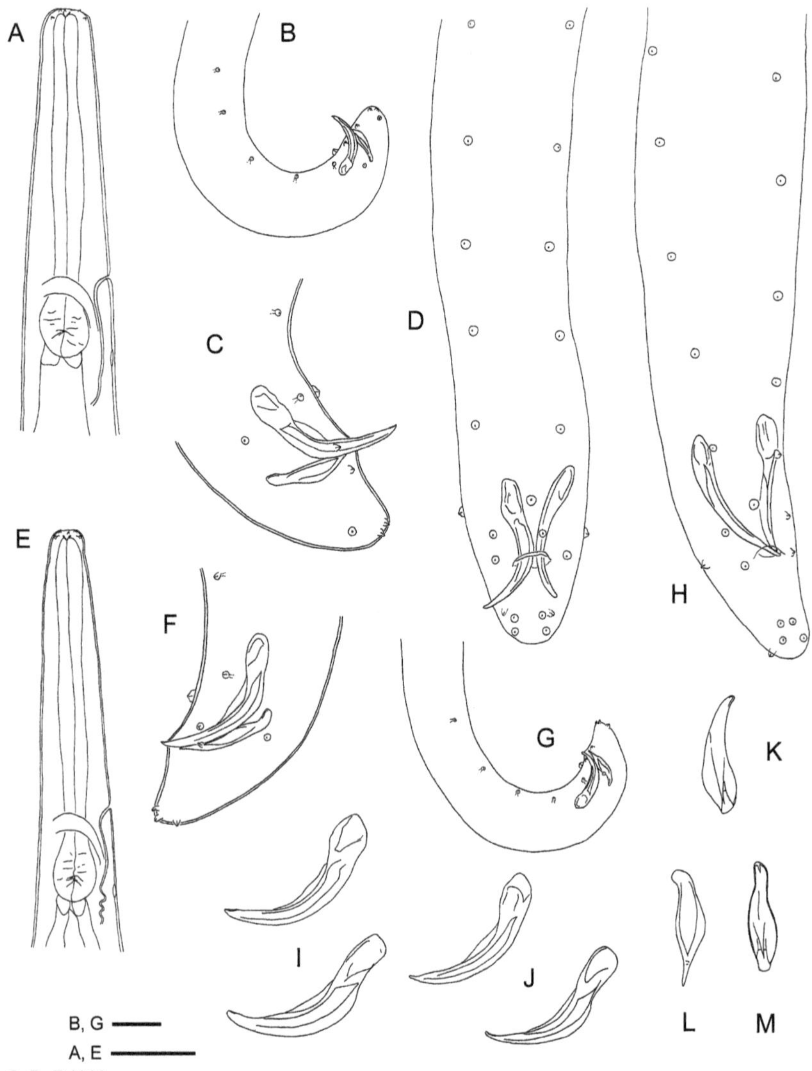

Fig. 4.5. Steinernema australe. *A-D, I, K: First generation male. A: Pharyngeal region; B: Tail region; C: Cloacal region; D: Tail region, ventral view; I: Spicules; K: Gubernaculum, subventral view. E-H, J, L, M: Second generation male. E: Pharyngeal region; F: Cloacal region; G: Tail region; H: Tail region, ventral view; J: Spicules; L: Gubernaculum, lateral view; M: Gubernaculum, ventral view. (Scale bars = 50 μm.) After Edgington* et al. *(2009a), courtesy* Nematology.

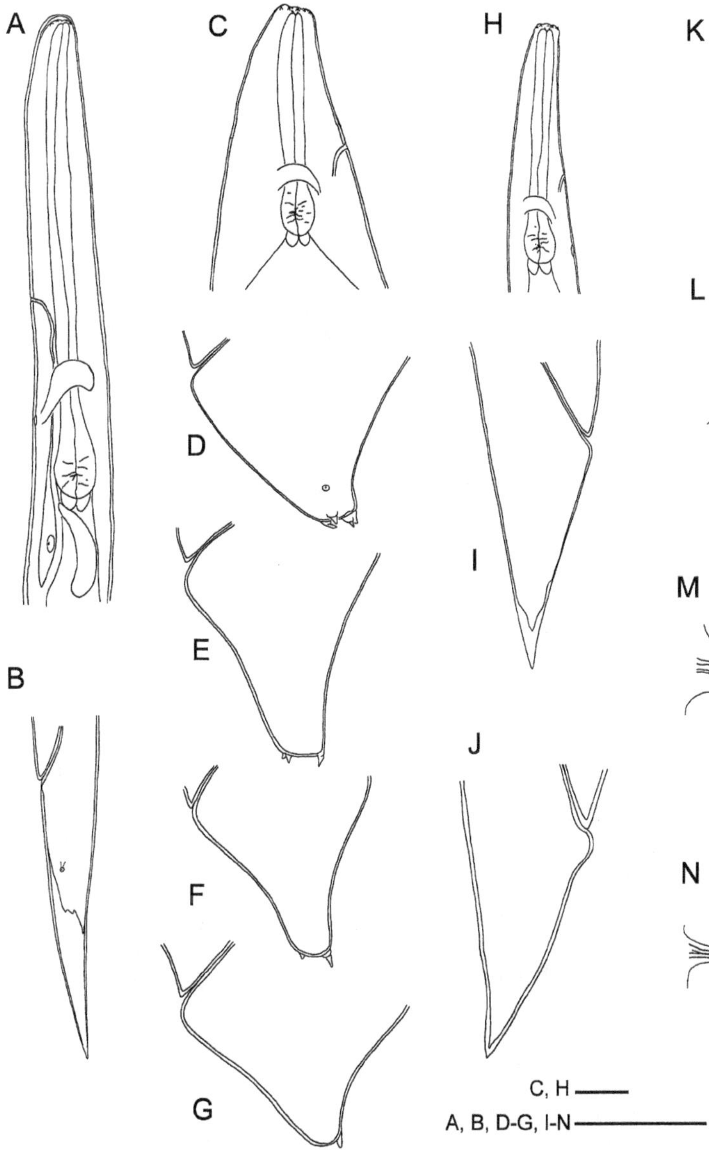

Fig. 4.6. Steinernema australe. *A, B: Infective juvenile. A: Pharyngeal region; B: Tail region. C-G, K, L: First generation female. C: Pharyngeal region; D-G: Tail region showing variation in mucron development; K, L: Epiptygmata. H-J, M, N: Second generation female. H: Pharyngeal region; I, J: Tail region showing variation in mucron development; M, N: Epiptygmata. (Scale bars = 50 μm.) After Edgington* et al. *(2009a), courtesy* Nematology.

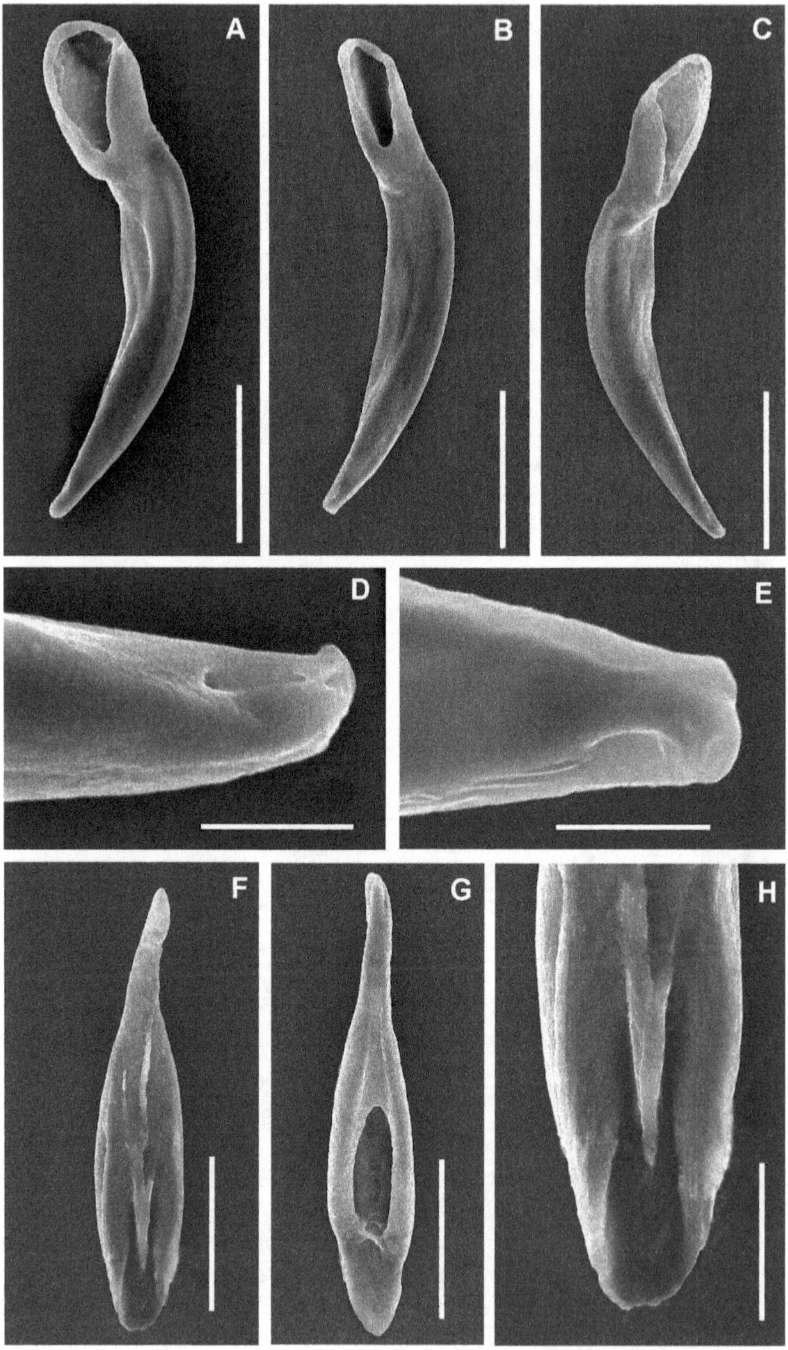
A
B
C
D
E
F
G
H

sinuate, less easily seen but debouching into a large excretory cell located *ca* one basal bulb length posterior to pharyngo-intestinal junction. Non-overlapping junction between pharynx and intestine, cardia present. Prominent papilliform deirid located laterally *ca* one basal bulb length posterior to pharyngo-intestinal junction. Genital tract monorchic, ventrally reflexed. Spicules paired, light brownish amber in colour. Manubrium shape distinctive, elongate (average length:width ratio = 1.7:1), slightly offset from shaft, shaft short, terminus of blade somewhat thickened with ventral notch in lateral view, low velum present, two internal ribs visible on each spicule. Gubernaculum boat-shaped in lateral view, corpus enlarged and closed posteriorly, cuneus V-shaped, large hole or lacuna present dorsally. Copulatory papillae 23 in number comprising a single precloacal midventral papilla and 11 pairs, six pairs of which are precloacal subventral, the anteriormost pairs being slightly more widely spaced, one pair adcloacal subventral ranging in position from just anterior to slightly posterior to cloacal aperture, one lateral pair located between midventral and cloacal level, two pairs postcloacal subventral arranged more or less in tandem and located subterminally and one pair postcloacal subdorsal, also subterminal. Tail more or less conoid, dorsally convex to a broad, usually bluntly rounded, terminus, terminal mucron absent.

Male (second generation)

Body shorter and narrower than that of first generation. Pharyngeal region, including deirid, as for first generation. Testis reflexion much shorter. Spicules shorter than first generation, manubrium slightly more elongate (average length:width ratio = 2.3:1). Tail similar in shape to first generation, terminal mucron absent.

Female (first generation)

Anterior region typical for *Steinernema* female. Body C-shaped when heat-relaxed, cuticle smooth. Labial region relatively low compared to its diam., rounded, with slight depression visible at junction with body

Fig. 4.7. Steinernema australe, *SEM, first generation male. A-C: Spicule, lateral view; D, E: Spicule tip; F: Gubernaculum, ventral view; G: Gubernaculum, dorsal view; H: Gubernaculum showing cuneus. (Scale bars: A = 17 μm; B, C = 20 μm; D, E = 3 μm; F, G = 15 μm; H = 6.7 μm.) After Edgington* et al. *(2009a), courtesy* Nematology.

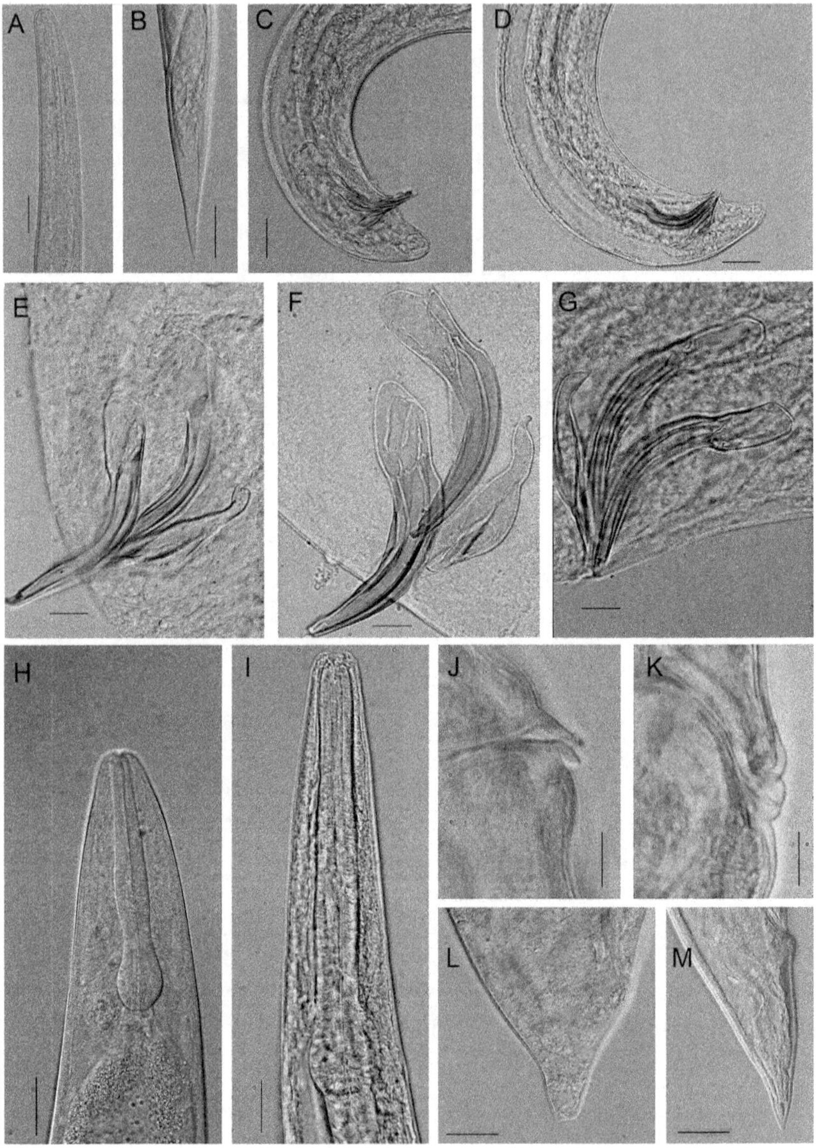
A
B
C
D
E
F
G
H
I
J
K
L
M

wall in most specimens. No lateral fields observed. Excretory pore usually located anterior to nerve ring but posterior to base of metacorpus, occasionally more posterior to level of anterior margin of basal bulb, duct cuticularised but excretory cell obscure. Deirid present laterally on each side *ca* one basal bulb length posterior to pharyngo-intestinal junction, not as distinct as in male, possibly because of size of female body and greater density of body contents. Non-overlapping junction between pharynx and intestine, cardia present. Genital system amphidelphic, ovaries reflexed. Vulva a median, transverse, slit. Vulval lips hypertrophied, protruding 16 (8-31) μm above body contour, epiptygmata double, well developed, closely adpressed, directed posteriorly. Vagina directed slightly anteriad. Postanal swelling absent in most specimens, weakly to moderately formed when present. Tail more or less conoid, shorter than anal body diam., usually dorsally directed, tapering strongly both dorsally and ventrally to a subcylindroid terminal section with a bluntly rounded terminus. Small mucron present in dorsal region of tail terminus, present on all specimens, occasionally with duplex termination. In addition, six of ten specimens observed with one or two small additional, subventral, mucrons (four of the six had two additional mucrons, the other two had but one).

Female (second generation)

Similar to first generation female in general morphology, but shorter, narrower and with a less prominent deirid. Body slightly arcuate when heat-relaxed. Isthmus poorly demarcated. Vulva protruding on most specimens by 4 (3-5) μm, epiptygmata double, closely adpressed, protruding and directed posteriad. Slight to pronounced postanal swelling on posterior lip on most specimens (nine of ten). Tail medium conoid, longer than anal body diam., tapering to a sharp, occasionally slightly offset, mucron-like terminus. No subsidiary mucrons visible.

Fig. 4.8. Steinernema australe, *light micrographs. A, B: Infective juvenile. A: Anterior region; B: Tail region. C, E, F: First generation male. C: Tail region; E, F: Spicules and gubernaculum. D, G: Second generation male. D: Tail region; G: Spicules and gubernaculum. H, J, L: First generation female. H: Pharyngeal region; J: Vulval region showing epiptygmata; L: Tail region. I, K, M: Second generation female. I: Pharyngeal region; K: Vulval region showing epiptygmata; M: Tail region. (Scale bars: A-D, I, L, M = 25 μm; E-G, J, K = 10 μm; H = 55 μm.) After Edgington* et al. *(2009a), courtesy* Nematology.

Table 4.2. *Morphometric data for* Steinernema australe. *All measurements are in* μm *and in the form: mean* $\pm$ *s.d. (range).*

Character	First generation			Second generation		Infective juvenile
	Male		Female	Male	Female	
	Holotype	Paratypes	Paratypes	Paratypes	Paratypes	Paratypes
n	–	20	20	20	20	20
L	1554	1606 ± 31.8 (1401-1937)	5133 ± 156.8 (4081-6468)	1368 ± 16.2 (1214-1499)	2305 ± 72.7 (1876-3087)	1316 ± 21.4 (1162-1484)
a	–	–	–	–	–	35 ± 0.5 (31-38)
b	–	–	–	–	–	7.8 ± 0.2 (6.6-9.2)
c	–	–	–	–	–	12.8 ± 0.2 (11-14.9)
c′	0.4	0.7 ± 0.02 (0.5-0.8)	0.9 ± 0.01 (0.7-1.2)	0.8 ± 0.03 (0.6-0.9)	1.7 ± 0.1 (1.5-2.1)	3.8 ± 0.1 (3.3-4.3)
V	–	–	51 ± 0.9 (48-64)	–	53 ± 0.5 (48-55)	–
Max. body diam.	66	75 ± 2.3 (61-97)	223 ± 8.8 (159-311)	64 ± 1.2 (56-71)	106 ± 5.0 (75-163)	38 ± 0.4 (34-41)
Excretory pore	121	132 ± 2.0 (119-148)	162 ± 4.9 (143-182)	138 ± 2.2 (123-154)	140 ± 2.4 (123-160)	110 ± 1.6 (95-125)
Nerve ring	138	145 ± 2.1 (129-155)	183 ± 4.3 (155-211)	147 ± 2.4 (121-159)	170 ± 2.0 (157-190)	126 ± 2.6 (102-143)
Pharynx	181	186 ± 2.4 (167-205)	242 ± 6.5 (210-308)	185 ± 2.2 (162-205)	222 ± 2.4 (203-243)	169 ± 2.9 (140-194)
Testis reflexion	354	533 ± 18.7 (408-661)	–	251 ± 15.4 (121-336)	–	–
Tail length	20	29 ± 0.9 (20-35)	66 ± 2.5 (53-88)	33 ± 1.1 (23-41)	87 ± 1.1 (80-95)	103 ± 1.6 (92-114)
Hyaline region	–	–	–	–	–	52 ± 1.3 (41-61)
Anal body diam.	38	43 ± 1.2 (37-59)	72 ± 1.9 (57-88)	40 ± 1.2 (31-57)	49 ± 1.2 (42-60)	27 ± 0.6 (25-30)
Spicule length	69	72 ± 1.2 (55-78)	–	67 ± 1.0 (61-75)	–	–
Spicule width	10	11 ± 0.3 (9-14)	–	10 ± 0.3 (9-13)	–	–
Gubernaculum length	43	45 ± 0.9 (36-51)	–	43 ± 1.0 (33-49)	–	–
Gubernaculum width	7	6 ± 0.4 (4-10)	–	8 ± 0.4 (5-11)	–	–
D%	70	71 ± 1.3 (59-87)	70 ± 3.5 (53-82)	74 ± 1.3 (64-82)	63 ± 1.4 (55-74)	65 ± 1.3 (57-78)
E%	–	–	–	–	–	107 ± 1.8 (94-122)
SW%	182	172 ± 5.1 (118-196)	–	173 ± 5.4 (152-204)	–	–
GS%	62	62 ± 1.3 (46-71)	–	64 ± 1.7 (44-77)	–	–
H%	–	–	–	–	–	51 ± 1.0 (42-61)

Infective juvenile

Anterior region typical for *Steinernema* IJ. Body almost straight when heat-relaxed, elongate, tapering gently towards anterior and posterior ends. Cephalic region gently rounded, continuous with body contour. Retained second-stage cuticle not usually observed after harvesting. Cuticle finely annulated with delicate transverse striations. Deirid distinct, located in middle of lateral field at basal bulb level, *i.e.*, at about hemizonid level. Hemizonid present, lens-like in shape, located between mid- and anterior-end of basal bulb. Hemizonion visible in most specimens, located posterior to basal bulb. Excretory pore prominent, situated anterior to nerve ring in isthmus region, excretory duct cuticularised. Lateral field with nine, equally spaced incisures (= eight ridges) at mid-body, 11-13 μm wide. Bacterial vesicle dorsally displaced, usually teardrop-shaped and tapering posteriorly, 12 (9-14) μm in diam., 34 (25-41) μm in length, containing rod-shaped bacteria. Tail long conoid, evenly attenuated, *ca* four times as long as anal body diam., prominent hyaline region occupying 51 (42-61)% of tail length. Phasmid distinct, tube-like, located at 31 (29-34)% of tail length posterior to anus, *i.e.*, at about beginning of hyaline region or slightly anterior (position varying from left- to right-hand side of nematode).

TYPE HABITAT AND LOCALITY

The natural host of *S. australe* is unknown as the species was isolated by *Galleria* baiting a soil sample taken on Isla Magdalena, an island situated 2 km off the west coast of Chile. The soil was a humid, loamy-sand located close to the sea, with dense, tall grass and moderate shading from trees.

TYPE MATERIAL

Holotype (first generation male), isolated from haemocoel of *G. mellonella*, deposited in the CABI Nematode Collection, Europe-UK Centre, Egham, UK. Five paratype males and females (first and second generations) and ten paratype IJ, all as permanent slide mounts, together with 20 paratypes of each stage in TAF, deposited in the nematode collections of CABI, Europe-UK Centre, Egham, UK, INIA-Quilamapu, Chillán, Chile, and the United States Department of Agriculture Nematode Collection (USDANC), Beltsville, MD, USA. In addition, numerous paratype IJ and other stages preserved in 70% ethyl alcohol deposited in the CABI Nematode Collection, Europe-UK Centre, Egham, UK.

RELATIONSHIPS

Steinernema australe is characterised by morphometrics of the IJ, which has a very long body of 1316 (1162-1484) μm, excretory pore located well posterior to the anterior end at 110 (95-125) μm, exceptionally long tail of 103 (92-114) μm, H% = 51 (42-61), E% = 107 (94-122) and a ratio = 35 (31-38). The first generation male is characterised by spicule and gubernaculum lengths of 72 (55-78) μm and 45 (36-51) μm, respectively, spicule and gubernaculum shape, the position of the excretory pore and SW% of 172 (118-196). The first generation female possesses well developed double epiptygmata, lacks a prominent postanal swelling, has a mucron on the tail tip and, in 60% of individuals, has one to two subsidiary mucrons on the tail tip. Based on the long IJ body length, *S. australe* is most similar to species of the *glaseri*-group.

MOLECULAR CHARACTERISATION

The sequence of the ITS regions of *S. australe*, including primers TW81 and AB28, is characterised by its length of 842 bp, ITS1 = 293 bp, and ITS2 = 315 bp. The sequence length of ITS1, at 293 bp, is shorter than that of *S. diaprepesi* (301 bp), *S. leizhouense* (301 bp) and *S. khoisanae* (314 bp) but longer than that of all other species in the *glaseri*-group. ITS2 (315 bp) is shorter than that of *S. scarabaei* (316 bp), *S. leizhouense* (325 bp) and *S. khoisanae* (330 bp) but longer than that of all other species. The distance between *S. australe* and *S. scarabaei* is the shortest (136 bp). For the D2-D3 regions, the sequence length is 835 bp. Pairwise distances showed that it differed from its closest taxon, *S. puertoricense*, by 16 bp and from the most divergent species, *S. karii*, by 60 bp. Further details are furnished by Edgington *et al.* (2009a).

BACTERIAL SYMBIONT

The bacterial symbiont was described as a new species, *Xenorhabdus magdalenensis* by Tailliez *et al.* (2012).

Bionomics

This species was isolated from soil with an *in situ* temperature of 13°C. As evaluated in the laboratory, the pH of the soil was 5.3.

Edgington *et al.* (2009a) provided details of the infectivity, development and reproduction of *S. australe* in *G. mellonella* at 7, 11, 16, 23, 28 and 33°C. Successful infectivity and reproduction occurred in larvae of *Cydia pomonella* (L.) (Lepidoptera: Tortricidae), *Orgyia antiqua* (L.) (Lepidoptera: Lymantriidae), *Brachysternus prasinus* (Guérin) (Coleoptera: Scarabaeidae) and *Phytoloema hermanni* (Germ.) (Coleoptera: Scolytidae) at 20°C.

Distribution

Known only from the type locality in Chile.

Steinernema balochiense[8,9] Fayyaz, Khanum, Ali, Solangi, Gulsher & Javed, 2015

(Fig. 4.9)

Measurements

See Table 4.3.

Diagnostic description

Male (first generation)

Anterior region typical for *Steinernema* male. Excretory pore mostly at base of metacorpus and always anterior to nerve ring. Cardia well developed, intestine with wide lumen up to mid-body. Testis single, well developed, ventrally reflexed, symmetrical and about half of testis length. Spicules paired dark yellow in colour. Spicule head region usually rounded, elongated or longer than width, slightly offset from shaft. Calomus moderately developed. Lamina moderately curved, tapering to a pointed terminus. Each spicule with two internal ribs. Velum thin, parallel to spicule edge. Gubernaculum boat-shaped with anterior part thin

[8] Named after the province of *Balochistan*, Pakistan, where the species was collected.

[9] Description after Fayyaz *et al.* (2015). Revised by Khuong B. Nguyen.

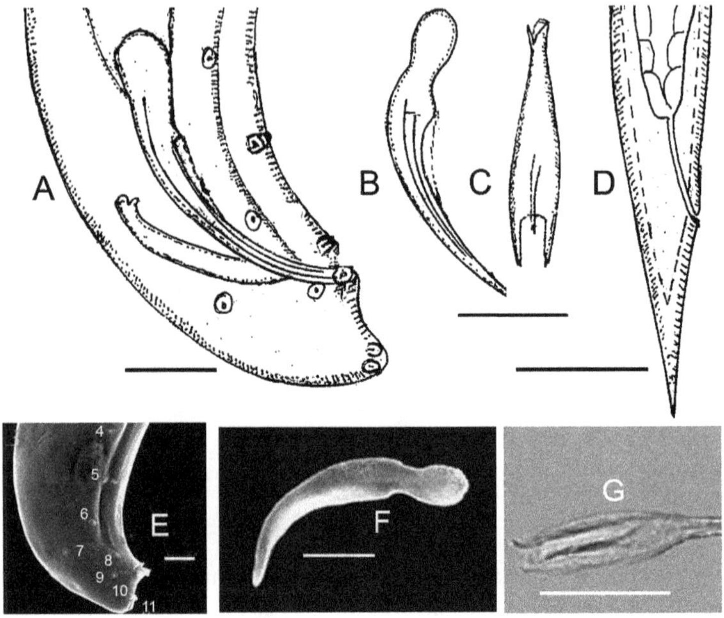

Fig. 4.9. Steinernema balochiense, *first generation males and infective juvenile. A: Posterior region showing genital papillae; B, C: Spicule and gubernaculum; D: Infective juvenile tail; E: SEM of posterior region showing genital papillae; F: SEM of spicule; G: Gubernaculum, ventral view showing Y-shaped cuneus. (Scale bars: A-C = 20 μm; D = 30 μm; E, F = 10 μm; G = 30 μm.) Modified after Fayyaz* et al. *(2015), with acknowledgement to* Zootaxa.

and ventrally curved. In ventral view, gubernaculum anteriorly tapering, funnel-shaped and distal end bifurcated with two wings (corpus). Cuneus Y-shaped. There are 23 genital papillae including 11 pairs and one single ventral precloacal papilla located just anterior to cloaca. Papillae arranged as follows: seven precloacal pairs (five subventral pairs and two ventrolateral pairs); one pair at cloacal, three post cloacal (one subventral and two subterminal pairs). Tail bluntly conoid with short mucron present in approximately 50% of first generation males.

Male (second generation)

Similar to the first generation but smaller and slender. Spicules and gubernaculum shorter and thinner. Tail with well developed mucron, *ca* 2-4 μm long.

Table 4.3. *Morphometrics of* Steinernema balochiense. *All measurements are in μm and in the form: mean ± s.d. (range).*

Character	First generation			Second generation		Infective juvenile
	Holotype (male)	Male	Female	Male	Female	
n	25	25	25	25	25	25
L	1135	1330 ± 199 (1135-1632)	3972 ± 825 (3245-5770)	943 ± 20.2 (927-1065)	2101 ± 564 (1350-2815)	455 ± 36 (415-528)
a	11.9	11 ± 0.8 (10-12.2)	20 ± 3.1 (17-25)	14 ± 1.9 (13-15)	13 ± 2.6 (9.9-17.5)	19 ± 1.35 (16-21)
b	8.73	9.5 ± 0.7 (8.7-11.2)	20 ± 3.1 (16-26)	6.0 ± 0.12 (5.9-7.6)	11.5 ± 2.7 (9-14)	4.9 ± 0.35 (4.4-5.8)
c	38.8	43.3 ± 3.6 (37.8-54)	84 ± 10.6 (60-115)	36 ± 2.4 (34-39)	39 ± 9.2 (25-49)	10 ± 0.7 (9.4-11.6)
c′	0.71	0.71 ± 0.2 (0.6-0.76)	0.79 ± 0.24 (0.6-1.33)	0.6 ± 0.2 (0.50-0.62)	1.2 ± 0.1 (1.1-1.4)	–
V	–	–	52 ± 5.1 (41-59)	–	53 ± 3.5 (48-58)	–
Body diam.	95	120 ± 24 (95-135)	199 ± 29 (145-240)	65 ± 5 (60-75)	160 ± 28.6 (120-193)	23 ± 1.13 (22-25)
Excretory pore	52	61 ± 9 (52-77)	77 ± 15.6 (65-100)	65 ± 2.5 (60-75)	73 ± 7.5 (65-90)	35 ± 2.0 (32-38)
Nerve ring	100	95 ± 5 (85-110)	118 ± 22.6 (80-145)	89 ± 7.6 (80-90)	110 ± 17 (100-134)	65 ± 4.6 (60-72)
Pharynx	130	136 ± 9 (130-152)	189 ± 16.5 (173-220)	134 ± 3.7 (117-138)	181 ± 18 (155-206)	91 ± 3 (85-98)
Tail length	30	31 ± 2.6 (28-35)	46 ± 7.2 (30-60)	26 ± 1.2 (23-28)	54 ± 2.5 (50-58)	44 ± 3.4 (40-51)
Anal body diam.	42	43 ± 2.7 (40-50)	59 ± 12.5 (45-80)	32 ± 2.5 (30-37)	43 ± 4.3 (38-50)	13.5 ± 0.6 (13-17)
Spicule length	76	70 ± 6.3 (60-76)	–	58 ± 2.8 (53-60)	–	–
Gubernaculum length	40	44 ± 2.6 (40-47)	–	33 ± 4 (30-42)	–	–
D%	40	44 ± 3.6 (40-51)	41 ± 4.8 (35-48)	42 ± 2.3 (40-48)	41 ± 5.1 (36-50)	39 ± 2.9 (36-44)
E%	173	200 ± 17.5 (173-256)	171 ± 31.3 (105-200)	204 ± 13 (200-260)	135 ± 13 (122-163)	80 ± 7.7 (70-92)
SW%	181	200 ± 12.9 (173-217)	–	140 ± 8 (132-150)	–	–
GS%	53	64 ± 7.7 (53-75)	–	57 ± 6.5 (50-63)	–	–
H%	–	–	–	–	–	52 ± 6.6 (39-60)

Female (first generation)

Anterior region typical for *Steinernema* female. Excretory pore mostly at base of metacorpus and always anterior to nerve ring. Cardia well developed. Gonads amphidelphic, reflexed dorsally. Vulva a transverse slit, slightly protruding from body contour, double-flapped epiptygmata present. Vulval opening mostly located near mid-body. Vagina short with muscular walls. Post-anal swelling absent. Egg development and hatching commonly inside reproductive system. Tail shape variable, bluntly conical to dome-shaped with a short spine on tip.

Female (second generation)

Similar in general aspect to first generation female, except for smaller length and diam. Vulva a transverse slit, generally protruding from body contour. Tail straight and longer than anal body diam., conoid, tapering evenly posteriorly and ending in a fine mucronate process.

Infective juvenile

Anterior region typical for *Steinernema* IJ. Excretory pore located in anterior region of pharynx, excretory duct long extending posteriorly to basal bulb. Hemizonid distinct, located at level of anterior margin of basal bulb. Bacterial vesicle present containing bacterial cells. Lateral field beginning anteriorly with one line at annule 6 or 7, two additional lines appearing to form two ridges at annule 12 or 13. At level of excretory pore number of ridges increasing from two to six. Maximum number of ridges in lateral field at mid-body is six. Intestine filled with numerous fat globules, lumen not visible due to fat globules. Genital primordium prominent. Rectum long and straight, anus distinct. Tail conoid with pointed terminus. Hyaline region occupying 39-60% of tail length.

TYPE HABITAT AND LOCALITY

Type host unknown. The nematode species was collected from the soil around the roots of *Psidium guajava* in Balochistan, Pakistan.

TYPE MATERIAL

Holotype male (first generation), 20 paratype males (first and second generation), 20 females (first and second generation) and 20 IJ deposited in the Nematode Collection of the National Nematological Research

Centre, University of Karachi, Karachi-75270, Pakistan. Twenty males, females and infective-stage juveniles deposited in the Nematode Collection of Guangdong Entomological Institute, Guangzhou, China.

Relationships

Steinernema balochiense belongs to the *carpocapsae*-group by having small juveniles. The new species can be recognised by IJ body length of 455 (415-528) μm, tail length = 44 (40-51) μm, and D% = 39 (36-44). The male is characterised by gubernaculum length = 44 (40-47) μm, D% = 43.8 (40-51), GS% = 63.8 (53-75), manubrium of spicule broad, wider than length, shaft small, mucron present in both generations. Proximal end of gubernaculum funnel-shaped [*sic*]. The female can be characterised by sclerotised cheilorhabdions, and mucron present in both generations. More information on the relationships is available in Fayyaz *et al.* (2015).

Molecular characterisation

Steinernema balochiense is characterised genetically by sequences of ITS (accession no. JX135547) and D2-D3 (JX068821) regions. More information about molecular characterisation and phylogenetic analysis is available in Fayyaz *et al.* (2015).

Symbiotic bacterium

No information.

Bionomics

No information.

Distribution

Steinernema balochiense was collected in Balochistan, Pakistan.

Steinernema bifurcatum[10,11] Fayyaz, Yan, Qui, Han, Gulsher, Khanum & Javed, 2014

(Figs 4.10-4.12)

MEASUREMENTS

See Table 4.4.

DIAGNOSTIC DESCRIPTION

Male (first generation)

Anterior region typical for *Steinernema* male. Deirids usually visible posterior to level of pharyngo-intestinal junction. Nerve ring just anterior to basal bulb. Excretory pore located anterior to nerve ring or at level of metacorpus. Gonad reflexed for about half of testis length. Spicules paired, light yellow in colour, manubrium with rounded broad anterior end, almost continuous with shaft, calomus very short. Lamina thick, variable in shape with two internal ribs, tapering gradually posteriorly to a point, velum prominent, posterior end of spicules blunt. Gubernaculum characteristically bifurcated at both proximal and distal ends. In ventral or dorsal view, anterior part shorter, curved anteriorly, proximal end with sharp pointed end. Genital papillae 23 in number, comprising 11 pairs and a single precloacal papilla. Six pairs precloacal (five subventral and one lateral), two adanal, three postcloacal (one dorsal, one ventral and one terminal) and a single ventral papilla located between fourth and fifth subventral papillae. Tail conoid, terminus lacking mucron.

Male (second generation)

Similar to first generation, but smaller. Tail terminus ending with fine mucron.

Female (first generation)

Anterior region typical for *Steinernema* female. Excretory pore located anterior to nerve ring at about mid-point of pharynx. Vulva a transverse slit, lips slightly protruding with short double-flapped epiptygmata. Tail shorter than anal body diam., mostly blunt, without mucron. Postanal swelling not developed.

[10] The specific epithet refers to the *bifurcate* nature of the gubernaculum.

[11] Description after Fayyaz *et al.* (2014). Revised by David J. Hunt.

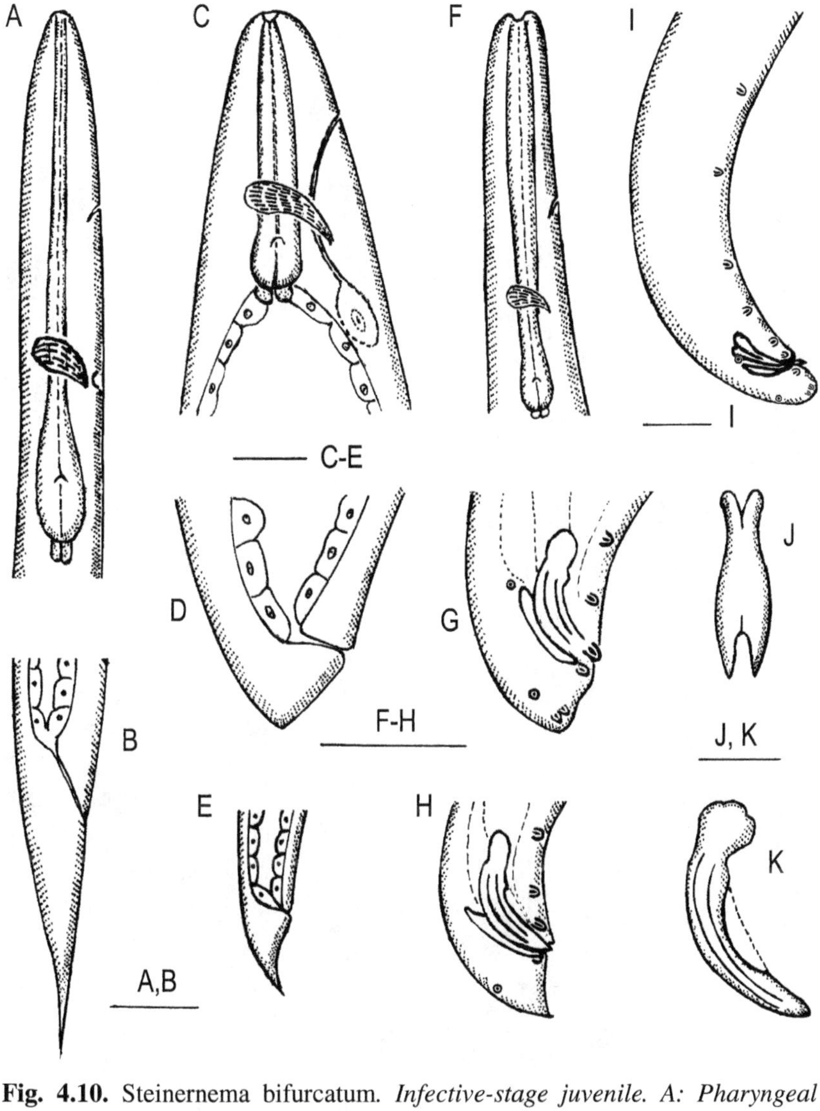

Fig. 4.10. Steinernema bifurcatum. *Infective-stage juvenile. A: Pharyngeal region; B: Tail. Female first generation. C: Pharyngeal region; D: Tail. Female second generation. E: tail. Male first generation. F: Pharyngeal region; G, I: Anal region showing papillae. Male second generation. H: Tail; J: Gubernaculum; K: Spicule. (Scale bars: A, B = 22 μm; C-I = 50 μm; J, K = 20 μm.) After Fayyaz* et al. *(2014), courtesy* Nematology.

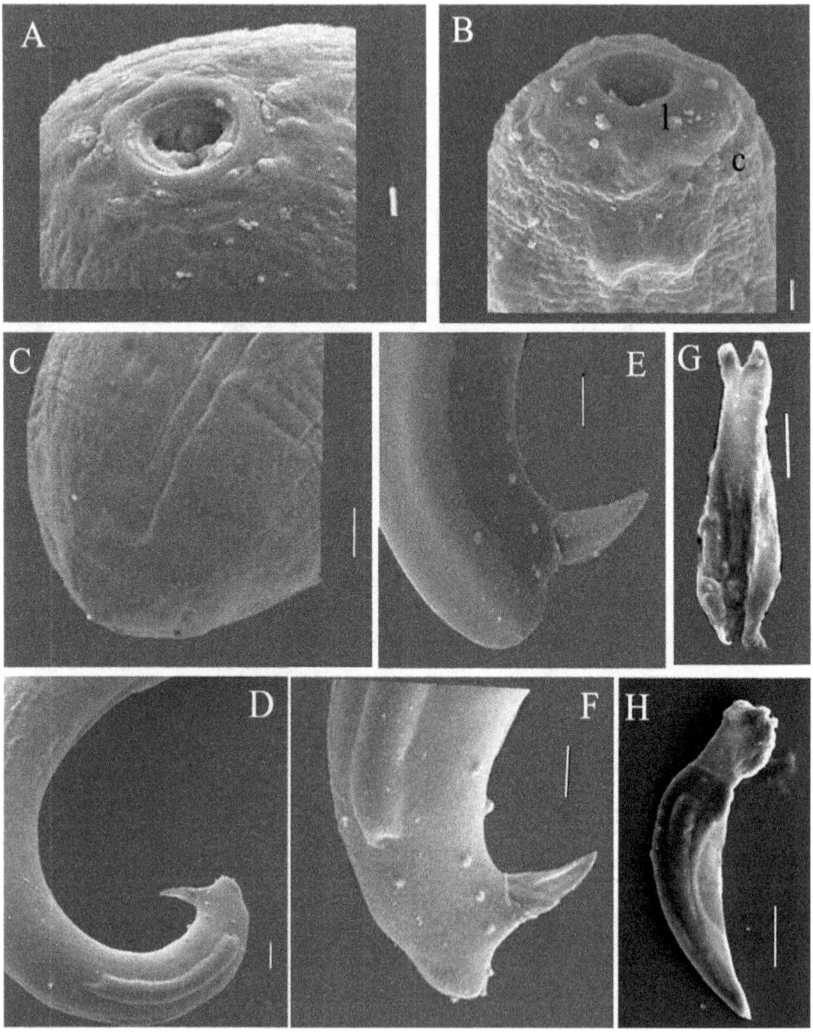

Fig. 4.11. Steinernema bifurcatum. *SEM of first generation female and male. Female. A, B: Anterior region showing labial (l) and cephalic papillae (c); C: Tail region. Male. D-F: Tail region showing papillae; G: Bifurcated gubernaculum; H: Spicule. (Scale bars: A, B = 1 μm; C-H = 10 μm.) After Fayyaz* et al. *(2014), courtesy* Nematology.

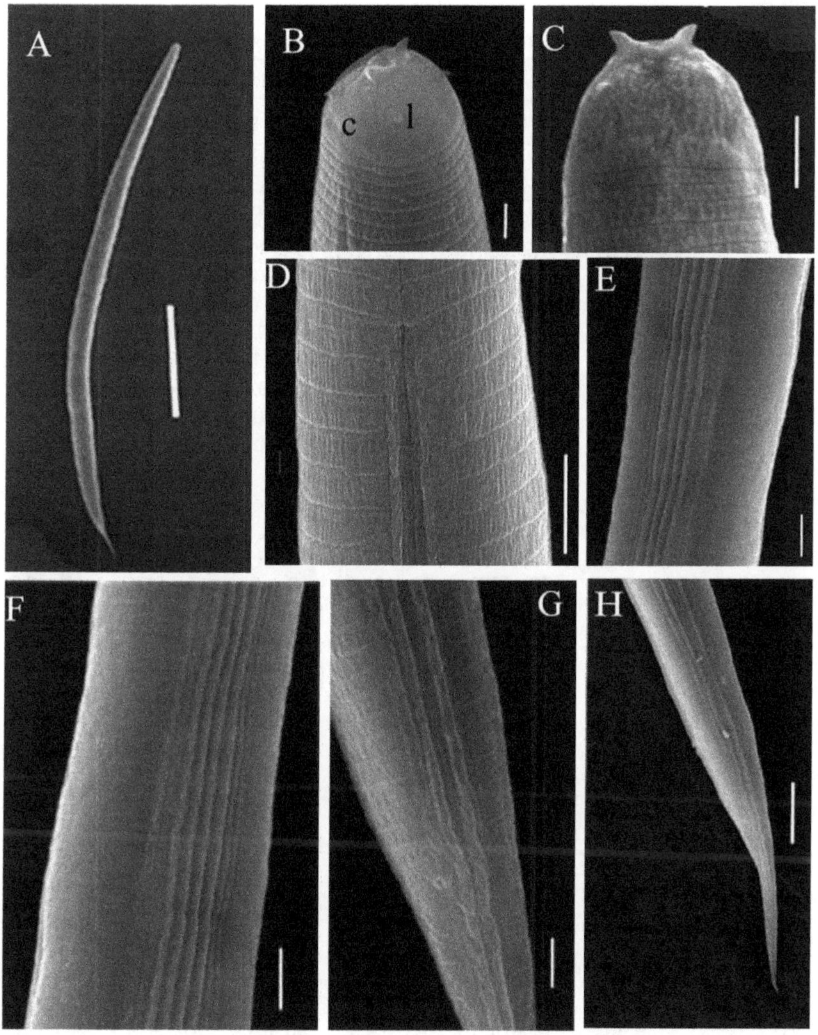

Fig. 4.12. Steinernema bifurcatum. *SEM of infective juvenile. A: Entire body; B, C: Anterior region showing labial (l), cephalic (c) papillae and horn-like structures; D: Origin of lateral field in anterior portion; E, F: Lateral field with eight ridges; G, H: Tail region showing gradual decrease in number of lateral ridges. (Scale bars: A = 100 μm; B-D, G = 2 μm; E, F = 5 μm; H = 10 μm.) After Fayyaz* et al. *(2014), courtesy* Nematology.

Table 4.4. *Morphometrics of* Steinernema bifurcatum. *All measurements are in μm and in the form: mean ± s.d. (range).*

Character	First generation			Second generation		Infective juvenile
	Male		Female	Male	Female	
	Holotype	Paratypes	Paratypes	Paratypes	Paratypes	Paratypes
n	–	25	25	25	25	25
L	1059	1192 ± 314 (1059-1454)	3655 ± 1850 (2292-7477)	621 ± 46 (560-685)	1241 ± 403 (994-1794)	521 ± 27.3 (460-590)
a	11.27	11.2 ± 3.7 (10.5-13.3)	17.6 ± 3.7 (12.7-24)	15 ± 1.78 (13-18)	16.5 ± 3 (13-20)	24 ± 1.14 (22-25)
b	7.4	8.6 ± 0.54 (7.4-9.2)	19.5 ± 7.4 (10-31)	4.8 ± 0.3 (4.5-5.5)	7.6 ± 1.7 (5.0-11.3)	4.5 ± 0.46 (3.8-5.6)
c	35.3	44.3 ± 5 (35-54)	78.6 ± 28.8 (42-159)	30 ± 1.6 (28-32)	25.8 ± 5.7 (14-30)	9.6 ± 0.4 (9.2-10.5)
c′	1.5	1.6 ± 0.12 (1.48-1.83)	0.78 ± 0.28 (0.5-0.83)	1.5 ± 0.06 (1.4-1.6)	1.0 ± 0.33 (0.6-1.5)	4.0 ± 0.3 (3.4-4.6)
V	–	–	56.6 ± 3 (53-62)	–	59.6 ± 3.6 (54-67.5)	–
Body diam.	94	108 ± 18.3 (85-117)	192 ± 51.4 (130-245)	40 ± 3 (35-44)	69 ± 18.4 (55-105)	22 ± 1.48 (20-24)
Excretory pore	72	70 ± 7.7 (58-86)	76 ± 21.7 (60-108)	50 ± 8.6 (49-61)	56 ± 7.3 (45-63)	45 ± 2.7 (40-49)
Nerve ring	100	117 ± 12 (100-130)	117 ± 11.8 (93-125)	98 ± 8.2 (88-100)	108 ± 11.3 (90-112)	69 ± 2.63 (66-80)
Pharynx	143	145 ± 7.8 (130-158)	174 ± 13.9 (158-200)	123 ± 3.6 (118-128)	143 ± 16 (120-182)	114 ± 8 (102-130)
Tail length	30	30 ± 2.9 (22-34)	43 ± 8 (38-60)	20 ± 0.6 (19-21)	43 ± 5.4 (33-56)	54 ± 2.45 (51-59)
Anal body diam.	45	48 ± 4.8 (45-56)	57 ± 12 (39-85)	30 ± 1.4 (28-34)	32 ± 3.9 (26-38)	13 ± 1.27 (11-17)
Spicule length	60	69 ± 7.7 (60-85)	–	48 ± 3.1 (44-52)	–	–
Gubernaculum length	38	39 ± 5.4 (30-49)	–	20 ± 2.4 (15-22)	–	–
Vulva	–	–	2071 ± 1033 (1010-4350)	–	658 ± 143 (452-992)	–
D%	50	48 ± 4.3 (42-58)	43 ± 9.8 (35-62)	46 ± 3.9 (45-52)	37.7 ± 9.1 (25-47)	39.7 ± 4.2 (33-47)
E%	240	239 ± 24 (197-286)	177 ± 32.9 (140-209)	275 ± 19.6 (245-295)	125 ± 34 (109-146)	85 ± 5.3 (77-94)
SW	1.3	1.4 ± 0.16 (1.2-1.7)	–	1.6 ± 0.1 (1.5-1.8)	–	–
GS	0.66	0.59 ± 0.08 (0.51-0.74)	–	0.40 ± 0.05 (0.29-0.45)	–	–
H%	–	–	–	–	–	54 ± 2.5 (50-57)

Female (second generation)

General morphology similar to first generation but smaller. Vulva with protruding lips. Tail conoid with postanal swelling. Terminal mucron present.

Infective juvenile

Anterior region typical for *Steinernema* IJ. Two horn-like structures present in labial region. Excretory pore situated at about mid-pharynx. Nerve ring distinct, encircling isthmus, anterior to basal bulb. Hemizonid distinct, located at anterior margin of basal bulb. At anterior portion of intestine, or just posterior to cardia, bacterial vesicle present containing bacterial cells. Lateral field beginning anteriorly with one line at annule 5 or 6. Two additional lines appearing to form two ridges at annule 14 or 15. At level of excretory pore number of ridges increases from two to six. Two submarginal ridges appearing more posteriorly to make a maximum of eight ridges (*i.e.*, nine lines). Marginal ridges less distinct. In posterior region submarginal ridges reducing to six ridges, near phasmid gradually reducing to two. With above arrangement, lateral field formula = 2, 6, 8, 6, 2. Tail elongate, conoid, tapering to a finely pointed terminus. Hyaline region occupying *ca* half of tail length.

Type habitat and locality

The type isolate, PAK. P. S. 37, was collected from soil around the roots of a rose plant (*Rosa indica* L.) at Sheikhupura Lahore district, Punjab, Pakistan.

Type material

Holotype male (first generation), 20 paratypes males (first and second generation), 20 paratype females (first and second generation) and 20 IJ deposited in the Nematode Collection of the National Nematological Research Centre, University of Karachi, Karachi-75270, Pakistan. Twenty specimens of first and second generation males, females and IJ deposited in the Nematode collection of Guangdong Entomological Institute, Guangzhou, China.

Relationships

Steinernema bifurcatum belongs to the *bicornutum*-group and is characterised by the male gubernaculum which is bifurcate at both proximal and distal ends, a key diagnostic feature. *Steinernema bifurcatum* has two horn-like structures on the labial region of the IJ and can be recognised by IJ body diam. = 22 (20-24) μm, pharynx = 114 (102-134) μm, ratio a = 24 (22-25) and D% = 39.7 (33-47). It can be further recognised by the male characters of D% = 48 (42-58), and genital papillae = 23 (22 + 1) in number and, for the first generation female, excretory pore = 75.7 (60-108) μm, pharynx = 174 (158-200) μm, and tail length = 43.2 (38-60) μm.

MOLECULAR CHARACTERISATION

Characterised by sequences of ITS (accession no. JX989267), D2-D3 regions (accession no. JQ838179) of rDNA and 12S (accession no. JX068826) mt DNA. The length of the ITS1 + 5.8S + ITS2 sequence = 824 bp, ITS1 = 267 bp and ITS2 = 309 bp and its composition is: A = 0.256, C = 0.160, G = 0.229, T = 0.353. The sequence of the D2-D3 region = 880 bp and its composition is: A = 0.240, C = 0.107, G = 0.380, T = 0.277. The 12S mtDNA sequence = 507 bp and its composition is A = 0.302, C = 0.1370, G = 0.193, T = 0.367. Further data are given by Fayyaz *et al.* (2014).

BACTERIAL SYMBIONT

According to Fayyaz *et al.* (2014) the bacterial symbiont showed 98% similarity with *Xenorhabdus indica* Somvanshi, Lang, Ganguly, Swiderski, Saxena & Stackebrandt, 2006 (accession no. KC020713; 1536 bp).

BIONOMICS

The nematode did not cross-hybridise with *S. abbasi* (Fayyaz *et al.*, 2014).

DISTRIBUTION

Known only from the Punjab, Pakistan (Fayyaz *et al.*, 2014).

Steinernema boemarei[12,13] Lee, Sicard, Skeie & Stock, 2009
(Figs 4.13, 4.14)

Measurements

See Table 4.5.

Diagnostic description

Male (first generation)

Anterior region typical for *Steinernema* male. Body slender, ventrally curved posteriorly, J-shaped when heat-killed. Cuticle appearing smooth with light microscopy. Lateral field and phasmids inconspicuous. Deirids conspicuous, located laterally in first third of body and posterior to excretory pore. *Vas deferens* with inconspicuous walls. Spicules paired, symmetrical, curved, ochre-brown. Manubrium rhomboidal. Calomus (shaft) distinct. Lamina with rostrum or retinaculum and two internal ribs. Velum present, narrow. Blade terminus blunt. Gubernaculum arcuate, *ca* 75% of spicule length long. Manubrium of gubernaculum curved ventrally. Tail conoid, mucron lacking. There are 23 genital papillae (11 pairs plus a single precloacal midventral papilla) arranged as six subventral precloacal pairs, a single midventral precloacal between precloacal pairs 5 and 6, one pair lateral (or slightly subdorsal) precloacal, one pair adcloacal subventral, and three pairs postcloacal, one of which is subdorsal and the other two subventral, terminal.

Male (second generation)

General morphology similar to that of first generation, but smaller in size. Tail with or without mucron. Spicules with manubrium morphology similar to that of first generation male but gubernaculum more slender and longer.

Female (first generation)

Anterior region typical for *Steinernema* female. Body C-shaped when heat-killed. Excretory pore located about mid-procorpus level. Ovaries opposed, reflexed dorsally, oviduct well developed, glandular

[12] Patronym honouring Dr Noel *Boemare* for his work on the bacterial symbionts of entomopathogenic nematodes.

[13] Description after Lee *et al.* (2009). Revised by David J. Hunt.

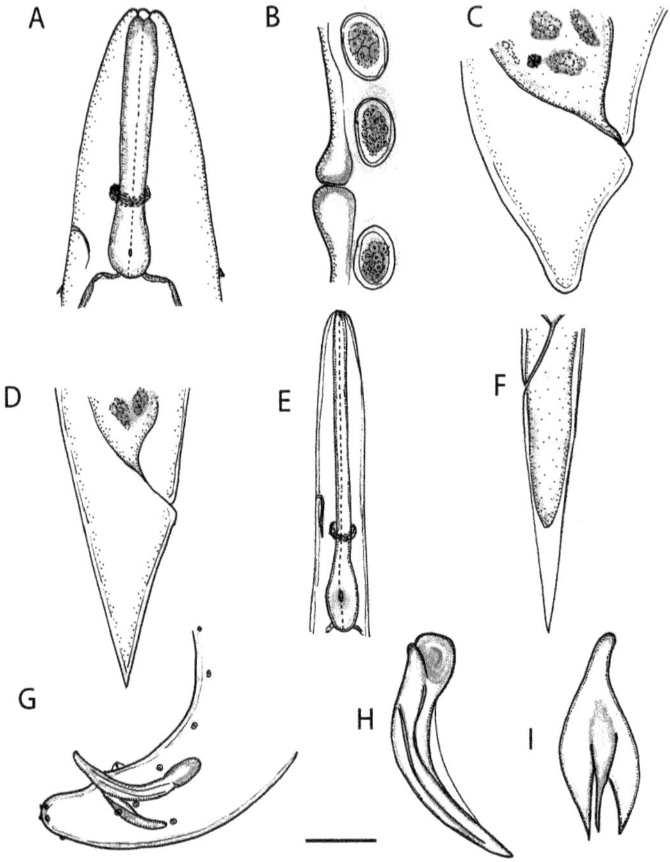

Fig. 4.13. Steinernema boemarei. *A: Anterior end of first generation male, dorsal view, showing position of deirids; B: Vulva of first generation female, lateral view; C: Tail of first generation female, lateral view; D: Tail of second generation female, lateral view; E: Anterior end of infective juvenile, lateral view; F: Tail of infective juvenile, lateral view; G: First generation male tail, lateral view, showing some of genital papillae; H: First generation male spicule, lateral view; I: First generation male gubernaculum, ventral view. (Scale bar: A = 64 μm; B-E = 37 μm; F = 33 μm; G = 45 μm; H = 15 μm; I = 19 μm.) After Lee* et al. *(2009), courtesy* Systematic Parasitology.

spermatheca and uterus lying ventrally. Vagina short with muscular walls. Vulva in form of a transverse slit, located near mid-body, lips protruding and slightly asymmetrical, posterior lip larger than anterior.

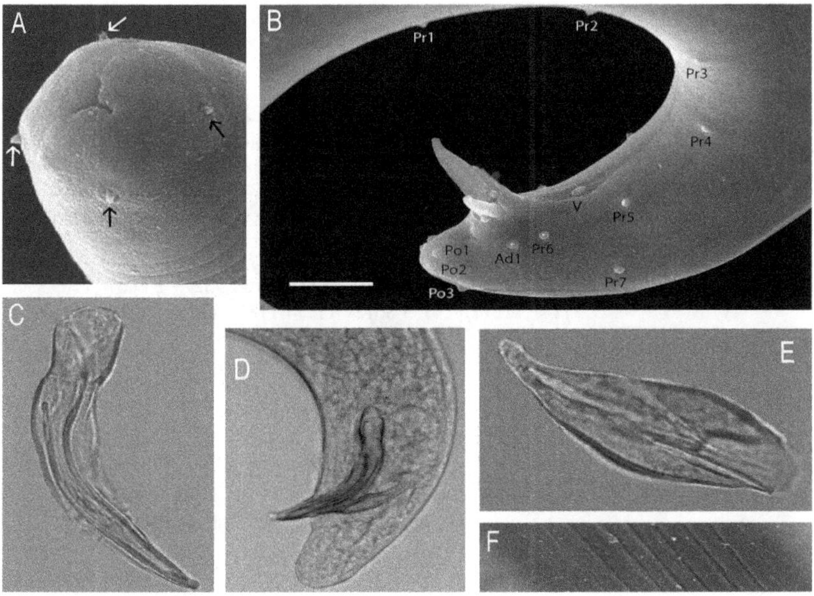

Fig. 4.14. *Light and scanning electron microscope photographs of* Steinernema boemarei. *A: SEM image of IJ labial region showing cephalic papillae (arrows); B: SEM image of male tail showing distribution of genital papillae (Pr = precloacal; Ad = adcloacal; Po = postcloacal; V = ventral); C: First generation male spicule, lateral view; D: Lateral view of first generation male tail, showing* in situ *spicule and gubernaculum; E: First generation male gubernaculum, dorsal view; F: Lateral field of IJ showing nine lines or incisures. (Scale bar: A = 2.35 μm; B = 16 μm; C = 29 μm; D = 18 μm; E = 13 μm, F = 3.2 μm.) After Lee* et al. *(2009), courtesy* Systematic Parasitology.

Tail blunt, conoid, lacking mucron. Postanal lips usually protruding and asymmetrical, anterior lip smaller than posterior.

Female (second generation)

Body open C-shape when heat-killed. Similar to first generation female but smaller. Vulva with protuberant symmetrical lips. Tail conoid, post-anal swelling present.

Infective juvenile

Anterior region typical for *Steinernema* IJ. Body of heat-relaxed specimens almost straight, slender, gradually tapering posteriorly. Cuticle with fine transverse striae. Lateral field distinct with nine distinct, evenly spaced lines or incisures (*i.e.*, eight ridges) in mid-body region. Labial

Table 4.5. *Morphometrics of* Steinernema boemarei. *All measurements are in μm and in the form: mean ± s.d. (range).*

Character	First generation			Second generation		Infective juvenile
	Male		Female	Male	Female	
	Holotype	Paratypes	Paratypes	Paratypes	Paratypes	Paratypes
n	–	24	25	24	25	25
L	2021	1850 ± 202 (1479-2214)	6394 ± 973 (4553-8612)	1351 ± 166 (1174-1782)	3273 ± 606 (2246-4630)	1103 ± 75 (1005-1323)
a	–	–	–	–	–	28 ± 2 (24-33)
b	–	–	–	–	–	8 ± 0.3 (7.0-8.4)
c	–	–	–	–	–	13 ± 0.5 (12-14)
V	–	–	51 ± 2 (48-56)	–	55 ± 2 (50-58)	–
Max body diam.	105	96 ± 16 (70-122)	316 ± 42 (241-427)	69 ± 8 (55-87)	205 ± 32 (151-259)	39 ± 3 (34-46)
Excretory pore	140	138 ± 15 (108-159)	197 ± 17 (165-226)	121 ± 15 (104-158)	165 ± 18 (137-202)	91 ± 6 (82-111)
Nerve ring	123	121 ± 13 (92-145)	164 ± 10 (146-183)	139 ± 13 (115-164)	182 ± 20 (151-223)	112 ± 8 (100-142)
Pharynx	169	162 ± 9 (140-176)	229 ± 10 (212-244)	173 ± 14 (142-202)	254 ± 27 (223-322)	143 ± 10 (131-182)
Tail length	46	40 ± 4 (32-46)	84.3 ± 13 (58-109)	39 ± 12 (26-87)	109 ± 11 (93-138)	86 ± 7 (77-109)
Anal body diam.	59	46 ± 5 (37-56)	90 ± 17 (64-131)	39 ± 5 (32-52)	57 ± 6 (47-69)	21 ± 2 (18-24)
Spicule length	77	79 ± 8 (64-96)	–	69 ± 6 (54-80)	–	–
Gubernaculum length	48	52 ± 5 (43-65)	–	42 ± 4 (34-48)	–	–
D%	83	85 ± 8 (68-99)	86 ± ? (76-101)	88 ± 9 (71-102)	65 ± 5 (56-81)	63 ± 3 (56-68)
E%	306	346 ± 42 (281-483)	239 ± 39 (188-360)	329 ± 79 (142-465)	153 ± 12 (132-183)	106 ± 6 (94-122)
SW%	150	170 ± 30 (120-240)	–	180 ± 30 (130-220)	–	–
GS%	60	70 ± 10 (50-90)	–	60 ± 10 (50-70)	–	–
H	–	–	–	–	–	35 ± 6 (23-52)
H%	–	–	–	–	–	41 ± 5 (26-49)

region continuous with body, slightly truncate. Nerve ring located at isthmus level, excretory pore situated about mid-corpus level. Anterior portion of intestine with small bacterial vesicle. Rectum long, straight, anus distinct. Tail conoid, with pointed terminus. Hyaline region comprising *ca* 41% of tail length.

Type-host and locality

The natural host is unknown, the new species being collected by *G. mellonella* baiting in sandy soil from a beach in Espiguette, Grand Travers, France.

Type material

First generation holotype male, five first generation paratype males, five first generation paratype females, five paratype IJ, one second generation paratype female and four second generation paratype males deposited in the USDA Nematode Collection, MD, USA; five first generation paratype males and females, and five paratype IJ deposited at the University of California Davis Nematode Collection, Davis, CA, USA.

Relationships

Steinernema boemarei is a member of the *glaseri*-group and has a very long IJ with a lateral field bearing nine equally developed lines (*i.e.*, eight ridges) arranged evenly in the mid-body region. According to Lee *et al.* (2009), most of the IJ morphometrics overlap with other species in the *glaseri*-group. The presence of deirids in IJ was not observed but their presence in the first generation male was regarded as diagnostic by Lee *et al.* (2009), although it is plausible that deirids are present in the male of most, if not all, species but have been overlooked, particularly in older descriptions. The first generation males were distinguished from other related species by spicule and gubernaculum morphology, the arrangement of the genital papillae, SW = 1.7 (1.2-2.4) and GS = 0.7 (0.5-0.9) (see Lee *et al.*, 2009 for more detail).

Molecular characterisation

Phylogenetic analyses of both LSU (accession no. FJ152415) and ITS (accession no. FJ152414) sequences placed *S. boemarei* in Clade V (see

Spiridonov *et al.*, 2004), the *glaseri*-group, characterised by IJ with long bodies (>1100 μm). Within this clade, *S. boemarei* was found to be closely related to *S. arenarium*, *S. glaseri*, *S. cubanum* and *S. diaprepesi*. The results of these two analyses placed *S. boemarei* close to *S. arenarium*, from which it differed by 135 of the compared 768 bp. Lee *et al.* (2009) present further details of the relationships.

SYMBIOTIC BACTERIUM

The bacterial symbiont was identified as *Xenorhabdus kozodoii* Talliez, Pagès, Ginibre, Boemare, 2006.

BIONOMICS

Cross-hybridisation tests between males and females of *S. boemarei* with *S. arenarium* and *S. glaseri* yielded no progeny. In the controls, offspring were produced in all self-crossed species. No progeny were observed in the single female control plates.

According to Lee *et al.* (2009), the life cycle of this new species is one of the shortest observed for the genus. At 25°C, IJ kill *G. mellonella* larvae at 16-24 h, J3 become J4 at 20-30 h and mature adult males and females develop *ca* 24-36 h. Gravid first generation females were observed as early as 36 h. Progeny from first generation adults were seen at 48-72 h. A massive emergence of J3 (mostly progeny of first generation adults) was observed at 96 h after initial exposure. Second generation adults also emerged at the same time as the IJ exited the cadaver (*ca* 96 h), which by then had largely disintegrated. Gravid second generation females may produce progeny outside the insect host or on the disintegrated cadaver.

DISTRIBUTION

Known only from southern France (Lee *et al.*, 2009).

Steinernema brazilense[14,15] Nguyen, Ginarte, Leite, Santos & Harakava, 2010
(Figs 4.15-4.17)

MEASUREMENTS

See Table 4.6.

DIAGNOSTIC DESCRIPTION

Male (first generation)

Anterior region typical for *Steinernema* male. Cuticle with faint transverse striae, lateral field present with one ridge. Excretory pore located posterior to mid-pharynx but anterior to nerve ring. Tail long, dorsally convex, ventrally concave, terminus bluntly rounded. No mucron-like projection observed at tail tip. Twenty-three genital papillae

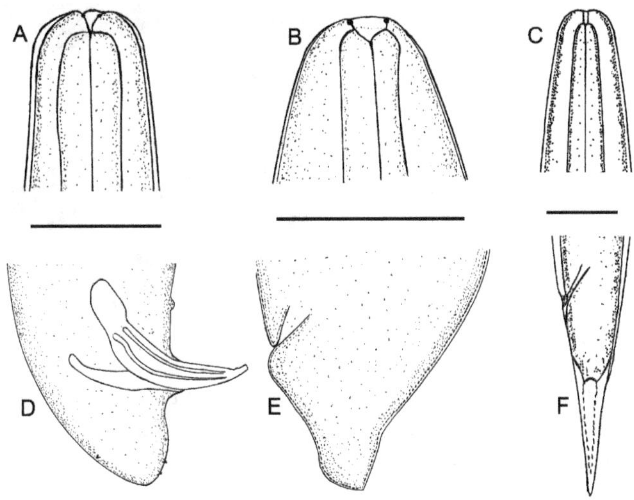

Fig. 4.15. Steinernema brazilense. *A-C: Anterior region of male, female and infective juvenile, respectively; D-F: Posterior region of male, female and infective juvenile, respectively. (Scale bars: A = 50 μm; B, E = 53 μm; C, F = 24 μm; D = 55 μm.) Modified after Nguyen* et al. *(2010), with acknowledgement to* Journal of Invertebrate Pathology.

[14] The specific epithet is derived from *Brazil*, the country from where the nematode was collected.

[15] Description after Nguyen *et al.* (2010). Revised by Khuong B. Nguyen.

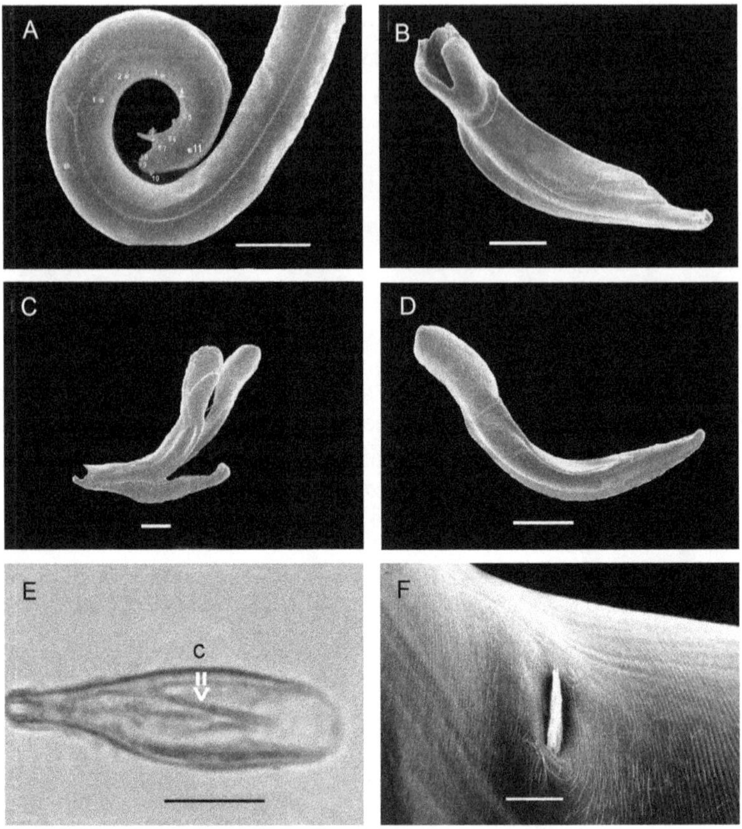

Fig. 4.16. *SEM of* Steinernema brazilense. *A: Posterior region of a male showing 11 pairs of genital papillae and lateral field with single ridge; B-D: Spicules and gubernaculum; E: Gubernaculum, ventral view showing arrowhead-shaped cuneus (c); F: Vulva with small epiptygmata. (Scale bars: A = 50 μm; B-D, F = 10 μm; E = 14.3 μm.) Modified after Nguyen* et al. *(2010), with acknowledgement to* Journal of Invertebrate Pathology.

visible with SEM (11 pairs and a single ventral precloacal papillae), including six subventral precloacal pairs, one pair dorsolateral, one pair adcloacal and subventral, two pairs subventral and subterminal, one pair subdorsal, a single midventral precloacal papilla. Spicules paired. Spicule head region longer than wide. Blade with dorsal lobe, lateral lobe well developed, ventral lobe poorly developed, sometimes not observed. An aperture sometimes observed on ventral side of spicule tip. Velum

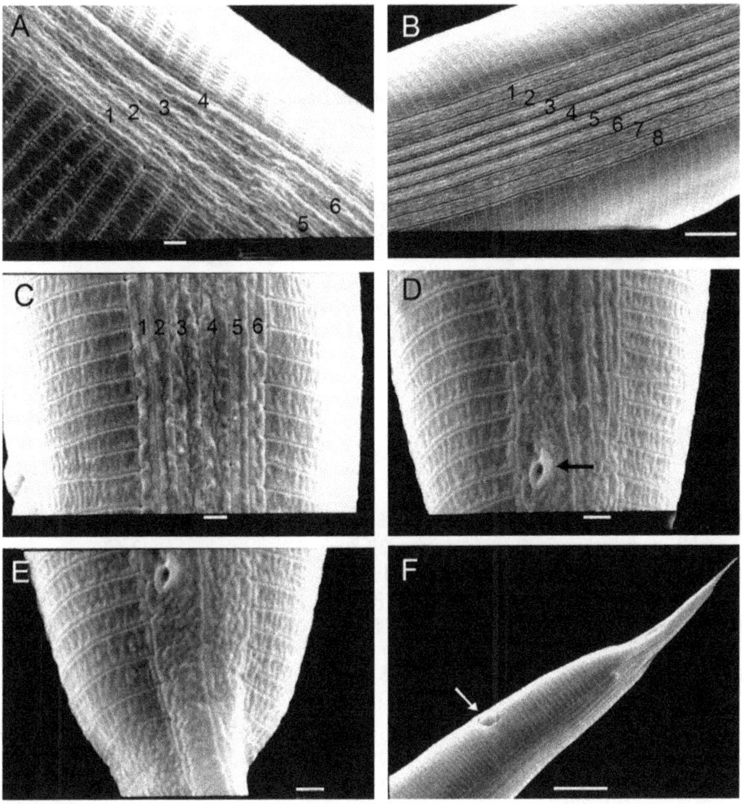

Fig. 4.17. *SEM of* Steinernema brazilense. *A-E: Lateral field showing arrangement of ridges and phasmid; F: Posterior end of infective juvenile. (Scale bars: A, C-E = 1 μm; B = 5 μm; F = 10 μm.) Modified after Nguyen* et al. *(2010).*

present. Gubernaculum with anterior part bending ventrally. Cuneus arrowhead-shaped.

Male (second generation)

Second generation male morphologically similar to first generation male but smaller. Mucron on tail terminus present.

Female (first generation)

Anterior region typical for *Steinernema* female. Excretory pore located posterior to mid-pharynx but anterior to nerve ring. Vulva a transverse slit situated on a protruding area with or without small epiptygmata. Tail usually shorter than anal body diam., mostly variable

Table 4.6. *Morphometrics of* Steinernema brazilense. *All measurements are in μm and in the form: mean ± s.d. (range).*

Character	First generation			Second generation		Infective juvenile
	Holotype (male)	Male	Female	Male	Female	
n	–	20	20	20	20	20
L	1692	1545 ± 98 (1382-1734)	5369 ± 2090 (3654-13108)	1169 ± 68 (1019-1280)	2642 ± 269 (1296-2654)	1157 ± 72 (1023-1284)
a	–	–	–	–	–	27.6 ± 1.9 (25.5-31)
b	–	–	–	–	–	7.8 ± 0.5 (7-9)
c	–	–	–	–	–	13 ± 0.8 (12-14.5)
V	–	–	53 ± 12 (37-80)	–	50 ± 12 (42-62)	–
Body diam.	80	93.6 ± 5.4 (78-99)	219 ± 83 (170-454)	64 ± 3.5 (58-70)	101 ± 14 (79-139)	42 ± 3 (39-49)
Excretory pore	117	115 ± 6.6 (102-128)	125 ± 30 (97-223)	111 ± 6.4 (101-124)	107 ± 5 (96-113)	94 ± 4.5 (87-102)
Nerve ring	130	131 ± 7.7 (117-144)	159 ± 24.5 (131-218)	136 ± 16 (114-168)	136 ± 9 (118-151)	114 ± 5 (108-121)
Pharynx	161	171 ± 9 (154-194)	214 ± 30.5 (178-280)	175 ± 21 (145-215)	185 ± 8 (174-200)	148 ± 5 (139-154)
Tail length	41	33 ± 1.7 (31-37)	44 ± 6.6 (33-59)	35 ± 6 (20-50)	71 ± 8 (62-82)	88 ± 7 (80-104)
Anal body diam.	39	43 ± 2.3 (41-47)	53 ± 12 (37-80)	39 ± 2 (35-41)	41 ± 4 (36-45)	24 ± 0.9 (22-25)
Spicule length	88	83 ± 4 (75-89)	–	73 ± 4 (67-80)	–	–
Gubernaculum length	61	47 ± 3.8 (41-56)	–	42 ± 2.6 (38-48)	–	–
D%	73	68 ± 5.3 (57-80)	58 ± 7.7 (52-80)	64 ± 7.7 (49-75)	58 ± 3 (53-64)	63 ± 3.3 (58-70)
E%	–	–	–	–	–	106 ± 6.7 (95-118)
SW%	225	192 ± 14 (158-208)	–	188 ± 10 (170-210)	–	–
GS%	69	56 ± 10 (48-65)	–	57 ± 10 (50-70)	–	–
H%	–	–	–	–	–	55 ± 3.3 (51-61)

with blunt tip, sometimes V-shaped. Ventral postanal swelling present or absent.

Female (second generation)

Similar to first generation female but much smaller. Tail tapering to a pointed end, longer than anal body diam., ventral postanal swelling present.

Infective juvenile

Anterior region typical for *Steinernema* IJ. Pharynx with slightly expanded corpus, isthmus distinct and surrounded by nerve ring. Excretory pore located anterior to nerve ring and anterior to mid-pharynx (D% = 63). Hemizonid distinct, lens-like in shape, located near end of basal bulb. Lateral field beginning anteriorly with one line from annule 3 or 4. At annule 7 or 8, two additional lines appearing to form two ridges. Near excretory pore number of ridges in lateral field increasing from two to four. A short distance posteriorly, this number becoming six then eight. Between anus and phasmid, number of ridges in lateral field becoming six. At level of phasmids ridges reducing to two. With this arrangement, lateral field formula = 2, 4, 6, 8, 6, 2. Phasmids pore-like, near mid-tail, located on ventral ridge. Tail attenuated, tapering gradually. Hyaline region occupying *ca* 55% of tail length.

TYPE HABITAT AND LOCALITY

The host of this nematode in nature is unknown as it was recovered from soil using *G. mellonella* larvae as bait. The soil sample was collected from a natural forest near Porto Murtinho city (22°10′16″S, 57°07″W), Mato Grosso state, Brazil.

TYPE MATERIAL

Holotype (male, first generation) isolated from haemocoel of *G. mellonella*. A vial of the mixtures of adults and IJ in 4% formalin are deposited in the United States Department of Agriculture Nematode Collection (USDANC), Beltsville, MD, USA. Several males, females and IJ in 4% formalin deposited in the Department of Entomology and Nematology, University of Florida, Gainesville, FL, USA.

Relationships

Steinernema brazilense is characterised morphologically by a combination of features of the various developmental stages of the nematode. For IJ, body length averaging 1157 (1023-1284) μm; distance from anterior end to excretory pore of 94 (87-102) μm, from anterior end to end of pharynx of 148 (139-153) μm; tail length of 84 (80-104) μm; and D% and E% values of 63.3 (58.0-70.0) and 106.0 (95.0-118.0), respectively. Lateral field pattern variable, formula = 2, 4, 6, 8, 6, 2. The portion with eight ridges is the longest. For male, the diagnostic characters include spicule length of 83 (75-89) μm, D% *ca* 65, and SW% *ca* 192. The length of the spicule head region is greater than its width and the lateral field has a single narrow ridge. First generation females are characterised by a vulva with or without epiptygmata. Ventral postanal swelling present. More information on relationships is available in Nguyen *et al.* (2010).

Molecular characterisation

The sequence of ITS regions (accession no. FJ410325) of *S. brazilense*, flanked by primers 18S and 26S is characterised by its sequence length 1019 bp, ITS1 = 284 bp, ITS2 = 319 bp. For the D2-D3 regions (accession no. FJ410326), the sequence length is 855 bp. More information about molecular characterisation and phylogenetic analysis is available in Nguyen *et al.* (2010).

Symbiotic bacterium

No information.

Bionomics

No information.

Distribution

Steinernema brazilense was collected in Mato Grosso state, Brazil.

Steinernema cameroonense[16,17] Kanga, Trinh, Waeyenberge, Spiridonov, Hauser & Moens, 2012

(Figs 4.18-4.20)

MEASUREMENTS

See Table 4.7.

DESCRIPTION

Male (first generation)

Anterior region typical for *Steinernema* male. Body curved ventrally posteriorly, broadly J-shaped when heat-killed. Cuticle with faint transverse striations, poorly seen under light microscope, but visible with SEM. Lateral fields present at mid-body with one narrow ridge. Perioral disc not observed. Excretory pore located anterior to nerve ring, slightly posterior to middle of pharynx. Excretory canal strongly cuticularised. Deirids not observed. Testis reflexed, comprising germinal zone with spermatocytes poorly distinguishable at level of flexure, growth zone posteriorly filled with spermatids, vacuolated *vas deferens*, and paired spicules. Spicules moderately curved, 15 μm long and labial region 12 μm wide. Velum 5 μm wide, ending 2.5-12 μm from rounded tip

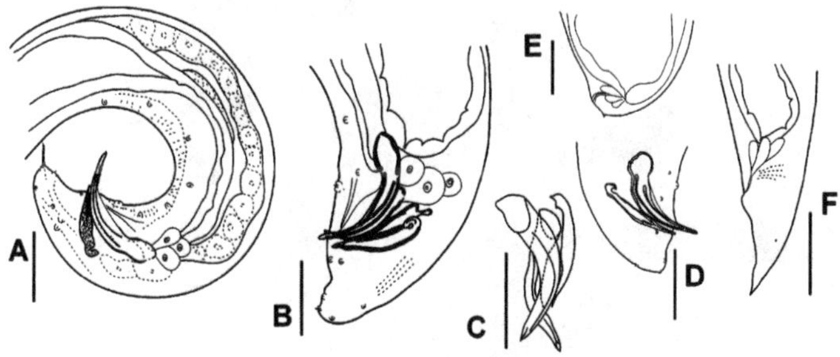

Fig. 4.18. Steinernema cameroonense. *A: First generation male, tail; B: Second generation male, posterior end; C, D: First generation male, spicules and gubernaculum; E: First generation female, tail; F: Second generation female, tail. (Scale bars = 50 μm.) After Kanga* et al. *(2014), courtesy* Russian Journal of Nematology.

[16] Specific epithet named after *Cameroon*, the country in which the species was found.

[17] Description after Kanga *et al.* (2012). Revised by Sergei E. Spiridonov.

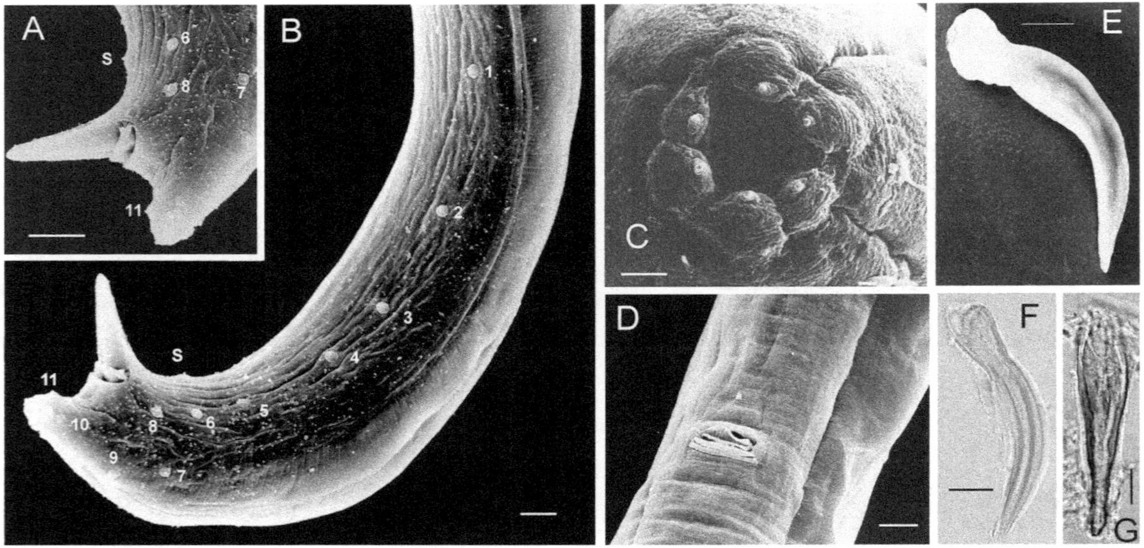

Fig. 4.19. Steinernema cameroonense. *A: First generation male, tail; B: First generation male, posterior end; C: First generation female, labial region, apical view; D: First generation female, vulva; E, F: First generation male, spicule; G: First generation male, gubernaculum. (Scale bars: A, B, D-G = 10 μm; C = 1 μm.) After Kanga* et al. *(2012), courtesy* Russian Journal of Nematology.

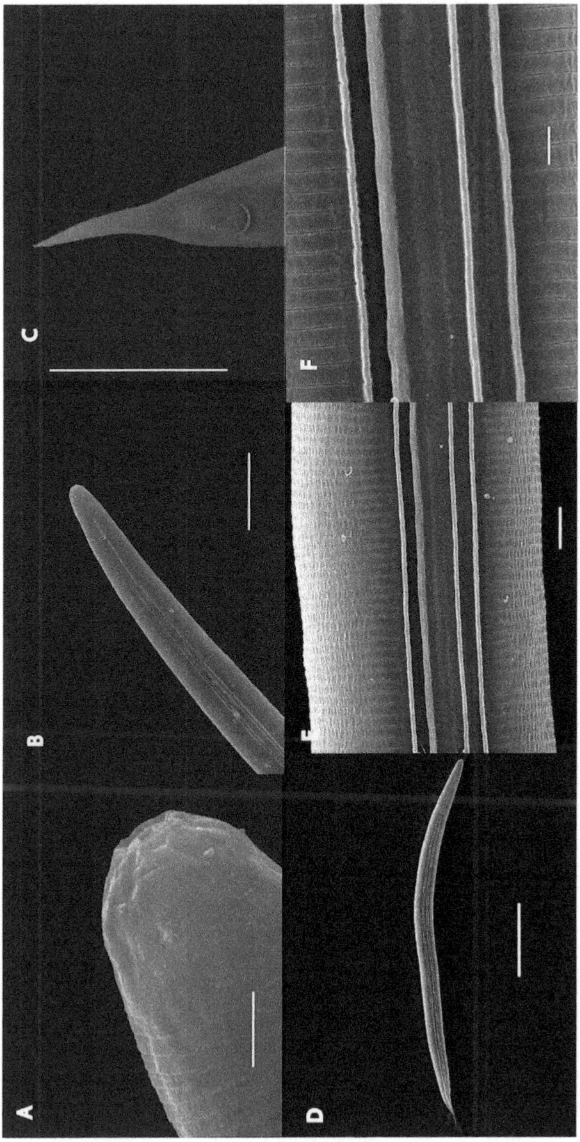

Fig. 4.20. Steinernema cameroonense, *infective juvenile. A: Labial region, lateral view; B: Anterior end, lateral view; C: Tail, ventral view; D: Entire view; E: Lateral field in middle of body; F: Ridges of lateral field. (Scale bars: A = 3 μm; B, C = 30 μm; D = 100 μm; E = 5 μm; F = 2 μm.) After Kanga* et al. (2014), *courtesy* Russian Journal of Nematology.

Table 4.7. *Morphometrics of type population of* Steinernema cameroonense. *All measurements are in μm and in the form: mean ± s.d. (range).*

Character	First generation			Second generation		Infective juvenile
	Holotype (male)	Male	Female	Male	Female	
n	–	20	20	20	20	20
L	1373	1331 ± 193 (1019-1718)	3168 ± 606 (2423-4607)	1018 ± 119 (813-1193)	2236 ± 223 (1790-2539)	622 ± 61 (490-694)
a	30	32 ± 4.1 (26-40)	18 ± 2.0 (15-22)	27 ± 2.7 (22-31)	22 ± 1.8 (18-25)	21 ± 1.9 (17-25)
b	9	9 ± 1.2 (7-12)	18 ± 2.7 (14-24)	8 ± 1.0 (7-10)	14 ± 1.4 (11-17)	6 ± 0.5 (5-6)
c	33	35 ± 4.5 (28-43)	85 ± 17.4 (58-112)	30 ± 3.6 (24-38)	43 ± 20 (25-107)	9 ± 1.6 (6-12)
V	–	–	52 ± 2.0 (48-57)	–	53 ± 2.5 (46-58)	–
Max. body diam.	103	90 ± 21 (65-124)	172 ± 25 (110-231)	60 ± 7.6 (51-84)	102 ± 9.7 (83-120)	30 ± 2.8 (24-35)
Excretory pore	102	93 ± 11.5 (65-109)	104 ± 21.0 (74-140)	85 ± 10.8 (69-115)	84 ± 11.6 (64-106)	54 ± 4.9 (45-64)
Pharynx	148	146 ± 8.8 (127-160)	173 ± 11.9 (152-198)	125 ± 11.4 (114-158)	155 ± 11.8 (131-179)	113 ± 5.6 (105-125)
Tail length	42	38 ± 2.9 (34-43)	38 ± 5.2 (30-52)	34 ± 2.7 (26-39)	58 ± 14.9 (19-81)	76 ± 18.7 (52-107)
Hyaline region	–	–	–	–	–	31 ± 4.8 (22-38)
Anal body diam.	46	41 ± 3.3 (36-48)	44 ± 7.8 (30-60)	38 ± 2.5 (35-45)	42 ± 5.4 (30-52)	15 ± 1.5 (12-18.7)
Spicule length	75	69 ± 7.9 (51-85)	–	70 ± 5.3 (56-82)	–	–
Gubernaculum length	50	45 ± 6.4 (37-57)	–	39 ± 5.7 (29-54)	–	–
D%	68.72	64 ± 6.3 (48-76)	60 ± 10.6 (43-79)	68 ± 6.5 (57-82)	54 ± 7.5 (41-66)	48 ± 4.0 (42-56)
E%	–	–	–	–	–	75 ± 18.3 (48-116)
SW%	162	170 ± 16.8 (131-201)	–	182 ± 10.3 (159-199)	–	–
GS%	65.87	64 ± 6.8 (47-76)	–	56 ± 5.0 (45-66)	–	–
H%	–	–	–	–	–	50 ± 4.7 (40-59)

of spicules. Manubrium slightly longer than wide (spicule head region length/width = 0.9-1.5), shaft very short. Gubernaculum boat-shaped in lateral view, anterior end rounded, needle-shaped cuneus present. Copulatory papillae totalling 22 + 1, comprising six pairs precloacal subventral papillae, single precloacal midventral papilla, one pair lateral, one pair adcloacal and three pairs postcloacal papillae, including two subterminal pairs and one subdorsal pair. Tail conoid with mucron usually present, as indicated in original description. In another population of this species (Kanga *et al.*, 2014) a mucron is absent, and when present, length not exceeding 3 μm. Phasmids inconspicuous.

Male (second generation)

Main morphological features similar to those of first generation male except for shorter body length and smaller body diam. Mucron 3-10 μm long, always present.

Female (first generation)

Anterior region typical for *Steinernema* female. Body C-shaped when killed by gentle heat. Cephalic end rounded, *ca* 26 μm diam. Cuticle smooth under light microscope but with fine annulation visible with SEM, well discernible near vulval lips. Corpus of pharynx increasing in diam. from 16 μm at stoma to 30 μm before isthmus. Isthmus 20 μm wide. Basal bulb 37 μm diam., with transparent 11 μm long lobes to cardia. Excretory duct 2-6 μm wide. Excretory gland not observed. Lateral field with one ridge. Vulva in form of a transverse slit with lips protruding up to 9-27 μm, unequal in size, with long double-flapped epiptygmata surrounded by two lamella-like structures on both sides and with surrounding annules. Tail conical, pointed with 3-9 μm long mucron on tip of type specimens, and 1-3 μm for specimens from another locality. Postanal swelling visible in fully mature females.

Female (second generation)

Similar to first generation female, but smaller and more slender. Tail conical, pointed or rounded with mucron *ca* 10 μm long.

Infective juvenile

Anterior region typical for *Steinernema* infective juvenile. Body elongated, on average 768 μm long, bow-shaped when heat-killed, usually exsheathed from second-stage cuticle. IJ of another population of this species with noticeably larger body size and longer pharynx (Kanga *et al.*, 2014). Cuticular striations (distinct under SEM) *ca* 1.5 μm wide at mid-body, but lip region and tail terminus smooth. Lateral fields consisting of six, unequally spaced and developed ridges in mid-body region (*i.e.*, seven lines). Only submarginal ridges prominent. Lateral field beginning with irregular cuticular depressions (lines) close to eight annules from anterior end; a short distance posteriorly, two ridges appearing and changing to five (six lines) at about excretory pore level; more posteriorly still central ridge dividing into two, thereby making a total of six ridges. In posterior part of body, number of ridges reducing gradually. Anterior to anus, two middle ridges merging to leave five ridges in lateral field. At phasmid opening, lateral field reduced to two, and at tail dorsal constriction to one ridge, extending almost to tail tip. With above arrangement, lateral field formula = 2, 5, 6, 5, 2, 1. Deirids not observed. Hemizonid visible as a 4-5 μm wide cuticular depression usually located between anterior of basal bulb and nerve ring. Pharynx collapsed, not well seen, corpus slender, cylindrical, isthmus indistinct, surrounded by nerve ring. Excretory pore located at level of mid-pharynx (D% = 46). Cardia *ca* 5-9 μm long, bacterial vesicle usually not well seen, ellipsoid, *ca* 30 × 10 μm. Rectum *ca* 75% of anal body diam. long. Tail conoid with characteristic dorsal constriction situated slightly anterior to proximal portion of hyaline region, in turn occupying *ca* 50% of tail length, terminus sharply pointed. Anus wide, sickle-shaped. Phasmids located at *ca* 45-50% of tail length posterior to anus, with aperture near beginning of dorsal constriction.

TYPE HABITAT AND LOCALITY

Isolated from soil collected from a teak plantation (latitude 4°30′N, longitude 12°01′E) at Obala, in the central region of southern Cameroon.

TYPE MATERIAL

Holotype male, five male paratypes, four female paratypes and five IJ paratypes deposited in the nematode collection of the Institute of Zoology, Ghent University, Ghent, Belgium.

RELATIONSHIPS

Steinernema cameroonense is characterised by a combination of features of the various developmental stages of the nematode. The IJ have a body length averaging 622 (490-693) μm, distance from anterior end to excretory pore of 54 (45-64) μm, from anterior end to the base of pharynx of 108 (100-114) μm, tail length of 76 (52-107) μm, D% and E% values of 48 (42-56) and 75 (48-116), respectively. The lateral field pattern of the new species is 2, 5, 6, 3 and 2. The first generation males of the new species can be recognised by a spicule length of 69 (51-85) μm, GS% = 64 (47-76), SW% = 170 (131-200), and D% = 64 (48-76). Females are recognised by a conical and pointed tail with a mucron on the tip and a protruding asymmetrical vulva with a long double-flapped epiptygmata surrounded by two lamella-like structures on both sides and with surrounding annules.

Steinernema cameroonense is similar to *S. nyetense*, *S. monticolum*, *S. robustispiculum*, *S. ashiuense* and *S. rarum*.

MOLECULAR CHARACTERISATION

The 769 bp long ITS rDNA sequence of *S. cameroonense* (JX985267) and 593 bp long sequence of D2-D3 expansion segment of LSU rDNA (JX985265) are deposited in NCBI GenBank. The sequences of this species together with that of another Cameroonian species (*S. nyetense*) form a well supported *Cameroonense*-clade with *S. sacchari* from South Africa (Nthenga *et al.*, 2014). The relationships of this clade with other groups of steinernematid species are not resolved.

SYMBIOTIC BACTERIUM

No information.

BIONOMICS

The possible hosts in nature are unknown. Cross-hybridisation tests with *S. feltiae*, *S. nyetense* and *S. everestense* (= junior synonym of *S. akhursti*) produced no progeny (Kanga *et al.*, 2012).

DISTRIBUTION

Another isolate of this species was also obtained from a soil sample collected near Obala, southern Cameroon (4°13′N, 11°35′E) (Kanga *et al.*, 2014).

Steinernema changbaiense[18,19] **Ma, Chen, De Clercq, Han & Moens, 2012**
(Figs 4.21, 4.22)

MEASUREMENTS

See Table 4.8.

DIAGNOSTIC DESCRIPTION

Male (first generation)

Anterior region typical for *Steinernema* male. Body curved ventrally, tapering toward anterior end, mostly C-shaped, rarely J-shaped in heat-relaxed specimens. Cuticle with very faint transverse striae. Excretory

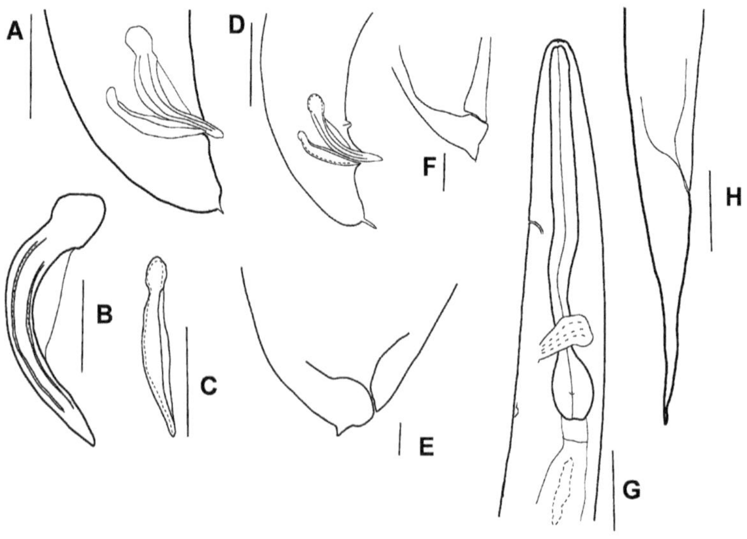

Fig. 4.21. Steinernema changbaiense. *A: First generation male, tail; B: First generation male, spicule; C: First generation male, gubernaculum; D: Second generation male, tail; E: First generation female, tail; F: First generation female, tail; G: Infective juvenile, anterior end; H: Infective juvenile, tail. (Scale bars: A, D-F = 50 μm; B, C, G, H = 25 μm.) After Ma* et al. *(2012a), courtesy* Russian Journal of Nematology.

[18] Specific epithet named after *Changbai* mountain range on the borders of China and North Korea, the locality from where the species was obtained.
[19] Description after Ma *et al.* (2012a). Revised by Sergei E. Spiridonov.

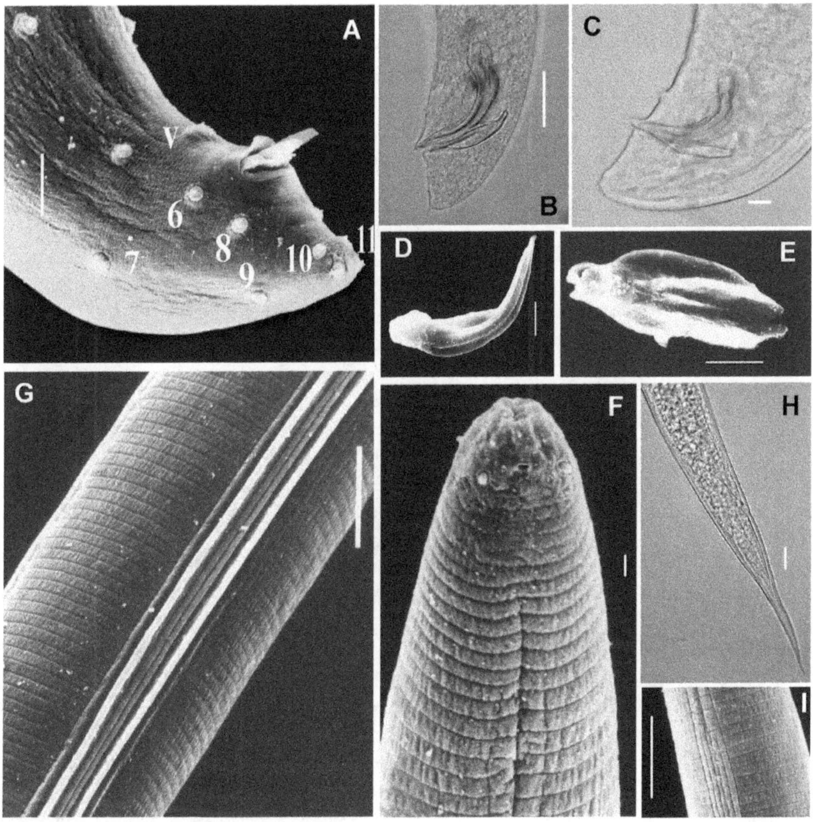

Fig. 4.22. Steinernema changbaiense. *A: First generation male, tail; B: Second generation male, tail; C: First generation male, tail; D: First generation male, spicule; E: First generation male, gubernaculum; F: Infective juvenile, labial region; G: Infective juvenile, lateral field; H: Tail of infective juvenile with hyaline region; I: Lateral field at excretory pore level. (Scale bars: A, C-I = 10 μm; B = 20 μm.) After Ma* et al. *(2012a), courtesy* Russian Journal of Nematology.

pore always located anterior to nerve ring (ranging from 52 to 66% of length from anterior body end to base of pharynx). Intestine generally with a wide lumen. Distinct, papilliform deirid situated at between pharyngeal bulb and nerve ring. Well developed and ventrally reflexed testis. Paired symmetrical, moderately curved and slender spicules, golden/yellowish in colour. Spicule length/width ratio (SL/SPW) = 7.0 (5.2-8.5). Manubrium region of spicules usually elongate (manubrium

Table 4.8. *Morphometrics of type population of* Steinernema changbaiense. *All measurements are in μm and in the form: mean ± s.d. (range).*

Character	First generation			Second generation		Infective juvenile
	Holotype (male)	Male	Female	Male	Female	
n	–	20	20	20	20	20
L	1452	1455 ± 165 (1140-1697)	4368 ± 439 (3273-5113)	1081 ± 49 (990-1300)	2470 ± 367 (2025-3240)	757 ± 34 (730-790)
a	12	13 ± 1.1 (11-14)	17 ± 2.1 (14-23)	13 ± 1.5 (11-16)	15 ± 1.6 (12-17)	22 ± 0.8 (21-23)
b	9.5	10 ± 1.0 (8.9-12)	22 ± 1.7 (19-24)	7.5 ± 0.7 (6.3-8.8)	12 ± 1.3 (10-14)	6.5 ± 0.3 (6.0-7.1)
c	43	45 ± 6.5 (33-54)	98 ± 25 (73-149)	39 ± 5.0 (32-47)	47 ± 7.2 (39-58)	9.8 ± 0.6 (8.7-11)
c′	0.8	0.8 ± 0.07 (0.6-0.9)	0.7 ± 0.2 (0.4-0.8)	0.9 ± 0.1 (0.7-1.0)	1.2 ± 0.2 (0.9-1.6)	4.1 ± 0.3 (3.7-4.6)
V	–	–	56 ± 2.4 (53-60)	–	57 ± 2.1 (55-61)	–
Max. body diam.	117	126 ± 20 (96-139)	261 ± 31 (201-308)	81 ± 9.6 (72-96)	175 ± 43 (140-230)	35 ± 1.6 (32-38)
Nerve ring	88	84 ± 4.9 (74-92)	97 ± 12 (80-110)	80 ± 6.2 (74-94)	104 ± 9 (90-112)	61 ± 3.7 (58-68)
Pharynx	153	145 ± 20 (130-159)	205 ± 8.9 (190-223)	144 ± 10 (131-160)	195 ± 15 (179-218)	117 ± 5.1 (109-125)
Tail length	34	34 ± 3.3 (27-41)	50 ± 8.7 (30-63)	28 ± 3.4 (23-33)	52 ± 5.4 (44-58)	77 ± 4.4 (71-85)
Hyaline region		–	–	–	–	35 ± 4.0 (30-40)
Anal body diam.	43	42 ± 4.3 (37-50)	69 ± 7.2 (57-80)	32 ± 2.3 (29-37)	43 ± 7.1 (35-56)	19 ± 1.0 (17-20)
Spicule length	66	63 ± 4.8 (56-70)	–	53 ± 6.2 (40-61)	–	–
Gubernaculum length	41	40 ± 3.7 (33-46)	–	31 ± 2.7 (27-34)	–	–
D%	60	58 ± 4.9 (52-66)	49 ± 7.2 (35-55)	56 ± 6.1 (46-66)	52 ± 5.7 (44-56)	53 ± 3.6 (48-58)
E%	274	269 ± 35 (203-318)	224 ± 44 (182-261)	276 ± 39 (220-360)	210 ± 29 (188-250)	80 ± 6.1 (73-88)
SW%	151	149 ± 15 (114-156)	–	169 ± 20 (130-203)	–	–
GS%	62	65 ± 4.7 (54-72)	–	59 ± 5.4 (51-66)	–	–
H%	–	–	–	–	–	45 ± 3.8 (41-49)

length/width = 1.2-1.5), slightly offset from shaft, calomus very short. Lamina slender and well curved, with two internal sclerotised ribs, tapering slightly posteriorly. Spicule terminus blunt. Rostrum not well developed, velum short. In lateral view, gubernaculum boat-shaped to almost straight, tapering gradually posteriorly, cuneus V-shaped, corpus with two wings opening posteriorly. Twenty-three genital papillae comprising 11 pairs and a single midventral papilla located just anterior to cloacal opening. Seven pairs located precloacal (six pairs subventral and one pair lateral), one pair adcloacal and three pairs postcloacal, consistently near tail terminus (one subdorsal, two at tail tip). Tail bluntly conoid, usually concave at ventral side, small mucron mostly present. Phasmids located just posterior to cloacal opening.

Male (second generation)

Similar to first generation, but smaller in size. Filamentous tail mucron much longer than in first generation, *ca* 4.4-9.5 μm long. Spicules and gubernaculum shorter and thinner.

Female (first generation)

Anterior region typical for *Steinernema* female. Body size variable, usually C-shaped on heat relaxation. Cuticle only faintly striated. Lateral field not observed. Cheilorhabdions well sclerotised. Morphology of pharynx similar to male. Excretory pore located anterior to nerve ring. Cardia prominent, protruding into intestinal lumen. Gonads amphidelphic, reflexed dorsally. Vulva mostly symmetrical, slightly protruding from body contour, occasionally forms observed where it is well protruding or non-protruding. Vulval opening slightly posterior to mid-body. Vagina short, leading into paired uteri. Rectum narrow, anal opening distinct. Tail blunt, conoid, shorter than anal body diam. Tail tip with a small pointed mucron. Postanal swelling not present.

Female (second generation)

Similar to first generation in general morphology, but shorter and narrower. Body arcuate when heat-relaxed. Vulva with protruding, symmetrical lips. Tail conical, bearing a terminal mucron.

Infective juvenile

Anterior region typical for *Steinernema* infective juvenile. Body of heat-relaxed specimens almost straight or slightly curved, slender, slightly tapering towards anterior and posterior ends. Body cuticle with fine transverse striae. Lateral field beginning with one cuticular depression or line at 7-8 annules from anterior end. At 12-13 annules two additional lines appearing posteriorly forming two thin ridges. Near excretory pore, number of ridges in lateral fields increasing from two to five. The five ridges remain unchanged for a short distance, then the central ridge dividing into two over a short distance after deirid, making a total of six, the maximum number of ridges in lateral field. Two submarginal ridges more distinct than the others. Number of ridges remaining unchanged over long distance. Two central ridges disappear forming four ridges between anus and phasmids. A very short distance posterior to phasmid only two prominent ridges are observed. With the above arrangement, the lateral field formula = 2, 5, 6, 4, 2. Excretory pore anterior to nerve ring. Hemizonid distinct, located anterior to base of basal bulb. Nerve ring encircling narrow isthmus. Bacterial vesicle dorsally displaced, just posterior to cardia, elongate (4-5 μm in width, 16-22 μm in length), containing rod-shaped bacteria. Tail long conoid with pointed terminus, hyaline region occupying less than half of tail length. Phasmid pore-like, located 30-40% of tail length posterior to anus.

TYPE HABITAT AND LOCALITY

The natural host is unknown as the species was isolated by *Galleria* baiting a soil sample taken in Benxi County, Liaoning province, China (40.89°N, 124.03°E). The type locality is characterised by hills with shrubs, grasses, locust tree and pines. The soil is humid, loamy, under dense grass and tree leaves with moderate shading by pines.

TYPE MATERIAL

Holotype male, four male paratypes, six female paratypes and ten IJ paratypes deposited in the nematode collection of the Institute of Zoology of Ghent University, K.L. Ledeganckstraat 35, Ghent, Belgium. Slides with ten first generation male paratypes, nine first generation female paratypes and 12 IJ paratypes are also deposited in the Institute of Plant Protection, Hebei Academy of Agricultural and Forestry Sciences, Baoding, Hebei, China.

RELATIONSHIPS AND MOLECULAR CHARACTERISATION

The 883 bp long sequence of ITSrDNA (JN865168) of this species is deposited in NCBI GenBank under the name '*Steinernema* sp. JM-2011'. The 909 bp long sequence of D2-D3 LSU rDNA of this species (JN830950) is deposited in NCBI GenBank under the name '*Steinernema* sp. 3 JM-2011'. *Steinernema changbaiense* is a member of the *monticolum*-group of species and demonstrates a high level of similarity with *S. monticolum*, *S. ashiunense* and *S. robustispiculum*. *Steinernema changbaiense* differs from *S. robustispiculum*, its closest taxon, by 14 bp in the ITS rDNA sequence. *Steinernema ashiunense* is the most similar species in the sequence of the D2-D3 region of LSU rDNA, differing by 4 bp.

BIONOMICS

Steinernema changbaiense has a life cycle comparable to that of other described species of the *krausseilfeltiaeloregonense*-group. *Galleria mellonella* larvae were killed after 2 days and first generation adults developed after 4-5 days at 25°C. Second generation adults developed after 10 days. It usually took more than 12 days for IJ to emerge from the insect cadavers.

DISTRIBUTION

Only known from Liaoning Province, Northeast China.

Steinernema cholashanense[20,21] Nguyen, Půža & Mráček, 2008
(Figs 4.23-4.25)

MEASUREMENTS

See Table 4.9.

[20] The specific epithet is derived from the *Chola Shan* Mountains, China, where the nematode was collected.

[21] Description after Nguyen *et al.* (2008). Revised by Khuong B. Nguyen.

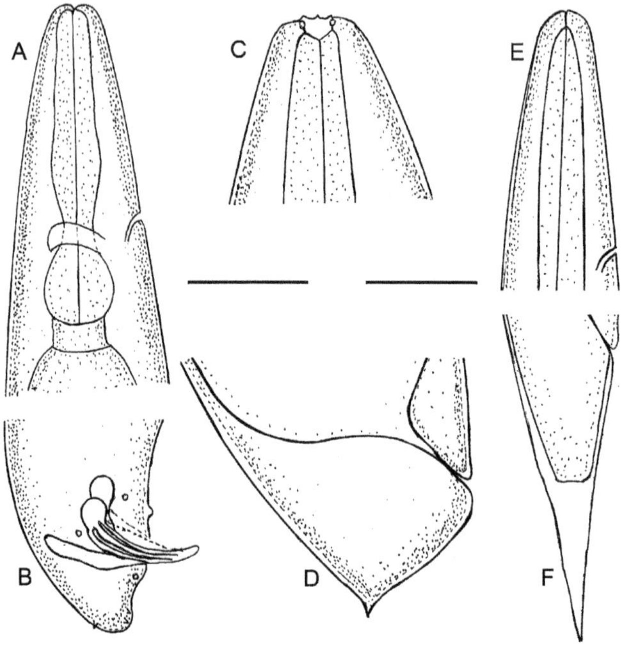

Fig. 4.23. Steinernema cholashanense. *Anterior and posterior regions of male (A, B), female (C, D) and infective juvenile (E, F). (Scale bars: A-D (left bar) = 50 μm; E, F (right bar) = 25 μm.) Modified after Nguyen* et al. *(2008), with acknowledgement to* Journal of Invertebrate Pathology.

Diagnostic description

Male (first generation)

Anterior region typical for *Steinernema* male. Excretory pore mostly at metacorpus and always anterior to nerve ring. Well developed and reflexed testis. Testis reflexion ventral in position and about half of testis length. Paired, symmetrical, moderately curved spicules, spicule tip bluntly pointed. Spicule manubrium usually rounded, occasionally tapering anteriorly with length:width ratio of 1:1; manubrium less than one-fourth of spicule length (*ca* 20-23%); calomus distinct. Lamina well curved anteriorly and dorsally with two internal ribs, slender and almost straight posteriorly. Rostrum sometimes well developed, slightly narrower than manubrium width, velum bow-like and distinct, extending from basal part of rostrum to distal part of lamina. In lateral view, gubernaculum boat-shaped, anterior part mostly straight, rarely hooked, sometimes slightly recurved, posterior part usually tapering abruptly. In

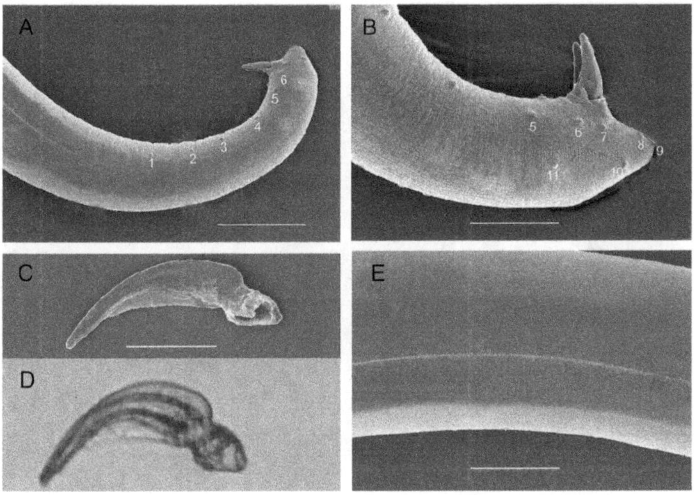

Fig. 4.24. Steinernema cholashanense. *A, B: SEM photographs of posterior regions of male showing 11 pairs of genital papillae and ends of spicules; C, D: SEM and LM photographs of spicule; E: SEM of male body showing lateral field with single narrow ridge (line). (Scale bars: A = 66.7 μm; B = 27.3 μm; C, D = 20 μm; E = 30 μm.) Modified after Nguyen* et al. *(2008), with acknowledgement to* Journal of Invertebrate Pathology.

ventral view, cuneus small, needle-shaped, or absent, corpus with two wings. Tail bluntly conoid, usually concave on ventral side, short mucron present, rarely not developed. There are 23 genital papillae comprising 11 pairs and a single ventral papilla located just anterior to cloacal opening. Of these, six pairs are located precloacal (five pairs subventral and one pair lateral), two pairs adcloacal, two pairs subterminal, and one pair subdorsal. Most posterior three pairs are consistently located near tail tip.

Male (second generation)

Similar to the first generation males, but slightly smaller. Max. body diam. much smaller. Filamentous tail mucron much longer than in first generation.

Female (first generation)

Anterior region typical for *Steinernema* female. Excretory pore located anterior to nerve ring. Genital tract amphidelphic with reflexed ovaries. Vulva mostly symmetrical and in form of a transverse slit,

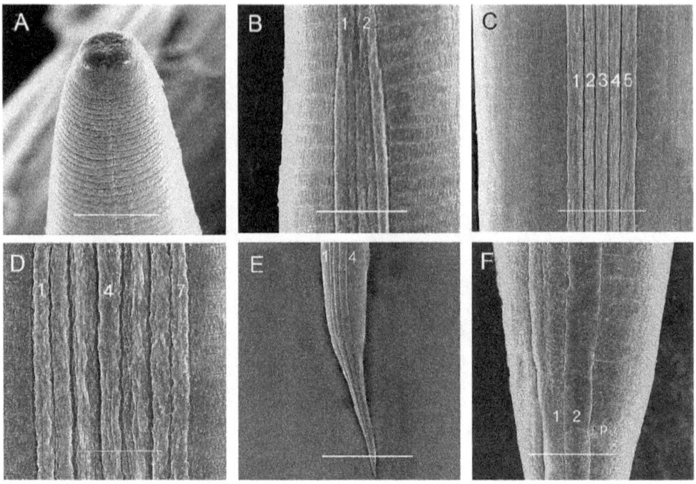

Fig. 4.25. Steinernema cholashanense. *A-F: SEM photographs of lateral field of infective juvenile showing arrangement of ridges from labial region to tail. (Scale bars: A = 5.6 μm; B, C = 6 μm; D, F = 5 μm; E = 23 μm.) Modified after Nguyen* et al. *(2008), with acknowledgement to* Journal of Invertebrate Pathology.

slightly protruding from body contour (depending on age), well protruding and non-protruding forms sometimes observed. Vulval opening slightly posterior to mid-body. Small epiptygmata present. Short vagina leading into paired uteri. Narrow rectum, distinct anal opening. Tail of mature females obtuse with a mucron at tip. Postanal swelling present on mature females, usually not well developed, absent or slightly developed on young females. Tail tip with a pointed mucron but without papilla-like projections. Pygmy forms not observed.

Female (second generation)

Similar to first generation females in general morphology, but smaller. Vulval opening posterior to mid-body. Tail conical with a mucron at the end. Postanal swelling not developed.

Infective juvenile

Anterior region typical for *Steinernema* IJ. Excretory pore in mid-pharynx region Lateral field beginning anteriorly with one line at first annule. At annule 14 or 15, two additional lines appearing to form two ridges. Near excretory pore, number of ridges in lateral fields increasing from two to five. Five ridges remaining unchanged for a short distance

Table 4.9. *Morphometrics of* Steinernema cholashanense *(holotype is the infective juvenile). All measurements are in μm and in the form: mean ± s.d. (range).*

Character	Holotype (IJ)	First generation Male	First generation Female	Second generation Male	Second generation Female	Infective juvenile
n	–	20	20	20	20	20
L	820	1428 ± 212 (1070-1778)	4692 ± 784 (3232-6363)	813 ± 34 (727-869)	2054 ± 241 (1555-2586)	843 ± 40 (727-909)
a	34	11 ± 3.5 (7.8-24)	19 ± 3.2 (13-23)	13 ± 1.8 (9.8-16)	15 ± 1.2 (12-17)	28 ± 2.5 (24-34)
b	7.5	9.3 ± 0.9 (7.5-11.2)	25 ± 3.4 (18-32)	6.3 ± 0.4 (5.8-7.1)	13 ± 1.1 (9.9-14)	6.8 ± 0.3 (6.1-7.2)
c	13	41 ± 4.4 (36-51)	83 ± 16 (62-119)	22 ± 2.8 (18-29)	37 ± 4.0 (29-47)	12 ± 1.0 (10-14)
c′	4.1	0.7 ± 0.1 (0.6-0.9)	0.8 ± 0.1 (0.6-1.0)	1.0 ± 0.2 (0.7-1.4)	1.3 ± 0.1 (1.1-1.6)	4.3 ± 0.4 (3.5-5.0)
V	–	–	53 ± 2.0 (50-57)	–	–	–
Body diam.	24	137 ± 39 (73-204)	255 ± 43 (156-332)	62 ± 9.2 (51-83)	137 ± 20 (107-173)	30 ± 1.9 (26-35)
Excretory pore	56	99 ± 19 (75-135)	94 ± 18 (53-130)	61 ± 6 (48-70)	82 ± 6.6 (73-100)	62 ± 1.9 (59-65)
Nerve ring	84	106 ± 9.1 (91-126)	129 ± 9.5 (111-148)	98 ± 7.3 (76-110)	121 ± 11 (95-140)	87 ± 5 (72-97)
Pharynx	110	152 ± 12 (135-173)	190 ± 11 (176-223)	129 ± 6.4 (111-138)	161 ± 13 (135-191)	125 ± 4.6 (110-138)
Testis reflexion	–	484 ± 102 (333-707)	–	229 ± 12 (207-266)	–	–
Tail length	63	35 ± 3.6 (29-43)	57 ± 7.3 (46-70)	37 ± 4.6 (25-45)	56 ± 4.9 (48-70)	73 ± 6 (60-80)
Hyaline region	21	–	–	–	–	24 ± 6.7 (11-35)
Anal body diam.	15	49 ± 6.4 (38-60)	77 ± 13 (54-95)	37 ± 4.7 (32-48)	43 ± 3.3 (35-48)	17 ± 0.9 (16-19)
Spicule length	–	66 ± 3.3 (60-71)	–	53 ± 3.3 (48-61)	–	–
Spicule width	–	13 ± 0.9 (11-14)	–	10 ± 0.8 (8-11)	–	–
Gubernaculum length	–	39 ± 3.4 (32-45)	–	30 ± 2.6 (24-35)	–	–
Gubernaculum width	–	5.3 ± 0.8 (4.8-6.4)	–	7 ± 1.2 (4.8-9.5)	–	–
D%	56	64 ± 9.5 (50-85)	50 ± 8.6 (29-65)	48 ± 4.3 (40-55)	51 ± 4.4 (44-59)	49 ± 1.9 (46-53)
E%	89	285 ± 45 (196-386)	167 ± 38 (92-250)	166 ± 21 (133-215)	148 ± 11 (132-175)	81 ± 3.9 (76-91)
SW%	–	115 ± 15 (92-144)	–	119 ± 15 (87-140)	–	–
GS%	–	71 ± 5.7 (61-85)	–	67 ± 3.8 (58-73)	–	–
H%	33	–	–	–	–	39 ± 1.9 (33-47)

then two external ridges gradually dividing to make seven. About two to three anal body diam. anterior to anus, ridges two, four and six becoming gradually smaller and then disappearing to leave four ridges. Between anus and phasmid, number of ridges in lateral field becoming two. With above arrangement, lateral field formula = 2, 5, 7, 4, 2. Deirid not observed. Phasmids pore-like, posterior to anus, approximately 41% of tail length from anus. Hemizonid distinct, lens-like in shape, located at beginning of basal bulb, cuticularised. Bacterial vesicle slender, subpyriform, *ca* 25 μm long. Intestine filled with numerous fat globules, lumen not visible. Rectum straight, anus sickle-shaped. Tail conoid with a slight dorsal constriction, terminus pointed. Hyaline region comprising less than half of tail length.

TYPE HABITAT AND LOCALITY

Host unknown. The type locality is characterised by an open agricultural area in a river basin in Sichuan province along the road from Garze to Manigango, approximately 2 km west of Garze, China. The soil sample was taken from the southern slope of the Chola Shan Mountains, which are covered by xerophytic vegetation.

TYPE MATERIAL

Holotype IJ, paratype IJ, males and females of first and second generation deposited in the EPN collection at the Laboratory of Insect Pathology, Institute of Entomology, Czech Academy of Sciences (CAS), Česke Budějovice, Czech Republic; paratype IJ in the nematode collection at the Department of Entomology and Nematology, University of Florida, Gainesville, FL, USA.

RELATIONSHIPS

Morphological and molecular studies showed that *S. cholashanense* is a member of the *feltiae*-group. *Steinernema cholashanense* is characterised by morphology and morphometrics of the IJ and male. For the IJ, the species can be recognised by the body length of 843 μm, pharynx = 125 μm, H% = 39% and E% = 81%. The lateral field pattern is 2, 5, 7, 4, 2. The male of the first generation is characterised by spicule shape and length and especially by the prominent velum and the presence of a mucron in both generations. More information on the relationships is available in Nguyen *et al.* (2008).

MOLECULAR CHARACTERISATION

Steinernema cholashanense is characterised genetically by the ITS and D2-D3 regions. The sequence of ITS regions of *S. cholashanense* (accession no. EF431959), flanked by primers 18S and 26S can be recognised by its length of 965 bp, ITS1 = 265 bp, and ITS2 = 303 bp. For D2-D3 regions, the sequence of *S. cholashanense* (accession no. EF520284) is 861 bp. More information on molecular characterisation and phylogenetic analysis is available in Nguyen *et al.* (2008).

SYMBIOTIC BACTERIUM

No information.

BIONOMICS

Steinernema cholashanense reproduced well on *G. mellonella* larvae in an incubator at temperatures of 15 and 25°C. At 15°C, nematode-infected *G. mellonella* larvae died after 2 days, first generation adults appeared after another 4 days. After 7 days second generation adults filled the cadavers and after 10 days the first IJ migrated into a water trap. At 25°C, the nematodes also developed as stated above except that the IJ emerged from cadavers after 8 days.

DISTRIBUTION

Steinernema cholashanense was collected from the Chola Shan Mountains, Sichuan province, China.

***Steinernema citrae*[22,23] Stokwe, Malan, Nguyen, Knoetze & Tiedt, 2011**

(Figs 4.26-4.29)

MEASUREMENTS

See Table 4.10.

[22] Specific epithet derived from *Citrus*, the genus of the trees growing in the type locality.

[23] Description after Stokwe *et al.* (2011). Revised by Khuong B. Nguyen.

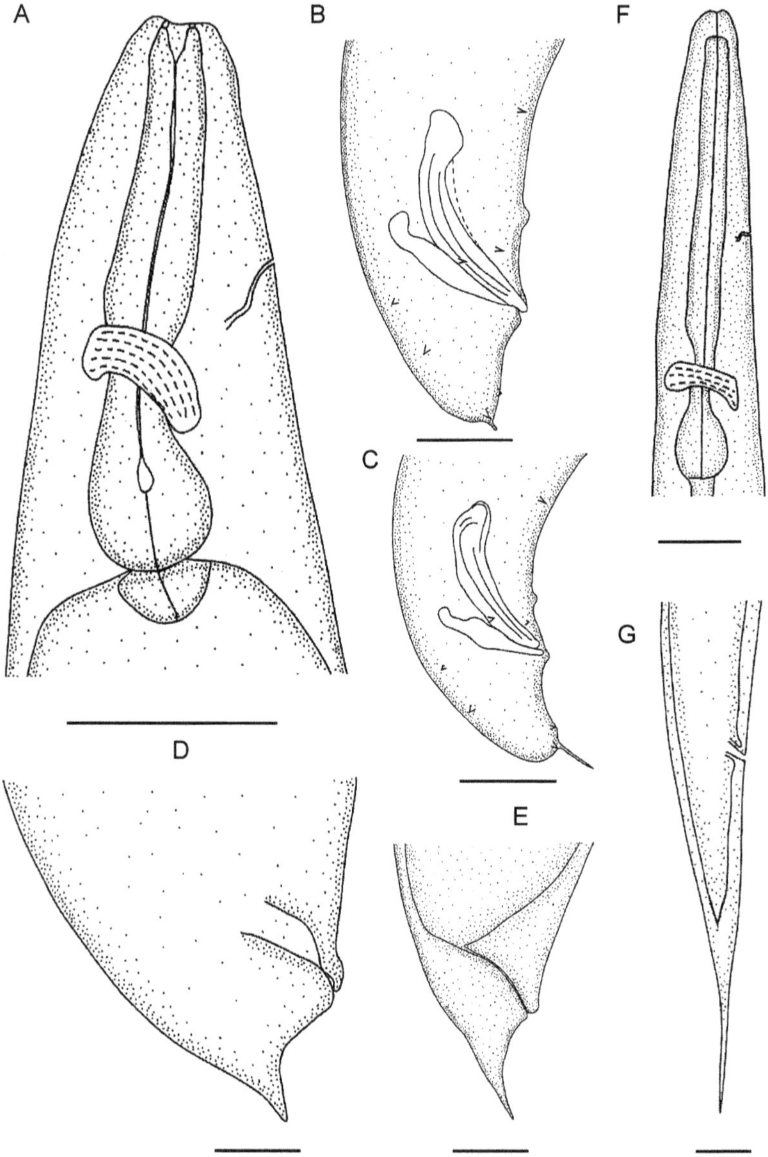

Fig. 4.26. Steinernema citrae. *A: Anterior region of first generation male; B: First generation male tail; C: Second generation male tail; D: Young first generation female tail; E: Second generation female tail; F: Pharynx region of infective juvenile; G: Tail region of infective juvenile. (Scale bars: A = 50 µm; B-E = 30 µm; F = 20 µm; G = 10 µm.) After Stokwe* et al. *(2011), courtesy* Nematology.

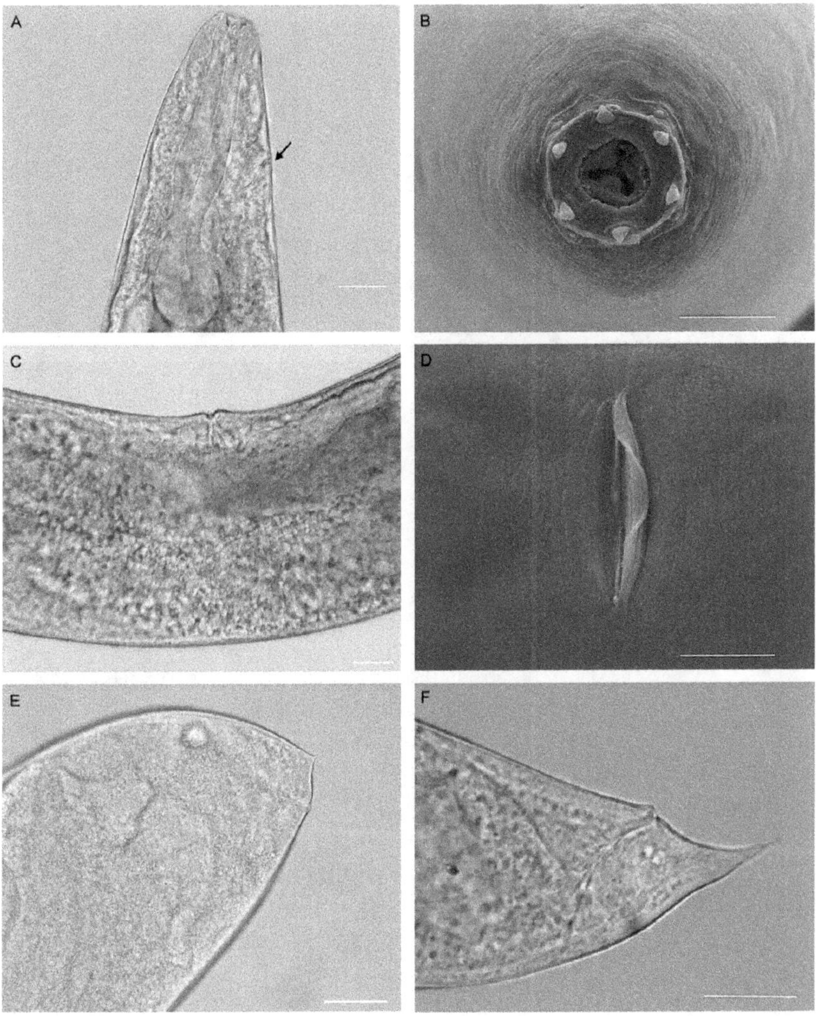

Fig. 4.27. *Female* Steinernema citrae. *A: Anterior region of first generation (excretory pore indicated by arrow); B:* En face *view; C: Light micrograph of first generation vulva; D: SEM of single-flapped epiptygmata; E: Light micrograph of first generation tail; F: Second generation tail. (Scale bars: A, C = 40 μm; B, D = 10 μm; E, F = 30 μm.) After Stokwe* et al. *(2011), courtesy* Nematology.

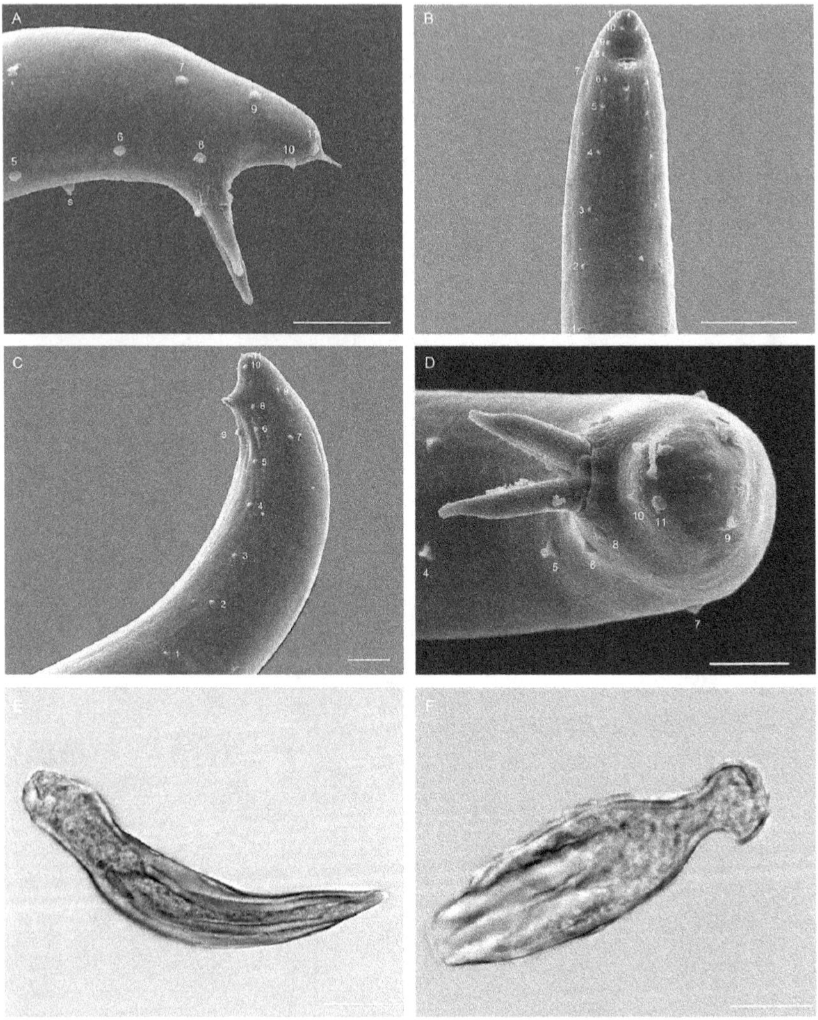

Fig. 4.28. *SEM of male* Steinernema citrae. *A, D: Posterior genital papillae (4-11) of second generation (s = midventral papilla); B, C: Posterior genital papillae of first generation; E: Spicule of first generation; F: Gubernaculum of first generation. (Scale bars: A, C = 20 μm; B = 50 μm; D = 10 μm; E, F = 15 μm.) After Stokwe* et al. *(2011), courtesy* Nematology.

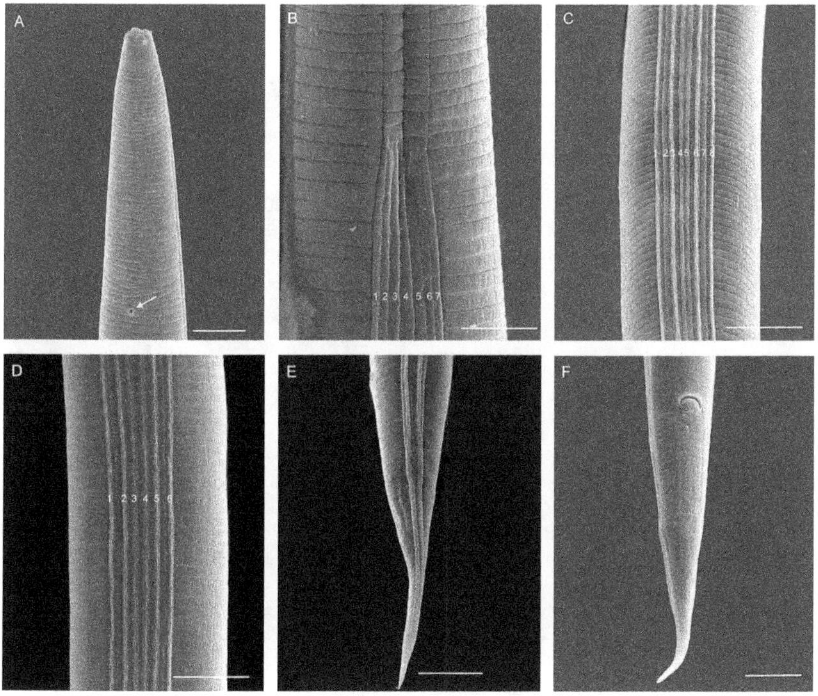

Fig. 4.29. *SEM of* Steinernema citrae *infective juvenile. A: Anterior region indicating position of excretory pore (arrow); B: Lateral field, expansion from two to seven ridges; C: Lateral field in mid-body, eight ridges; D: Lateral field posterior to mid-body, six ridges; E: Lateral field, two ridges shrinking towards posterior end; F: Ventral region of tail showing anus. (Scale bars: A, C-F = 10 μm; B = 5 μm.) After Stokwe* et al. *(2011), courtesy* Nematology.

DIAGNOSTIC DESCRIPTION

Male (first generation)

Anterior region typical for *Steinernema* male. Lateral field present in mid-body, with one narrow ridge. Excretory pore anterior to nerve ring, excretory duct well cuticularised. Genital system monorchic, reflexed. Testis reflexed, comprising germinal zone, growth zone, *vas deferens*. Copulatory papillae 23, comprising six pairs precloacal subventral papillae, single precloacal midventral papilla, one pair lateral, one pair adcloacal; postcloacal papillae including two pairs subterminal and one pair subdorsal. Spicule paired. Manubrium longer than wide (3:2), shaft short, blade moderately curved, rostrum not observed. In light

Table 4.10. *Morphometrics of* Steinernema citrae. *All measurements are in μm and in the form: mean ± s.d. (range).*

Character	First generation			Second generation		Infective juvenile
	Holotype (male)	Male	Female	Male	Female	
n	–	20	20	20	20	25
L	1099	1154 ± 80 (1028-1402)	3087 ± 575 (2038-4019)	770 ± 64 (640-859)	1732 ± 291 (1232-2152)	754 ± 48 (623-849)
a	–	–	–	–	–	30 ± 2.2 (25-34)
b	–	–	–	–	–	6.0 ± 0.4 (5.1-7.1)
c	–	–	–	–	–	15 ± 1.3 (13-17)
c′	–	–	–	–	–	–
V	–	–	54 ± 2.5 (50-59)	–	54 ± 2.1 (50-57)	–
Body diam.	92	103 ± 7.2 (87.1-113.7)	175 ± 22 (137-212)	51 ± 34 (43-60)	121 ± 16 (90-140)	26 ± 1.5 (23-28)
Excretory pore	82	81 ± 7 (64-92)	75 ± 10 (54-90)	57 ± 5 (46-67)	71 ± 8 (59-84)	56 ± 4.6 (49-64)
Nerve ring	97	106 ± 7 (92-119)	151 ± 14 (130-179)	96 ± 6 (83-108)	115 ± 9 (101-130)	98 ± 5.9 (88-110)
Pharynx	126	139 ± 7 (123-155)	206 ± 16 (182-238)	130 ± 5 (122-141)	172 ± 11 (152-187)	125 ± 5.3 (118-137)
Testis reflexion	67	91 ± 19 (59-130)	–	104 ± 24 (68-145)	–	–
Tail length	25	25 ± 4 (17-31)	44 ± 7 (33-60)	25 ± 5 (14-33)	48 ± 4 (41-55)	71 ± 4.3 (63-81)
Anal body diam.	34	33 ± 4 (28-43)	62 ± 11 (43-79)	27 ± 3 (22-33)	35 ± 4 (29-39)	14 ± 0.9 (13-17)
Mucron	–	8.5 ± 1.7 (7.3-9.7)	–	12.4 ± 3.4 (7.9-18.3)	–	–
Hyaline	–	–	–	–	–	30 ± 2.4 (25-34)
Spicule length	63	65 ± 6 (57-80)	–	49 ± 4 (42-56)	–	–
Spicule width	15	14 ± 2.8 (9.5-19.4)	–	10 ± 1.7 (7.0-12.5)	–	–
Gubernaculum length	44	44 ± 6.8 (32-59)	–	28 ± 5.2 (18-35)	–	–
D%	65	58 ± 5 (47-67)	37 ± 6.0 (27-46)	44 ± 4 (35-52)	41 ± 3.8 (34-48)	44 ± 3.8 (39-58)
E%	–	–	–	–	–	110 ± 10.5 (85-132)
SW%	188	198 (156-233)	–	181 ± 23 (138-224)	–	–
GS%	70	68 ± 12 (48-89)	–	58 ± 11 (41-83)	–	–
H%	–	–	–	–	–	43 ± 3.6 (37-50)

microscopy, spicule blade with three lobes, one dorsal, one lateral and one ventral. Gubernaculum boat-shaped in lateral view, anterior end rounded, arrow-shaped cuneus present. Tail conoid, tail terminus with small mucron in 10% of specimens.

Male (second generation)

Body of second generation male similar to that of first generation male, except shorter and narrower. Conoid tail with mucron.

Female (first generation)

Anterior region typical for *Steinernema* female. Excretory pore anterior to nerve ring, well cuticularised; excretory gland not observed. Lateral field absent. Vulva with transverse slit on a slightly protruding area. Low, single-flapped epiptygmata present. Tail tapering to a pointed terminus. Slight pre-anal swelling in fully mature females. Tail shorter than anal body diam.

Female (second generation)

Similar to first generation female, but smaller and more slender. Body an open C when heat-killed and fixed with 80°C TAF. Vulva on a slight protuberance. No postanal swelling present. Tail tapering gently to a sharp point.

Infective juvenile

Anterior region typical for *Steinernema* IJ. Excretory pore anterior to nerve ring. Lateral field beginning anteriorly with one line on the tenth annule, two additional lines appearing at annules 30 to 31. On third annule posterior to excretory pore, number of lateral line increasing to form seven ridges. In mid-body, number of ridges increasing to eight, toward posterior third of body number of ridges decreasing to six. In region of anus, ridges decreasing to four, then two, ending in one toward tail end. In mid-body, two sublateral and central pair of eight ridges less prominent. More posterior to mid-body, ridges reduced to six, with strongly reduced submarginal ridges still slightly visible. Lateral field formula = 2, 7, 8, 6, 4, 2. Cardia inconspicuous. Bacterial chamber prominent, located only in anterior intestine. Hemizonion and hemizonid not observed. Tail three anal body diam. long, attenuate, tapering evenly. Hyaline region occupying posterior 41 (30-54)% of tail length.

TYPE HABITAT AND LOCALITY

Natural host unknown as the species was collected by baiting. *Steinernema citrae* (isolate 141-C) was collected from the soil of a citrus orchard (32°35.285′S, 18°39.383′E) on the Rietkloof farm near the town of Piketberg in the Western Cape Province, South Africa.

TYPE MATERIAL

Holotype (male, first generation), isolated from the haemocoel of *G. mellonella*, deposited in the National Collection of Nematodes, Biosystematics Division, Plant Protection Research Institute, Agricultural Research Council, Pretoria, South Africa. Many specimens of the first and second generation males and females and IJ in TAF deposited in the same collection as the holotype. Glycerin-mounted paratypes of first and second generation males and females and IJ, as well as many specimens of the different stages in TAF (vial nos T-555p-T559p), deposited in the United States Department of Agriculture Nematode Collection (USDANC), Beltsville, MD, USA.

RELATIONSHIPS

Steinernema citrae is characterised by morphometrics of the IJ, with body length of 754 (623-849) μm, distance from anterior end to excretory pore of 56 (49-64) μm, tail length = 71 (63-81) μm, ratio a = 30 (25-34), H% = 43 (37-50) and E% = 110 (85-132). Lateral field formula = 2, 7, 8, 6, 4, 2. The male of the first generation can be recognised by spicule and gubernaculum length and shape, excretory pore position, D% = 58 (47-67) and GS% = 68 (48-89). The female is recognisable by the vulva having low, single-flapped epiptygmata. More information on relationships is available in Stokwe *et al.* (2011).

MOLECULAR CHARACTERISATION

Steinernema citrae is characterised genetically by sequences of the ITS region (accession no. EU740970) and D2-D3 domain (accession no. GU004534) of rDNA. The sequence of ITS regions of *S. citrae*, including ITS1 + 5.8S + ITS2 can be recognised by its length of 714 bp. Sequence length of each transcribed spacer is ITS1 = 265 bp, ITS2 = 292 bp. For the D2-D3 domain, the sequence of *S. citrae*, flanked by primers D2F and 536, is 908 bp. More information about molecular characterisation and phylogenetic analysis is available in Stokwe *et al.* (2011).

SYMBIOTIC BACTERIUM

No information.

BIONOMICS

Crossbreeding between males and females of *S. citrae* with *S. feltiae* produced no progeny (Stokwe *et al.*, 2011). In the control, when each of the species was self-crossed, males and females produced offspring normally.

DISTRIBUTION

Steinernema citrae was collected near the town of Piketberg in the Western Cape Province, South Africa.

Steinernema colombiense[24,25] López-Núñez, Plichta, Góngora-Botero & Stock, 2008
(Figs 4.30, 4.31)

MEASUREMENTS

See Table 4.11.

DIAGNOSTIC DESCRIPTION

Male (first generation)

Anterior region typical for *Steinernema* male. Body slender, ventrally curved posteriorly, J-shaped when heat-relaxed. Cuticle smooth under light microscopy. Lateral field and phasmids inconspicuous. Excretory pore opening circular, located anterior to nerve ring at anterior one-third of metacorpus. Testis single, reflexed, consisting of germinal

[24] Specific epithet named after *Colombia*, the country in which the species was found.

[25] Description after López-Núñez *et al.* (2008). Revised by David J. Hunt.

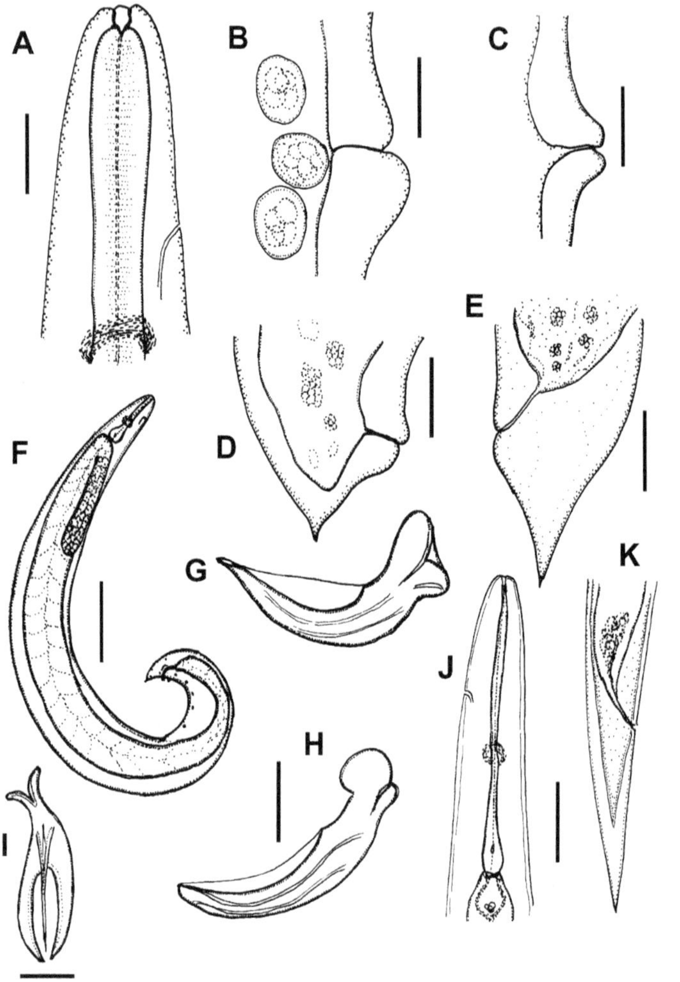

Fig. 4.30. Steinernema colombiense. *A: Anterior end of first generation female, lateral view; B: Vulva of first generation female, lateral view; C: Vulva of second generation female, lateral view; D: Tail of first generation female, lateral view; E: Tail of second generation female, lateral view; F: First generation male,* in toto, *lateral view; G: First generation male spicule, lateral view; H: Second generation male spicule, lateral view; I: First generation male gubernaculum, dorsal view; J: Anterior end of infective juvenile, lateral view; K: Tail of infective juvenile, lateral view. (Scale bars: A, J = 23 μm; B-D = 35 μm; E, G-I, K = 15.5 μm; F = 170 μm.) After López-Núñez* et al. *(2008), courtesy* Nematology.

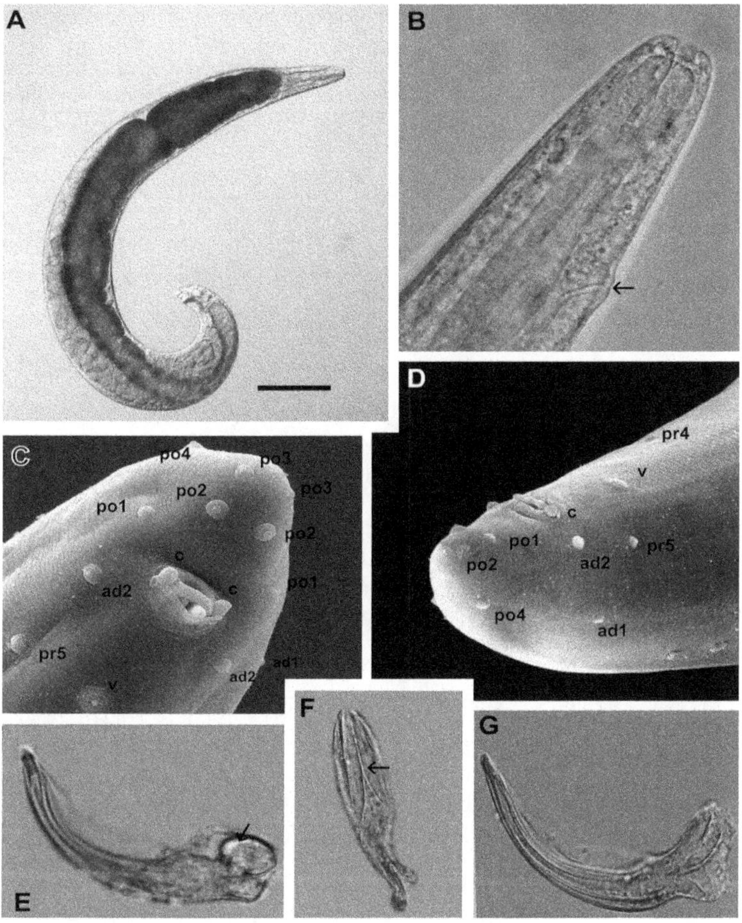

Fig. 4.31. *Light and scanning electron microscope photographs of male specimens of* Steinernema colombiense. *A: First generation male,* in toto*; B: Anterior end, showing excretory pore location (arrow); C, D: Scanning electron microscopy images of tail showing distribution of several genital papillae (pr = precloacal, c = cloacal, ad = adcloacal, po = postcloacal, v = length from anterior end to vulval opening); E: First generation male spicule, lateral view showing 'knob-like' manubrium (arrow); F: First generation male gubernaculum, ventral view showing bifurcate head region and cuneus (arrow); G: Second generation male spicule, lateral view. (Scale bar: A = 100 μm; B, D, E = 16 μm; C, F, G = 12 μm.) After López-Núñez* et al. *(2008), courtesy* Nematology.

Table 4.11. *Morphometrics of* Steinernema colombiense. *All measurements are in μm and in the form: mean ± s.d. (range).*

Character	First generation			Second generation		Infective juvenile
	Male		Female	Male	Female	
	Holotype	Paratypes	Paratypes	Paratypes	Paratypes	Paratypes
n	–	19	19	20	20	20
L	1678	1542 ± 226 (1209-2040)	4608 ± 102 (2796-6752)	870 ± 61 (762-980)	1470 ± 211 (996-2123)	636 ± 15 (549-732)
a	–	–	–	–	–	21 ± 3 (18-28)
b	–	–	–	–	–	5.4 ± 0.6 (4.4-6.4)
V	–	–	52 ± 2 (49-55)	–	59 ± 2 (55-64)	–
Max. body diam.	95	102 ± 20 (58-121)	140 ± 15 (115-163)	56 ± 6 (46-68)	75 ± 18 (51-116)	31 ± 4 (22-36)
Pharynx length	158	156 ± 15 (123-199)	185 ± 17 (155-215)	122 ± 7 (105-140)	147 ± 12 (111-179)	118 ± 6 (106-128)
Head to excretory pore	70	67 ± 5 (56-76)	76 ± 7 (62-89)	54 ± 4 (46-63)	67 ± 11 (52-115)	35 ± 4 (31-40)
Head to nerve ring	134	121 ± 14 (96-155)	137 ± 14 (112-173)	92 ± 6 (80-105)	120 ± 12 (102-155)	83 ± 5 (73-92)
Tail length	40	31 ± 6 (23-41)	48 ± 10 (33-74)	25 ± 4 (21-35)	45 ± 5 (28-60)	41 ± 5 (32-53)
Anal body diam.	38	37 ± 6 (26-49)	64 ± 11 (47-81)	32 ± 5 (23-43)	36 ± 7* (38-62)	11.5 ± 2 (9-14)
Spicule length	70	71 ± 4 (64-77)	–	54 ± 5 (43-62)	–	–
Gubernaculum length	61	50 ± 8 (47-63)	–	35 ± 5 (23-43)	–	–
SW%	180	200 ± 30 (150-270)	–	170 ± 40 (100-240)	–	–
GS%	90	150 ± 30 (110-180)	–	160 ± 20 (130-230)	–	–
D%	44	40 ± 4 (30-50)	–	44 ± 4 (39-49)	–	29 ± 2 (25-33)
E%	57	52 ± 9 (35-69)	–	218 ± 3 (166-263)	–	205 ± 31 (138-284)
H	–	–	–	–	–	14 ± 2 (11-19)
H%	–	–	–	–	–	35 ± 7 (27-50)

* Data as per original description.

growth zone leading to seminal vesicle. *Vas deferens* with inconspicuous walls. Spicules paired, symmetrical, curved, ochre-brown in colour. Manubrium rhomboidal, with a rounded, 'knob-like', protrusion. Calomus (shaft) distinct. Lamina with rostrum or retinaculum and two internal ribs. Velum present, narrow. Blade terminus blunt. Gubernaculum arcuate, *ca* 75% of spicule length. Manubrium of gubernaculum curved ventrally and bifurcate. Tail conoid and non-mucronate. Gubernaculum more slender and longer than that of first generation male. Twenty-five genital papillae (12 pairs and single midventral) arranged as follows: five subventral precloacal pairs, single midventral precloacal papilla (located between precloacal pairs 4 and 5); one pair of cloacal papillae (located laterally on cloacal opening); two pairs adcloacal (one pair subventral, one pair subdorsal); four pairs postcloacal (one pair subventral, two pairs ventral, one pair subdorsal).

Male (second generation)

General morphology similar to that of first generation males, but smaller in size. Tail with or without mucron. Spicules with manubrium morphology slightly different to that of first generation male. Manubrium wider than long, with two small ribs. Calomus indistinct, continuous with shaft. Lamina with a small rostrum, velum present.

Female (first generation)

Anterior region typical for *Steinernema* female. Cuticle, lip region, stoma and pharyngeal region as in male. Body C-shaped when heat-relaxed. Excretory pore located about mid-procorpus level. Ovaries opposed, reflexed dorsally; oviduct well developed; glandular spermatheca and uterus in ventral position. Vagina short, with muscular walls. Vulva located near mid-body, with protruding asymmetric lips, posterior lip larger than anterior. Tail blunt, conoid, lacking mucron. Anal lips usually protruding, asymmetric with anterior lip smaller than posterior.

Female (second generation)

Body open C-shape when heat-relaxed. Similar to first generation female, but smaller. Vulva with protruding, symmetrical lips. Tail conoid, with post-anal swelling.

Infective juvenile

Anterior region typical for *Steinernema* IJ. Body of heat-relaxed specimens almost straight, slender, gradually tapering posteriorly. Cuticle with fine transverse striae. Lateral field distinct with six longitudinal ridges (*i.e.*, seven lines/incisures) in mid-body region. Excretory pore located *ca* mid-corpus level. Hemizonid not observed. Anterior portion of intestine with bacterial receptacle. Intestine filled with numerous fat globules, narrow lumen. Rectum long, straight, anus distinct. Genital primordium evident. Tail conoid with pointed terminus. Hyaline region occupying *ca* 25-50% of tail length.

TYPE HABITAT AND LOCALITY

The natural host is unknown as *S. colombiense* (isolate SNI-0198) was collected at 1402 m a.s.l. by *Galleria* baiting from soil samples from a coffee plantation (*Coffea arabica* var. Castillo) in Maracay Experimental Station, Quimbaya County, Quindío department, Colombia.

TYPE MATERIAL

Holotype first generation male, five first generation paratype males, five first generation paratype females and five paratype IJ deposited in the USDA Nematode Collection, Beltsville, MD, USA, and five first generation paratype males, five first generation paratype females and five paratype IJ deposited in the University of California Davis Nematode Collection, Davis, CA, USA.

RELATIONSHIPS

Steinernema colombiense is characterised by the morphometrics of the IJ which has a small body length of 636 (549-732) μm, narrow body diam. (31 (22-36) μm), the position of the excretory pore (35 (31-40) μm) and a short tail (41 (32-53) μm). Additional diagnostic traits include D% = 29 (25.5-33), E% = 205 (138-284) and H% = 35 (27-50). First generation males can be distinguished by the morphology of the spicules and gubernaculum, the values of ratios SW = 2.0 (1.5-2.7) and GS = 1.5 (1.1-1.8) and the number and arrangement of the genital papillae. First generation females can be recognised by the presence of protruding and asymmetric vulval lips.

Phylogenetic analysis of LSU sequences data placed *S. colombiense* in a clade that includes four species: *S. carpocapsae*, *S. monticolum*, *S. siamkayai* and *S. scapterisci*, all characterised by having the IJ with a short body length (average = 572 μm). However, *S. colombiense* can be differentiated from these taxa by several morphometric differences of the IJ and first generation male. For example, IJ of the new species have a larger body size (average = 636 μm) and higher value of E% (average = 205) than *S. carpocapsae*, *S. scapterisci* and *S. monticolum*. First and second generation males of *S. colombiense* also differ from those of *S. carpocapsae*, *S. scapterisci* and *S. monticolum* by the number and distribution of genital papillae and by the overall morphology of the spicules and gubernaculum. In the new species the spicules are characterised by having a rhomboidal manubrium, with a rounded, 'knob-like' protrusion. The gubernaculum is also unique in its morphology with the manubrium curved ventrally and bifurcate. In addition to these qualitative differences, males of the new species can be differentiated by the SW value (average = 2.0 and 1.7 for first and second generation males, respectively). Both first and second generation females of *S. colombiense* are characterised by having protruding vulval lips. This feature is shared with females of *S. carpocapsae*. Females of *S. monticolum* lack protruding vulval lips and *S. siamkayai* and *S. scapterisci* have only slightly protruding vulval lips. Both first and second generation females of the new species lack a tail mucron, whereas *S. carpocapsae*, *S. scapterisci* and *S. monticolum* may or may not have a mucron.

Molecular characterisation

Sequences were deposited in GenBank as follows: 28S: EU345420; ITS: EU345421. Maximum parsimony (MP) analysis placed *S. colombiense* as a relative of Clade I. This clade comprises *Steinernema* spp. known to have IJ with small body size (average < 600 μm). Association of the new species with this clade was strongly supported by bootstrap resampling (100%). Pairwise distances also shows that *S. colombiense* differs from closely related species by 16-20 bp. Analysis of ITS rDNA sequences placed *S. colombiense* as a member of a clade of species with small IJ, including *S. tami*, *S. carpocapsae*, *S. scapterisci* and *S. siamkayai*. Further details are provided by López-Núñez *et al.* (2008).

Bionomics

Cross-hybridisation assays between males and females of *S. colombiense* with *S. carpocapsae*, *S. siamkayai* and *S. scapterisci* yielded no progeny. In the controls, offspring were produced in all self-crossed species. No progeny were observed in the single female control plates.

DISTRIBUTION

Known only from Colombia, the type locality (López-Núñez *et al.*, 2008).

Steinernema costaricense[26,27] Uribe-Lorío, Mora & Stock, 2007
(Figs 4.32, 4.33)

MEASUREMENTS

See Table 4.12.

DIAGNOSTIC DESCRIPTION

Male (first generation)

Anterior region typical for *Steinernema* male. Body slender, ventrally curved posteriorly, J-shaped when heat-killed. Big and small forms were reported for the Missouri population. Lateral field and phasmids inconspicuous. Labial region truncated to slightly round, continuous with body. Deirids at mid-level of bulb. Cardia 5-9 μm long, distinct, projecting into intestine. Excretory pore located anterior to nerve ring in anterior third of metacorpus. Single reflexed testis consisting of germinal growth zone leading to seminal vesicle. *Vas deferens* with inconspicuous walls. Spicules paired, symmetrical, curved, ochre-brown in colour. Manubrium rhomboid, continuous dorsally with lamina. Calomus (shaft) inconspicuous. Lamina with rostrum or retinaculum and two internal ribs. Velum present. Distal tip of spicule with ventral aperture. Gubernaculum arcuate, *ca* half length of spicules. Manubrium of gubernaculum straight or curved ventrally in some specimens. Twenty-three genital papillae (11 pairs and one single) arranged as follows: six precloacal subventral pairs, two subventral precloacal pairs (usually located on both sides of single ventral papillae), a single

[26] Specific epithet named after *Costa Rica*, the country from which the species was described.

[27] Description after Uribe-Lorío *et al.* (2007) and Ivanova *et al.* (2013). Revised by Sergei E. Spiridonov.

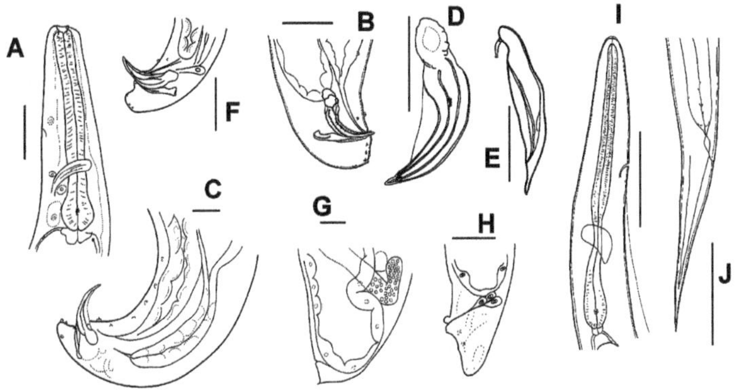

Fig. 4.32. *Morphology of* Steinernema costaricense. *A: Male, second generation, labial region; B: Male, first generation, tail; C: Male first generation, posterior end; D: Spicule, first generation male; E: Gubernaculum, first generation male; F: Male, second generation, tail; G: female, first generation, tail; H: Female, second generation, tail; I: Infective juvenile, anterior end; J: Infective juvenile, tail. (Scale bars = 50 μm.)*

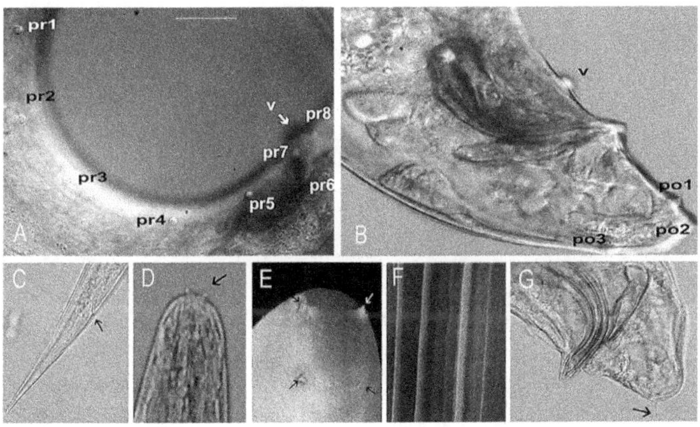

Fig. 4.33. *A: First generation male tail (lateral view) showing position of precloacal (pr1-8) and ventral (v) genital papillae; B: First generation male tail showing ventral (v) and postcloacal genital papillae (po1-po3); C: Infective juvenile tail (lateral view, anus arrowed); D: Infective juvenile anterior end (lateral view) showing horn-like cephalic papillae (arrow); E: SEM of infective juvenile showing cephalic papillae (arrows); F: SEM of infective juvenile lateral field pattern; G: Second generation male tail (lateral view) showing mucron (arrow). (Scale bar: A, B, D, G = 22.5 μm; C = 32 μm; E = 10 μm; F = 15 μm.) Modified after Stock* et al. *(2007), courtesy* Systematic Parasitology.

Table 4.12. *Morphometrics of holotype of* Steinernema costaricense *with other data from a USA population described by Ivanova* et al. *(2013). All measurements are in μm and in the form: mean ± s.d. (range).*

Character	First generation			Second generation		Infective juvenile (measured alive)
	Holotype (male)	Male	Female	Male	Female	
n	–	15	10	5	10	26
L	2349	2335 ± 374 (1456-2786)	5637 ± 2238 (2504-9072)	1203 ± 454 (733-1906)	1198 ± 59 (1093-1292)	855 ± 94 (680-1032)
a	18.4	12 ± 1.6 (9.5-14.8)	16.4 ± 3.9 (10.7-21.1)	10.6 ± 3.8 (7-16)	18 ± 7.1 (12.7-37.7)	25.6 ± 2 (21.3-29.2)
b	10.9	11.5 ± 1.3 (7.7-12.7)	22.9 ± 7.3 (12.3-33.1)	7.5 ± 2.1 (5-11)	7.4 ± 0.5 (6.8-8.1)	5.9 ± 0.5 (5.2-6.7)
c	55.9	62.1 ± 14.2 (41.6-92.9)	143.4 ± 56.3 (75.5-226.8)	47.7 ± 22.6 (27-84)	21.3 ± 3.2 (17.6-28.8)	10.2 ± 0.6 (8.9-11.4)
c′	–	0.59 ± 0.1 (0.4-0.8)	0.35 ± 0.1 (0.23-0.49)	0.53 ± 0.2 (0.22-0.67)	1.5 ± 0.3 (0.9-1.9)	4.6 ± 0.7 (3.6-7.0)
V	–	–	54.9 ± 3.3 (51-60.3)	–	56.9 ± 2.5 (51.9-59.1)	–
Max. body diam.	127	200 ± 49 (140-280)	337 ± 87 (230-480)	113 ± 6 (105-120)	72 ± 15 (32-86)	34 ± 3.9 (26-42)
Excretory pore	120	112 ± 15 (72-135)	95 ± 17 (70-112)	99 ± 13 (82-112)	83 ± 3 (80-88)	63 ± 4.3 (54-72)
Pharynx	214	202 ± 16 (174-230)	240 ± 24 (203-290)	157 ± 15 (140-175)	162 ± 8 (148-171)	144 ± 10.8 (120-167)
Tail length	42	38 ± 5 (30-47)	42 ± 9 (32-57)	28 ± 12 (11-40)	57 ± 9 (38-69)	84 ± 7.8 (68-98)
Hyaline region	–	–	–	–	–	26 ± 3.9 (20-32)
Anal body diam.	56	66 ± 7 (54-75)	118 ± 33 (68-172)	52 ± 7 (42-60)	40 ± 10 (34-67)	17 ± 3 (12-24)
Spicule length along chord	101	91 ± 6 (71-100)	–	75 ± 6 (69-82)	–	–
Gubernaculum length	48	53.7 ± 6.9 (40-62)	–	40 ± 8 (32-50)	–	–
D%	56	55 ± 0 (38-66)	39.4 ± 0.1 (24.1-47.7)	63 ± 0 (59-69)	51.6 ± 0 (48.5-54)	43.9 ± 2.1 (37-47)
E%	285	297 ± 0.5 (206-383)	228 ± 0.5 (140-287)	417 ± 2 (270-818)	149 ± 0.3 (123-224)	76 ± 6.2 (70-93)
SW%	0.55	1.4 ± 0.2 (1.2-1.8)	–	1.4 ± 0.2 (1.2-1.7)	–	–
GS%	0.48	0.59 ± 0.1 (0.47-0.7)	–	0.54 ± 0.1 (0.46-0.68)	–	–
H%	–	–	–	–	–	57 ± 4 (50-67)

ventral precloacal papilla, two pairs subventral postcloacal (with one pair sometimes in terminal position) and a subdorsal postcloacal pair. Phasmids inconspicuous. Mucron absent.

Male (second generation)

General morphology similar to that of first generation. Not significantly shorter than small form of first generation, with similar size of spicules and gubernaculum. Mucron present or absent.

Female (first generation)

Anterior region typical for *Steinernema* female. Body sausage-like, gradually tapering from pharynx base towards rounded anterior extremity, C-shaped when heat-killed. Cuticle smooth, faintly annulated. Lateral fields indistinct. Stoma 14 ± 3 (10-18) μm long and 12 ± 2 (8-14) μm wide. Procorpus expanded, 31 ± 4 (26-38) μm, isthmus 23 ± 2 (19-26) μm wide, bulb 42 ± 5 (32-50) μm diam. Excretory pore located at about mid-procorpus level. Ovaries opposed, reflexed in dorsal position, oviduct well developed, spermatheca glandular, uterus in ventral position. Ovaries extending for some distance from anterior and posterior extremities, reflexing at mid-body, with several loops. Spermathecae located at junction of ovaries. Oviducts containing spermatozoa arranged into chains of 5-10 cells *ca* 6 μm diam. Vagina short, with muscular walls. Vulva located near middle of body, with slightly protruding lips. Gelatinous mating plug often observed in fertilised females. Tail blunt, conoid, with or without postanal swelling. About one in ten specimens lacking a mucron.

Female (second generation)

Viviparous. Tail conoid, with or without postanal swelling.

Infective juvenile

Anterior region typical for *Steinernema* IJ. Body of heat-relaxed specimens almost straight, slender, gradually tapering posteriorly. About 50% IJ retaining sheath (second-stage cuticle) after 5 months storage. Cuticle with fine transverse striae. Lateral field extending from 35-40 μm from anterior extremity to level adjacent to tail tip. Two low ridges present at anterior extremity, transforming into six equally developed and closely arranged ridges, at level of nerve ring and posterior to it, each lateral field provided with four prominent ridges thus forming three bands, with central one wider than others. Complete lateral field formula = 2, 6, 2. Two additional, very low, poorly defined ridges stretching along central, internal part of lateral field, becoming more pronounced closer to mid-body, only two marginal ridges present at posterior extremity. Labial region continuous with body, slightly truncate. Excretory pore located about middle of corpus. Cardia up to 7-8 μm long. Deirids situated between marginal and submarginal ridges at level of bulb base or cardia. Hemizonid appearing as cuticular depression, 4 μm long, at level of deirids or just anterior. Genital primordium comprised of several dozens of cells, 62-90 μm long Anterior portion of intestine with dorsally displaced vesicle containing symbiotic bacterium. Poorly defined oval to bean-shaped vesicle with walls less than 1 μm thick present in intestine. Two- to 3-week-old IJ with few rod-like bacterial cells *ca* 6 μm long in a vesicle measuring 10-15 × 5-7 μm; 5-month-old IJ with vesicle measuring 18-35 × 7-15 μm, containing numerous rods. Intestine filled with numerous fat globules, lumen of intestine narrow. Rectum long and straight, anus distinct. Tail conoid, with pointed terminus. Phasmids distinct, located 25-42 μm posterior to anus. Hyaline region occupying *ca* half of tail length.

Type habitat and locality

Natural host unknown, the isolate being recovered by *Galleria* baiting soil from a premontane transitional tropical rainforest along a valley slope between the Cacao and Rincón de la Vieja volcanoes, Costa Rica, at 1200 m a.s.l.

Type material

Holotype male, first generation (T-605t); five paratype males, first generation (T-5483p); five paratype females, first generation (T-5484p, T-5485p); five paratype IJ (T-5486p) deposited in the USDA Nematode Collection, MD, USA.[28]

Relationships

According to Uribe-Lorío *et al.* (2007), *S. costaricense* was characterised by the IJ body size of 1696 (1600-1774) μm [*sic*], excretory pore position of 77 (75-82) μm, D% = 53 (45-60), E% = 85 (82-91.5),

[28] But apparently unavailable according to Ivanova *et al.* (2013).

and presence of six ridges in the lateral field. Ivanova *et al.* (2013) redescribed what they believed to be this species using an isolate from the USA. They discussed in detail problems with the original description, including the erroneous attribution of the species to the *bicornutum*-group on the basis of horn-like cephalic papillae (the *bicornutum*-group is characterised by the presence of two lateral horn-like protuberances adjacent to the oral aperture, these being in addition to the four cephalic papillae), and also pointed out that the great length of the IJ given in the original description was likely to be a calibration error. First generation males can be distinguished by the position of the excretory pore, cloacal body diam., tail length, D%, morphology of the spicules and gubernaculum, and disposition and number of the genital papillae. First generation females have a blunt tail, which may have or lack a mucron, and non-protruding to slightly protruding vulval lips.

MOLECULAR CHARACTERISATION

Uribe-Lorío *et al.* (2007), in the original description, stated that, on the basis of the LSU rDNA sequence and what appears to be a misinterpretation of the sensilla of the IJ labial region, *S. costaricense* belonged to the *bicornutum*-group yet, according to the very long body of the IJ, it was a member of the *glaseri*-group. It seems clear that the original description involved a number of errors/misinterpretations, these being discussed in detail by Ivanova *et al.* (2013) who redescribed the species from an isolate from Missouri, USA, type material being apparently not available.

In their analysis, Ivanova *et al.* (2013) stated that the LSU rDNA sequences of the different Costa Rican isolates of *S. costaricense* (GU569047, FJ165551, EF187017) were identical, and differed by only 3 bp from the 'Bush-Augusta' isolate (Missouri, USA). Phylogenetic analysis of the D2-D3 LSU rDNA sequence of *S. costaricense* (both Costa Rican and Missouri isolates) did not reveal strongly supported links with known clades of steinernematids (Ivanova *et al.*, 2013).

The ITS rDNA sequence of the Missouri isolate (AY171276) is most similar to the sequence of *S. scarabaei* Stock & Koppenhöfer, 2003 (GU569047), and formed a weakly supported clade with the following three steinernematid species from the Americas: *S. rarum*, *S. scarabaei* and *S. unicornum*. No intraspecific variability between five studied clones was detected in ITS rDNA sequences of 'Bush-Augusta' isolate (Ivanova *et al.*, 2013).

SYMBIOTIC BACTERIUM

No specific information, although sequences of 16S, *Rec*A and *Ser*C have been deposited.

BIONOMICS

Cross-hybridisation tests with *S. glaseri*, *S. cubanum* and *S. puertoricense* produced no progeny (Uribe-Lorío *et al.*, 2007).

DISTRIBUTION

In addition to the type locality in Costa Rica, this species has also been reported from Bush Augusta State Park, Missouri, USA. Samples were collected in November 1999 in deciduous woodland near a stream (Ivanova *et al.*, 2013).

Steinernema ethiopiense[29,30] Tamiru, Waeyenberge, Hailu, Ehlers, Půža & Mráček, 2012

(Figs 4.34-4.36)

MEASUREMENTS

See Table 4.13.

DIAGNOSTIC DESCRIPTION

Male (first generation)

Anterior region typical for *Steinernema* male. Body J-shaped in heat-relaxed specimens, curved ventrally posteriorly, tapering toward labial region. Stoma *ca* 3-4 μm long and 6-8 μm broad. Excretory pore located posterior to mid-pharynx but anterior to nerve ring (55-65% of distance from anterior body end to base of pharynx). Spicules paired, symmetrical, slightly curved, golden-brown in colour, *ca* 70-80 μm long, spicule tip pointed, Spicule manubrium slightly variable, elongated to ellipsoid, with an average length:width ratio of 1.45:1. Manubrium

[29] The new species is named after *Ethiopia*, the country from which it was isolated.

[30] Description after Tamiru *et al.* (2012). Revised by David J. Hunt.

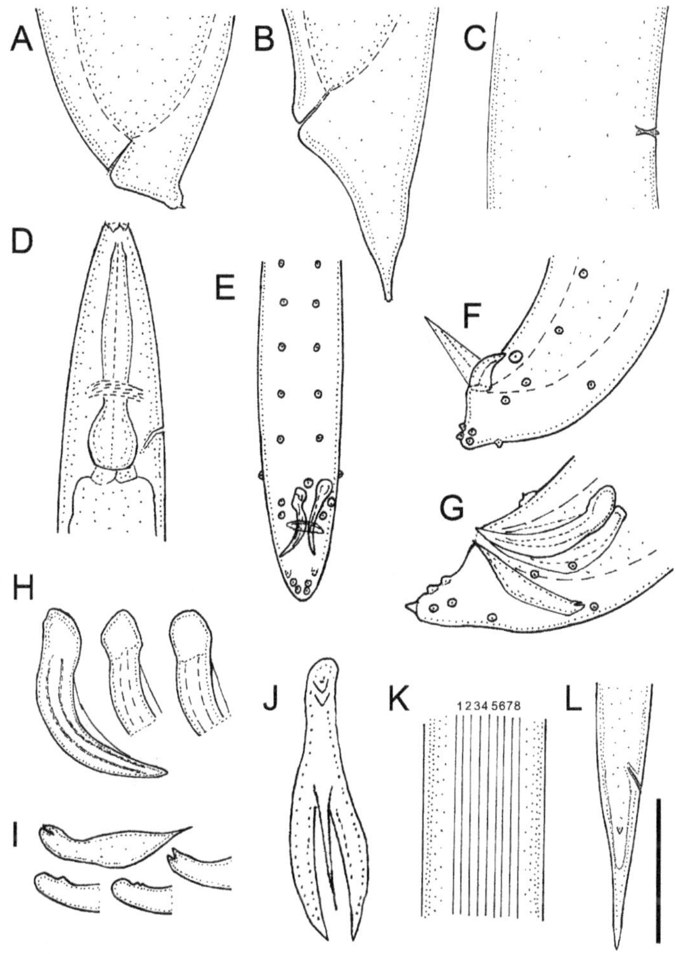

Fig. 4.34. Steinernema ethiopiense. *A-C: Female. A: First generation tail; B: Second generation tail; C: First generation vulva. D-J: Male. D: Labial region and pharynx; E: First generation tail with genital papillae (ventral); F: First generation tail with genital papillae (slightly rolled); G: Second generation tail with genital papillae and mucron (slightly rolled); H: First generation spicule and variability of proximal end; I: First generation gubernaculum and variability of proximal end; J: Gubernaculum (ventral). K, L: Infective juvenile. K: Lateral field with eight ridges; L: Tail with hyaline region and phasmids. (Scale bar: A = 145 μm; B, H = 50 μm; C = 240 μm; D = 105 μm; E = 100 μm; F = 70 μm; G = 55 μm; I = 45 μm; J = 25 μm; K = 40 μm; L = 65 μm.) After Tamiru* et al. *(2012), courtesy* Nematology.

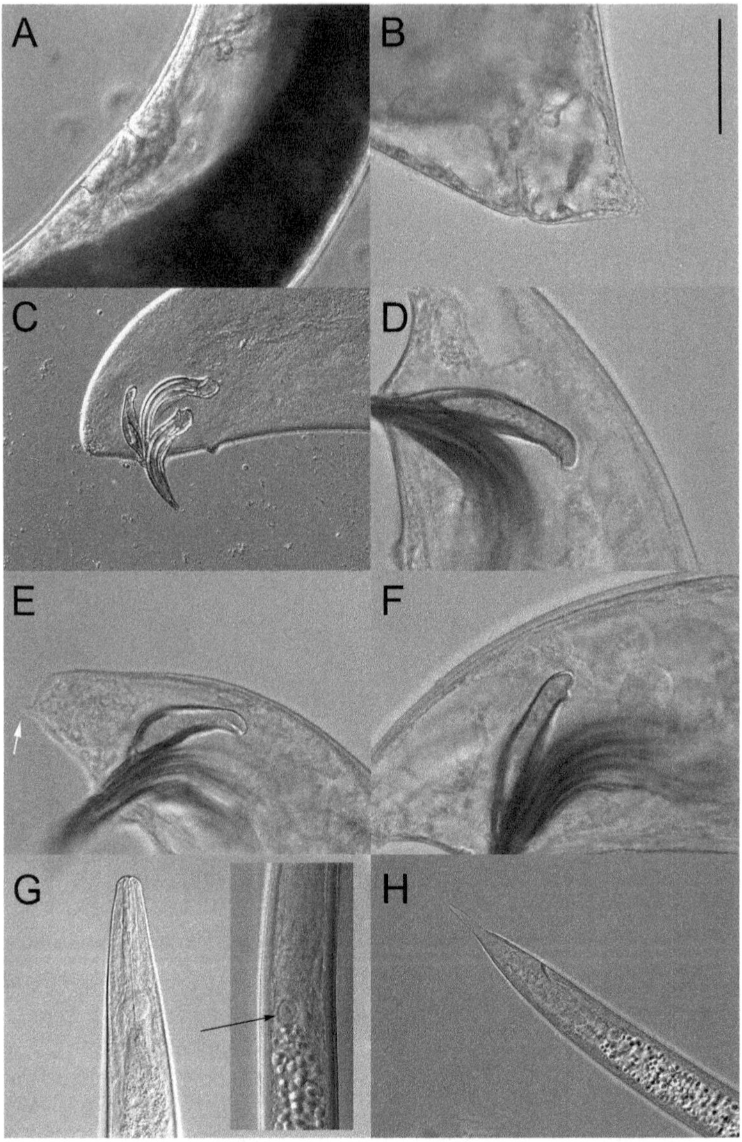

Fig. 4.35. Steinernema ethiopiense. *DIC microscopy. A: First generation female vulva; B: First generation female tail showing minute protuberance; C: First generation male tail with spicules and gubernaculum; D: First generation male tail with two protuberances on gubernaculum; E: Second generation male tail with a short mucron (arrow) and irregular proximal end of gubernaculum;*

forming slightly more than 25% of spicule length, calomus distinct. Lamina with two ribs, finely curved. Rostrum not developed, velum bow-like, distinct, extending from calomus part to distal part of lamina. In lateral view, gubernaculum boat-shaped, short proximal part bent, usually bearing one or two protuberances or irregular. Posterior part usually tapering abruptly to distal tip. In ventral view, cuneus small, needle-shaped, or absent, corpus with two wings. Tail bluntly conoid, usually concave on ventral surface, mucron absent. Genital papillae totalling 23, comprising 11 pairs and a single midventral papilla located just anterior to cloacal opening. Of the paired papillae, eight pairs located precloacal (five pairs subventral, one pair subdorsal and two lateral adcloacal pairs) and three pairs postcloacal and consistently near tail terminus with one pair subdorsal and two pairs subventral near tail tip.

Male (second generation)

Similar to first generation, but smaller. Spine-like tail mucron developed as a short, 2-3 μm long, projection, never filamentous. Spicule manubrium slightly variable in shape, similar to first generation male.

Female (first generation)

Anterior region typical for *Steinernema* female. Body variable in length, usually C-shaped on heat relaxation. Labial region bluntly rounded, slightly tapering anteriorly, not offset from body contour. Stoma *ca* 5-7 μm long and 10-13 μm broad. Genital tract amphidelphic with reflexed ovaries. Vulva indistinct mostly symmetrical, rarely slightly asymmetrical and in form of a transverse slit, not or only slightly (3 μm), protruding from body contour. Vulval opening slightly posterior to mid-body. Small epiptygmata developed. Vagina short, leading into paired uteri. Rectum narrow, anal opening distinct. Tail of mature females obese with a short conical tip bearing only one minute protuberance. Postanal swelling not developed in mature females. Pygmy forms not observed. *Endotokia matricida* rarely observed.

F: First generation male tail with single protuberance on gubernaculum; G: Infective juvenile with rounded labial region and pharynx (left) and bacterial vesicle (right; arrow); H: Infective juvenile tail. (Scale bar: A = 145 μm; B = 57 μm; C = 80 μm; D-F = 27 μm; G = 47 μm; H = 70 μm.) After Tamiru et al. *(2012), courtesy* Nematology.

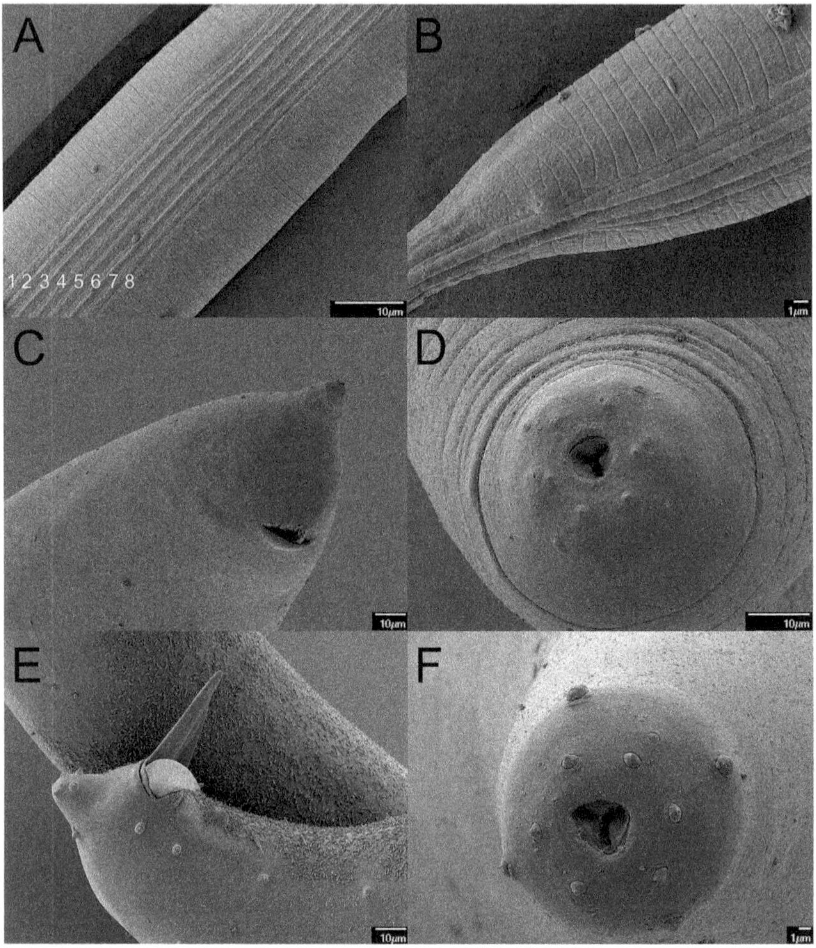

Fig. 4.36. Steinernema ethiopiense. *SEM microscopy. A: Lateral field of infective juvenile with eight ridges (1-8) at mid-body; B: Lateral field of infective juvenile showing gradual disappearance of ridges on tail, four ridges reducing to two at phasmid level; C: First generation female tail; D: First generation female labial region with two circles of papillae and pair of amphidial apertures; E: First generation male tail with genital papillae and spicules; F: First generation male labial region with two circles of papillae and pair of amphidial apertures. After Tamiru* et al. *(2012), courtesy* Nematology.

Female (second generation)

Similar to first generation in general morphology, but smaller. Vulval opening slightly posterior to mid-body, slightly to moderately protrud-

Table 4.13. *Morphometric characters of* Steinernema ethiopiense, *type isolate Dero-8. All measurements are in μm and in the form: mean ± s.d. (range).*

Character	First generation			Second generation		Infective juvenile
	Male		Female	Male	Female	
	Holotype	Paratypes	Paratypes	Paratypes	Paratypes	Paratypes
n	–	20	20	20	20	20
L	1394	1081 ± 40 (1028-1232)	4836 ± 880 (2889-6262)	980 ± 22 (906-1011)	2129 ± 264 (1677-2525)	898 ± 50 (768-1010)
a	12.8	12 ± 0.6 (11-13)	16 ± 1.7 (12-19)	12 ± 0.4 (11-13)	17 ± 1.9 (14-20)	27 ± 2.6 (20-30)
b	8.9	7.1 ± 0.2 (6.3-7.4)	23 ± 3.9 (15-29)	6.9 ± 0.1 (6.0-7.4)	11 ± 1.1 (8.7-13)	6.4 ± 0.3 (5.7-7.1)
c	44	42 ± 3.4 (35-49)	80 ± 18 (52-108)	40 ± 1.1 (36-41)	33 ± 2.9 (26-37)	12 ± 0.6 (11-14)
c′	–	–	0.8 ± 0.2 (0.6-1.5)	–	1.4 ± 0.1 (1.2-1.7)	3.2 ± 0.2 (2.9-3.8)
V	–	–	51 ± 3.4 (45-57)	–	54 ± 1.6 (51-56)	–
Max. body diam.	109	92 ± 4.8 (81-99)	301 ± 44 (232-374)	85 ± 3.4 (79-91)	127 ± 17 (109-176)	34 ± 3.0 (32-43)
Anterior end to excretory pore	86	88 ± 1.7 (85-92)	165 ± 17 (129-192)	82 ± 2.1 (79-87)	102 ± 15 (82-125)	78 ± 4.6 (65-84)
Pharynx length	157	153 ± 3.1 (149-161)	208 ± 13 (180-227)	143 ± 4.0 (137-152)	192 ± 15 (164-215)	140 ± 6.9 (127-153)
Tail length	32	29 ± 1.9 (25-33)	62 ± 12 (47-94)	25 ± 0.5 (23-25)	65 ± 6.0 (55-74)	73 ± 3.9 (64-80)
Anal body diam.	45	45 ± 0.8 (43-46)	78 ± 14 (51-102)	41 ± 1.7 (39-46)	47 ± 4.8 (39-59)	23 ± 1.3 (19-24)
Spicule length	72	73 ± 2.0 (69-77)	–	67 ± 3.9 (64-80)	–	–
Gubernaculum length	52	49 ± 1.3 (46-57)	–	41 ± 0.9 (40-43)	–	–
Anterior end to vulva	–	–	2489 ± 492 (1656-3454)	–	1147 ± 141 (909-1353)	–
Anterior end to nerve ring	–	–	–	–	–	104 ± 7.7 (83-113)
Bulb length	–	–	–	–	–	24 ± 1.6 (19-27)
Hyaline region	–	–	–	–	–	38 ± 3.7 (29-43)
D%	57	57 ± 1.6 (54-61)	–	57 ± 1.8 (54-61)	–	56 ± 2.2 (51-58)
E%	269	304 ± 23 (264-357)	–	333 ± 11 (316-354)	–	107 ± 5.5 (91-116)
SW%	160	164 ± 5.8 (154-175)	–	164 ± 7.4 (155-185)	–	–
GS%	72	67 ± 2.1 (63-70)	–	62 ± 3.4 (53-66)	–	–
H%	–	–	–	–	–	52 ± 3.0 (43-56)

ing. Tail conical, sharply pointed, longer than anal body diam., rarely extending to form a mucron. Postanal swelling slightly developed.

Infective juvenile

Anterior region typical for *Steinernema* IJ. Body elongate, on average 898 μm long, almost straight or slightly C-shaped when heat-killed, often enclosed in second-stage cuticle sheath. Cuticular striations (distinct under SEM) *ca* 1.5 μm wide at mid-body, but lip region and tail terminus smooth. Lateral fields consisting of eight, equally spaced and equally developed ridges in mid-body region (*i.e.*, nine lines). Lateral field beginning with one cuticular depression (line) close to fourth annule from anterior end; a short distance posteriorly, two ridges appearing and changing to six at about excretory pore level; more posteriorly still central ridge dividing into two, making eight ridges in all. Number of ridges reducing gradually in posterior part of body. *Circa* two anal body diam. anterior to anus, two middle ridges merging to leave six ridges in lateral field. Close to anus, lateral field reduced to four and closer to anus to two ridges, extending almost to tail tip. With above arrangement, lateral field formula = 2, 6, 8, 6, 4, 2. Deirids not observed. Hemizonid visible as a 5-6 μm wide cuticular depression usually located between anterior to middle of basal bulb. Excretory pore located at level of mid-pharynx (D% = 56). Cardia *ca* 7-10 μm long, with rounded posterior end. Bacterial vesicle usually well developed, ellipsoid to rounded. Rectum *ca* 0.5 anal body diam. long. Tail attenuated, tapering gradually, contour depressed both dorsally and ventrally at about commencement of hyaline region. Hyaline region occupying *ca* 50% of tail length. Anus wide, sickle-shaped. Phasmids located *ca* 35-40% of tail length posterior to anus, with aperture near point where lateral field ridges reduce from four to two.

TYPE HABITAT AND LOCALITY

The type isolate, Dero-8, was recovered from brown clay soil at 1850 m a.s.l. from a forest near Mendi (9°48′N, 35°6′E) in the Western Welega Province, western Ethiopia.

OTHER LOCALITIES

Two other isolates of this nematode (Dero-1, Mosisa-1) were recorded very close to the type isolate at Mendi, although precise data are not available.

TYPE MATERIAL

Holotype first generation male, paratype males (12 slides with 32 specimens), paratype females (11 slides with 16 specimens) and juveniles (four slides with 22 specimens) deposited at the EPN collection at the Laboratory of Insect Pathology, Institute of Entomology, Biology Centre, Czech Academy of Sciences (CAS), Česke Budějovice, The Czech Republic. Three slides with paratype IJ, males and females deposited at the USDA Nematode Collection, Beltsville, MD, USA. Numerous paratype IJ also deposited in liquid nitrogen in the EPN collection at the Department of Biotechnology & Biological Control, Institute for Phytopathology, Christian-Albrechts-University, Kiel, Germany.

RELATIONSHIPS

The species is characterised morphologically by a combination of features of the IJ and adults. For IJ, the relatively short body length of 898 (768-1010) μm, distance from anterior end to excretory pore of 78 (65-84) μm, anterior end to end of pharynx of 140 (127-153) μm; tail length of 73 (64-80) μm, D% = 56 (51-58) and E% = 107 (91-116) are characteristic. For first generation males, the diagnostic characters include the spicule length of 73 (69-77) μm, D% = 57 (54-61), SW% = 164 (154-175) and GS% = 67 (63-70). Both generations possess characteristic protuberances or irregularities at the tip of the proximal part of the gubernaculum. Females have an indistinct, symmetrical, rarely slightly asymmetrical, vulva. The tail of mature first generation females is obese with a short conical tip (peg) bearing a minute protuberance. A postanal swelling is not developed.

MOLECULAR CHARACTERISATION

All three isolates (Dero-1, Dero-8 and Mosisa-1) of *S.ethiopiense* revealed a PCR-product of *ca* 800 bp for the ITS-region whilst the 28S rRNA gene containing the D2 and D3 expansion fragments had a length of *ca* 650 bp (accession nos JN651409-JN651414 (Dero-1 D2-D3 sequence: JN651409, Dero-1 ITS sequence: JN651412; Mosisa-1 D2-D3 sequence: JN651410, Mosisa-1 ITS sequence: JN651413; Dero-8 D2-D3 sequence: JN651411, Dero-8 ITS sequence: JN651414)). BLAST analysis of the ITS sequences for all studied isolates indicated 93% similarity with *S. karii* (AY230173) and 85% or lower with any other *Stein-*

ernema species/isolate. Within the D2-D3 sequences, a similarity percentage of 98% with *S. karii* AF331902 and 95% or lower with other *Steinernema* isolates was recorded.

The isolates of *S. ethiopiense* formed a subgroup within the *glaseri*-group (*sensu* Spiridonov *et al.*, 2004), and were most closely related to *S. karii*. For D2-D3, the MP tree was calculated based on 246 parsimony informative characters; for the ME tree 499 characters were used. In the MP and ME trees, the *S. ethiopiense* isolates were closest to *S. karii* but clustered within the *glaseri*-group, although as a separate subgroup (see Tamiru *et al.*, 2012 for further detail).

Bacterial symbiont

No information.

Bionomics

No information.

Distribution

Known only from the Western Welega Province, western Ethiopia (Tamiru *et al.*, 2012).

Steinernema huense[31,32] Phan, Mráček, Půža, Nermut' & Jarošová, 2014
(Figs 4.37-4.39)

Measurements

See Table 4.14.

Diagnostic description

Male (first generation)

Anterior region typical for *Steinernema* male. Stoma *ca* 4-5 μm long and 7-9 μm broad. Excretory pore located posterior to mid-pharynx but

[31] Named after *Hue* province, from where the type population was recovered.
[32] Description after Phan *et al.* (2014). Revised by David J. Hunt.

anterior to nerve ring at 41-61% of distance from anterior body end to base of pharynx. Spicules paired, symmetrical, finely curved, slender, brown, *ca* 60-72 μm long, spicule tip pointed, spicule manubrium slightly variable, finely rounded, with an average length:width ratio of *ca* 1.2:1. Manubrium forming slightly more than 25% of spicule length, calomus and rostrum distinct. Lamina slender with two ribs, finely curved. Velum distinct and short, slightly bow-like, extending from rostrum to one-third of lamina. In ventral view, cuneus small, needle-shaped, corpus with two wings. Genital papillae totalling 23, comprising 11 pairs and a single midventral papilla located just anterior to cloacal opening. Six pairs located precloacal (five subventral, one lateral), five pairs located postcloacal, two pairs adcloacal subventrally and three pairs consistently near tail terminus, one pair being subdorsal and two at tail tip. Tail bluntly conoid, concave on ventral surface with a minute 1-2 μm long mucron.

Male (second generation)

Similar to first generation but smaller. Mucron short, 2-3 μm long. Spicules slender, manubrium mostly oblong, with an average length: width ratio of *ca* 1.6:1, slightly variable in shape, different from first generation.

Female (first generation)

Anterior region typical for *Steinernema* female. Lateral fields absent. Stoma *ca* 6-8 μm long and 12-19 μm wide. Vulva distinct, mostly symmetrical, rarely slightly asymmetrical, in form of a transverse slit and slightly (3-5 μm), protruding from body contour, double-flapped epiptygmata present. Vulval opening posterior to mid-body, mean V = 53. Tail obese, bearing small peg or single minute protuberance. Post-anal swelling not developed. Pygmy forms not observed.

Female (second generation)

Similar to first generation in general morphology, but smaller. Vulval opening slightly posterior to mid-body, moderately protruding (10-15 μm). Tail obese, sharply pointed, longer than anal body diam., extending to form a 5-8 μm long filamentous mucron. Post-anal swelling slightly developed.

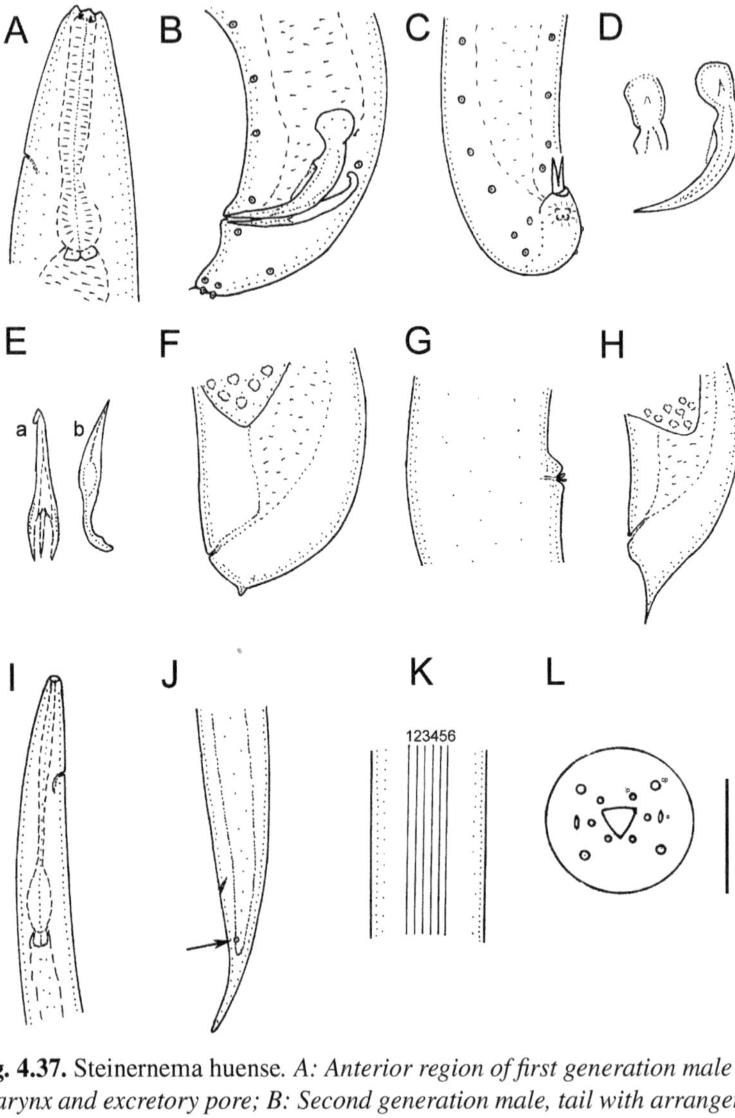

Fig. 4.37. Steinernema huense. *A: Anterior region of first generation male with pharynx and excretory pore; B: Second generation male, tail with arrangement of genital papillae and mucron, lateral; C: Tail of first generation male with arrangement of genital papillae, dorso-ventral; D: First generation male spicules and manubrium variation; E: First generation male gubernaculum, ventral (a) and lateral (b). F, G: First generation female. F: Tail with terminal peg; G: Vulval region with double-flapped epiptygmata. H: Second generation female, tail with mucron. I-K: Infective juvenile. I: Anterior region with excretory pore; J: Tail with anus and hyaline region (arrow indicates phasmid*

Infective juvenile

Anterior region typical for *Steinernema* IJ. Body short, on average 527 μm long, straight or slightly bow-shaped when heat-killed, usually ensheathed in second-stage cuticle. Labial region continuous, rounded, not offset from body contour. Cuticular striations (distinct under SEM) *ca* 1.5 μm wide at mid-body, but lip region and tail terminus smooth. Lateral fields consisting of six equally spaced and developed ridges (*i.e.*, seven lines) in mid-body region. Sub-marginal ridges not developed. Lateral field beginning with irregular cuticular depressions (lines) close to eighth annule from anterior end; a short distance posteriorly, two indistinct ridges appearing and changing to six ridges (seven lines) at about excretory pore level. In posterior part of body, number of ridges gradually reducing. Anterior to anus, two middle ridges disappearing to leave four ridges in lateral field. At phasmid opening, lateral field reducing to two ridges extending almost to tail tip. With above arrangement, lateral field formula = 2, 6, 4, 2. Deirids not observed. Hemizonid visible as a 3-4 μm wide cuticular depression usually located between anterior of basal bulb and nerve ring. Excretory pore located at level of mid-pharynx (D% = 37). Cardia *ca* 3-6 μm long, bacterial vesicle usually not well seen, oval, *ca* 13 $\times$ 7 μm in size. Rectum *ca* 80-120% of anal body diam. long. Tail conoid, with slight ventral constriction/bent situated slightly anterior to proximal portion of hyaline region. Hyaline region occupying *ca* 44% of tail length. Tail terminus sharply pointed, claw-like. Phasmids located at *ca* 65-75% of tail length posterior to anus, with aperture before beginning of ventral constriction.

TYPE HABITAT AND LOCALITY

Brown clay soil at an elevation of 100 m a.s.l., in soil from beneath a canopy of broad-leaved trees in Bach Ma national park, Vietnam (16°13′53″N, 107°49′02″E). The natural insect host is unknown.

---

opening); K: Lateral field with ridges numbered. L: Adult arrangement of labial (lp), cephalic (cp) papillae, and amphidial apertures (a). (Scale bar: A = 70 μm; B, C = 130 μm; D, E = 45 μm; F = 85 μm; G = 220 μm; H = 65 μm; I = 52 μm; J = 40 μm; K = 25 μm; L = 20 μm.) After Phan et al. *(2014), courtesy* Nematology.

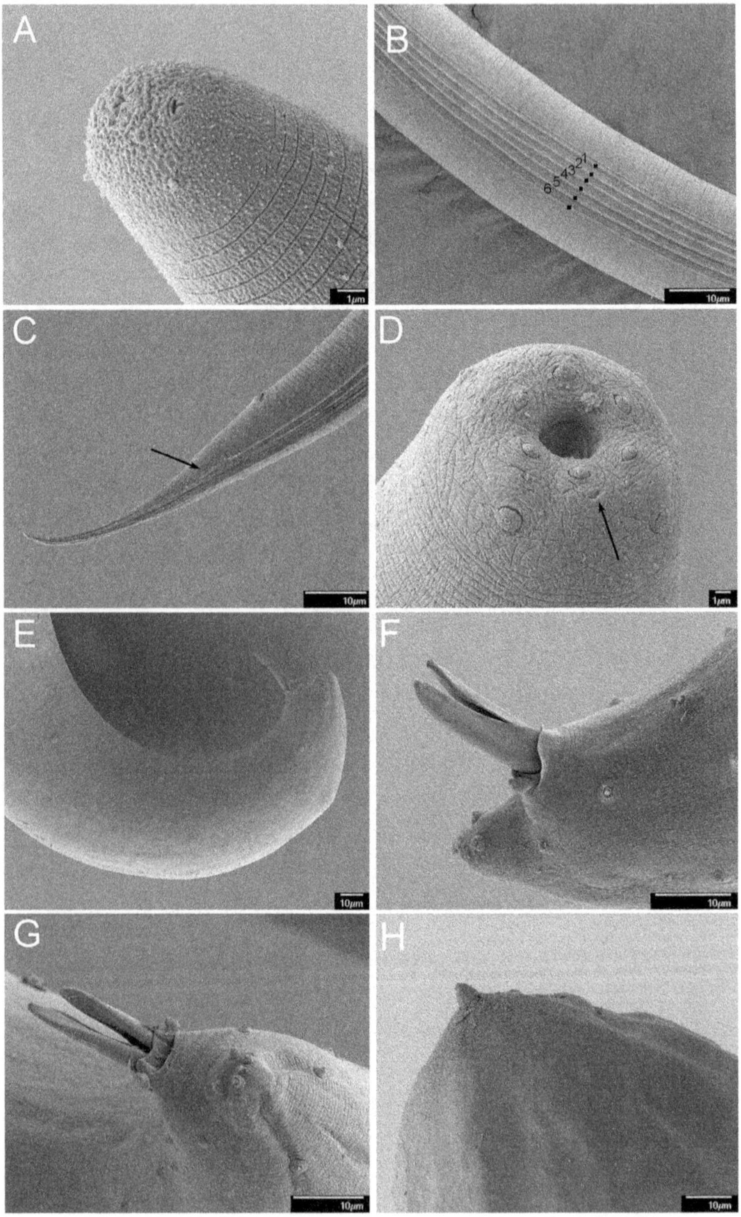

Fig. 4.38. Steinernema huense. *SEM of infective juvenile, male and female. A-C: Infective juvenile. A: Labial region with four papillae and amphid openings; B: Lateral field in mid-body (ridges numbered 1-6); C: Lateral field in tail region with phasmid opening (arrow). D-G: First generation male. D: Labial*

TYPE MATERIAL

Holotype first generation male, paratype males (M1 – ten slides with 24 specimens, M2 – five slides with 30 specimens), paratype females (F1 – ten slides with 26 specimens, F2 – three slides with ten specimens) and IJ (five slides with 67 specimens) deposited at the EPN collection in the Laboratory of Entomopathogenic Nematodes, Institute of Entomology, Biology Centre CAS, České Budějovice, Czech Republic. Nine slides with paratype IJ (one slide with ten specimens), males and females (F1 – five slides with five specimens, F2 – one slide with three specimens, M1 – one slide with one specimen and M2 – one slide with five specimens) deposited at the Vietnam National Museum of Nature, Vietnam Academy of Science and Technology, Hanoi, Vietnam.

RELATIONSHIPS

Steinernema huense belongs to the *carpocapsae*-group and is characterised morphologically by a combination of features of the IJ and adults. For IJ, the very short body length of 527 (444-571) μm, anterior end to excretory pore = 43 (38-46) μm, anterior end to end of pharynx = 116 (103-129) μm, tail length = 50 (43-56) μm, D% = 37 (34-39), E% = 85 (79-93) and presence of six ridges (seven lines) in the lateral field are characteristic. For first generation males, the diagnostic characters include the spicule length of 67 (60-72) μm, D% = 48 (41-61), SW% = 195 (168-238) and GS% = 76 (68-81). Both generations possess a minute filamentous mucron. First generation females have a non-protruding, symmetrical vulva, with a double-flapped epiptygmata. The tail of mature females is obese with a short conical tip (peg) and lacks minute protuberances. A postanal swelling is not developed.

region with labial papillae, cephalic papillae and amphidial aperture (arrow); E: Posterior body region, lateral; F: Tail tip with genital papillae, lateral, showing one terminal pair, two adcloacal papillae and dorso-lateral papilla; G: Tail tip, dorso/ventral, showing two terminal pairs, one subdorsal and one pair adcloacal papillae. H: First generation female tail with terminal peg. After Phan et al. *(2014), courtesy* Nematology.

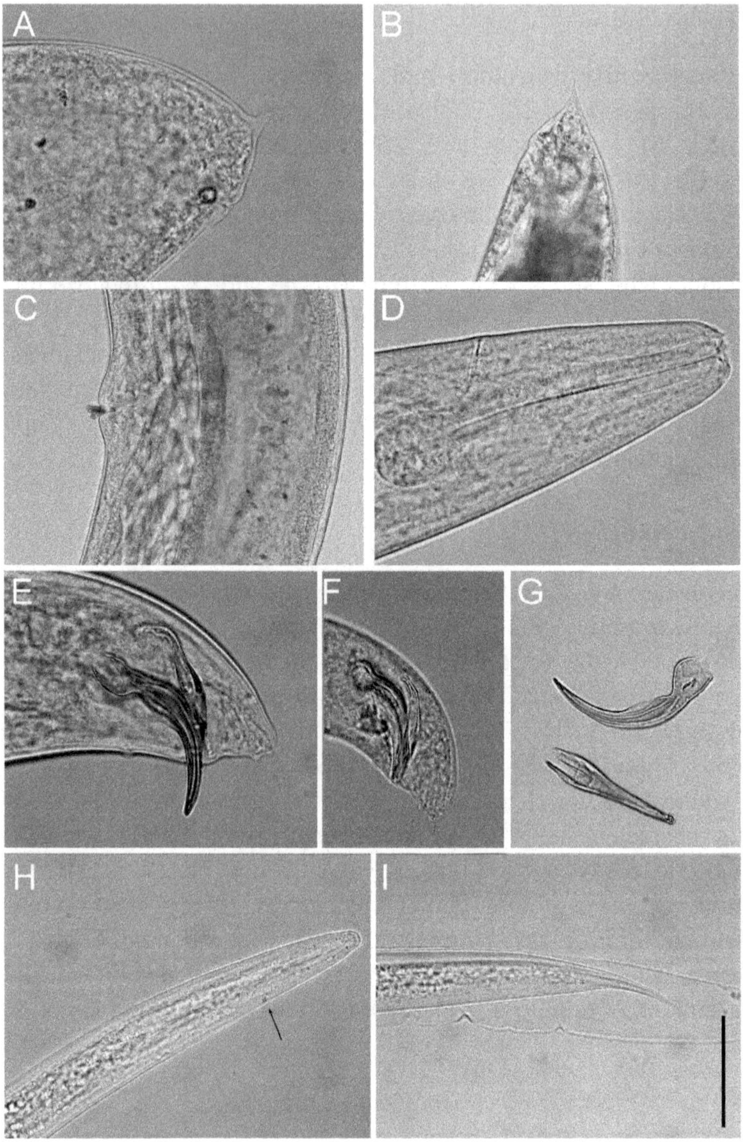

Fig. 4.39. Steinernema huense. *LM of infective juvenile, male and female. A, C: First generation female. A: Tail region; C: Vulval region. B: Second generation female, tail with spine-like mucron. D, E, G: First generation male. D: Labial region with excretory pore and pharynx; E: Tail with spicules and gubernaculum; G: Spicule lateral and gubernaculum, ventral. F: Second generation male, tail with filamentous mucron, spicules and gubernaculum.*

MOLECULAR CHARACTERISATION

Length of the ITS regions = 951 bp (accession no. KF857581). Length of D2-D3 regions = 871 bp (accession no. KF857582). More information is available in Phan *et al.* (2014).

SYMBIOTIC BACTERIUM

All three genetic markers in a BLAST search showed the closest match to be with *Xenorhabdus stockiae* (99% similarity for 16S rDNA and *recombinase* A gene and 98% for *gyrase* B gene). This bacterium was originally isolated from the closely related nematodes *S. siamkayai* (Tailliez *et al.*, 2006) and *S. minutum* (Maneesakorn *et al.*, 2010).

BIONOMICS

No information.

DISTRIBUTION

Known only from Vietnam (Phan *et al.*, 2014).

Steinernema ichnusae[33,34] Tarasco, Mráček, Nguyen & Triggiani, 2008
(Figs 4.40-4.43)

MEASUREMENTS

See Table 4.15.

H, I: Infective juvenile. H: Anterior portion showing rounded labial region and excretory pore (arrow); I: Tail with anus and hyaline region. (Scale bar: A-C = 65 μm; D, E, H, I = 40 μm; F = 50 μm; G = 45 μm.) After Phan et al. *(2014), courtesy* Nematology.

[33] The specific epithet is derived from the ancient Greek name for Sardinia: Hyknusa or *Icnussa*.

[34] Description after Tarasco *et al.* (2008). Revised by Khuong B. Nguyen.

Table 4.14. *Morphometric characters of* Steinernema huense. *All measurements are in μm and in the form: mean ± s.d. (range).*

Character	First generation			Second generation		Infective juvenile
	Male		Female	Male	Female	
	Holotype	Paratypes	Paratypes	Paratypes	Paratypes	Paratypes
n	–	20	20	20	–	20
L	1313	1516 ± 170 (1151-1717)	6132 ± 1358 (4505-9171)	1024 ± 137 (747-1232)	1754 ± 178 (1454-2040)	527 ± 33 (444-571)
a	11	11 ± 1.2 (8.9-14)	24 ± 3.6 (19-32)	14 ± 1.5 (12-18)	13 ± 1.2 (11-16)	21 ± 1.2 (19-23)
b	9.9	11 ± 1.0 (8.8-12)	27 ± 5.5 (20-41)	6.6 ± 0.7 (4.9-7.8)	8.8 ± 0.9 (7.8-11)	4.5 ± 0.2 (4.2-5.1)
c	63	55 ± 7.1 (44-72)	181 ± 39 (128-260)	43 ± 4.8 (34-53)	37 ± 5.3 (28-45)	11 ± 0.5 (10-12)
c′	0.7	0.7 ± 0.1 (0.5-0.7)	0.4 ± 0.1 (0.3-0.5)	0.8 ± 0.1 (0.7-0.9)	1.2 ± 0.2 (0.8-1.6)	3.4 ± 0.2 (2.9-3.9)
V	–	–	53 ± 5.6 (46-74)	–	56 ± 2.2 (52-60)	–
Max. diam.	119	138 ± 11 (121-152)	253 ± 28 (202-303)	73 ± 11 (56-89)	136 ± 19 (102-167)	26 ± 2.3 (21-30)
Anterior end to excretory pore	86	86 ± 7.8 (72-99)	100 ± 13 (74-125)	74 ± 6.1 (65-91)	84 ± 10 (72-107)	43 ± 1.9 (38-46)
Pharynx length	132	144 ± 10.5 (122-159)	225 ± 16 (192-258)	156 ± 8.2 (137-169)	201 ± 20 (167-231)	116 ± 6.7 (103-129)
Tail length	21	28 ± 4.1 (24-37)	34 ± 4.6 (27-43)	24 ± 2.4 (21-32)	48 ± 6.0 (35-56)	50 ± 3.5 (43-56)
Anal body diam.	32	43 ± 5.8 (35-56)	96 ± 12 (78-121)	30 ± 2.4 (25-35)	41 ± 5.3 (32-48)	15 ± 1.0 (13-16)
Spicule length*	68	67 ± 3.2 (60-72)	–	58 ± 2.8 (52-64)	–	–
Gubernaculum length	56	53 ± 2.7 (46-56)	–	44 ± 2.9 (40-49)	–	–
Anterior end to vulva	–	–	3225 ± 661 (2424-5111)	–	977 ± 103 (808-1131)	–
Anterior end to nerve ring	97	–	–	112 ± 10 (97-135)	139 ± 9 (118-154)	83 ± 4.2 (72-87)
Hyaline region	–	–	–	–	–	22 ± 1.6 (19-25)
D%	65	60 ± 5.8 (46-69)	44 ± 6.1 (31-54)	48 ± 4.8 (41-61)	42 ± 3.3 (37-48)	37 ± 1.6 (34-39)
E%	409	312 ± 43 (222-413)	294 ± 50 (211-371)	318 ± 37 (235-407)	180 ± 31 (141-260)	85 ± 4.3 (79-93)
SW%	213	159 ± 22 (111-196)	–	195 ± 18 (168-238)	–	–
GS%	82	79 ± 4.6 (73-88)	–	76 ± 3.3 (68-81)	–	–
H%	–	–	–	–	–	44 ± 3.8 (39-52)

* Measured along the chord.

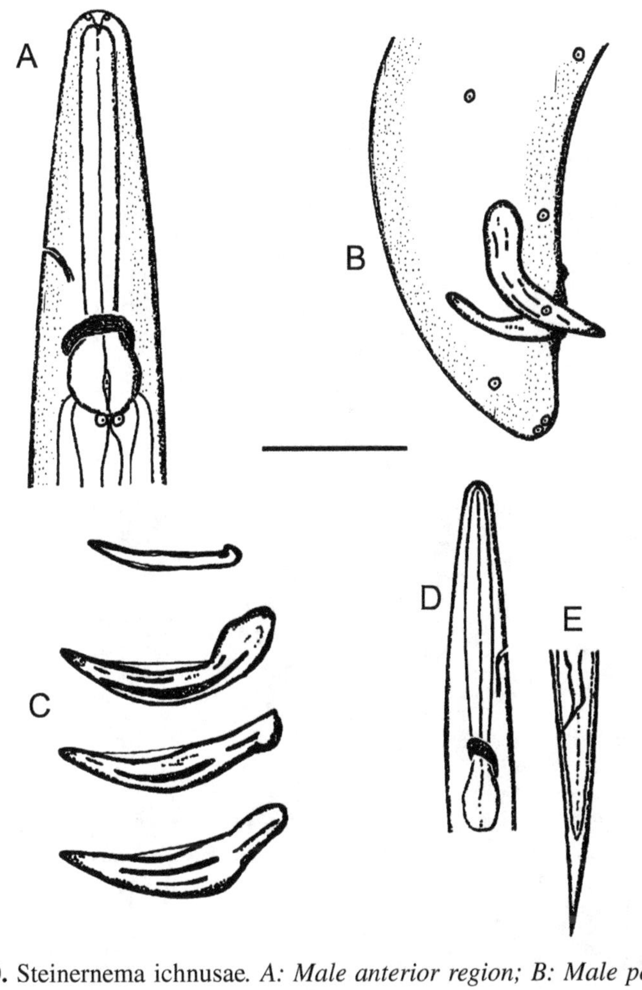

Fig. 4.40. Steinernema ichnusae. *A: Male anterior region; B: Male posterior region; C: Gubernaculum and variation of spicules; D: Anterior region of an infective juvenile; E: Posterior region of an infective juvenile. (Scale bar: A = 60 μm; B = 100 μm; C: gubernaculum = 44 μm, spicule = 40 μm; D = 54 μm; E = 58 μm.) Modified after Tarasco* et al. *(2008), with acknowledgement to* Journal of Invertebrate Pathology.

Diagnostic description

Male (first generation)

Anterior region typical for *Steinernema* male. Excretory pore mostly at metacorpus and always anterior to nerve ring. Testis reflexed, well

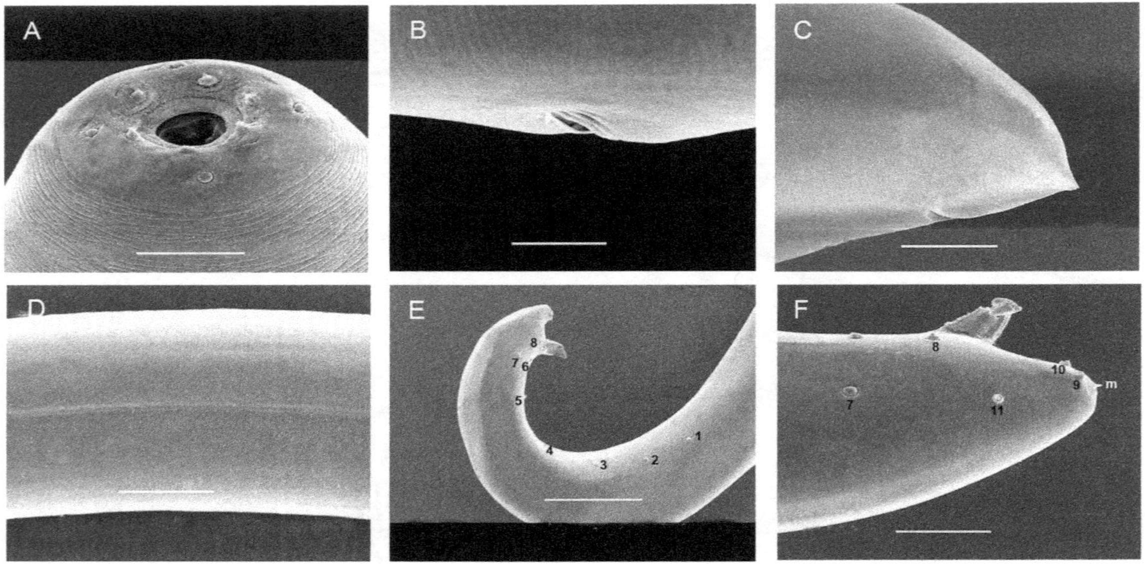

Fig. 4.41. *SEM of* Steinernema ichnusae. *A-C: Female labial region, vulva and tail; D-F: Male lateral field with single ridge and posterior regions showing arrangement of genital papillae. (Scale bars: A, B = 10 μm; C = 27.3 μm; D = 13 μm; E = 60 μm; F = 16.7 μm.) Modified after Tarasco* et al. *(2008), with acknowledgement to* Journal of Invertebrate Pathology.

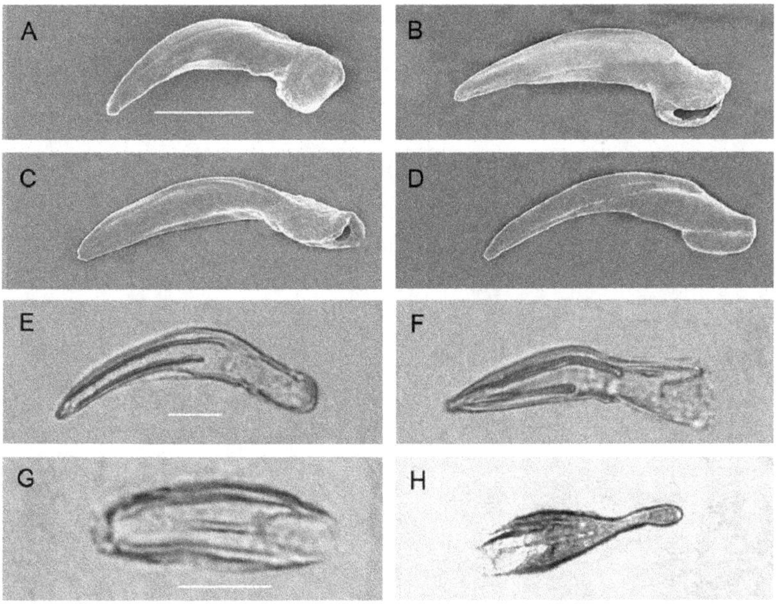

Fig. 4.42. Steinernema ichnusae. *Spicule and gubernaculum, SEM and LM. (Scale bars: A = 20 μm; B-D = 23.1 μm; E, F = 11.1 μm; G = 14 μm; H (scale bar in G) = 19.3 μm.) Modified after Tarasco* et al. *(2008), with acknowledgement to* Journal of Invertebrate Pathology.

developed. Testis reflexion ventral in position and about half of testis length. Paired symmetrical, moderately curved spicules, colourless or slightly golden/yellowish in a light microscope. Spicule manubrium always oblongate in lateral view, on average approximately one-third of spicule length with average length:width ratio of 2.1:1; spicule tip blunt; calomus distinctly separating manubrium from lamina. Lamina slightly curved, supported by two internal ribs, wider and almost straight posteriorly. Rostrum, if present, slightly to moderately developed, velum bow-like and distinct. In lateral view, gubernaculum boat-shaped. Proximal portion in most specimens characteristically finger- or club-like. In ventral view, cuneus small, needle-shaped, or absent, corpus with two wings. Tail bluntly conoid, usually concave on ventral side. Mucron mostly absent, if developed, very short. In some males, tail with finger-like projection. Genital papillae 21 or 23 comprising ten or 11 pairs and a single midventral papilla located just anterior to cloacal opening. Of these, 6-7 pairs are located precloacal (five or six pairs subventral and

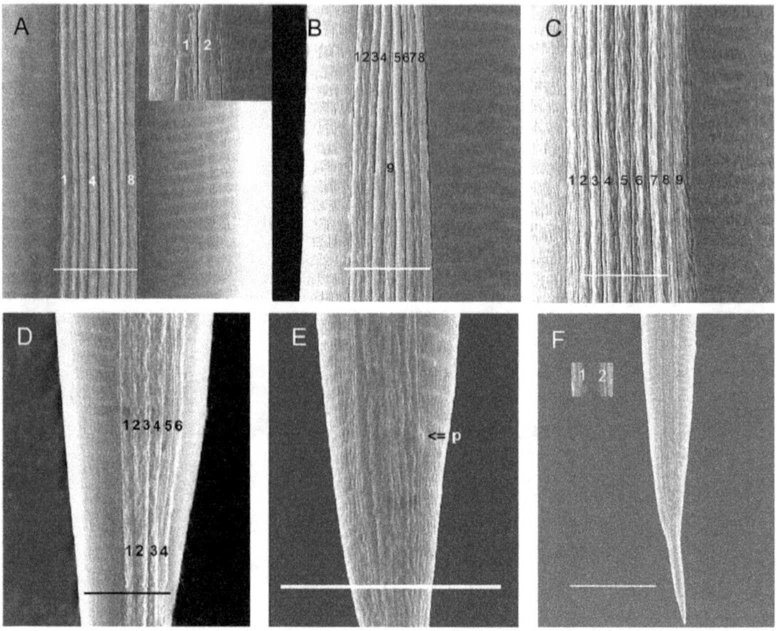

Fig. 4.43. Steinernema ichnusae. *Arrangement of ridges in lateral field of infective juvenile. A: Lateral field with eight ridges, inset of anterior part with two ridges; B: Lateral field with eight ridges, becoming nine posteriorly; C: Lateral field with nine ridges; D: Lateral field in posterior part reducing to six then four ridges; E, F: Lateral field in posterior region (p = phasmid), inset of posterior part with two ridges. (Scale bars: A = 6.67 μm; B, C = 5 μm; D = 6 μm; E = 12.1 μm; F = 16.7 μm.) Modified after Tarasco* et al. *(2008), with acknowledgement to* Journal of Invertebrate Pathology.

one pair lateral), one pair adcloacal, two pairs subterminal and one pair subdorsal. Most posterior pairs consistently located near tail tip.

Male (second generation)

Similar to first generation males, but slightly smaller. Filamentous or spine-like tail mucron always developed and much longer than in mucronated specimens of first generation.

Female (first generation)

Anterior region typical for *Steinernema* female. Stoma *ca* 5-7 μm long and 10-13 μm wide. Excretory pore located anterior to nerve ring. Genital tract amphidelphic with reflexed ovaries. Indistinct vulva mostly symmetrical and transverse slit, not or rarely slightly (3 μm) protruding

Table 4.15. *Morphometrics of* Steinernema ichnusae *(holotype is the infective juvenile). All measurements are in μm and in the form: mean ± s.d. (range).*

Character	First generation			Second generation		Infective juvenile
	Holotype (IJ)	Male	Female	Male	Female	
n	–	20	20	20	20	20
L	859	1341 ± 146 (1151-1494)	5514 ± 425 (4547-6186)	954 ± 86 (808-1110)	2054 ± 829 (1474-2725)	866 ± 61 (767-969)
a	31	22 ± 3.6 (20-29)	21 ± 1.9 (17-24)	17 ± 1.3 (15-19)	16 ± 1.8 (13-20)	28 ± 2.1 (24-32)
b	6.1	8.2 ± 1.1 (7.0-9.4)	23 ± 1.5 (21-26)	6.8 ± 0.6 (5.9-8.2)	11 ± 1.5 (9.0-14)	6.3 ± 0.3 (5.6-6.9)
c	11	34 ± 4.3 (29-39)	93 ± 13 (68-113)	31 ± 2.0 (27-35)	34 ± 5.8 (24-43)	11 ± 0.8 (8.8-12.0)
c′	4.5	0.84 ± 0.03 (0.81-0.88)	0.8 ± 0.11 (0.6-1.0)	0.9 ± 0.1 (0.7-1.0)	1.5 ± 0.2 (1.3-1.9)	4.6 ± 0.2 (4.2-5.1)
V	–	53 ± 1.7 (51-57)	–	54 ± 1.7 (51-57)	–	–
Body diam.	28	137 ± 39 (73-204)	269 ± 24 (242-323)	57 ± 4.4 (51-72)	129 ± 13 (110-149)	31 ± 2.2 (27-35)
Excretory pore	65	101 ± 6.2 (94-108)	126 ± 13 (106-156)	69 ± 5.8 (57-80)	73 ± 4.6 (67-82)	63 ± 2.7 (59-68)
Nerve ring	99	–	–	–	–	102 ± 4.7 (94-108)
Pharynx	140	165 ± 6.0 (135-173)	239 ± 13 (215-262)	140 ± 9.6 (124-153)	180 ± 10 (160-196)	138 ± 6.1 (119-148)
Testis reflexion	–	333 ± 157 (252-475)	–	132 ± 22 (98-152)	–	–
Tail length	81	40 ± 6.4 (33-48)	60 ± 7.6 (51-79)	31 ± 2.1 (27-35)	60 ± 5.4 (55-70)	81 ± 3.2 (76-89)
Anal body diam.	–	48 ± 6.1 (41-56)	80 ± 6.7 (70-94)	35 ± 4.3 (29-43)	42 ± 4.3 (35-51)	–
Spicule length	–	66 ± 1.4 (64-67)	–	54 ± 2.3 (51-59)	–	–
Gubernaculum length	–	44 ± 1.4 (43-46)	–	38 ± 3.3 (32-45)	–	–
D%	46	62 ± 2.3 (59-65)	53 ± 4.0 (47-63)	50 ± 4.1 (43-58)	41 ± 2.1 (36-44)	46 ± 1.4 (42-49)
E%	80	256 ± 50 (200-320)	213 ± 32 (153-257)	227 ± 18 (200-260)	123 ± 12 (100-143)	77 ± 3.4 (68-83)
SW%	–	139 ± 18 (120-162)	–	159 ± 22 (127-200)	–	–
GS%	–	67 ± 1.9 (64-69)	–	71 ± 6.3 (61-84)	–	–
H%	49	–	–	–	–	48 ± 1.8 (44-50)

from body contour (depending on age). Vulval opening located slightly posterior to mid-body. Small epiptygmata rarely observed. Short vagina leading into paired uteri. Narrow rectum, distinct anal opening. Tail of mature females obtuse with a conical projection on tip. In older females (these forming *endotokia matricida*) conus tip slightly rounded or slightly bifurcate. Postanal swelling not developed in mature females. Pygmy forms not observed. *Endotokia matricida* rarely observed.

Female (second generation)

Similar to first generation females in general morphology, but smaller. Vulval opening posterior to mid-body. Tail conical extended into a mucron at its end. Postanal swelling not developed.

Infective juvenile

Anterior region typical for *Steinernema* IJ. Lateral field beginning anteriorly with one line at annules 7 to 10. About three body diam. from anterior end, two additional lines appearing to form two ridges. Near excretory pore, number of ridges in lateral fields increasing from two to eight. If lateral field has nine ridges, a short distance after that, ninth ridge appearing at centre of lateral field. Nine ridges remaining unchanged to *ca* two-thirds of body and then central ridge disappearing to make eight. Near anus, eight ridges becoming six, then four. Between anus and phasmid, number of ridges in lateral field decreasing to two. Lateral field formula = either 2, 8, 9, 8, 6, 4, 2, or 2, 8, 6, 4, 2. Deirid not observed. Phasmids pore-like at *ca* 45% of tail length posterior to anus. Hemizonid distinct, lens-like in shape, located from beginning to middle of basal bulb. Excretory pore in mid-pharynx region, cuticularised, *ca* 1 μm wide. Cardia present. Bacterial vesicle slender, subpyriform, *ca* 24 $\times$ 13 μm in size. Rectum straight, anus sickle-shaped. Tail sharply conoid, terminus pointed. Hyaline region comprising *ca* half of tail length.

TYPE HABITAT AND LOCALITY

Natural host unknown. *Steinernema ichnusae* was collected from a sandy beach without any vegetation at Platamona Beach, north-west coast of Sardinia, Italy.

Type material

Holotype IJ, paratype IJ, males and females of first and second generation deposited in the EPN collection at the Laboratory of Insect Pathology, Institute of Entomology, Biology Centre CAS, Česke Budějovice, The Czech Republic; paratype IJ and adults in the nematode collection at the Department of Entomology and Zoology, University of Bari, Italy, and Department of Entomology and Nematology, University of Florida, Gainesville, FL, USA.

Relationships

This species belongs to the *feltiae*-group and is closely related to *S. feltiae*. It is characterised by morphological characters of the IJ, males and females. The IJ has a body length of 866 (767-969) μm, distance from labial region to excretory pore of 63 (59-68) μm, tail length of 81 (76-89) μm, ratio E% = 77 (68-83), hyaline region forming *ca* half of the tail length, and lateral field with eight to nine ridges, the lateral field formula = either 2, 8, 9, 8, 6, 4, 2, or 2, 8, 6, 4, 2. The male tail has a mucron in the second generation, rarely in the first generation. The first generation male spicule length = 66 (64-67) μm and gubernaculum = 44 (43-46) μm. The first generation female has no, or only a slight, vulval protrusion, no postanal swelling, characteristic mucrons on the tail tip, and D% = 53 (47-63). More information on relationships is available in Tarasco *et al.* (2008).

Molecular characterisation

Steinernema ichnusae is characterised genetically by the ITS (accession no. EU421129) and D2-D3 regions (accession no. EU421130). The sequence of ITS regions of *S. ichnusae* including primers TW81 and AB28 can be recognised by its length of 794 bp, ITS1 = 265 bp, ITS2 = 318 bp. For the D2-D3 regions, the sequence length of *S. ichnusae* is 840 bp. More information about molecular characterisation and phylogenetic analysis is available in Tarasco *et al.* (2008).

Symbiotic bacterium

No information.

Bionomics

In the laboratory, *S. ichnusae* reproduced well on *G. mellonella* larvae in an incubator at temperatures of 15 and 25°C. At 15°C, nematode-infected *G. mellonella* larvae died after 2 days, first generation adults appearing 4 days after that. After 7 days, second generation adults filled the cadavers and after 10 days the first IJ migrated into a water trap. At 25°C, the nematodes also developed as stated above except that the IJ emerged from cadavers after 8 days.

Distribution

Known only from Sardinia, Italy.

Steinernema innovationi[35,36] Çimen, Lee, Hatting, Hazir & Stock, 2015

(Fig. 4.44)

Measurements

See Table 4.16.

Diagnostic description

Male (first generation)

Anterior region typical for *Steinernema* male. Excretory pore opening circular, located anterior to nerve ring. Testis monorchic, ventrally reflexed. Spicules paired, symmetrical, curved. Spicule head region rhomboid. Shaft distinct. Blade with rostrum, two internal ribs prominent. Velum present, narrow. Blade terminus blunt. Gubernaculum boat-shaped or arcuate, 75% length of spicules. Head region of gubernaculum curved ventrally. Tail conoid and non-mucronate, with no bursa. There are 23 genital papillae (11 pairs and one single) arranged as follows: six precloacal pairs, of which five pairs are subventral and one pair is lateral,

[35] The specific epithet is derived from the *Innovation* Foundation that provided financial support for the survey of entomopathogenic nematodes in South Africa.

[36] Description after Çimen *et al.* (2015). Revised by Khuong B. Nguyen.

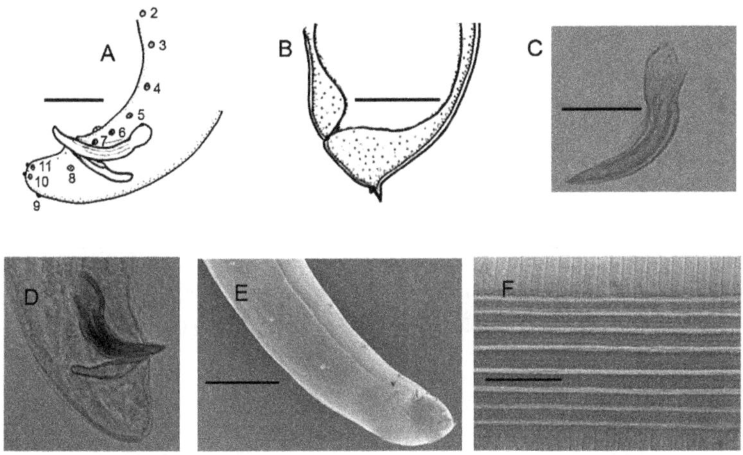

Fig. 4.44. Steinernema innovationi. *A: Posterior region of first generation male showing distribution of genital papillae; B: First generation female tail; C: First generation male spicule; D: Posterior region of first generation male showing spicule and gubernaculum; E: SEM of posterior region of first generation male showing distribution of genital papillae; F: SEM of lateral field showing eight ridges. (Scale bars: A = 50 μm; B = 55 μm; C, D = 40 μm; E = 100 μm; F = 10 μm.) Modified after Çimen* et al. *(2015), with acknowledgement to* Journal of Helminthology.

one pair in cloacal region, one single midventral papilla, and four pairs postcloacal, of which one pair subventral, one pair subdorsal and two pairs subventral and subterminal.

Male (second generation)

General morphology similar to that of first generation males, but smaller in size. Tail with or without mucron. Gubernaculum more slender than that of first generation male.

Female (first generation)

Anterior region typical for *Steinernema* female. Excretory pore located near mid-procorpus or surrounding isthmus. Gonads didelphic, amphidelphic, reflexed dorsally; oviduct well developed, glandular. Spermatheca and uterus in ventral position. Vagina short, with muscular walls. Vulva located near middle of body with slightly protruding lips and mostly symmetrical. Postanal lips non-protruding or slightly protruding. Tail blunt, conoid and with a digitate tip.

Table 4.16. *Morphometrics of* Steinernema innovationi. *All measurements are in μm and in the form: mean ± s.d. (range).*

Character	First generation			Second generation		Infective juvenile
	Holotype (male)	Male	Female	Male	Female	
n	–	24	25	25	25	25
L	1845	1896 ± 259 (1664-2773)	4070 ± 474 (3398-5423)	1322 ± 54 (1220-1435)	2063 ± 152 (1837-2281)	1054 ± 27 (1000-1103)
a	–	–	–	–	–	29 ± 3 (21-34)
b	–	–	–	–	–	7 ± 0.3 (6.5-7.6)
V	–	–	–	–	–	–
Head to vulva	–	–	2167 ± 228 (1729-2761)	–	1124 ± 84 (981-1743)	–
Body diam.	125	139 ± 21 (100-171)	173 ± 15 (144-202)	80 ± 7 (70-94)	149 ± 14 (123-175)	37 ± 5 (31-50)
Excretory pore	138	140 ± 12 (115-164)	147 ± 10 (125-168)	116 ± 6 (105-125)	127 ± 5 (116-136)	88 ± 3 (82-91)
Nerve ring	127	127 ± 10 (112-158)	135 ± 8 (126-154)	103 ± 6 (93-114)	122 ± 5 (113-129)	108 ± 4 (97-117)
Pharynx	166	170 ± 6 (162-190)	192 ± 6 (179-202)	148 ± 3 (142-154)	174 ± 2 (170-179)	150 ± 5 (142-160)
Tail length	50	51 ± 8 (42-76)	58 ± 3 (52-66)	39 ± 2 (34-41)	65 ± 6 (57-74)	76 ± 3 (66-80)
Anal body diam.	54	58 ± 7 (45-73)	65 ± 6 (49-83)	45 ± 2 (42-51)	50 ± 4 (41-57)	23 ± 3 (19-30)
Spicule length	80	81 ± 5 (74-91)	–	59 ± 0.8 (58-61)	–	–
Gubernaculum length	59	58 ± 3 (54-63)	–	41 ± 1 (39-43)	–	–
D%	83	82 ± 7 (67-95)	–	79 ± 4 (71-88)	–	58 ± 2 (54-63)
E%	277	279 ± 40 (196-325)	–	301 ± 15 (277-344)	–	115 ± 7 (104-137)
SW%	150	140 ± 20 (110-190)	–	130 ± 10 (120-140)	–	–
GS%	70	70 ± 10 (60-80)	–	70 ± 0* (70)	–	–
H%	–	–	–	–	–	43 ± 3 (37-46)

* Data cited as in original description.

Female (second generation)

Body open C-shape when heat-killed. Morphological characteristics similar to first generation female but smaller. Tail conoid, without postanal swelling.

Infective juvenile

Anterior region typical for *Steinernema* IJ. Excretory pore located near mid-corpus. Lateral field beginning anteriorly with one line at fourth or fifth annule, and posteriorly two additional lines appearing forming two ridges. More posteriorly, number of ridges increasing to six then eight at mid-body, evenly spaced. Ridges reducing to six near anus and to two near phasmid. With this arrangement, lateral field formula = 2, 6, 8, 6, 2. Anterior portion of intestine with small bacterial vesicle. Rectum long, straight, anus distinct. Genital primordium evident. Tail conoid with pointed terminus. Hyaline region occupying 32% of tail length.

Type habitat and locality

The natural host of this species is unknown. Isolate SGI-60 was collected from a grain field at the Agricultural Research Center, Small Grain Institute in Bethlehem, South Africa.

Type material

Holotype first generation male, five first generation paratype males, five first generation paratype females, and five paratype IJ deposited in the USDA Nematode Collection, MD, USA. Five first generation paratype males, five first generation paratype females, and five paratype IJ deposited at the University of California Davis Nematode Collection, Davis, CA, USA.

Relationships

Based on the body length, this species is a member of the *glaseri*-group. *Steinernema innovationi* is characterised by having long IJ with a body length of 1054 (1000-1103) μm, tail length = 108 (97-117) μm, location of the excretory pore 88 (82-91) μm, D% = 58 (54-63), E% = 115 (104-137), H% = 43 (37-46) and by the lateral field formula = 2, 6, 8, 6, 2. This species differs from other species by morphological characteristics and the shape of the spicules and gubernaculum. More information on relationships is available in Çimen *et al.* (2015).

MOLECULAR CHARACTERISATION

Steinernema innovationi is characterised genetically by sequences of ITS regions (accession no. KJ578793) and D2-D3 (28S) (accession no. KJ578794). More information about molecular characterisation and phylogenetic analysis is available in Çimen *et al.* (2015).

SYMBIOTIC BACTERIUM

No information.

BIONOMICS

No information.

DISTRIBUTION

Steinernema innovationi was collected in Bethlehem, South Africa.

Steinernema jeffreyense[37,38] **Malan, Knoetze & Tiedt, 2015**
(Fig. 4.45)

MEASUREMENTS

See Table 4.17.

DIAGNOSTIC DESCRIPTION

Male (first generation)

Anterior region typical for *Steinernema* male. Excretory pore anterior of nerve ring. Cardia prominent. Testis monorchic, reflexed; spicules paired; gubernaculum 65% of spicule length. Spicule with manubrium longer than wide, ratio of head region length:head region width 2:1, shaft almost absent, blade moderately curved, with two internal ribs,

[37] The specific epithet is derived from *Jeffrey* Bay, South Africa, where the nematode was collected.

[38] Description from Malan *et al.* (2015). Revised by Khuong B. Nguyen.

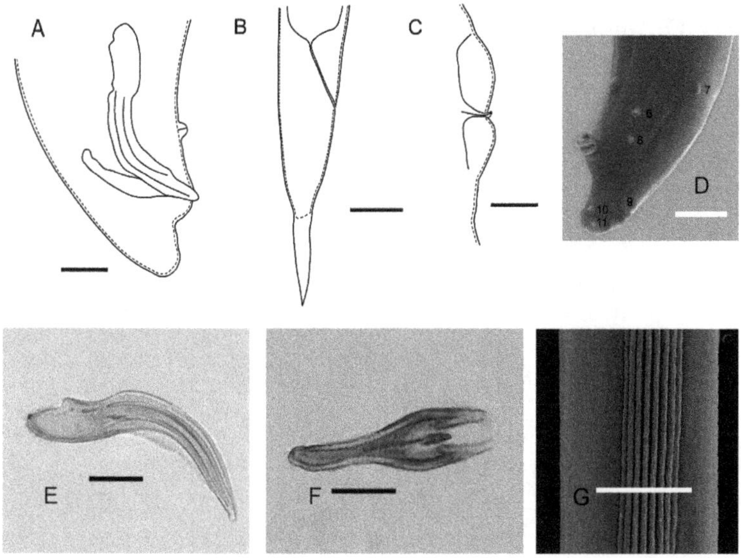

Fig. 4.45. Steinernema jeffreyense. *A: First generation male tail; B: Infective juvenile tail; C: First generation female vulva showing small epiptygmata; D: First generation male tail region showing posterior genital papillae; E, F: First generation male spicule and gubernaculum, respectively; G: Infective juvenile lateral field. (Scale bars: A, B, D-F = 20 μm; C = 50 μm; G = 10 μm.) Modified from Malan* et al. *(2015), with acknowledgement to* Journal of Helminthology.

velum prominent, not reaching tail tip. Gubernaculum with short cuneus, not reaching tip of wings. Six pairs of genital papillae, with a single precloacal midventral papilla, two pairs adcloacal, one pair lateral, two pairs subterminal, and one pair subdorsal. Tail tip either rounded or digitate, without a mucron.

Male (second generation)

Body smaller than first generation male, tightly curled posteriorly, sometimes 6-shaped. Tail of second generation male similar to that of first generation male, and always without a mucron.

Female (first generation)

Anterior region typical for *Steinernema* female. Excretory pore located just anterior to, or at middle of, basal bulb. Gonads didelphic, amphidelphic, reflexed, containing many eggs. Vulva at mid-body, with an asymmetrical pronounced protuberance in older females, with short

Table 4.17. *Morphometrics of* Steinernema jeffreyense. *All measurements are in μm and in the form: mean ± s.d. (range).*

Character	First generation			Second generation		Infective juvenile
	Holotype (male)	Male	Female	Male	Female	
n	–	20	20	20	20	24
L	1711	1634 ± 130* (1740-1899)	3863 ± 817 (1943-5115)	1212 ± 53 (1119-1301)	1508 ± 123 (1316-1687)	926 ± 83 (784-1043)
a	10	12 ± 1.7 (9.9-15.2)	16 ± 2.7* (19-22)	17 ± 12 (15-21)	16 ± 1.4 (13-19)	26 ± 3.2 (20-35)
b	11	11 ± 1.0 (9.4-13)	26 ± 3.8 (19-34)	8.5 ± 0.5 (7.5-9.4)	13 ± 1.8 (9.6-16)	7.0 ± 0.5 (6.1-7.9)
c	58	61 ± 6.2 (49-73)	47 ± 12 (35-74)	32 ± 3.0 (27-38)	20 ± 3.0 (16-28)	12 ± 2.0 (9.7-19)
c′	0.67	0.65 ± 0.05 (0.55-0.71)	0.62 ± 0.2 (0.24-0.99)	0.9 ± 0.1 (0.7-1.1)	1.3 ± 0.2 (0.6-1.6)	2.7 ± 0.4 (2.1-3.5)
V	–	–	53 ± 2 (51-58)	–	53 ± 2.5 (48-58)	–
Body diam.	147	139 ± 22 (94-167)	244 ± 40 (135-337)	72 ± 4.3 (64-80)	93 ± 5.2 (85-102)	35 ± 4.4 (23-43)
Excretory pore	71	71 ± 10 (50-99)	148 ± 21 (101-184)	112 ± 6.0 (100-120)	114 ± 10 (95-138)	87 ± 9.3 (78-107)
Nerve ring	122	121 ± 8.3 (100-137)	136 ± 14 (113-159)	102 ± 3.1 (94-106)	120 ± 5.8 (110-128)	100 ± 7.0 (85-114)
Pharynx	153	151 ± 8.5 (136-165)	190 ± 12 (171-209)	142 ± 5.6 (129-153)	163 ± 5.3 (155-171)	133 ± 7.2 (116-149)
Testis reflexion	473	451 ± 49 (366-574)	–	238 ± 33 (205-285)	–	–
Tail length	30	27 ± 3.01 (21-33)	56 ± 15 (29-89)	27 ± 3.5 (22-35)	78 ± 7.8 (52-88)	81 ± 10.1 (50-96)
Anal body diam.	44	42 ± 3.5 (35-49)	89 ± 26 (38-126)	44 ± 3.8 (36-51)	62 ± 6.6 (54-85)	29 ± 4.6 (19-39)
Spicule length	92	88 ± 3.6 (79-95)	–	72 ± 3.5 (66-81)	–	–
Spicule width	12	12 ± 0.8 (11-13)	–	10 ± 1.0 (9.3-14)	–	–
Gubernaculum length	59	57 ± 2.8 (51-61)	–	43 ± 2.1 (39-48)	–	–
D%	46	47 ± 8.4 (34-68)	78 ± 10 (59-99)	78 ± 4.7 (71-87)	70 ± 7.3 (57-85)	66 ± 6.1 (57-85)
E%	46	47 ± 8.4 (34-68)	286 ± 111 (193-616)	78 ± 4.7 (71-87)	149 ± 22 (124-223)	109 ± 19 (86-169)
SW%	211	215 ± 27 (171-295)	–	165 ± 14 (140-189)	–	–
GS%	64	65 ± 2.57 (61-71)	–	60 ± 3.3 (51-66)	–	–
H%	–	–	–	–	–	43 ± 6.3 (20-52)

* Data as per original description.

double-flapped epiptygmata always forward anteriad. Anus without postanal swelling. Tail shorter than anal body diam., conoid, bluntly pointed with dorsal projection on rounded tail tip.

Female (second generation)

Body straight to slightly curved at vulva. Vulva with asymmetrical protuberance, double-flapped epiptygmata present, anteriorly directed. Tail tapering gently to a sharp point, no postanal swelling.

Infective juvenile

Anterior region typical for *Steinernema* IJ. Position of excretory pore in vicinity of nerve ring just anterior to basal bulb. Hemizonion distinct, posterior to excretory pore. Stoma closed. Lateral fields of exsheathed larvae with following formula: 2, 4, 6, 8, 7, 6, 2. Rectum long, straight, anus distinct. Tail conoid, hyaline region occupying 54% of tail length.

TYPE HABITAT AND LOCALITY

The host of this nematode in nature is unknown as it was recovered from soil using *G. mellonella* larvae as bait. The species was collected around a guava tree in sandy soil, close to the ocean, in the town of Jeffrey's Bay in the Eastern Cape province, South Africa.

TYPE MATERIAL

Holotype male (mounted) and 20 paratypes each of females, males and many IJ in TAF, isolated from the haemocoel of *G. mellonella*, were deposited in the United States Department of Agriculture, Nematode Collection (USDANC), Beltsville, MD, USA. In addition, 20 paratypes each of hermaphrodites, males, females and many IJ in TAF were deposited in the National Collection of Nematodes, Biosystematics Division, Plant Protection Research Institute, Agricultural Research Council, Pretoria, South Africa. Slides of one male, one female of the second generation and several IJ were deposited in the California Collection of Nematodes, University of California Davis Nematode Collection, Davis, CA, USA. Several slides of hermaphrodites, males and females, with all stages in TAF, are maintained in the Department of Entomology at Stellenbosch University, South Africa.

Relationships

Steinernema jeffreyense is characterised by morphometrics of the IJ with body length 926 (784-1043) μm, distance from anterior end to excretory pore = 87 (78-107) μm, tail length = 81 (52-96) μm, ratio a = 26 (20-35), H% = 43 (20-52) and E% = 109 (86-169). The first generation male can be recognised by the length and shape of the spicule and gubernaculum, position of the excretory pore, D% = 47 (34-68) and SW% = 215 (171-295). The female can be recognised by the vulva having low epiptygmata and the presence of a dorsal projection on the rounded tail tip. More information on relationships is available in Malan *et al.* (2015).

Molecular characterisation

The sequence of ITS regions (accession no. KC897093) of *S. jeffreyense*, flanked by primers 18S and 26S, is characterised by its sequence length of 1050 bp, ITS1 = 278 bp, ITS2 = 323 bp. For the D2-D3 regions, the sequence, flanked by the primers D2F and 536 (accession no. KP164866), is 586 bp. More information about molecular characterisation and phylogenetic analysis is available in Malan *et al.* (2015).

Symbiotic bacterium

No information.

Bionomics

No information.

Distribution

Known only from the type locality, Eastern Cape Province, South Africa.

Steinernema lamjungense[39,40] Khatri-Chhetri, Waeyenberge, Spiridonov, Manandhar & Moens, 2011
(Figs 4.46, 4.47)

MEASUREMENTS

See Table 4.18.

DIAGNOSTIC DESCRIPTION

Male (first generation)

Anterior region typical for *Steinernema* male. Body curved ventrally posteriorly, tapering toward anterior end, mostly C-shaped, rarely J-shaped in heat-relaxed specimens. Cuticle with very faint transverse striae, lateral field present with one ridge. Labial region rounded, slightly offset from rest of body. Six lips bearing labial papilla. Four cephalic papillae present. Amphids distinct, located between labial and cephalic circle of papillae. Excretory pore always anterior to nerve ring, mostly located at level of isthmus, sometimes at metacorpus. Well developed and reflexed testis. Testis reflexion ventral in position and less than half of testis length. Paired, symmetrical, moderately curved, yellowish-brown coloured spicules. Spicule manubrium usually rounded, occasionally tapering anteriorly with length:width ratio of 1:1, manubrium almost 25% of spicule length, very short calomus separating manubrium from lamina. Lamina well curved anteriorly and dorsally, with two internal ribs, slender and almost straight posteriorly. Rostrum not well developed, velum bow-like, strong and distinct, extending from base part not covering distal part of lamina. Spicule terminus blunt with ventral notch. In lateral view, gubernaculum boat-shaped to almost straight, anterior part slightly upturned, posterior part usually tapering. In ventral view, cuneus narrow V-shaped, post-forked, corpus with two wings. Tail bluntly conoid, mucron absent. Twenty-three genital papillae, comprising 11 pairs and a single midventral papilla located just anterior to cloacal opening. Of these, seven pairs located precloacal (six pairs subventral and one pair lateral), one pair adcloacal, two

[39] Specific epithet named after the *Lamjung* district of Nepal, the area where the species was found.
[40] Description after Khatri-Chhetri *et al.* (2011a). Revised by Sergei E. Spiridonov.

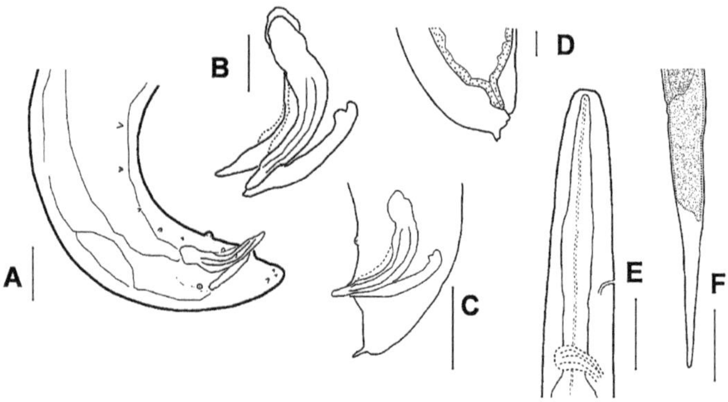

Fig. 4.46. Steinernema lamjungense. *A: First generation male, tail; B: First generation male, spicules and gubernaculum; C: Second generation male, posterior end; D: First generation female, posterior end; E: Infective juvenile, anterior end; F: Infective juvenile, tail. (Scale bars: A, C, D = 50 μm; B, E, F = 25 μm.) After Khatri-Chhetri* et al. *(2011a), courtesy* Nematology.

pairs subterminal, and one pair subdorsal. Posteriormost three pairs consistently located near tail tip.

Male (second generation)

Second generation male similar to the first generation except for length and body diam., spicule and gubernaculum, and presence of tail mucron.

Female (first generation)

Anterior region typical for *Steinernema* female. Body variable in length, usually C-shaped, sometimes coiled on heat relaxation. Cuticle faintly striated. Lateral field not observed in light microscopy. Labial region very small, bluntly rounded, sometimes abruptly tapering anteriorly. Cardia prominent. Excretory pore located anterior to nerve ring, duct well cuticularised. Genital tract amphidelphic with reflexed ovaries. Vulva in form of transverse slit, lips protruding, lips mostly unequal, anterior more protruding than posterior, small epiptygmata present. Sometimes, forms observed with well protruding or non-protruding vulval lips. Vulval opening mostly at mid-body. Short vagina leading into paired uteri. Postanal swelling usually not present, if present usually not well developed. Tail shape variable, mostly nipple-shaped with blunt terminus sometimes a small mucron-like structure present on dorsal side of

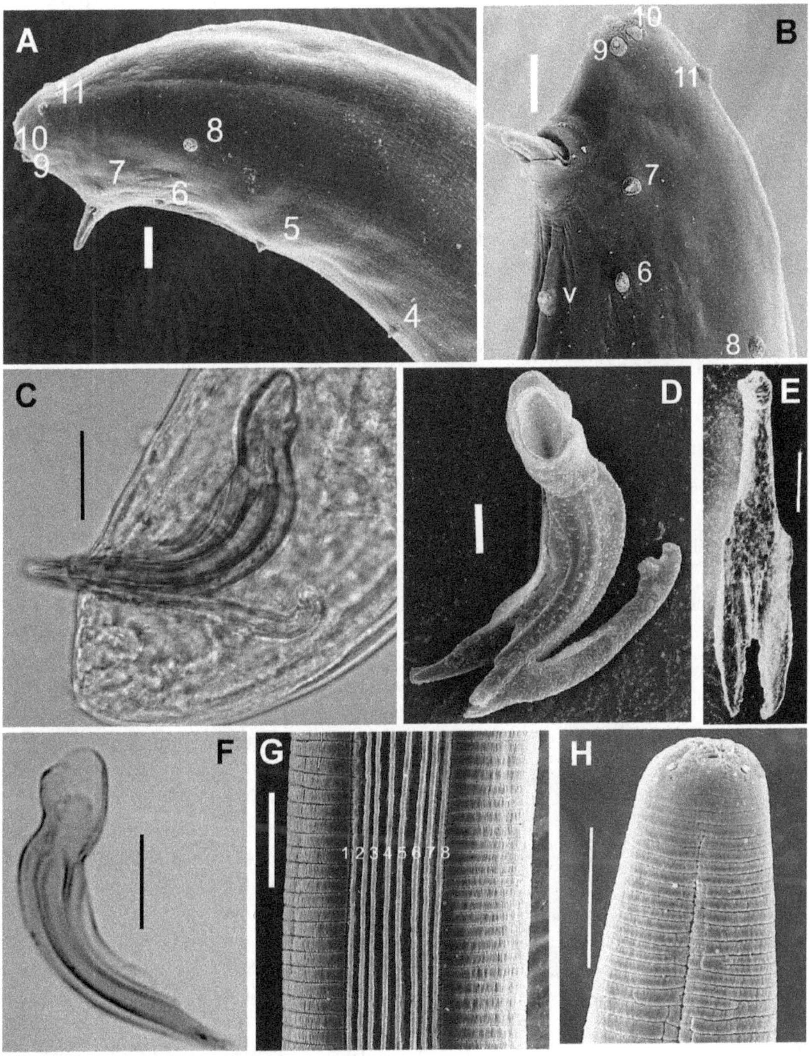

Fig. 4.47. Steinernema lamjungense. *A, B: First generation male, tail (papillae numbered); C: First generation male, tail; D: First generation male, spicule and gubernaculum; E: First generation male, gubernaculum; F: First generation male, spicule (light microscopy); G: Infective juvenile, lateral field (ridges numbered 1-8); H: Infective juvenile, labial region. (Scale bars: A, B, D-H = 10 μm; C = 20 μm.) After Khatri-Chhetri* et al. *(2011a), courtesy* Nematology.

Table 4.18. *Morphometrics of type population of* Steinernema lamjungense. *All measurements are in* μ*m and in the form: mean* $\pm$ *s.d. (range).*

Character	First generation			Second generation		Infective juvenile
	Holotype (male)	Male	Female	Male	Female	
n	–	20	20	20	20	20
L	2306	1865 ± 191 (1567-1963)	5260 ± 1156 (2755-6939)	1199 ± 149 (1013-1679)	1136 ± 109 (968-1315)	832 ± 88 (690-950)
a	12	12 ± 1.3 (9.5-14.6)	18 ± 3.2 (11-22)	16 ± 2.4 (11-20)	15 ± 2.2 (12-20)	30 ± 3.1 (23-40)
b	13	11 ± 1.3 (8.4-13)	22 ± 4.9 (10-28)	8 ± 2.0 (7-14)	8.3 ± 1.0 (6.8-10.3)	6.6 ± 0.5 (5.8-7.2)
c	70	61 ± 11 (47-91)	216 ± 87 (97-408)	47 ± 8.0 (31-60)	33 ± 4.1 (27-42)	9.6 ± 0.7 (7.7-10.8)
c′	–	–	–	0.7 ± 0.1 (0.5-0.8)	56 ± 1.2 (54.1-58.1)	–
V	–	–	–	50 ± 9.5 (41-88)	55 ± 3.9 (46-61)	–
Max. body diam.	185	154 ± 22 (116-205)	315 ± 49 (242-386)	77 ± 21 (57-138)	76 ± 9.3 (59-93)	27 ± 2.4 (23-31)
Excretory pore	112	105 ± 18 (80-172)	82 ± 18 (52-110)	79 ± 6.2 (68-88)	61 ± 6.6 (50-81)	68 ± 3.7 (61-73)
Pharynx	183	173 ± 11 (155-190)	239 ± 39 (190-380)	143 ± 15 (110-167)	138 ± 9.2 (125-162)	127 ± 8.8 (111-142)
Tail length	33	31 ± 4.3 (21-38)	27 ± 7.5 (14-40)	26 ± 3.1 (21-34)	35 ± 4.1 (28-43)	88 ± 11 (67-103)
Hyaline region	–	–	–	–	–	53 ± 8.2 (37-71)
Anal body diam.	58	50 ± 3.8 (43-58)	64 ± 12 (50-93)	36 ± 3.8 (31-44)	31 ± 6.7 (22-52)	17 ± 2.5 (14-25)
Spicule length along chord	94	87 ± 3.1 (81-94)	–	73 ± 6.1 (66-93)	–	–
Gubernaculum length	56	57 ± 3.5 (50-66)	–	46 ± 4.4 (40-58)	–	–
D%	61	61 ± 11 (44-91)	33 ± 11 (21-47)	56 ± 6.7 (47-71)	44 ± 3.9 (38-52)	54 ± 3.8 (48-63)
E%	341	345 ± 75 (237-554)	313 ± 133 (193-707)	308 ± 45 (217-400)	179 ± 26 (128-246)	79 ± 8.7 (69-97)
SW%	161	173 ± 12 (151-193)	204 ± 21 (163-251)	–	–	–
GS%	60	65 ± 4.3 (56-77)	64 ± 4.9 (55-73)	–	–	–
H%	–	–	–	–	–	61 ± 3.7 (55-69)

tail terminus. Pygmy forms not observed. *Endotokia matricida* rarely observed.

Female (second generation)

Similar to first generation female in general morphology, but smaller. Body slightly curved on heat relaxation. Vulval opening slightly posterior to mid-body. Tail tapering to pointed end. Swollen vulva mostly with small epiptygmata. Postanal swelling mostly developed.

Infective juvenile

Anterior region typical for *Steinernema* IJ. Heat-relaxed specimens with medium long, slightly curved, slender body, slightly tapering towards anterior and posterior ends. IJ enclosed in sheath (second-stage cuticle), sheath usually lost during storage or in fixative. Cuticular striation indistinct under light microscope, but distinct with SEM, 1.5 μm thick in mid-body region. Lateral field beginning anteriorly with a single ridge at third annule. At annule 6 or 7, two additional ridges, more prominent from annule 18 or 19. Each ridge dividing into three, ridges increasing from two to six over a very short distance (10-12 annules). One extra ridge (number seven) appearing in middle, also over very short distance. Middle ridge dividing and number of ridges becoming eight. Number of ridges remaining unchanged over long distance. At phasmid level, ridges merging to three over few annules up to constriction of tail where middle ridge disappearing and number of ridges becoming two on tail. With above arrangement, lateral field formula = 2, 6, 7, 8, 3, 2. Deirid not observed. Phasmid pore-like, posterior to anus, located at about first quarter of tail. Labial region from rounded to slightly truncate, usually slightly offset from body contour, not annulated. Excretory pore in mid-pharynx region, cuticularised, *ca* 1 μm wide. Cardia present. Bacterial vesicle subpyriform, *ca* 17 $\times$ 7 μm in size. Intestine filled with numerous fat globules, lumen not visible. Genital primordia prominent. Rectum straight, anus sickle-shaped. Tail, sometimes with slight dorsal constriction situated between anus and proximal portion of hyaline region. Hyaline region occupying slightly more than half of tail length. Tail attenuate with blunt terminus.

Type habitat and locality

Collected from soil samples taken from mixed deciduous forest dominated by *Schima wallichii* and *Costanopsis indica* in Tarku (84°23′E,

28°09′N, 710 m a.s.l.), Lamjung district in western mid-hills region of Nepal. The soil was loamy, acidic (pH 4.2) with moderate organic matter content.

TYPE MATERIAL

Holotype male (UGMD 104182), 12 male paratypes (four in UGMD 104182) and eight in (UGMD 104183) and five female paratypes (UGMD 104184) deposited in the nematode collection of the Institute of Zoology, Ghent University, K.L. Ledeganckstraat 35, Ghent, Belgium. Five female paratypes and six male paratypes also deposited in the nematode collection of Nepal Agricultural Research Council, Nepal.

RELATIONSHIPS

The IJ of can be recognised by the body length of 832 (690-950) μm, pharynx length of 127 (110-142) μm, H% = 61 (55-69) and E% = 79 (69-97). The lateral field pattern is 2, 6, 7, 8, 3, 2. The males have well curved, yellowish-brown spicules with a prominent velum. The spicule terminus is blunt with a longitudinal ventral notch. In lateral view, the gubernaculum is boat-shaped to almost straight, cuneus narrow V-shaped, forked, corpus with two wings. Spicule length of the first generation males is 87 (81-94) μm and gubernaculum length is 57 (50-66) μm. A tail mucron is absent in the first generation male and present in the second generation. Genital papillae total 23, comprising 11 pairs and a single midventral papilla. Of these, seven pairs are located precloacal (six pairs subventral and one pair lateral), one pair adcloacal, two pairs subterminal, and one pair subdorsal. The vulva is usually slightly protruding with unequal lips and epiptygmata. A postanal swelling is absent in the first generation, but well developed in the second generation, female.

Steinernema lamjungense is a member of the *arenarium*/*glaseri*/*karii*/*longicaudum*-group (Spiridonov *et al.*, 2004) and morphologically similar to *S. guangdongense*.

MOLECULAR CHARACTERISATION

ITS rDNA sequence of *S. lamjungense* (HM000101) reveals its closeness to *S. khoisanae*, *S. guangdongense* and *S. longicaudum* (*i.e.*, the *longicaudum*-subgroup of the *glaseri*-group). Phylogenetic analysis of D2-D3 sequence of LSU rDNA (deposited as HM000102) revealed only that *S. lamjungense* is a species of the *glaseri*-group (Khatri-Chhetri *et al.*, 2011a).

SYMBIOTIC BACTERIUM

No information.

BIONOMICS

Cross-hybridisation tests with *S. glaseri* produced no progeny. In addition to *G. mellonella*, *S. lamjungense* reproduced well on rice moth larvae, *Corcyra cephalonica*, lesser wax moth larvae, *Achroia grisella*, silk worm, *Bombyx mori*, and *Spodoptera littoralis* (Khatri-Chhetri *et al.*, 2011a). The natural hosts are unknown.

DISTRIBUTION

In addition to the type population, three other isolates of this species were reported by Khatri-Chhetri *et al.* (2011a) from the neighbourhood of the type locality (84°23′E, 28°09′N, 690-710 m a.s.l.), and another from Swarketari, Syangja (83°53′E, 28°09′N, 1133 m a.s.l.) district. These are distant locations (>100 km) but have similar mild temperatures and vegetation type.

Steinernema minutum[41,42] Maneesakorn, Grewal & Chandrapatya, 2010

(Fig. 4.48)

MEASUREMENTS

See Table 4.19

DIAGNOSTIC DESCRIPTION

Male (first generation)

Anterior region typical for *Steinernema* male. Body slender, ventrally curved posterior, J-shaped when heat-relaxed. Cuticle smooth with

[41] Specific epithet refers to the small body size of the species.

[42] Description after Maneesakorn *et al.* (2010). Revised by David J. Hunt.

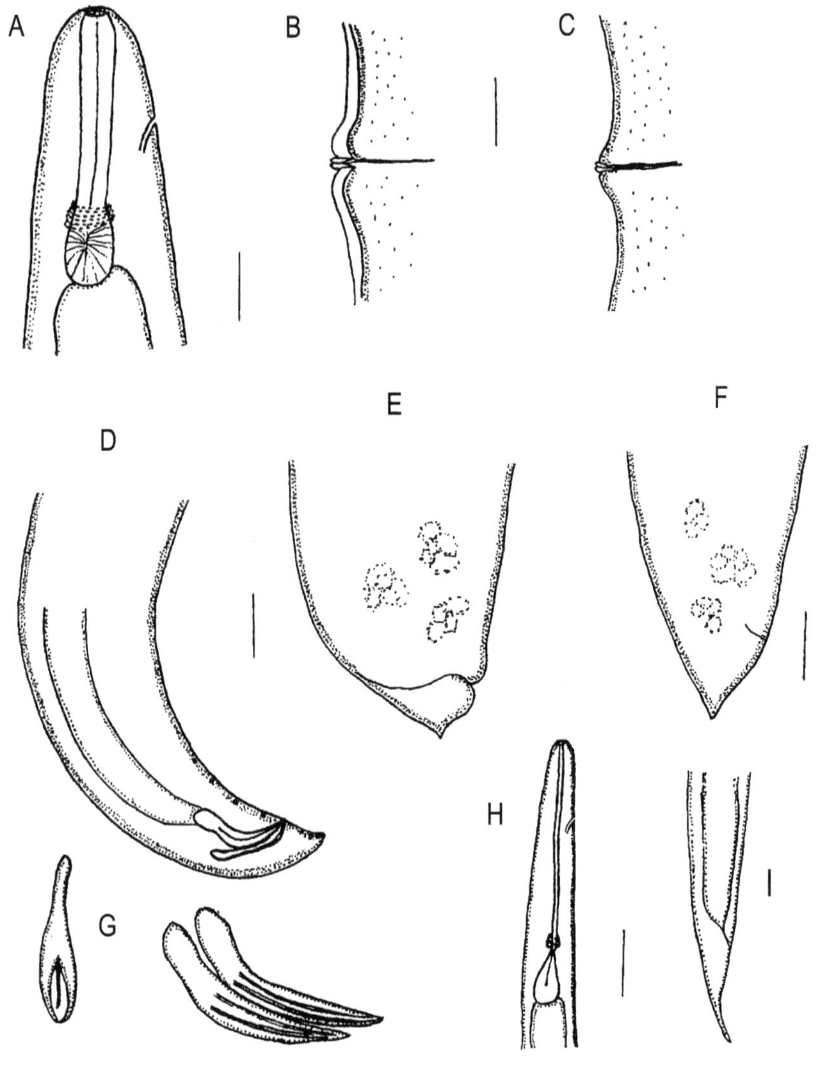

Fig. 4.48. Steinernema minutum. *A: Anterior of first generation female; B, C: Vulval region of first and second generation female; D: Posterior region of first generation male; E, F: Posterior region of first and second generation female; G: Spicule and gubernaculum of first generation male; H, I: Anterior and posterior of infective juvenile. (Scale bars: A, D, F = 50 μm; B, C, E, G, H, I = 25 μm.) Modified after Maneesakorn* et al. *(2010), courtesy* International Journal of Nematology.

Table 4.19. *Morphometrics of* Steinernema minutum. *All measurements are in μm and in the form: mean ± s.d. (range).*[*]

Character	First generation		Second generation		Infective juvenile
	Male	Female	Male	Female	
	Paratypes	Paratypes	Paratypes	Paratypes	Paratypes
n	20	20	20	20	20
L	1566 ± 100 (1394-1737)	6390 ± 763 (5252-8040)	816 ± 54 (737-909)	1600 ± 145 (1414-1980)	403 ± 22.2 (364-455)
a	–	–	–	–	19.2 ± 1.1 (17.4-21.6)
b	–	–	–	–	4.3 ± 0.3 (3.9-4.9)
c	–	–	–	–	9.9 ± 0.8 (8.5-11.2)
Max. body diam.	124 ± 10.3 (104-150)	215 ± 20.1 (185-258)	57 ± 5.9 (48-68)	101 ± 15.3 (80-139)	21.0 ± 1.0 (19.6-22.7)
Excretory pore	71 ± 4.6 (65-81)	83 ± 7.6 (67-96)	54 ± 3.9 (45-62)	53 ± 5.6 (40-62)	34 ± 1.4 (31-36)
Nerve ring	102 ± 7.2 (87-116)	155 ± 10.1 (140-171)	84 ± 4 (79-93)	102 ± 6.4 (89-133)	66 ± 2.7 (61-73)
Pharynx	140 ± 6.7 (122-152)	209 ± 13 (190-234)	114 ± 3.1 (108-119)	145 ± 6.5 (131-155)	94 ± 3.4 (85-99)
Tail	25 ± 1.9 (21-28)	23 ± 4.3 (18-35)	17 ± 1.3 (14-19)	27 ± 3.1 (21-31)	41 ± 2.6 (37-46)
Anal body diam.	40 ± 4.0 (32-46)	68 ± 7.9 (56-83)	31 ± 3.5 (23-35)	35 ± 5.0 (25-45)	11.5 ± 0.6 (10.4-12.9)
Spicule length	75 ± 4 (68-84)	–	65 ± 5.2 (57-78)	–	–
Spicule width	11.5 ± 1.2 (8.4-13.2)	–	9.5 ± 1.4 (7.8-12.9)	–	–
Gubernaculum length	58 ± 4.2 (50-66)	–	46 ± 5.1 (37-54)	–	–
Gubernaculum width	6.4 ± 0.7 (5.2-7.5)	–	5.4 ± 0.9 (3.7-6.7)	–	–
D%	50.7 ± 3.4 (46.7-61.2)	39.7 ± 3.7 (32.2-46.2)	47.4 ± 3.7 (40.7-54.4)	36.8 ± 4 (28.7-45.7)	35.8 ± 1.4 (33.3-39.1)
E%	–	–	–	–	82.8 ± 5 (76.4-94.6)
SW%	189 ± 20.9 (156-232)	–	212 ± 27.3 (171-260)	–	–
GS%	77.2 ± 6 (66.1-87.1)	–	71.6 ± 7.4 (59.2-85.1)	–	–
H%	–	–	–	–	49.4 ± 4.8 (37.9-57.9)

[*] Holotype data not cited in original paper.

very faint transverse striae. Lateral field and phasmids inconspicuous. Excretory pore opening located in anterior half of pharynx. Testis single, reflexed. Spicules paired, symmetrical, strongly curved, ochre-brown in colour, head region (manubrium) length (46 μm), longer than wide (27 μm), calomus (shaft) distinct, lamina with rostrum or retinaculum and two internal ribs, velum present. Gubernaculum arcuate, *ca* 77% of spicule length, boat-shaped in lateral view, swollen in middle with ventrally curved knob at proximal end, cuneus long, bifurcate, not reaching end of corpus, corpus separated posteriorly. Tail conoid with an inconspicuous mucron. Genital papillae totalling 19, comprising nine pairs and single midventral precloacal papilla. Paired papillae arranged as five pairs subventral precloacal, one lateral adcloacal pair, one postcloacal subdorsal pair, and two pairs subventral arranged close together near tail terminus.

Male (second generation)

Similar to first generation, but smaller. Tail mucron prominent, 3-4 μm long.

Female (first generation)

Anterior region typical for *Steinernema* female. Body robust, C-shaped when heat-killed. Excretory pore cuticularised, excretory gland swollen. Cardia prominent, protruding into intestinal lumen. Genital system amphidelphic, reflexed, filled with eggs. Vulva located in mid-body region, in form of a transverse slit, protruding from body, double epiptygmata present. Vagina short, oblique, with muscular walls. Slight postanal swelling. Tail shorter than anal body diam., mucron always present.

Female (second generation)

Similar to first generation female in general morphology, but smaller. Vulva symmetrical, lips slightly protruding, double epiptygmata present. Excretory pore positioned further anteriad. Tail conoid, sharply pointed, longer than anal body diam., postanal swelling absent.

Infective juvenile

Anterior region typical for *Steinernema* IJ. Body almost straight when heat-killed, tapering regularly from base of pharynx to anterior end and from anus to terminus. Mouth and anus closed. Cuticle striated. Pharynx long, narrow, isthmus distinct, nerve ring located just anterior to basal bulb, basal bulb elongate with valve. Cardia prominent. Excretory pore located at 36% of pharynx length. Hemizonid distinct, anterior to basal bulb. Unfortunately, characteristics of lateral field and incisures neither mentioned nor figured in original description – a major omission. Tail long, constricted in hyaline region, especially on dorsal side. Hyaline region comprising *ca* 49% of tail length.

TYPE HABITAT AND LOCALITY

The natural host is unknown, the species being collected by *G. mellonella* baiting soil collected from beneath a pine tree at Amphoe Mueang, Chumphon Province, southern Thailand, in April 2007. The soil was a sandy-loam and acidic (pH 4.4).

TYPE MATERIAL

Holotype, first generation male, five first generation paratype males, five first generation paratype females and five paratype IJ were deposited in the USDA Nematode Collection, Beltsville, MD, USA; five first generation paratype males, five first generation paratype females and five paratype IJ deposited in the University of California Davis Nematode Collection, Davis, CA, USA; five first generation paratype males, five first generation paratype females and five paratype IJ were deposited in the Department of Entomology, Kasetsart University, Bangkok, Thailand.

RELATIONSHIPS

Steinernema minutum belongs to a group of species related to *S. tami*, which are in the 'small IJ' or *carpocapsae*-group, and characterised by an IJ body length of <600 μm.

MOLECULAR CHARACTERISATION

The ITS sequence of 852 bp (accession no. GU647156) was compared with that of *Steinernema* species deposited in the NCBI GenBank database and showed 97% similarity to the most closely related species, *S. tami*. Further details of the ITS phylogeny are available in Maneesakorn *et al.* (2010).

Bacterial symbiont

The 16S rDNA sequence of 916 bp for the bacterium isolated from *S. minutum* species showed 100% similarity with *Xenorhabdus stockiae*, a recognised symbiont of *S. siamkayai*, another species originally isolated from Thailand.

Bionomics

Cross-hybridisation studies between *S. minutum* and *S. siamkayai* Stock, Somsook & Reid, 1998 produced no progeny according to Maneesakorn *et al.* (2010). Live material of other closely related species was not available for testing.

Distribution

Known only from Thailand (Maneesakorn *et al.*, 2010).

Note

This species is not well described by contemporary standards, lacks important information, including morphometric data for the holotype and the lateral field pattern of the IJ, and is in urgent need of redescription. However, the molecular data, although based only on ITS, appear to indicate that it is a valid species, hence its inclusion here rather than being regarded as *species inquirenda*.

Steinernema nepalense[43,44] Khatri-Chhetri, Waeyenberge, Spiridonov, Manadhar & Moens, 2011

(Figs 4.49, 4.50)

Measurements

See Table 4.20.

[43] Specific epithet derived from *Nepal*, the country of origin.
[44] Description after Khatri-Chhetri *et al.* (2011b). Revised by Sergei E. Spiridonov.

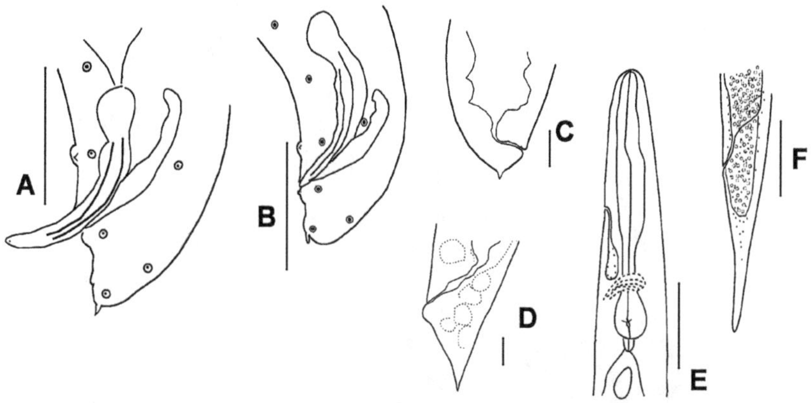

Fig. 4.49. Steinernema nepalense. *A: First generation male, tail; B: Second generation male, tail; C: First generation female, tail; D: Second generation female, tail; E: Infective juvenile, anterior end; F: Infective juvenile, tail. (Scale bars: A-D = 50 μm; E, F = 25 μm.) After Khatri-Chhetri* et al. *(2011b), courtesy* Russian Journal of Nematology.

Diagnostic description

Male (first generation)

Anterior region typical of *Steinernema* male. Body mostly C-shaped, sometimes J-shaped in heat-relaxed specimens. Cuticle with faint transverse striae, lateral differentiation of cuticle in shape of single rounded or flattened ridge. Labial region rounded, mostly offset from rest of body. Six lips, each lip bearing a labial papilla. Four cephalic papillae present. Amphids distinct, crescent-shaped, located laterally between labial and cephalic circle of papillae. Excretory pore mostly at base of metacorpus and always anterior to nerve ring. Well developed reflexed testis. Testis reflexion ventral in position, forming less than half of testis length. Paired, symmetrical, moderately curved, yellowish-brown, slender spicules. Spicule manubrium shape variable, usually elongated with length longer than width, manubrium almost one-fourth of spicule length, long calomus separating manubrium from lamina. Lamina well curved anteriorly and dorsally with two thin internal ribs. Rostrum not well developed, velum thin, distinct. Spicule terminus round. Gubernaculum slender boat-shaped in lateral view, with anterior part curved upwards ventrally, posterior part usually tapering. In ventral view cuneus long, Y-shaped, corpus with two wings. Tail bluntly conoid, short mucron present (2.8 ± 0.93 μm). Twenty-five genital papillae

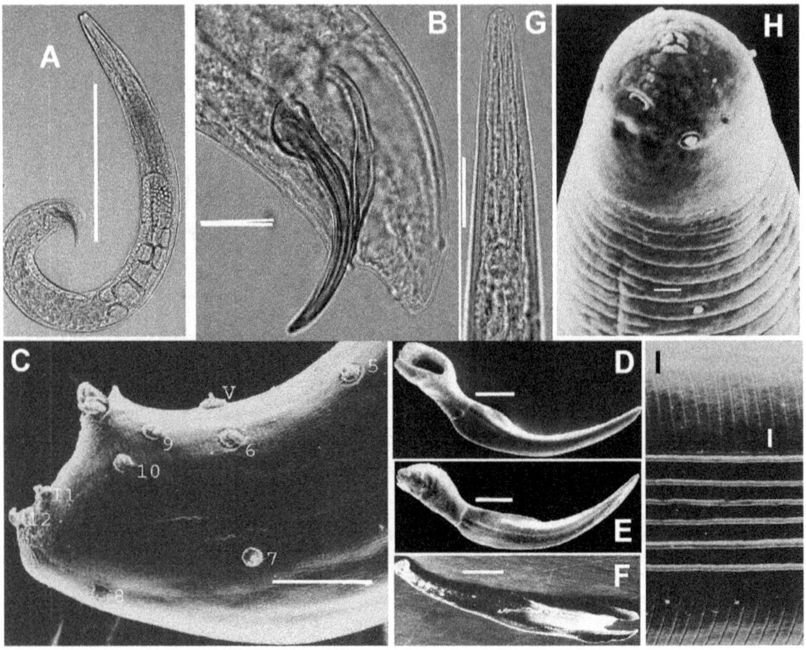

Fig. 4.50. Steinernema nepalense. *A: First generation male, entire view; B: First generation male, tail; C: First generation male, tail; D, E: First generation male, spicule; F: First generation male, gubernaculum; G: Infective juvenile, anterior end; H: Infective juvenile, anterior end; I: Infective juvenile, lateral field. (Scale bars: A = 500 μm; B, C, G = 25 μm; D-F, H, I = 10 μm.) After Khatri-Chhetri* et al. *(2011b), courtesy* Russian Journal of Nematology.

comprising 12 pairs and a single midventral papilla located just anterior to cloacal opening. Of these, seven pairs located precloacal (six pairs subventral and one pair lateral) and five pairs postcloacal (two pairs adcloacal, two pairs subterminal, and one pair subdorsal). Posteriormost two pairs consistently located near tail tip. Next to these genital papillae, a pair of sensory/papillae-like structures protruding from edges of cloacal aperture are also present.

Male (second generation)

Second generation male similar to that of first generation except body, spicule and gubernaculum shorter and thinner.

Table 4.20. *Morphometrics of type population of* Steinernema nepalense. *All measurements are in μm and in the form: mean ± s.d. (range).*

Character	First generation			Second generation		Infective juvenile
	Holotype (male)	Male	Female	Male	Female	
n	–	20	20	20	20	20
L	1555	1295 ± 155 (1050-1556)	4763 ± 901 (2846-6308)	769 ± 46 (707-861)	1050 ± 141 (839-1440)	479 ± 21 (424-513)
a	11.51	12 ± 0.9 (11-14)	25 ± 2.8 (19-28)	13 ± 1.0 (11-15)	14 ± 2.2 (10-19)	20 ± 1.5 (17-23)
b	10.87	9.6 ± 0.9 (8-11)	24 ± 3.2 (17-30)	6.0 ± 0.4 (5.3-6.9)	7.2 ± 0.8 (6.1-9.2)	4.8 ± 0.3 (4.4-5.4)
c	77.75	63 ± 8.3 (49-79)	196 ± 80 (77-440)	49 ± 4.9 (39-58)	26 ± 3.9 (21-35)	9.1 ± 0.8 (7-10)
V	–	–	52 ± 3.2 (46-58)	–	58 ± 5.8 (41-70)	–
Max. body diam.	135	105 ± 15.0 (76-135)	192 ± 31 (148-267)	59 ± 4.1 (51-66)	74 ± 12 (52-95)	24 ± 1.7 (21-27)
Excretory pore	72	62 ± 7.9 (43-75)	77 ± 14 (41-96)	54 ± 5.2 (46-69)	61 ± 7.3 (53-86)	38 ± 4.4 (32-46)
Pharynx	143	135 ± 10 (120-163)	198 ± 24 (157-229)	127 ± 5.9 (118-136)	146 ± 5.4 (135-157)	100 ± 5.6 (91-115)
Tail length	20	21 ± 2.2 (16-25)	27 ± 9.4 (11-45)	16 ± 1.3 (14-18)	41 ± 4.5 (28-49)	53 ± 3.5 (48-61)
Hyaline region	–	–	–	–	–	33 ± 3.7 (28-40)
Anal body diam.	32	33 ± 3.3 (28-41)	69 ± 17 (37-107)	26 ± 1.8 (24-30)	32 ± 12 (25-84)	14 ± 1.1 (13-17)
Spicule length along chord	73	71 ± 5.7 (56-78)	–	63 ± 4.6 (55-69)	–	–
Gubernaculum length	56	57 ± 3.5 (50-66)	–	46 ± 4.4 (40-58)	–	–
D%	50.34	46 ± 5.2 (34-54)	39 ± 7.1 (18-49)	43 ± 5.1 (37-57)	41 ± 4.3 (35-55)	38 ± 5.6 (31-51)
E%	360	297 ± 35 (243-365)	310 ± 107 (136-609)	346 ± 48 (283-460)	149 ± 23 (120-210)	71 ± 7.9 (58-87)
SW%	228	218 ± 25 (165-260)	–	240 ± 21 (207-283)	–	–
GS%	82.19	79 ± 5.9 (70-93)	–	67 ± 4.2 (58-74)	–	–
H%	–	–	–	–	–	63 ± 6.0 (52-71)

Female (first generation)

Anterior region typical of *Steinernema* female. Body variable in length, usually C-shaped, sometimes coiled on heat relaxation. Labial region rounded, sometimes slightly offset from body. Morphology of labial region and pharynx similar to male. Excretory pore anterior to nerve ring. Genital tract amphidelphic with reflexed ovaries. Vulva transverse slit-like, with slightly swollen, or sometimes well swollen, equal lips, double-flapped epiptygmata present. Vulval opening at 58% of body length. Well cuticularised short vagina leading into paired uteri. Narrow rectum, distinct anal opening. Postanal swelling mostly present, sometimes not well developed or slightly developed. Tail shape variable, mostly dome-shaped with small mucron. Pygmy forms not observed. *Endotokia matricida* rarely observed.

Female (second generation)

Similar to first generation female in general morphology, but smaller. Body slightly curved on heat relaxation. Gymnostom with thick, well sclerotised walls. Vulval opening, posterior to mid-body. Tail conical with mucron. Uniformly swollen vulval lips with small epiptygmata. Postanal swelling mostly developed.

Infective juvenile

Anterior region typical of *Steinernema* IJ. Heat-relaxed specimens short (*ca* 479 μm), body slightly curved sometimes straight, slightly tapering towards anterior and posterior ends. Cuticular striation indistinct under light microscope, but distinct with SEM, 1-1.25 μm thick in mid-body region. Lateral field beginning anteriorly with one line at fourth or fifth annule. Two additional lines starting after some distance and forming two ridges. Twelve annules after starting point of two ridges, each ridge dividing into three. Both ridges sharing middle single ridge and thus number of ridges increasing to seven with two marginal ridges more prominent than others. Five annules posterior, middle ridge further dividing into two and ridge number reaching eight. Two submarginal ridges thinner than others, which become invisible after some distance with number of ridges becoming six. These six ridges are prominent and uniform for a long distance. Anterior to anus, widths of ridges are irregular and not all ridges are visible. Here, number of ridges reducing to five for short distance, one submarginal line being more prominent than others. At level of anus one more ridge disappearing, four ridges remaining up to level of phasmid. Thereafter, only two prominent marginal lines remaining on tail. With above arrangement, lateral field formula = 2, 7, 8, 6, 5, 4, 2. Deirid not observed. Phasmid pore-like, posterior to anus, in anterior half of tail. Labial region rounded, not annulated. Stoma slightly opened. Labial papillae not observed, amphids crescent-like, four distinct cephalic papillae located posterior to amphids. Hemizonid distinct, lens-like, located near nerve ring level. Excretory pore in mid-pharynx region, cuticularised. Cardia present. Bacterial vesicle oval to elongated, *ca* 20 × 11 μm in size. Intestine filled with numerous fat globules, lumen not visible. Genital primordia sometimes prominent. Rectum straight. Hyaline region occupying slightly more than half (63%) of tail length, terminus finely pointed.

Type habitat and locality

Recovered from a soil sample collected from sub-tropical low altitude area, at Itahari (87°16′E, 26°39′N, 96 m a.s.l.), Sunsari district of eastern Nepal. The soil sample was taken from a lentil field in December 2007. The soil was silt loam, acidic (pH 3.6) with moderate organic matter content.

Type material

Holotype male (UGMD 104194), nine male paratypes (UGMD 104194) and six female paratypes (UGMD 104195) deposited in the nematode collection of the Institute of Zoology, Ghent University, K.L. Ledeganckstraat 35, Ghent, Belgium. Eight male paratypes and five female paratypes also deposited in the nematode collection of Centre of Parasitology of A.N. Severtsov Institute of Ecology and Evolution of Russian Academy of Sciences, Moscow, Russia.

Relationships

The IJ of *S. nepalense* can be distinguished from those of *S. surkhetense* by slightly longer body (479 (424-513) *vs* 415 (393-450) μm), longer EP distance (38 (32-46) *vs* 32 (28-34) μm), slightly longer tail length (53 (48-61) *vs* 45 (38-53) μm) and lateral line formula = 2, 7, 8, 6, 5, 4, 2 *vs* 2, 8, 7, 5, 2. First generation males of both species can be distinguished by the structure of the lateral field (presence of single flattened ridge *vs* absence), lower SW% value of 218

(165-260) *vs* 230 (168-330), and longer gubernaculum of 56 (47-60) *vs* 52 (42-63) μm, respectively.

MOLECULAR CHARACTERISATION

An analysis of the 814 bp long ITS rDNA sequence (HQ190044) and 614 bp long D2-D3 LSU rDNA (HQ190045) sequences indicates that *S. nepalense* and *S. surkhetense* are the closest species in the *carpocapsae*-group.

SYMBIOTIC BACTERIUM

No information.

BIONOMICS

Cross-hybridisation tests with *S. carpocapsae* and *S. siamkayai* produced progeny in 10% of cases, but these were non-fertile (Khatri-Chhetri *et al.*, 2011b).

Steinernema nepalense reproduced well on *G. mellonella* larvae at 22 ± 3°C. Nematode-infected *G. mellonella* larvae died after 1 or 2 days. First generation adults appeared after 3 days. Second generation adults appeared 7 days after inoculation. After 12 days the first cohort of IJ migrated into a water trap. Besides *G. mellonella*, the species reproduced well on rice moth, *Corcyra cephalonica* Stainton, lesser wax moth larvae, *Achroia grisella* Fabricius, and silk worm *Bombyx mori* L. (Khatri-Chhetri *et al.*, 2011b).

Steinernema nyetense[45,46] Kanga, Trinh, Waeyenberge, Spiridonov, Hauser & Moens, 2012

(Figs 4.51, 4.52)

MEASUREMENTS

See Table 4.21.

[45] Specific epithet named after *Nyété*, the place in Cameroon near where the species was found.

[46] Description after Kanga *et al.* (2012). Revised by Sergei E. Spiridonov.

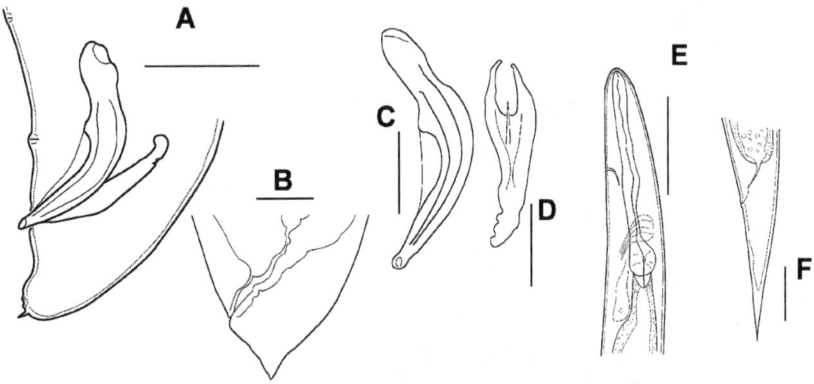

Fig. 4.51. Steinernema nyetense. *A: First generation male, tail; B: First generation female, tail; C: Spicule; D: Gubernaculum; E: Infective juvenile, anterior end; F: Infective juvenile, tail. (Scale bars: A, B, E, F = 50 μm; C, D = 25 μm.) After Kanga* et al. *(2012), courtesy* Russian Journal of Nematology.

DESCRIPTION

Male (first generation)

Anterior region typical of *Steinernema* male. Body curved ventrally posteriorly, broadly C-shaped when heat-killed. Cuticle with faint transverse striations, poorly seen under light microscope, visible with SEM. Lateral field present at mid-body with one narrow ridge. Phasmids not observed. Labial region rounded, gently tapering toward anterior end, slightly swollen, with six lips each with one labial papilla at tip and four more protruding cephalic papillae at its base, former papillae smaller than latter. Two pore-like amphidial apertures located posterior to labial lateral papillae. Anterior end with a perioral disc. Cheilostom cuticle connected to a thick strongly cuticularised 1.9-2.8 μm long ring 1.6-2.6 μm thick, forming middle part of stoma walls. Additional thinner cuticular ring situated posterior to thicker one, embedded in pharynx tissue. Excretory pore anterior to nerve ring, slightly posterior to middle of pharynx. Excretory duct strongly cuticularised. Deirids not observed. Intestinal lumen open throughout body length. Testis reflexed. Germinal zone with poorly distinguishable spermatocytes at the level of flexure. Growth zone posteriorly to flexure filled with spermatids. *Vas deferens* vacuolated. Spicules paired, moderately curved 19 μm long and 12 μm wide labial region. Velum 6 μm wide, ending 7-13 μm from rounded tip of spicules. Manubrium longer than wide (spicule head region

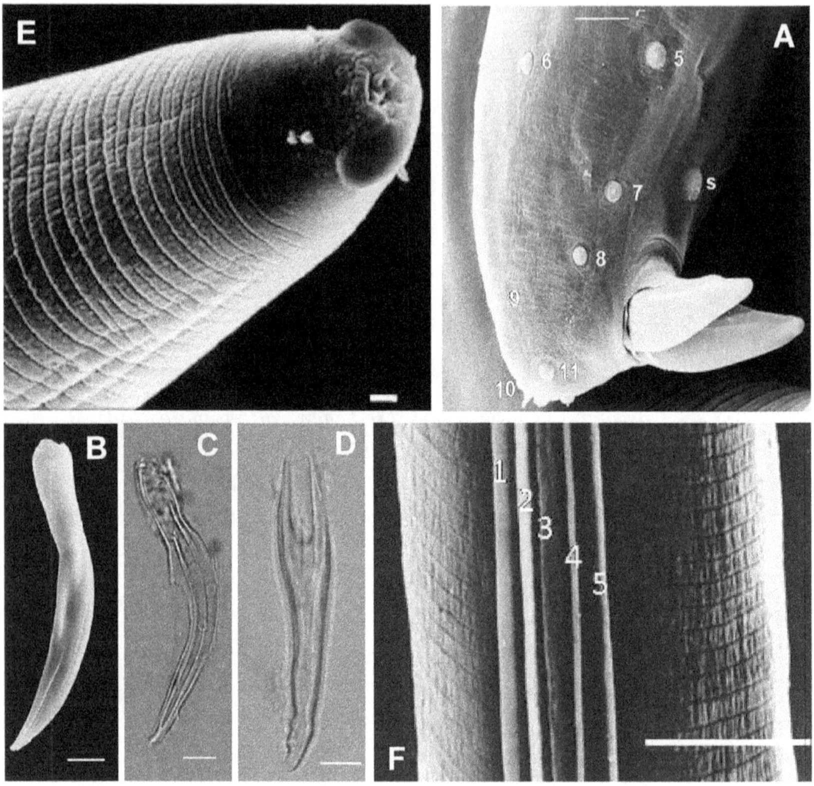

Fig. 4.52. Steinernema nyetense. *A: First generation male, tail (papillae numbered, s = single midventral papilla); B: First generation male, spicule; C: First generation male, spicule, light microscopy; D: First generation male, gubernaculum, light microscopy; E: Infective juvenile, anterior end; F: Infective juvenile, lateral field (ridges numbered). (Scale bars: A-D, F = 10 μm; E = 1 μm.) After Kanga* et al. *(2012), courtesy* Russian Journal of Nematology.

length/width = 1.4-2.0), shaft elongate. Gubernaculum boat-shaped in lateral view, anterior end rounded with two to three projections from neck toward labial region, needle-shaped cuneus present. Copulatory papillae totalling 22 + 1, comprising seven pairs precloacal subventral papillae, single precloacal midventral papilla, one pair adcloacal; three pairs postcloacal papillae including two subterminal and one subdorsal pairs. Tail conoid with mucron usually present. Phasmids inconspicuous.

Table 4.21. *Morphometrics of type population of* Steinernema nyetense. *All measurements are in* μm *and in the form: mean* ± *s.d. (range).*

Character	First generation			Second generation		Infective juvenile
	Holotype (male)	Male	Female	Male	Female	
n	–	20	20	20	20	20
L	1887	1460 ± 264 (1081-1887)	3615 ± 856 (2570-5257)	678 ± 149 (414-941)	1296 ± 249 (825-1758)	648 ± 36 (565-708)
a	55	36 ± 7.8 (22-55)	16 ± 2.6 (11-22)	18 ± 4.3 (10-28)	12 ± 2.6 (8-20)	21 ± 1.5 (19-26)
b	11	9 ± 1.1 (8-11)	14 ± 2.6 (11-19)	4 ± 0.7 (3-6)	7 ± 0.9 (5-8)	6 ± 0.4 (5-6)
c	39	42 ± 6.0 (28-56)	87 ± 26.5 (53-150)	20 ± 4.3 (14-31)	24 ± 7.1 (14-44)	8 ± 1.6 (6-11)
V	–	–	53 ± 3 (47-58)	–	53 ± 3 (47-58)	–
Max. body diam.	151	106 ± 27.4 (62-159)	219 ± 41.7 (153-310)	51 ± 6.3 (37-66)	113 ± 29.1 (42-164)	32 ± 2.5 (25-37)
Excretory pore	103	88 ± 16.6 (57-121)	124 ± 29.4 (72-205)	68 ± 12.2 (40-85)	80 ± 12.5 (58-103)	52 ± 2.8 (46-57)
Pharynx	177	158 ± 22.7 (98-203)	251 ± 37.5 (207-384)	156 ± 23.9 (103-196)	194 ± 22.5 (141-221)	114 ± 6.6 (104-128)
Tail length	48	35 ± 6.0 (26-48)	42 ± 5.9 (33-56)	35 ± 5.3 (25-49)	55 ± 9.0 (33-77)	82 ± 19.1 (54-113)
Hyaline region	–	–	–	–	–	28 ± 3.9 (20-35)
Anal body diam.	34	42 ± 6.9 (30-63)	63 ± 6.3 (54-79)	37 ± 4.0 (31-44)	41 ± 6.0 (29-54)	16 ± 1.5 (14-20)
Spicule length	89	80 ± 7.6 (67-98)	–	57 ± 9.4 (36-75)	–	–
Gubernaculum length	57	53 ± 6.4 (40-62)	–	34 ± 7.9 (24-52)	–	–
D%	58.07	55 ± 6.8 (40-70)	–	43 ± 5 (35-52)	42 ± 8.0 (28-55)	46 ± 3.1 (37-50)
E%	–	–	–	–	–	75 ± 18.3 (48-116)
SW%	262	199 ± 41.2 (125-283)	–	154 ± 26.3 (115-211)	–	–
GS%	64.22	66 ± 7.3 (51-77)	–	61 ± 9.1 (43-79)	–	–
H%	–	–	–	–	–	50 ± 5.5 (35-57)

Male (second generation)

Main morphological features similar to those of first generation male except for shorter body length and smaller body diam.

Female (first generation)

Anterior region typical of *Steinernema* female. Body C-shaped when killed by gentle heat. Labial region rounded, *ca* 40 μm diam., continuous with body, with six labial papillae and four cephalic papillae. Cuticle smooth under light microscope, fine annulations visible with SEM. Stoma shallow with buccal cavity 11-20 μm broad. Cheilorhabdions prominent, well sclerotised, posterior part funnel-shaped. Excretory pore anterior to nerve ring, well cuticularised, excretory gland not observed. Lateral field with one ridge. Vulva a transverse slit with lips protruding up to 9-25 μm, equal in size, with long double-flapped epiptygmata. Tail conoid and pointed with 3-8 μm long mucron on tip. Prominent postanal swelling in fully mature females. Tail shorter than anal body diam.

Female (second generation)

Similar to first generation female, but smaller. Tail, tapering gradually to tail terminus, longer than anal body diam.

Infective juvenile

Anterior region typical of *Steinernema* IJ. Body elongate, almost straight or slightly curved, gradually tapering towards anterior and posterior ends. Sheath (second-stage cuticle) present immediately after harvesting, but many IJ losing sheath during storage. Exsheathed juvenile with four cephalic papillae. Labial region smooth, continuous with body. Amphidial apertures prominent and sometimes swollen. Cuticle marked with prominent transverse striations. Excretory pore anterior to nerve ring. Hemizonid distinct, located anterior to base of basal bulb. Lateral field beginning anteriorly with one line at sixth annule and two additional lines appearing at annule 15 or 16. At fourth annule posterior to excretory pore, number of lateral lines increasing to form four ridges, a fifth additional line appearing at seven or eight annules posterior to excretory pore. Fifth ridge appearing in middle of two marginal ones making a total of five equally developed ridges at mid-body. Near region of anus, ridges decreasing to three, ending with two towards tail end. With above arrangement, lateral field formula = 2, 4, 5, 4, 3, 2. Hemizonid distinct. Hemizonion not observed. Tail five times as long as anal body diam., attenuated. Phasmid present near mid-tail, just ventral to lateral field. Hyaline region occupying 36 (20-54) μm of tail length.

TYPE HABITAT AND LOCALITY

Steinernema nyetense was collected from a soil sample taken in a forest at Nyété in the southern region of south Cameroon (2°49′N, 10°07′E). The possible hosts in nature are unknown.

TYPE MATERIAL

Holotype male, five male paratypes, four female paratypes and five IJ paratypes deposited in the nematode collection of the Institute of Zoology, Ghent University, Ghent, Belgium.

RELATIONSHIPS

Steinernema nyetense is characterised by combination of features of the various developmental stages of the nematode. The IJ are characterised by an average body length of 648 (565-708) μm, the distance from anterior end to excretory pore of 52 (46-57) μm, distance from anterior end to the base of pharynx of 114 (104-127) μm, a tail length of 82 (54-112) μm, and D% and E% values of 46 (37-50) and 66 (43-89), respectively. The lateral field pattern of the species is 2, 4, 5, 4, 3, and 2. The first generation males of the new species have a spicule length of 80 (67-98) μm, GS% = 66 (51-77), SW% = 198 (124-283) and D% = 55 (40-70). Females are recognised by a conoid and pointed tail with a mucron on the tip and a protruding and symmetrical vulva with a double-flapped epiptygmata.

Steinernema nyetense is similar to *S. cameroonense*, *S. monticolum*, *S. robustispiculum*, *S. ashiuense* and *S. rarum*.

MOLECULAR CHARACTERISATION

Two sequences of *S. nyetense* are deposited in NCBI GenBank: a 735 bp long one for ITS rDNA (partial 18S + ITS1 + 5.8S + ITS2 + partial 28S) – JX985266; and a 592 bp long sequence of D2-D3 expansion segment of 28S deposited as JX985264. The sequences of this species, together with that of another Cameroonian species

(*S. cameroonense*), form a well supported clade with *S. sacchari* from South Africa (Nthenga *et al.*, 2014) – the *Cameroonense*-clade. The relationships of this clade with other groups of steinernematid species are not resolved.

SYMBIOTIC BACTERIUM

No information.

BIONOMICS

The possible hosts in nature are unknown. Cross-hybridisation tests with *S. cameroonense*, *S. feltiae* and *S. everestense* (= junior synonym of *S. akhursti*) produced no progeny (Kanga *et al.*, 2012).

DISTRIBUTION

Known only from Cameroon (Kanga *et al.*, 2012).

Steinernema papillatum[47,48] San-Blas, Portillo, Nermut', Půža & Morales-Montero, 2015
(Figs 4.53-4.57)

MEASUREMENTS

See Table 4.22.

DIAGNOSTIC DESCRIPTION

Male (first generation)

Anterior region typical for *Steinernema* male. Body curved ventrally posteriorly, G-shaped when heat-killed, tapering toward labial region. Excretory pore always anterior to isthmus and nerve ring (ranging from 48-65% of distance from anterior body end to pharynx base), excretory duct cuticularised. Pharyngo-intestinal valve present, usually weakly developed. Deirids located laterally at level of basal bulb base, visible

[47] Specific epithet referring to the very distinct pair of deirids located at the base of the basal bulb of the male.

[48] Description after San-Blas *et al.* (2015). Revised by David J. Hunt.

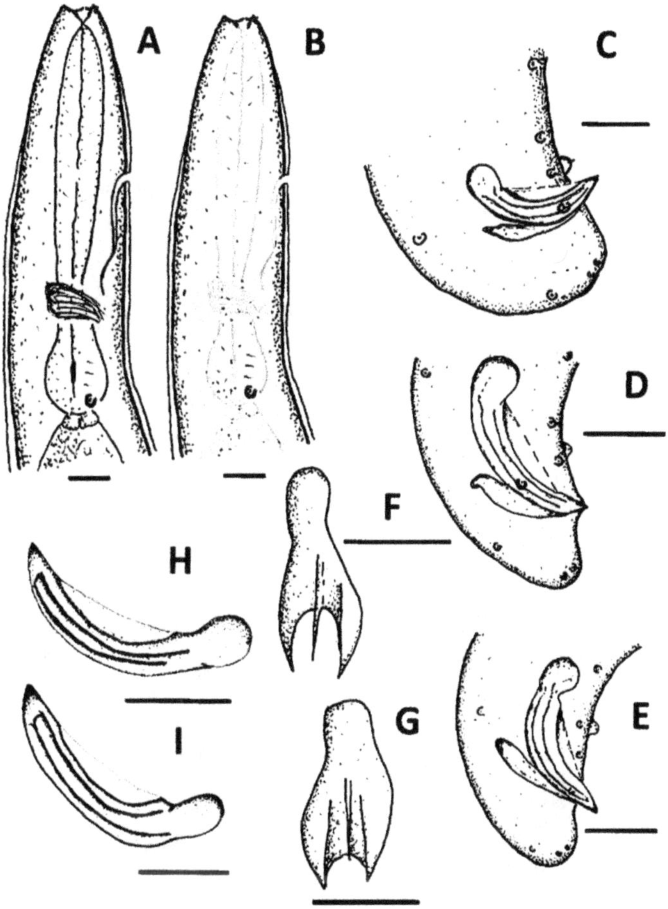

Fig. 4.53. *Male of* Steinernema papillatum. *A: Labial and pharyngeal region of first generation; B: Anterior region showing deirids; C-E: Tails showing different gubernaculum shapes; F, G: Ventral view of gubernaculum; H, I: Different spicule shapes. (Scale bars = 20 μm.) After San-Blas* et al. *(2015), courtesy* Nematology.

under both light and electronic microscopy. Gonad monorchic, ventrally reflexed. Spicules paired, curved, light yellow-brown in colour, two internal ribs visible on each spicule. Manubrium rounded in shape, continuous with dorsal side of lamina. Length and width of spicule manubrium similar. Distinctive flat area on ventral side of spicule tip. Thin velum present. Calomus short, sometimes indistinct. Two internal ribs visible on both spicules. Gubernaculum boat-shaped with

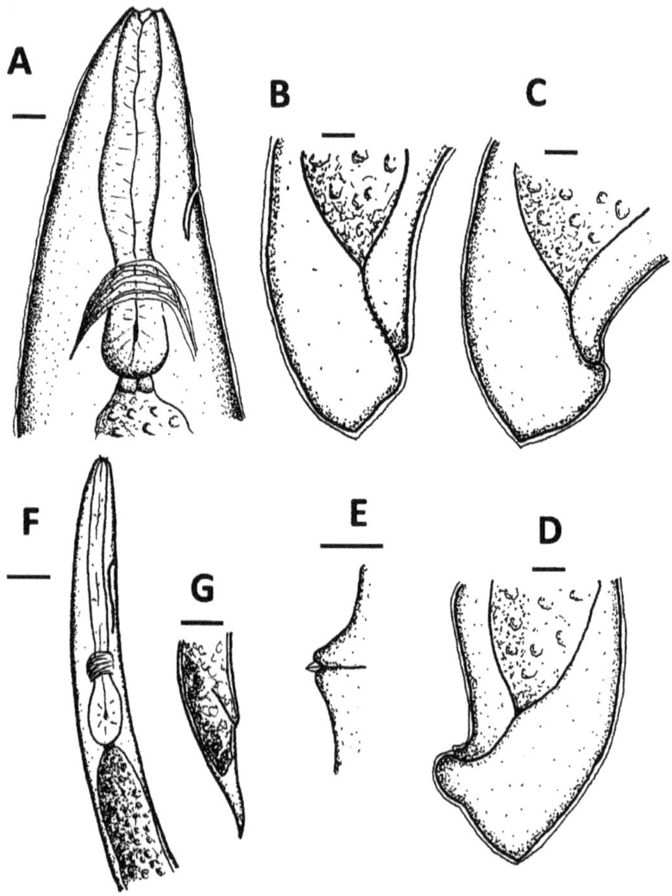

Fig. 4.54. *Female and infective juvenile of* Steinernema papillatum. *A: Labial and pharyngeal region of first generation female; B-D: Different female tails showing postanal swelling; E: Vulva showing epiptygmata; F: Labial and pharyngeal region of infective juvenile; G: Tail of infective juvenile. (Scale bars = 20 μm.) After San-Blas* et al. *(2015), courtesy* Nematology.

anterior part tapering (lateral view) or hooked in 77% of examined specimens, cuneus needle-shaped. Twenty-three genital papillae visible with light microscopy and ESEM (11 pairs and single midventral precloacal papilla), including six subventral precloacal pairs, one pair dorsolateral, one pair adcloacal, one pair postcloacal subdorsal and two pairs subterminal. Tail short and rounded, mucron absent.

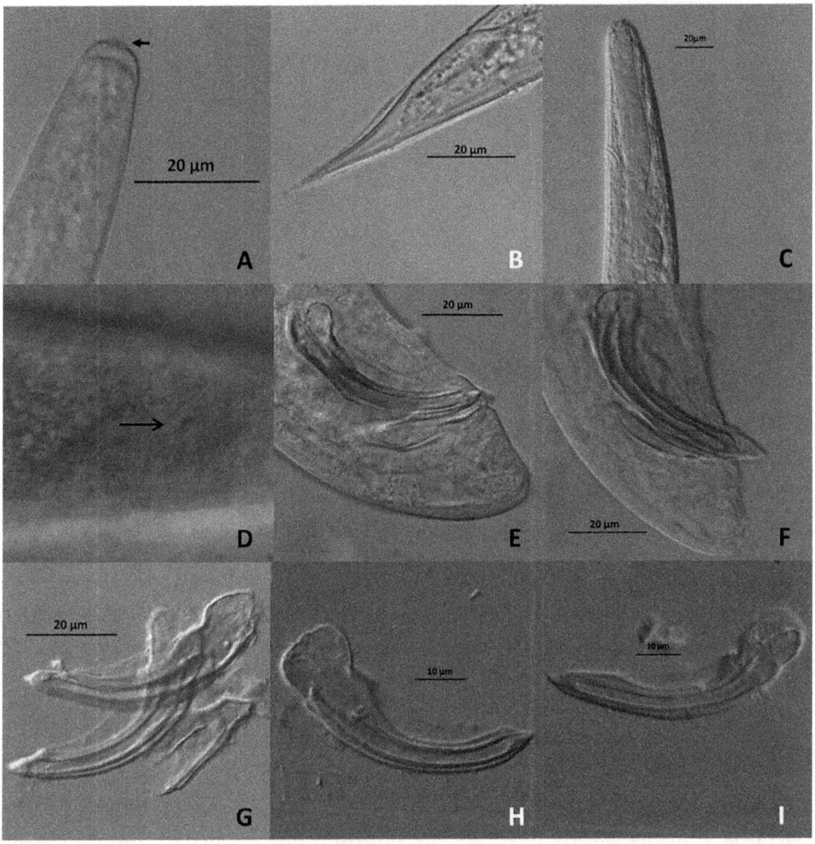

Fig. 4.55. *Microphotography of male and infective juvenile of* Steinernema papillatum. *A: Labial region of infective juvenile showing 'horn-like' structures on labial region (arrow); B: Tail of infective juvenile; C: Pharyngeal region of first generation male; D: Deirid of first generation male (arrow); E: Tail of first generation male; F: Tail of second generation male; G: Spicules and gubernaculum; H, I: Spicules. After San-Blas* et al. *(2015), courtesy* Nematology.

Male (second generation)

General morphology similar to first generation male but smaller in size with thinner spicules and shorter testis reflexion. Pharyngeal region, including deirids, as for first generation.

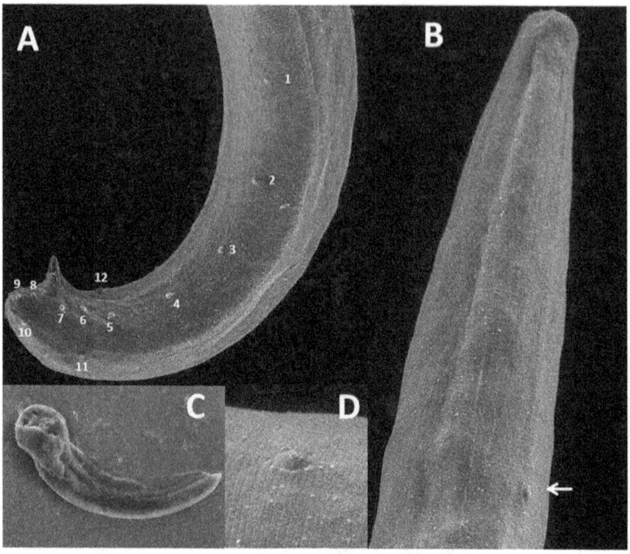

Fig. 4.56. *Environmental scanning electron microscope photography of male* Steinernema papillatum. *A: Posterior region showing genital papillae distribution (numbered); B: Anterior region showing position of deirid (arrow); C: Spicule; D: Deirid. After San-Blas* et al. *(2015), courtesy* Nematology.

Female (first generation)

Anterior region typical for *Steinernema* female. Body length highly variable. Cuticle faintly striated. No lateral fields observed. Labial region short and slightly rounded. Excretory pore prominent with cuticularised duct always anterior to nerve ring and at level of, or anterior to, metacorpus base. Cardia prominent. Ovaries amphidelphic and reflexed, vulva in form of transverse slit located near mid-region of body (52%), well developed, lips protruding. Small but distinct double epiptygmata present. Tail short, conoid with no terminal mucron, postanal swelling present in 50% of studied specimens. *Endotokia matricida* in mature females observed.

Female (second generation)

General morphology similar to first generation female but smaller in size. Excretory pore more anterior and anal body diam. smaller. Postanal swelling present but variable in shape and size.

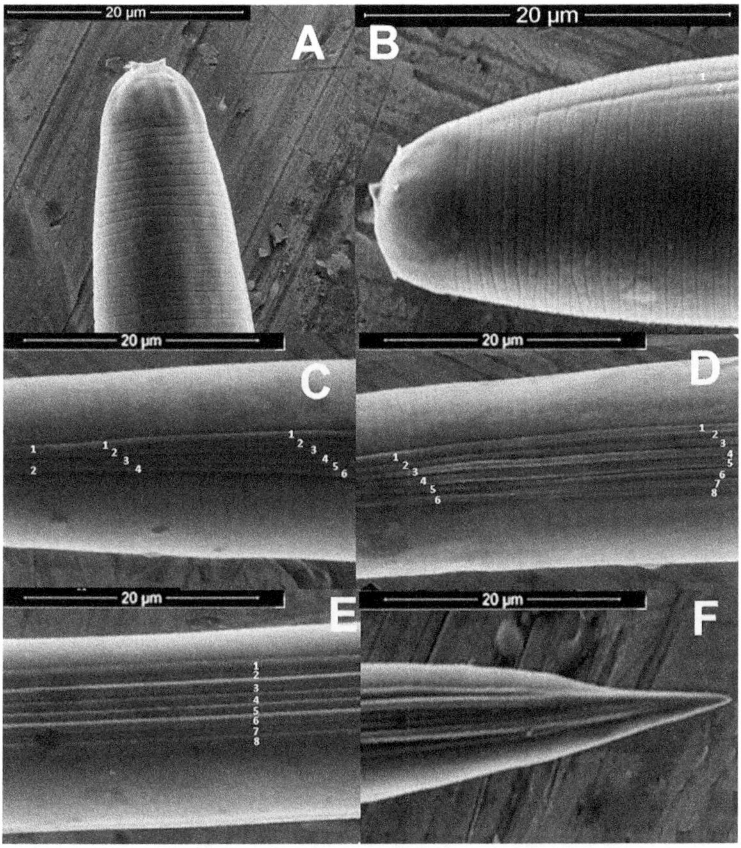

Fig. 4.57. *Environmental scanning electron microscope photography of infective juvenile of* Steinernema papillatum. *A: Labial region showing 'horn-like' structures on labial region; B: Lateral fields showing two ridges; C: Lateral fields showing two ridges and the transition to four and six ridges; D: Lateral fields showing six ridges and the transition to eight ridges; E: Lateral fields showing eight ridges at mid-body; F: Lateral fields showing the third and sixth ridges merging into the last two ridges and tail. After San-Blas* et al. *(2015), courtesy* Nematology.

Infective juvenile

Anterior region typical for *Steinernema* IJ apart from two horn-like protuberances on labial region. Body straight to slightly arcuate when heat-relaxed. Gradually tapering anteriorly and posteriorly. Body elongated and enclosed in sheath (second-stage cuticle) after emergence, 100% of IJ losing their sheath after 2 weeks of storage. Labial region

Table 4.22. *Morphometrics of* Steinernema papillatum *(LPV023 type strain). All measurements are in μm and in the form: mean ± s.d. (range).*

Character	First generation			Second generation		Infective juvenile
	Male		Female	Male	Female	
	Holotype	Paratypes	Paratypes	Paratypes	Paratypes	Paratypes
n	–	20	20	20	20	20
L	977	1146 ± 127 (851-1321)	6728 ± 2617 (3333-12568)	996 ± 211 (551-1344)	2489 ± 571 (1582-3464)	652 ± 39.6 (572-720)
a	–	–	–	–	–	27 ± 2.5 (22-30)
b	–	–	–	–	–	5.9 ± 0.4 (5.0-6.4)
c	–	–	–	–	–	12.1 ± 1.3 (8.3-15)
V	–	–	52 ± 3.3 (45-57)	–	54 ± 2.5 (47-58)	–
Max. body diam.	70	69 ± 9.9 (54-87)	210 ± 52.8 (132-312)	63 ± 10.9 (44-82)	117 ± 19.7 (81-167)	24.4 ± 3.2 (21-31)
Excretory pore	70	73 ± 9.1 (54-96)	101 ± 17.7 (66-129)	72 ± 7.2 (57-84)	71 ± 11.3 (45-88)	50 ± 3.3 (44-58)
Nerve ring	105	104 ± 11.5 (74-125)	138 ± 17.5 (112-174)	113 ± 22.4 (67-168)	134 ± 6.4 (117-155)	88 ± 4.1 (81-96)
Pharynx	132	136 ± 15.6 (94-163)	194 ± 24.5 (160-242)	140 ± 13 (121-160)	177 ± 12.5 (153-206)	110 ± 4.9 (103-121)
Testis reflexion	227	340 ± 54.3 (238-454)	–	182 ± 43.5 (72-258)	–	–
Tail length	26	25 ± 4.9 (16-35)	39 ± 9 (20-51)	21 ± 3.7 (14-28)	32 ± 6.4 (21-49)	54 ± 6.7 (40-78)
Hyaline region	–	–	–	–	–	31 ± 3.7 (24-43)
Anal body diam.	33	34 ± 4.7 (26-44)	62 ± 18.5 (29-97)	30 ± 7.3 (20-44)	46 ± 6.1 (35-62)	–
Spicule length (chord)	56	52 ± 4.3 (42-62)	–	52 ± 4.4 (37-56)	–	–
Spicule length (arc)	63	58 ± 4.6 (47-66)	–	60 ± 6.6 (39-69)	–	–
Spicule width	7	12 ± 1.3 (9-14)	–	7 ± 1.2 (4-8)	–	–
Gubernaculum length	23	31 ± 3.2 (23-36)	–	26 ± 4.5 (13-31)	–	–
Gubernaculum width	5	7 ± 1.2 (4-9)	–	6 ± 0.8 (5-8)	–	–
D%	53	54 ± 5.3 (43-65)	52 ± 6.3 (37-64)	55 ± 9.1 (45-85)	40 ± 6.6 (27-50)	46 ± 3.3 (40-52)
E%	–	–	–	354 ± 52.3 (262-436)	–	93 ± 10.8 (66-121)
SW%	169	156 ± 21.2 (125-194)	–	174 ± 32.8 (112-227)	–	–
GS%	41	59 ± 6.5 (48-70)	–	49 ± 8.2 (34-66)	–	–
H%	–	–	–	–	–	57 ± 5.3 (47-75)

continuous with body, exsheathed IJ with two horn-like structures on labial region, very distinct by light microscopy, four cephalic papillae visible with Environmental Scanning Electron Microscope (ESEM). Stoma closed. Excretory pore prominent, located anterior to nerve ring and almost at mid-pharynx (D% = 45.6). Lateral field with eight ridges evenly spaced at mid-body. Lateral field beginning with a single line around fifth annule, becoming two ridges between annuli 17-20. Around annule 42, two ridges splitting into four, then in six and eight (of which submarginal ridges are less distinct), at mid-body. At this point, third and sixth ridges becoming wider than others, which are still present when ridges reduce to six and then forming the last two ridges. Lateral field formula = 2, 4, 6, 8, 6, 2. Anus distinct, semi-circular with ESEM. Tail conoid with pointed terminus, hyaline portion occupying 57% of tail length.

TYPE HABITAT AND LOCALITY

The host of the type strain nematode (LPV023) in nature is unknown as it was recovered from soil using *G. mellonella* larvae as bait. The type locality is cultivated grassland at Finca El Tigre (10°48′01″N, 72°18′23″W) close to the Cachirí river margins in Zulia State, Venezuela. Soil texture was sandy-loam (8.3 ± 0.83% clay, 22.1 ± 1.98% loam and 69.6 ± 2.36% sand) with a pH value of 7.2 ± 0.22. The altitude of the sampling point was 109 m a.s.l.

OTHER LOCALITY

The second population (LPV723) was isolated from a natural grassland in the eastern side of Maracaibo Lake, 5 km northeast from La Mona (10°31′10.73″N, 70°49′35.06″W) in Falcon State, Venezuela. Soil texture was sandy-loam (7.9 ± 0.83% clay, 24.1 ± 1.98% loam and 68.0 ± 1.82% sand) with a pH value of 6.7 ± 0.33. The altitude of the sampling point was 540 m a.s.l.

TYPE MATERIAL

Holotype male (first generation), five paratype males (first generation), five paratype females (first generation) and two slides containing six paratype IJ each deposited in the University of California Riverside Nematode Collection (UCRNC), CA, USA. The rest of the paratypes remain in the Laboratorio de Protección Vegetal (LPV), at Instituto Venezolano de Investigaciones Cientificas, Maracaibo, Venezuela. Living material is available at LPV and at the Institute of Entomology, Branišovská, České Budějovice, Czech Republic.

RELATIONSHIPS

In having two horn-like structures on the labial region of the IJ, *S. papillatum* belongs to the *bicornutum*-group. It is characterised by morphometrics of the IJ, with body length 652 (572-720) μm, pharynx length (ES) of 110 (103-121) μm, a short tail 54 (40-78) μm long, c = 12.1 (8.3-15) and E% = 93 (66-121). Male specimens of the first generation can be recognised by the shorter spicule length (SL) of 52 (42-62) μm and the gubernaculum length (GL) of 31 (23-36) μm, both being the shortest within the *bicornutum*-group. The males possess a very distinct pair of deirids at basal bulb level in both first and second generations.

MOLECULAR CHARACTERISATION

Gel electrophoresis revealed a PCR-product of 926 bp for the ITS-region, comprising the transcribed spacer regions ITS1 and ITS2, the 5.8S rRNA gene and flanking region of the 18S rRNA gene. *Steinernema papillatum* is separated by 125-288 bp from other species in the *bicornutum*-group and its ITS1 region with 314 bp is 23-47 bp longer. The D2 and D3 expansion fragments of the 28S rRNA gene, generated by PCR, were 836 bp long and separated by 41-97 bp from other species in the *bicornutum*-group.

Phylogenetic analyses of the *bicornutum*-group, based on both ITS and D2-D3 regions, showed a clear separation of *S. papillatum* from all other species. In both analyses, *S. papillatum* formed a strongly supported group of American species with *S. riobrave* but the higher-level topology differed depending on the marker used. In the ITS tree, *S. papillatum-S. riobrave* formed a group with *S. bicornutum-S. ceratophorum* whereas *S. abbasi-S. yirgalemense* were sister to this clade. In the D2-D3 tree, *S. papillatum-S. riobrave* grouped with *S. abbasi-S. bifurcatum-S. yirgalemense*.

Steinernema phyllophagae[49,50] Nguyen & Buss, 2011
(Figs 4.58-4.61)

MEASUREMENTS

See Table 4.23.

DIAGNOSTIC DESCRIPTION

Male (first generation)

Anterior region typical for *Steinernema* male. Excretory pore anterior to nerve ring, anterior part of excretory duct well cuticularised, excretory cells not observed. Gonad monorchic, reflexed, mostly reduced in size. Testis with short ventral reflexion, germinal zone, growth zone and *vas deferens*. Distance from base of pharynx to anterior end of testis variable, longer than pharynx length. Spicules paired, dark brown in colour. Spicule head region (manubrium) somewhat elongate (head region length/width = 1.20-1.40), shaft (calomus) very short or absent, rostrum small, blade (lamina) tapering slightly posteriorly, blade terminus blunt with or without a small hook, velum not observed. Each spicule with two internal ribs. Three lobes on spicule not well defined with SEM. An unusual short form of spicule (present in 5% of males) also observed. Gubernaculum boat-shaped in lateral view, anterior end curved ventrally. Cuneus Y-shaped in ventral view. Copulatory papillae numbering 21, comprising a single precloacal midventral papilla and ten pairs (five subventral precloacal pairs, one pair dorsolateral, one pair adcloacal and subventral, two pairs subventral and subterminal, one pair subdorsal). Tail conoid, bursa absent, phasmid absent, tail terminus without mucron.

Male (second generation)

Similar to first generation except for shorter body and smaller body diam. Tail terminus without mucron.

Female (first generation)

Anterior region typical for *Steinernema* female. Excretory pore position variable, usually just posterior to, or at same level of, nerve

[49] Specific epithet derived from *Phyllophaga*, the genus of insect from which the nematode was collected.

[50] Description after Nguyen & Buss (2011). Revised by Khuong B. Nguyen.

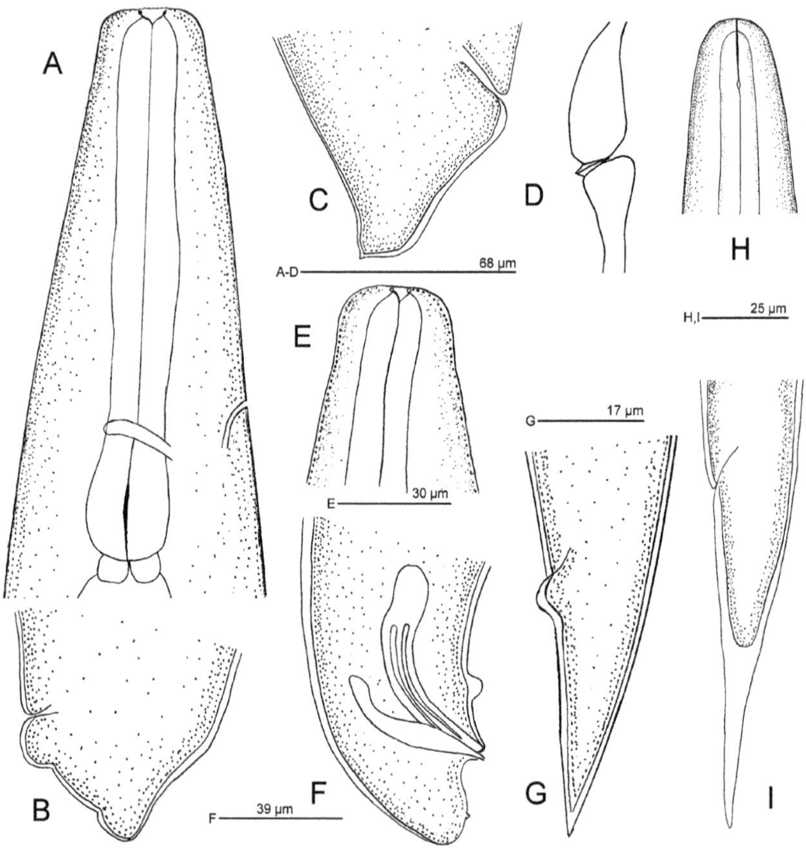

Fig. 4.58. Steinernema phyllophagae. *A-D: First generation female. A: Anterior region; B, C: Tail; D: Epiptygmata. E, F: First generation male. E: Labial region; F: Tail. G: Second generation female tail; H, I: Infective juvenile. H: Anterior region; I: Tail. After Nguyen & Buss (2011), courtesy* Nematology.

ring. Vulva a transverse slit located on a protuberant area, small double-flapped epiptygmata present. Postanal swelling present in most females, more pronounced in fully mature females. Tail shorter than anal body diam., tapering to a blunt end. Tail terminus with three or more papilla-like structures.

Female (second generation)

Similar to first generation female but smaller. Body diam. greater anterior to vulva than posterior. Vulva on asymmetrical protuberance, situated near mid-body, epiptygmata present. Postanal swelling present.

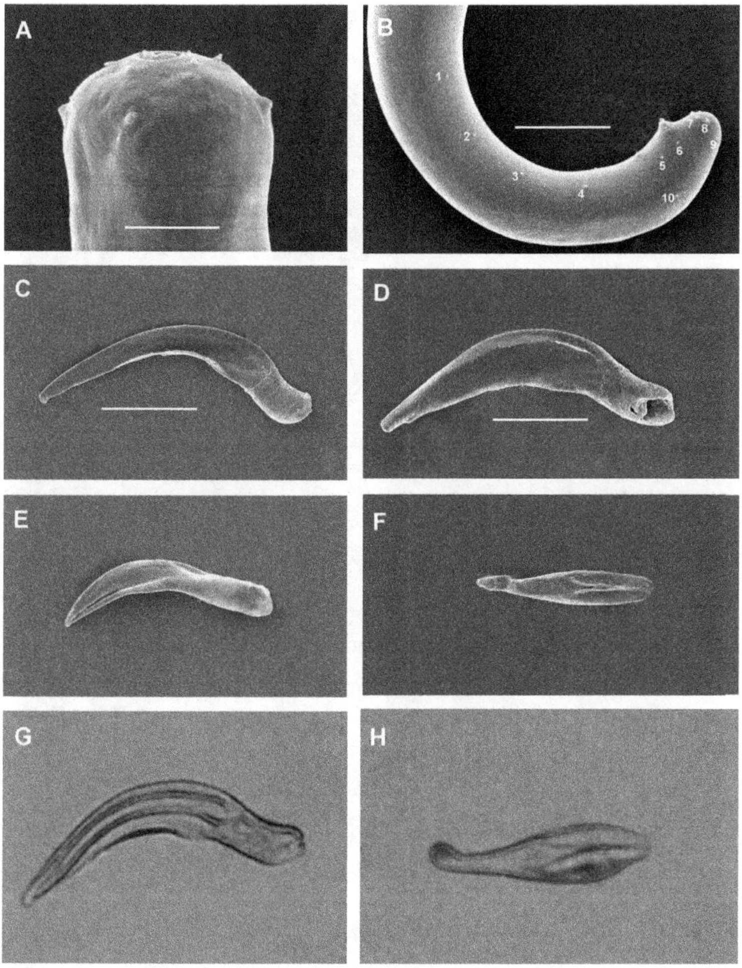

Fig. 4.59. *SEM of* Steinernema phyllophagae *first generation male. A: Male labial region showing labial and cephalic papillae; B: Posterior region showing genital papillae from pair 1 to pair 10; C, D: Normal spicules; E: Abnormal spicule; F: Gubernaculum, ventral view showing Y-shaped cuneus; G, H: Spicule and gubernaculum (light microscopy). (Scale bars: A = 7.52 μm; B = 42.5 μm; C, E-H = 23.1 μm; D = 16.7 μm.). After Nguyen & Buss (2011), courtesy* Nematology.

Tail mostly longer than anal body diam., tapering gently to a sharp point. For some large females (*ca* 5% of population), tail shorter than anal body diam.

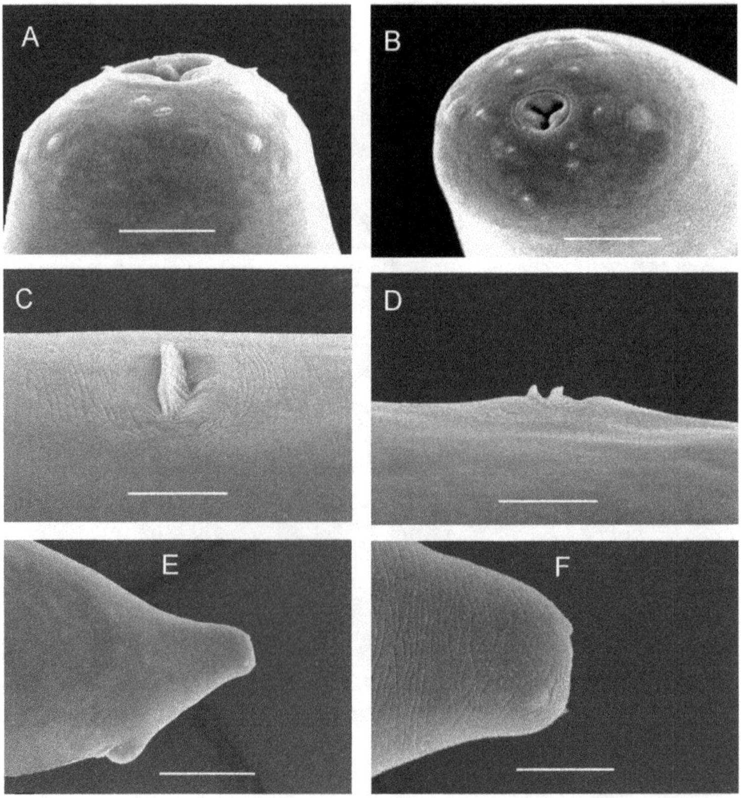

Fig. 4.60. *SEM of* Steinernema phyllophagae *female. A, B: First generation female face view showing mouth, six labial papillae and four cephalic papillae; C, D: First generation female vulva with epiptygmata; E, F: Tail with post-anal swelling and papilla-like structures on tip. (Scale bars: A, C, D = 7.5 μm; B = 12.2 μm; E = 15.0 μm; F = 5.0 μm.) After Nguyen & Buss (2011), courtesy* Nematology.

Infective juvenile

Anterior region typical for *Steinernema* infective juvenile. Cuticle marked with prominent transverse striations. Excretory pore anterior to nerve ring. Lateral field beginning anteriorly with one line at fourth or fifth annule. A short distance posteriorly, two additional lines appearing to form two ridges. Near excretory pore level, number of ridges in lateral fields increasing from two to five. More posteriorly, number of ridges increasing to six, then eight. Near anus all ridges becoming smaller, reducing to six in number. Near phasmid, only two ridges observed in

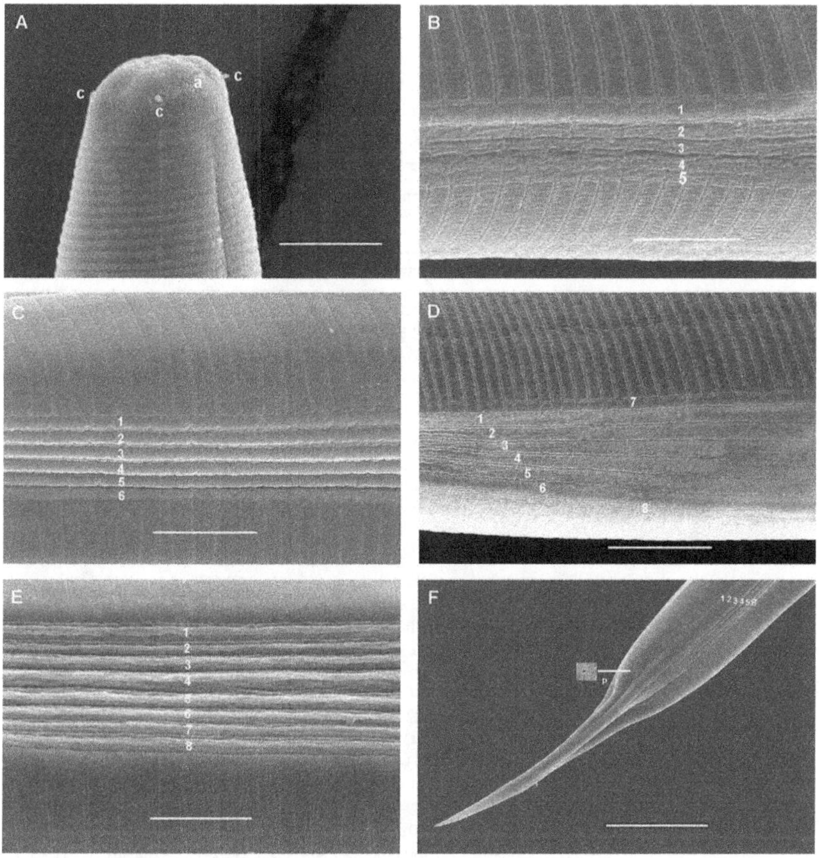

Fig. 4.61. *SEM of* Steinernema phyllophagae *infective juvenile. A: Anterior region showing one of two amphidial apertures (a), three of four cephalic papillae (c) and single line in lateral field; B: Lateral field showing five ridges; C: Lateral field showing six ridges; D: Lateral field with six ridges and beginning of ridges seven and eight; E: Lateral field with eight ridges; F: Posterior region of IJ showing lateral field with six and two ridges near anus and phasmid (p), respectively. (Scale bars: A = 7.5 μm; B = 5.0 μm; C = 6.65 μm; D = 8.57 μm; E = 4.3 μm; F = 16.7 μm.) After Nguyen & Buss (2011), courtesy* Nematology.

lateral field. With above arrangement, lateral field formula = 2, 5, 6, 8, 6, 2. Bacterial vesicle obscure. Tail *ca* three anal body diam. long, attenuate. Phasmid present near mid-tail, just ventral to lateral field. Hyaline region occupying 57 (51-60)% of tail length.

Table 4.23. *Morphometrics of* Steinernema phyllophagae. *All measurements are in* μm *and in the form: mean* $\pm$ *s.d. (range).*

Character	First generation				Second generation		Infective juvenile reared in	
	Male	Males reared in		Female	Male	Female		
		Galleria	*Phyllophaga*				*Galleria*	*Phyllophaga*
	Holotype	Paratypes	Paratypes	Paratypes	Paratypes	Paratypes	Paratypes	Paratypes
n	–	20	20	20	20	20	20	20
L	1524	1380 ± 89 (1253-1529)	1469 ± 136 (1259-1606)	4871 ± 726 (3559-6118)	1266 ± 379 (1118-2423)	2502 ± 159 (2294-2882)	1289 ± 76 (1133-1395)	1251 ± 104 (1017-1446)
a	–	–	–	–	–	–	31 ± 1.6 (27.6-35.5)	32 ± 2 (26.8-35.5)
b	–	–	–	–	–	–	8.9 ± 0.9 (7.6-11.0)	9.3 ± 0.9 –
c	–	–	–	–	–	–	14.6 ± 1.1 (12.4-15-9)	14.7 ± 0.8 (12.1-15.9)
c′	–	–	–	–	–	–	3.3 ± 0.4 (2.7-4.3)	3.1 ± 0.3 (2.9-3.8)
V	–	–	–	51 ± 3 (45-55)	–	55 ± 1.8 (54-59)	–	–
Body diam.	72	65 ± 6 (58-79)	61 (55-65)	193 ± 14 (170-214)	58 ± 14 (53-81)	56 ± 4.6) (48-64)	41 ± 1.6 (37-43)	39 ± 1.8 (16-41)
Excretory pore	113	116 ± 5 (108-123)	132 ± 18 (102-148)	162 ± 20 (128-202)	108 ± 12 (77-120)	68 ± 3.4 (62-74)	99 ± 11 (84-120)	100 ± 7 (86-107)
Nerve ring	124	122 ± 7 (108-139)	140 ± 11 (121-156)	177 ± 29 (122-250)	141 ± 16 (113-163)	95 ± 6.5 (87-107)	112 ± 14 (101-135)	107 ± 8 (96-117)
Pharynx	153	155 ± 7 (147-181)	167 ± 12 (141-179)	244 ± 45 (193-348)	172 ± 19 (136-201)	118 ± 6 (108-132)	146 ± 13 (120-178)	135 ± 7.8 (120-143)
Tail length	29	28 ± 1.8 (25-34)	26 ± 3.4 (21-33)	51 ± 8 (39-60)	28 ± 2.6 (24-33)	41 ± 2.5 (36-44)	89 ± 7 (77-109)	85 ± 7 (72-96)
Anal body diam.	41	39 ± 4.7 (34-55)	38 ± 3.6 (30-46)	70 ± 6 (60-80)	38 ± 2.8 (34-43)	21 ± 2.2 (19-27)	28 ± 2.8 (22-30)	25 ± 2.7 (19-31)
Spicule length	74	72 ± 3.7 (65-77)	67 ± 4.3 (59-73)	–	72 ± 6.5) (59-83)	–	–	–
Spicule width	12	12.4 ± 1.2 (7.4-14.8)	11.0 ± 1.7 (8.8-11.9)	–	11.4 ± 2.5 (10.4-15.0)	–	–	–
Gubernaculum length	55	51 ± 2.7 (46-56)	48 ± 4 (41-55)	–	52 ± 2.7 (46-56)	–	–	–
Gubernaculum width	10	8.7 ± 1.4 (7.4-10.4)	8.2 ± 1.3 (5.9-8.8)	–	8.7 ± 1.2 (7.0-10.4)	–	–	–
D%	74	75 ± 3.6 (68-80)	78 ± 6 (72-88)	–	63 ± 7 (52-79)	–	70 ± 9 (58-100)	73 ± 5.8 (61-83)
E%	–	–	–	–	–	–	110 ± 18 (70-139)	115 ± 7.6 (103-130)
SW%	180	187 ± 23 (136-223)	180 ± 18 (152-226)	–	182 ± 31 (151-206)	–	–	–
GS%	74	71 ± 4.3 (61-77)	71 ± 3.9 (59-85)	–	74 ± 8 (64-93)	–	–	–
H%	–	–	–	–	–	–	57 ± 2.6 (51-60)	56 ± 3.1 (53-64)

Type habitat and locality

Isolated from a larva of *Phyllophaga* sp. collected in a live oak (*Quercus virginiana*) nursery in Levy County (29°14′13″N, 82°28′75″W), FL, USA.

Type material

Holotype (male, first generation), isolated from haemocoel of *G. mellonella*, deposited in the United States Department of Agriculture Nematode Collection (USDANC), Beltsville, MD, USA. Three paratype males, three paratype females and ten paratype IJ (mounted and with same data as holotype) and several males and females of the first generation and many IJ in 4% formalin in a vial deposited in USDANC. Several males and females of the first and second generations and IJ in 4% formalin in two vials deposited in the Department of Entomology and Nematology, University of Florida, Gainesville, FL, USA.

Diagnosis and relationships

Steinernema phyllophagae is characterised by morphometrics of the IJ with body length 1289 (1133-1395) μm, distance from anterior end to excretory pore 99 (84-120) μm, tail length = 89 (77-109) μm, ratio a = 31 (27.6-35.5), H% = 57 (51-60) and E% = 110 (70-139). The lateral field formula = 2, 5, 6, 8, 6, 2. The first generation male can be recognised by the length and shape of the spicule and gubernaculum, position of the excretory pore, D% = 75 (68-80) and GS% = 71 (61-77). The female can be recognised by the vulva having low epiptygmata and the presence of several papilla-like structures that are always present near the tail tip. More information on relationships is available in Nguyen & Buss (2011).

Molecular characterisation

The sequence of ITS regions (accession no. FJ410327) of *S. phyllophagae*, flanked by primers 18S and 26S, is characterised by its sequence length of 958 bp, ITS1 = 268 bp, ITS2 = 284 bp. For the D2-D3 regions (accession no. FJ666054), the sequence of *S. phyllophagae* flanked by primers D2F and 536, is 855 bp. More information about molecular characterisation and phylogenetic analysis is available in Nguyen & Buss (2011).

SYMBIOTIC BACTERIUM

No information.

BIONOMICS

No information.

DISTRIBUTION

Steinernema phyllophagae was collected in Florida, USA.

Steinernema poinari[51,52] Mráček, Půža & Nermut', 2014
(Figs 4.62, 4.63)

MEASUREMENTS

See Table 4.24.

DIAGNOSTIC DESCRIPTION

Male (first generation)

Anterior region typical for *Steinernema* male. Body characteristically J- or C-shaped in heat-relaxed specimens, tapering toward labial region, tail curved ventrally. Cuticle with distinct transverse striation, lateral fields absent. Excretory pore located at level of mid-procorpus. Stoma *ca* 3-4 μm long and 5-8 μm broad. Excretory pore located posterior to mid-pharynx but anterior to nerve ring (42-64% of distance from anterior body end to base of pharynx). Pharynx muscular with procorpus extending into slightly swollen non-valvate metacorpus, narrower isthmus region surrounded by nerve ring, followed by rounded basal bulb. Distinct, papilliform deirid situated slightly anterior to nerve ring. Cardia rather long, intestine generally with a wide lumen. Spicules paired, symmetrical, finely curved, robust, colourless, *ca* 62-78 μm long, spicule tip pointed, spicule manubrium slightly variable, finely rounded, with an average length:width ratio approximate of 1:1. Manubrium forming

[51] Patronym in honour of Prof. George *Poinar* Jr for his contribution to entomopathogenic nematology.

[52] Description after Mráček *et al.* (2014). Revised by Sergei E. Spiridonov.

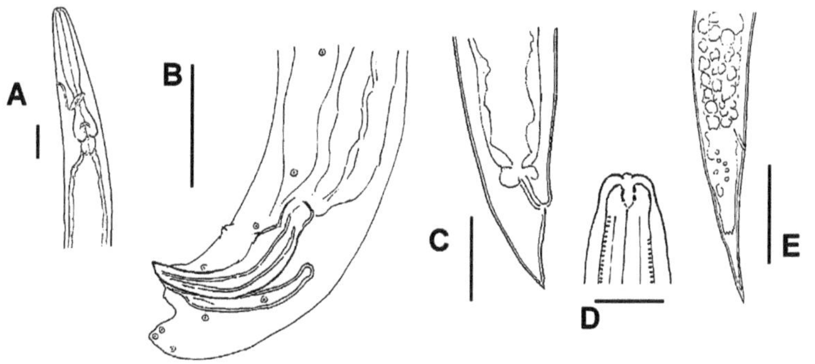

Fig. 4.62. Steinernema poinari. *A: Second generation male, anterior end; B: Second generation male, tail; C: Second generation female, tail; D: Infective juvenile, labial region; E: Infective juvenile, tail. (Scale bars: A-C, E = 50 μm; D = 10 μm.) Original drawing of Estonian isolate 'Polva'.*

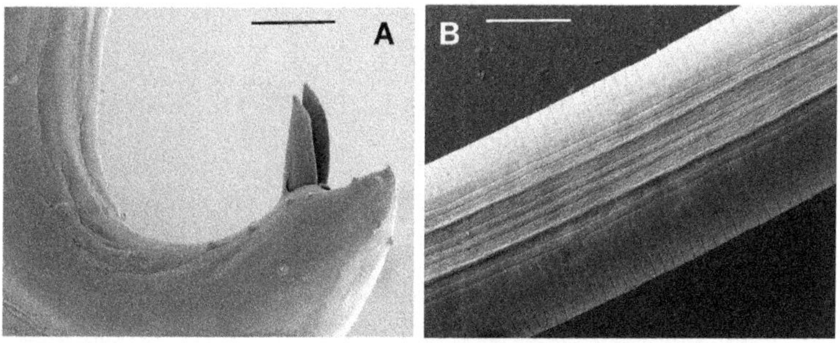

Fig. 4.63. Steinernema poinari. *A: First generation male, tail; B: Infective juvenile, lateral field. (Scale bars = 10 μm.) SEM courtesy Dr Vladimír Půža (Institute of Entomology, Czech Academy of Sciences, České Budějovice, Czech Republic).*

slightly less than 25% of spicule length, calomus and rostrum distinct. Lamina robust with two ribs, finely curved. Dorso-proximal portion of lamina often with a distinct hump. Velum bow-like, distinct, extending from rostrum for two-thirds of lamina length. Rostrum developed or fused with calomus. In lateral view gubernaculum boat-shaped, short proximal part straight, never bent as in many steinernematids, from *ca* 75% of its length strongly attenuating to distal point. In ventral view cuneus small, needle-shaped, or absent, corpus with two wings. Tail bluntly conoid, concave on ventral surface with papillate protuberance on tip, filamentous mucron absent. Genital papillae totalling 23, comprising 11 pairs and a single midventral papilla located just anterior

Table 4.24. *Morphometrics of type population of* Steinernema poinari. *All measurements are in μm and in the form: mean ± s.d. (range).*

Character	First generation			Second generation		Infective juvenile
	Holotype (male)	Male	Female	Male	Female	
n	–	20	20	20	20	20
L	2020	1815 ± 224 (1475-2363)	3244 ± 331 (2424-3858)	1069 ± 119 (909-1374)	2272 ± 203 (2040-2707)	768 ± 41 (687-848)
a	13	12 ± 1.4 (9.4-14)	14 ± 2.7 (10-22)	14 ± 2 (10-17)	18 ± 0.9 (15-19)	24 ± 1.6 (21-28)
b	12	10 ± 1.0 (8.5-12)	22 ± 10 (13-38)	6.7 ± 0.6 (6.0-8.6)	13 ± 0.9 (11-15)	5.5 ± 0.3 (4.8-6.3)
c	44	38 ± 4.1 (31-46)	60 ± 6.9 (47-76)	60 ± 6.9 (47-76)	46 ± 5.2 (38-57)	10 ± 0.6 (8.7-12)
c′	–	–	0.7 ± 0.1 (0.5-0.8)	–	1.2 ± 0.2 (1.0-1.3)	–
V	–	–	55 ± 2.1 (52-59)	–	57 ± 1.4 (54-59)	–
Max. body diam.	156	158 ± 30 (118-219)	240 ± 44 (137-313)	76 ± 11 (65-100)	130 ± 13 (117-156)	32 ± 2.0 (29-37)
Excretory pore	102	99 ± 9.1 (78-114)	140 ± 56 (78-227)	75 ± 8 (64-94)	91 ± 5.7 (78-98)	64 ± 3.7 (56-70)
Nerve ring	164	182 ± 16 (151-211)	174 ± 61 (86-254)	158 ± 10 (140-178)	181 ± 15 (160-219)	141 ± 8.4 (119-151)
Tail length	46	48 ± 3.6 (43-59)	55 ± 7.2 (43-70)	36 ± 2.4 (33-41)	49 ± 4.4 (39-59)	77 ± 6.3 (68-87)
Hyaline region	–	–	–	–	–	38 ± 3.9 (32-45)
Anal body diam.	62	65 ± 6.7 (56-78)	74 ± 8.8 (59-90)	44 ± 5.3 (37-56)	42 ± 3.8 (35-47)	19 ± 0.9 (17-21)
Spicule length along chord	72	70 ± 3.8 (62-78)	–	56 ± 4.6 (51-67)	–	–
Gubernaculum length	54	49 ± 3.3 (45-57)	–	42 ± 4.4 (33-49)	–	–
D%	62	54 ± 6.1 (42-64)	108 ± 87 (41-252)	48 ± 5.3 (40-61)	51 ± 5.0 (38-60)	46 ± 3.2 (40-55)
E%	221	207 ± 22 (167-246)	252 ± 84 (167-414)	211 ± 22 (182-262)	187 ± 18 (162-230)	84 ± 5.0 (76-95)
SW%	115	109 ± 8.8 (98-123)	–	130 ± 14 (94-146)	–	–
GS%	76	70 ± 7.2 (58-87)	–	74 ± 5.8 (66-91)	–	–
H%	–	–	–	–	–	50 ± 3.4 (43-56)

to cloacal opening. Of the paired papillae, six pairs located precloacal (five pairs subventral, one pair lateral). Two pairs adcloacal subventrally. Three pairs on the tail tip and consistently near tail terminus with one pair subdorsal and two at tail tip.

Male (second generation)

Similar to first generation, but smaller in size of body and internal organs. Mucron short, either 2-3 μm filamentous or papillate. Spicules slightly variable in shape, similar to first generation male.

Female (first generation)

Anterior region typical for *Steinernema* female. Body variable in length, usually C-shaped on heat relaxation. Labial region bluntly rounded, slightly tapering anteriorly, not offset from body contour. Stoma *ca* 4-7 μm long and 10-14 μm broad. Genital tract amphidelphic with reflexed ovaries. Vulva distinct, mostly symmetrical, rarely slightly asymmetrical and in form of a transverse slit, not, or only slightly (3-4 μm), protruding from body contour. Vulval opening always posterior to mid-body, on average 55%. Vulva slightly protruding (*ca* 12 μm), epiptygmata not developed. Vagina short, leading into paired uteri. Rectum narrow, anal opening distinct. Tail of mature females obese, bearing two minute protuberances. Postanal swelling slightly (2-8 μm) or not developed in mature females. Pygmy forms observed. *Endotokia matricida* rarely seen.

Female (second generation)

Similar to first generation in general morphology, but smaller. Vulval opening slightly posterior to mid-body, slightly to moderately protruding. Tail conical, sharply pointed, longer than anal body diam., extending to form a conical/filamentous mucron (6-10 μm long). Postanal swelling absent or slightly developed. *Endotokia matricida* observed.

Infective juvenile

Anterior region typical for *Steinernema* IJ. Body elongate, on average 768 μm long, bow-shaped when heat-killed, usually exsheathed from second-stage cuticle. Cuticular striations (distinct under SEM) *ca* 1.5 μm wide at mid-body, but lip region and tail terminus smooth. Lateral fields consisting of six unequally spaced and developed ridges in mid-body region (*i.e.*, seven lines). Only submarginal ridges prominent.

Lateral field beginning with irregular cuticular depressions (lines) close to eighth annule from anterior end. A short distance posteriorly, two ridges appearing and changing to five ridges (six lines) at about excretory pore level; more posteriorly still, central ridge dividing into two, making a total of six ridges. In posterior part of body, number of ridges reducing gradually. Anterior to anus, two middle ridges merging to leave five ridges in lateral field. At phasmid opening, lateral field reduced to two and at tail dorsal constriction to one ridge, extending almost to tail tip. With above arrangement, lateral field formula = 2, 5, 6, 5, 2, 1. Deirids not observed. Hemizonid visible as a 4-5 μm wide cuticular depression usually located between anterior of basal bulb and nerve ring. Excretory pore located at level of mid-pharynx (D% = 46). Cardia *ca* 5-9 μm long, bacterial vesicle usually not well seen, ellipsoid, *ca* 30 × 10 μm in size. Rectum *ca* 75% of anal body diam. long. Tail conoid with characteristic dorsal constriction that is situated slightly before proximal portion of hyaline region, itself occupying *ca* half of tail length (50%), terminus sharply pointed. Anus wide, sickle-shaped. Phasmids located at *ca* 45-50% of tail length posterior to anus, with aperture near point where dorsal constriction begins.

Type habitat and locality

The type isolate, CZ 1152, was recovered from brown clay soil at an elevation of 450 m a.s.l. from the bank of an unnamed creek covered by willow and alder brush in west Bohemia near Plzeň city, Czech Republic (49°50′54.162″N, 13°8′8.154″E).

Type material

Holotype first generation male, paratype males (51 specimens), paratype females (24 specimens) and juveniles (87 specimens) deposited at the EPN collection in the Laboratory of Entomopathogenic Nematodes, Institute of Entomology, Biology Centre CAS, Česke Budějovice, Czech Republic. Three slides with paratype IJ, males and females deposited at the USDA Nematode Collection, Beltsville, MD, USA.

Relationships

The IJ mean body length of *S. poinari* is shorter than 800 μm, the species belonging to the *affine*/*intermedium*-group and being most similar to *S. affine*, *S. intermedium*, *S. sichuanense* and *S. beddingi*. The IJ of *S. poinari* clearly differ from those of *S. affine* in lacking a refractile spine in the tail terminus, whilst the tail form of the first generation female, which bears two papilla-like protuberances, is an important diagnostic character. Other comparisons are given by Mráček *et al.* (2014).

Molecular characterisation

Several ITS rDNA sequences of this species (973-1057 bp long) are deposited in GenBank. In addition to sequences of the three Czech isolates (1093, 1160 and 1152: KF241752-241754), there are also data for an isolate from Tomsk (Western Siberia). Sequences of the D2-D3 segment of LSU rDNA for all of the above-mentioned isolates were deposited with accession numbers KF241749-KF241751 and KF145174. GenBank also contains several ITS rDNA sequences (AY171297, AY171294, AY171295 and AY171299) that were deposited as '*Steinernema* sp.' before the formal description was published. From its closest relatives, *S. affine*, *S. intermedium*, *S. sichuanense* and *S. beddingi*, *S. poinari* differs by 37-52 bp of ITS rDNA. Further details can be obtained from Mráček *et al.* (2014).

Symbiotic bacterium

No information.

Bionomics

Cross-hybridisation studies with *S. affine* and *S. intermedium* produced no progeny (Mráček *et al.*, 2014).

Distribution

This species was also recorded from another locality in the Czech Republic (Novohradské Hory Mts, 48°38′55.221″N, 14°41′23.277″E, 825 m a.s.l.), and in south Bohemia (48°59′22.723″N, 14°19′47.778″E, 441 m a.s.l.). In the original description the authors state that isolates of this species, identical at the 99-100% level with the type isolate, were also recorded from Belgium, Estonia, Russia (Moscow, Tomsk region and the Altai Republic) and the UK (Mráček *et al.*, 2014).

Steinernema pui[53,54] Qiu, Zhao, Wu, Lv & Pang, 2011
(Figs 4.64, 4.65)

MEASUREMENTS

See Table 4.25.

DIAGNOSTIC DESCRIPTION

Male (first generation)

Anterior region typical for *Steinernema* male. Excretory pore opening located slightly anterior to nerve ring. Lateral fields and phasmids not observed. Spicule head region longer than wide with length/width ratio *ca* 1.7 (1.5-2.0). Two internal ribs distinct. Spicule tip blunt, bearing a circular aperture located on tip and a slit-like concave located on ventral side close to tip. Velum moderately developed, starting and ending at same level as ventral rib. Gubernaculum boat-shaped in lateral view. In ventral view, gubernaculum tapering anteriorly and ending in a slightly enlarged, ventrally bent capitulum. Cuneus needle-shaped, short. Spicules of young male slightly different from those of mature. Genital papillae 25, comprising a single precloacal midventral papilla and 12 pairs of papillae (seven pairs precloacal, one pair adcloacal, one pair dorsolateral and three pair postcloacal). Tail conoid, mucron lacking.

Male (second generation)

Similar to first generation male but smaller. Excretory pore located much more posteriorly. Tail conoid without mucron.

Female (first generation)

Anterior region typical for *Steinernema* female. Excretory pore opening anterior to nerve ring. Lateral fields and phasmids not observed. Gonads didelphic, reflexed. Vagina muscular and short. Vulva a transverse slit, slightly protruding from body surface and with short double-flapped epiptygmata. Tail conoid, pointed, with a short projection.

---

[53] Patronym in honour of Zhelong *Pu*, first director of State Key Laboratory for Biocontrol, Sun Yat-sen University.

[54] Description after Qiu *et al.* (2011). Revised by Khuong B. Nguyen.

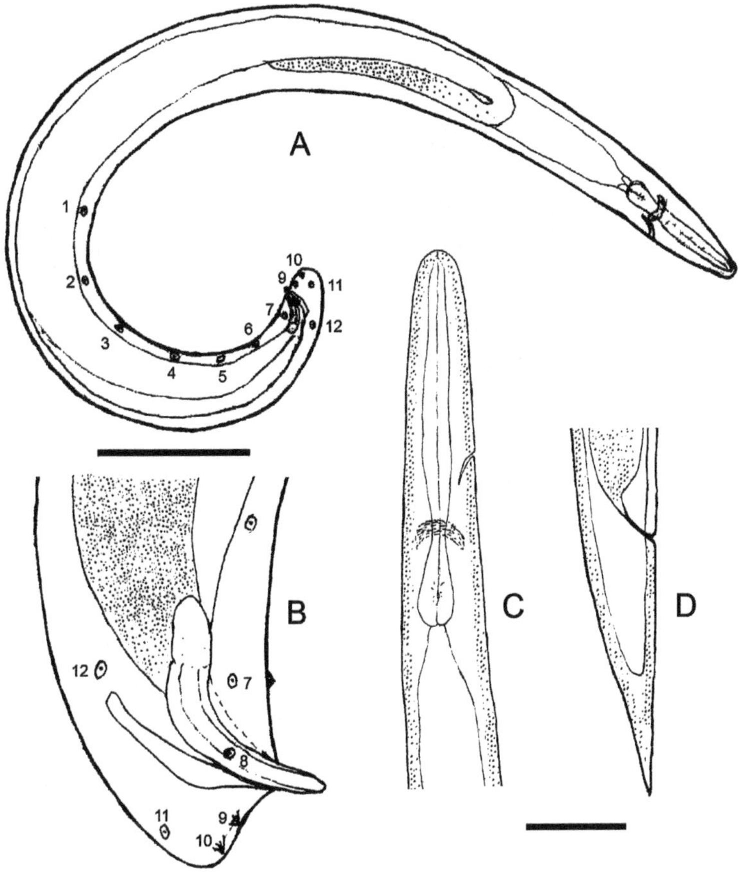

Fig. 4.64. Steinernema pui. *A, B: Male body and posterior region showing 12 pairs of genital papillae and spicules and gubernaculum; C, D: Anterior and posterior region of infective juvenile. (Scale bars: left bar: A = 320 μm; B = 65 μm; right bar: C = 32 μm; D = 28 μm.) Modified after Qiu* et al. *(2011), with acknowledgement to* Zootaxa.

Female (second generation)

Similar to first generation female but smaller. Vulva asymmetrical, protruding from body surface. Epiptygmata indistinct or absent. Tail longer than anal body diam., tapering to a pointed end.

Infective juvenile

Anterior region typical for *Steinernema* IJ. Excretory pore located anterior to nerve ring. Bacterial vesicle located in anterior portion

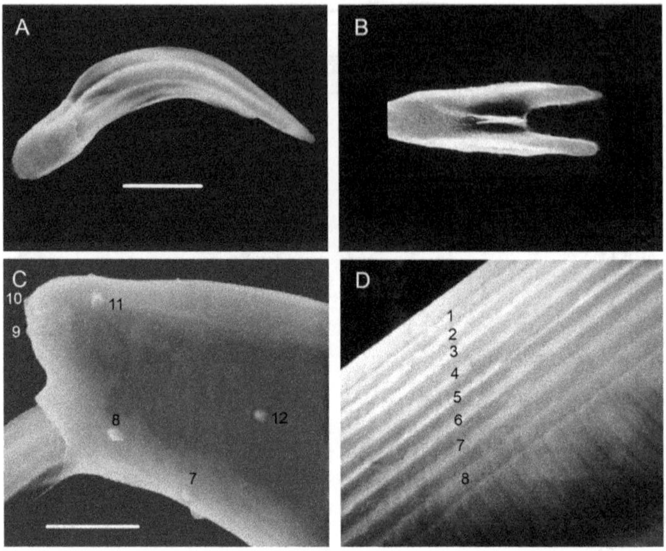

Fig. 4.65. *SEM of* Steinernema pui. *A: Spicule; B: Gubernaculum showing needle-shaped cuneus; C: Posterior region of a male showing pairs 7-12 of genital papillae; D: SEM photograph of a part of female body showing lateral field with eight ridges. (Scale bars: A = 15 μm; B = 12 μm; C = 25 μm; D = 8 μm.) Modified after Qiu* et al. *(2011), with acknowledgement to* Zootaxa.

of intestine. Cuticle with prominent transverse striations. Lateral field beginning anteriorly with one line and increasing to nine lines posterior to base of pharynx, forming eight ridges. Number of ridges reducing to six near anus then to five near phasmids and to two after that, then finally disappearing. With this arrangement, lateral field formula = 2, 8, 6, 5, 2. Phasmid distinct, located on posterior end of first ridge from ventral side. Tail short, attenuated, tapering evenly. Hyaline region occupying 46% of tail length.

TYPE HABITAT AND LOCALITY

Natural host unknown as it was recovered using *Galleria* larvae as bait from a soil sample collected from a rubber plantation at Xiao-jie town (22.01°N, 100.47°E), Yunnan province, China.

Table 4.25. *Morphometrics* Steinernema pui. *All measurements are in μm and in the form: mean ± s.d. (range).*

Character	First generation			Second generation		Infective juvenile
	Holotype (male)	Male	Female	Male	Female	
n	–	20	20	10	10	20
L	2025	2059 ± 124 (1800-2350)	6081 ± 1151 (4850-8875)	1425 ± 67 (1300-1500)	2495 ± 278 (2070-2950)	1004 ± 75 (900-1120)
a	15	15 ± 1.3 (14-16)	27 ± 5 (20-35)	17 ± 3 (14-20)	18 ± 4 (14-22)	28 ± 1.7 (26-31)
b	10	11 ± 1.4 (9.5-12)	26 ± 3 (22-31)	10.1 ± 0.2 (9.6-11)	14.3 ± 3 (11-18)	7.03 ± 0.74 (6.0-8.2)
c	60	64 ± 6 (58-70)	107 ± 14 (96-118)	53 ± 4 (48-66)	42 ± 3 (37-46)	15 ± 0.8 (13-16)
c′	0.62	0.58 ± 0.06 (0.53-0.61)	0.58 ± 0.07 (0.53-0.61)	0.71 ± 0.06 (0.53-0.61)*	1.07 ± 0.1 (0.53-0.61)*	3.1 ± 0.2 (2.9-3.3)
V	–	–	50 ± 5 (37-55)	–	52 ± 2.3 (50-57)	–
Body diam.	130	137 ± 15 (118-180)	228 ± 32 (155-275)	84 ± 7 (70-90)	138 ± 10 (125-150)	36 ± 2 (33-40)
Excretory pore	140	152 ± 13 (130-180)	147 ± 17 (125-188)	138 ± 10 (120-150)	126 ± 6.1 (120-137)	85 ± 4 (80-95)
Nerve ring	150	157 ± 10 (143-188)	166 ± 17 (150-218)	110 ± 12 (93-125)	137 ± 9.3 (125-152)	117 ± 7 (100-125)
Pharynx	193	196 ± 12 (175-228)	236 ± 21 (213-288)	141 ± 13 (120-162)	183 ± 16 (162-207)	144 ± 8 (130-150)
Anal body diam.	55	55 ± 3 (48-63)	98 ± 13 (85-125)	38 ± 4 (35-48)	56 ± 5 (50-60)	22 ± 1 (20-25)
Spicule length	80	84 ± 4 (78-88)	–	65 ± 4.3 (58-70)	–	–
Gubernaculum length	65	62 ± 2 (58-65)	–	44 ± 4.4 (35-50)	–	–
D%	73	77 ± 6 (70-91)	–	98 ± 12 (77-114)	69 ± 6.7 (57-81)	60 ± 3 (55-66)
E%	412	465 ± 48 (403-512)	–	511 ± 48 (389-621)	210 ± 45 (139-272)	125 ± 8 (109-142)
SW%	145	152 ± 10 (140-184)	–	171 ± 20 (132-193)	–	–
GS%	81	74 ± 4 (69-81)	–	68 ± 8 (52-78)	–	–
H%	–	–	–	–	–	45 ± 3.7 (40-50)

* Data range as cited in original description, but clearly in error.

Type material

Slides of holotype (first generation male), allotype (first generation female) and paratypes of about 60 IJ on three slides, ten each of the first and second generation males and females on 20 slides (one male and one female from the same generation on each slide) of *S. pui* were deposited in the State Key Lab for Biocontrol, School of Life Sciences, Sun Yat-sen University, Guangzhou 510275, China.

Relationships

Steinernema pui is characterised by morphometrics of the IJ with body length 1004 (900-1120) μm, distance from anterior end to excretory pore 85 (80-95) μm, tail length = 69 (60-80) μm, E% = 125 (109-142) and hyaline region occupying 40-50% of tail length. The first generation male can be recognised by spicule length = 84 (78-88) μm, presence of an aperture on its tip, gubernaculum with needle-shaped cuneus, D% = 77 (70-91), SW% = 152 (140-184) and GS% = 74 (69-81). The female can be recognised by the presence of a short mucron on the tail tip and short double-flapped epiptygmata. More information on relationships is available in Qiu *et al.* (2011).

Molecular characterisation

Steinernema pui is characterised molecularly by the sequences of ITS (accession no. GU395618) and D2-D3 (accession no. GU395636) regions. More information on molecular characterisation and phylogenetic analysis is available in Qiu *et al.* (2011).

Symbiotic bacterium

No information.

Bionomics

No information.

Distribution

Steinernema pui was collected in Yunnan province, China.

Steinernema puntauvense[55,56] Uribe-Loría, Mora & Stock, 2007
(Figs 4.66, 4.67)

MEASUREMENTS

See Table 4.26.

DIAGNOSTIC DESCRIPTION

Male (first generation)

Anterior region typical for *Steinernema* male. Body slender, curved ventrally, J-shaped when heat-killed. Cuticle smooth under light microscopy. Lateral field and phasmids inconspicuous. Labial region truncate to slightly rounded, continuous with body. Excretory pore located anterior to nerve ring in anterior third of metacorpus. Testis single, reflexed, consisting of germinal growth zone leading to seminal vesicle. *Vas deferens* with inconspicuous walls. Spicules paired, symmetrical, curved, ochre-brown in colour. Manubrium subtriangular or rhomboid, continuous dorsally with lamina. Calomus (shaft) inconspicuous. Lamina with rostrum or retinaculum and two internal ribs. Velum thin, extending from rostrum to spicule terminus. Gubernaculum *ca* half spicule length long, arcuate in lateral view. Manubrium elongate (in lateral view), lamina wide. Genital papillae numbering 23 (11 pairs + 1) arranged as follows: six precloacal subventral pairs, a single midventral precloacal papilla (located between precloacal pairs 5 and 6), one pair subventral adcloacal (or precloacal in some specimens), one pair subventral postcloacal, one pair subdorsal postcloacal, and two pairs terminal postcloacal. Tail conoid, mucronate.

Male (second generation)

Smaller than first generation (847 *vs* 1591 μm). Tail with or without mucron.

Female (first generation)

Anterior region typical for *Steinernema* female. Body large, C-shaped when heat-relaxed. Cuticle, lip region, stoma and pharyngeal region as

[55] Specific epithet named after *Punta Uva* beach, the locality of the original isolation.

[56] Description after Uribe-Loría *et al.* (2007). Revised by David J. Hunt.

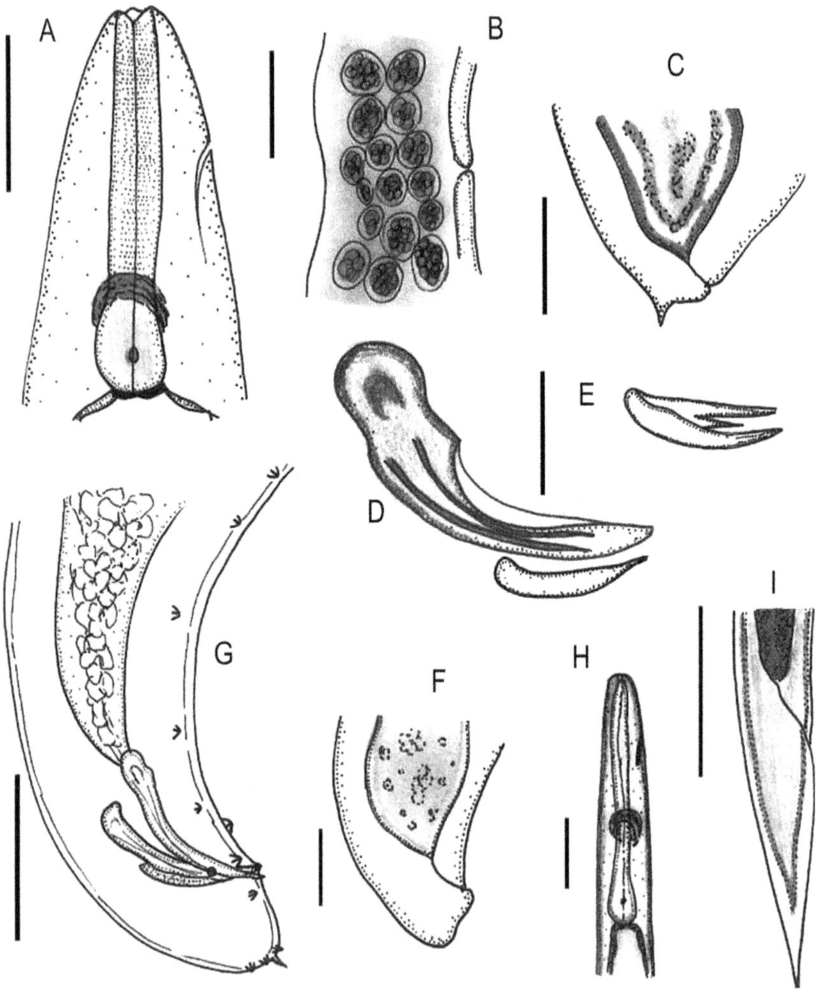

Fig. 4.66. Steinernema puntauvense *(lateral view). A: First generation female anterior end; B: First generation female vulva; C: First generation female tail; D: First generation male spicule and gubernaculum; E: First generation male gubernaculum; F: Second generation female tail; G: First generation male tail; H: Infective juvenile anterior end; I: Infective juvenile tail. (Scale bars: A-C, G = 65 μm; D, E = 23 μm; F = 11 μm; H = 58 μm; I = 33 μm.) Modified after Uribe-Lorío* et al. *(2007), courtesy* Systematic Parasitology.

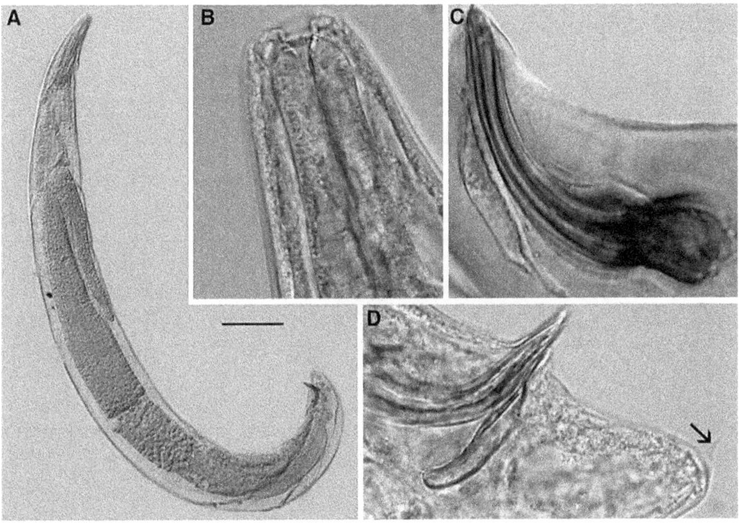

Fig. 4.67. Steinernema puntauvense. *A: First generation male entire; B: First generation male anterior end showing stoma; C: First generation male spicule and gubernaculum; D: First generation male tail showing mucron (arrow). (Scale bar: A = 105 μm; B, D = 10 μm; C = 11.5 μm.) Modified after Uribe-Lorío* et al. *(2007), courtesy* Systematic Parasitology.

in male. Excretory pore located about mid-level of procorpus. Ovaries opposed, reflexed in dorsal position, oviduct well developed, glandular spermatheca and uterus in ventral position. Vagina short, with muscular walls. Vulva located near mid-body, lips non-protruding. Tail blunt, conoid, with mucron, no post-anal swelling.

Female (second generation)

Smaller than first generation (1465 *vs* 6198 μm). Vulval lips slightly protruding. Tail conoid, with or without post-anal swelling.

Infective juvenile

Anterior region typical for *Steinernema* IJ. Body of heat-relaxed specimens almost straight, slender, gradually tapering posteriorly. Cuticle with transverse striae. Lateral field distinct, bearing eight longitudinal ridges (*i.e.*, nine lines) in mid-body region, with two middle ridges more distantly spaced than others. Excretory pore located at about mid-point of corpus. Hemizonid not observed. Anterior portion of intestine with dorsally displaced vesicle containing symbiotic bacteria. Intestinal

Table 4.26. *Morphometrics of* Steinernema puntauvense. *All measurements are in μm and in the form: mean ± s.d. (range). Some data have been adjusted from the original where several errors were evident.*

Character	First generation			Second generation		Infective juvenile
	Male		Female	Male	Female	
	Holotype	Paratypes	Paratypes	Paratypes	Paratypes	Paratypes
n	–	19	19	20	20	20
L	1066	1591 ± 306 (1010-1931)	6198 ± 802 (3687-8335)	847 ± 64 (734-983)	1465 ± 147.5 (1231-1844)	670 ± 30.5 (631-728)
a	–	–	–	–	–	20 ± 1.4 (17-23)
b	–	–	–	–	–	7.1 ± 0.5 (6.1-7.9)
c	–	–	–	–	–	12 ± 0.5 (11-13)
V	–	–	53 ± 1.5 (51-55)	–	54 ± 1.1 (52-56)	–
Max body diam.	110	119 ± 10 (101-139)	198 ± 12.1 (181-221)	76 ± 6.9 (52-77.5)	133 ± 9.9 (108-145)	33 ± 2.1 (31-37.5)
Excretory pore	177	94 ± 12.5 (68-114)	70 ± 9.5 (51-84)	76.5 ± 6.9 (60-89)	75 ± 5.1 (68-84)	25 ± 2.3 (20-30)
Nerve ring	104	115 ± 6 (104-128)	135 ± 5.8 (123-146)	89 ± 5.9 (80-98)	82 ± 6.7 (69-93)	54 ± 8.4 (46-69)
Pharynx	139	140 ± 8.5 (130-159)	192 ± 8.8 (141-206)	117 ± 2.1 (111-121)	148 ± 8.5 (132-169)	93 ± 5.3 (81-103)
Tail length	31	33 ± 3.6 (28-40)	49 ± 6.5 (41-66)	29 ± 2.4 (22-30)	54 ± 5.5 (45-62)	54 ± 2.3 (51-59)
Anal body diam.	42	46 ± 5 (40-55)	76 ± 9.6 (57-102)	34 ± 2.2 (30-38)	40 ± 4.5 (32-47)	17 ± 0.8 (15-18)
Spicule length	72	77 ± 3 (71-81)	–	64 ± 6.2 (52-76)	–	–
Gubernaculum length	51	34 ± 2.5 (30-40)	–	30 ± 2.5 (25-35)	–	–
D%	55	67 ± 9.8 (45-85)	37 ± 5 (25-45)	65 ± 5.8 (53-75)	50 ± 5.8 (43-56)	42 ± 6.8 (25-50)
E%	250	286 ± 45 (230-399)	146 ± 30.2 (99-202)	293 ± 29 (239-366)	140 ± 16.7 (115-178)	44 ± 5 (35-56)
SW%	170	170 ± 20 (140-200)	–	190 ± 20 (150-220)	–	–
GS%	70	65 ± 9.8 (55-75)	–	50 ± 10 (40-70)	–	–
H	–	–	–	–	–	30 ± 0.8 (28-33)
H%	–	–	–	–	–	54 ± 1.5 (52-55)

lumen narrow. Rectum long, straight, anus distinct. Tail conoid, with pointed terminus. Hyaline region occupying *ca* one-third of tail length.

Type habitat and locality

Isolated from a rainforest strip along the Punta Uva coast in the Caribbean region of Costa Rica. The natural host is unknown as the species was recovered by *Galleria* baiting.

Type material

Holotype male first generation (T-606t), five paratype males first generation (T-5487p), five paratype males first generation (T-5489p, T-5490p), five paratype IJ (T-5488p) deposited in the USDA Nematode Collection, MD, USA, all types from isolate Li 6.

Relationships

Morphologically, the IJ of *S. puntauvense* is most similar to a group of species that belong to the *intermedium*-group. This group is characterised by IJ body length of 600-800 μm. Phylogenetic analysis of 28S rDNA sequences placed *S. puntauvense* as a sister taxon of *S. feltiae*, although morphological/morphometric differences exist between these two taxa. For example, IJ of *S. puntauvense* have the excretory pore more anteriorly located at 25 (20-30) *vs* 62 (53-67) μm from the anterior end, and have a shorter tail of 54 (51-59) *vs* 81 (70-92) μm. The morphology of the first generation male spicules and size of gubernaculum also separate *S. puntauvense* from *S. feltiae*. Additionally, first generation females lack protruding vulval lips, a feature present in *S. feltiae*.

Molecular characterisation

According to Uribe-Lorío *et al.* (2007), maximum parsimony analysis of the D2-D3 data (accession no. EF187018; 880 bp) set yielded 387 parsimony informative characters and produced six equally parsimonious trees with a tree length of 998 steps (CI = 0.57). *Steinernema puntauvense* was found to be most closely related (bootstrap support 94%) to *S. feltiae* and within Clade III, a group comprising species with an IJ body length of 800-1000 μm. An adjusted distance matrix showed that *S. puntauvense* and *S. feltiae* differed by 3 bp, indicating a very close relationship.

SYMBIOTIC BACTERIUM

Steinernema puntauvense naturally hosts a strain of *Xenorhabdus bovienii*. In a study of the bacterial symbiont, eight of nine strains of *X. bovienii* successfully colonised *S. puntauvense*, the only one that did not being the *intermedium* strain (Lucero, 2011). Murfin *et al.* (2015) molecularly characterised and compared *X. bovienii* strain *puntauvense*.

BIONOMICS

Attempts to cross *S. puntauvense* with *S. feltiae*, *S. kraussei*, *S. riobrave* and *S. oregonense* yielded no progeny. Control crosses using individuals of the same species produced viable offspring. When crossing *S. puntauvense* with *S. feltiae*, progeny production was observed in *ca* 20% of replicates. However, when pre-adults of the resulting hybrids were crossed with each other, no offspring were produced (Uribe-Lorío *et al.*, 2007).

DISTRIBUTION

Reported only from the type locality in Costa Rica (Uribe-Lorío *et al.*, 2007).

Steinernema sacchari[57,58] **Nthenga, Knoetze, Berry, Tiedt & Malan, 2014**
(Figs 4.68-4.71)

MEASUREMENTS

See Table 4.27.

DIAGNOSTIC DESCRIPTION

Male (first generation)

Anterior region typical for *Steinernema* male. Body curved posteriorly, mostly J-shaped when heat-relaxed. Cuticle smooth under light microscope, but with striation under SEM. Lateral field absent. Excretory

[57] Specific epithet derived from *Saccharum*, the genus of plant growing in the type locality.
[58] Description after Nthenga *et al.* (2014). Revised by David J. Hunt.

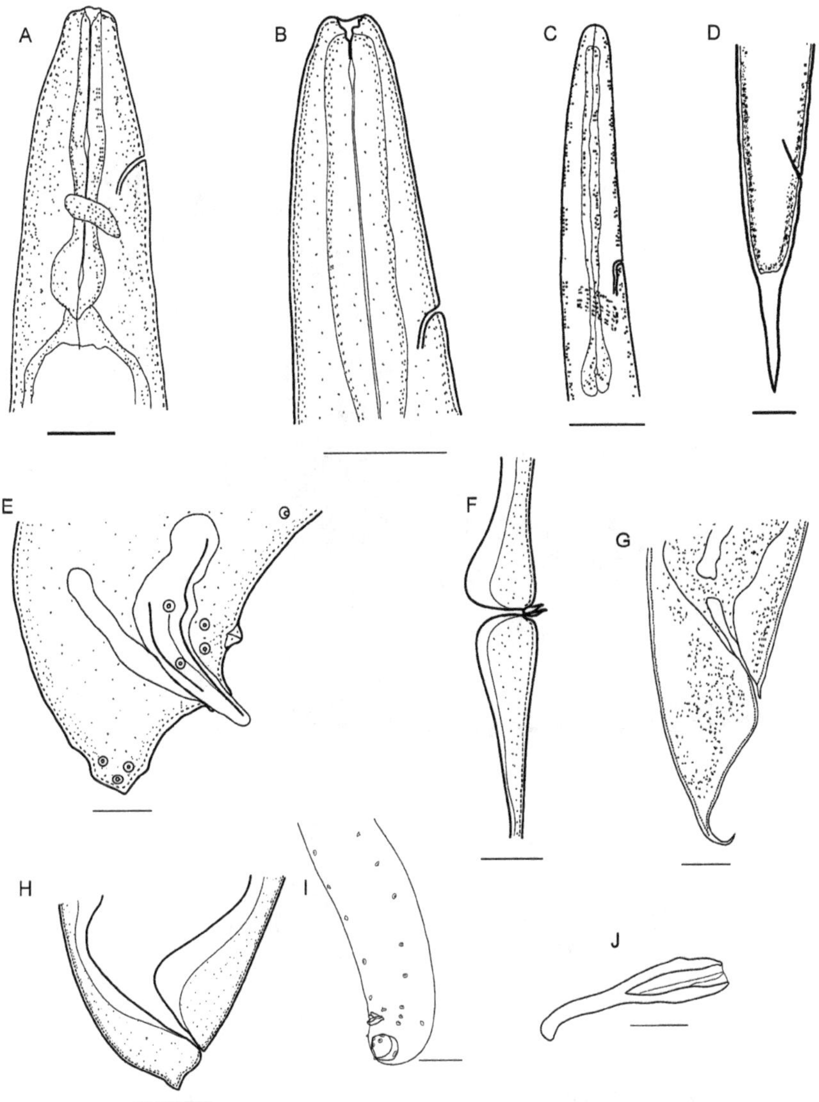

Fig. 4.68. Steinernema sacchari. *First generation male. A: Anterior region; E: Lateral view of tail region; I: Ventral view of tail; J: Gubernaculum. First generation female. B: Anterior region; F: Vulva with double-flapped epiptygmata; H: Tail region. Infective juvenile. C: Anterior region; D: Tail region. G: Tail of second generation female. (Scale bars: A = 50 μm; B-H, J = 20 μm; I = 25 μm.) After Nthenga* et al. *(2014), courtesy* Nematology.

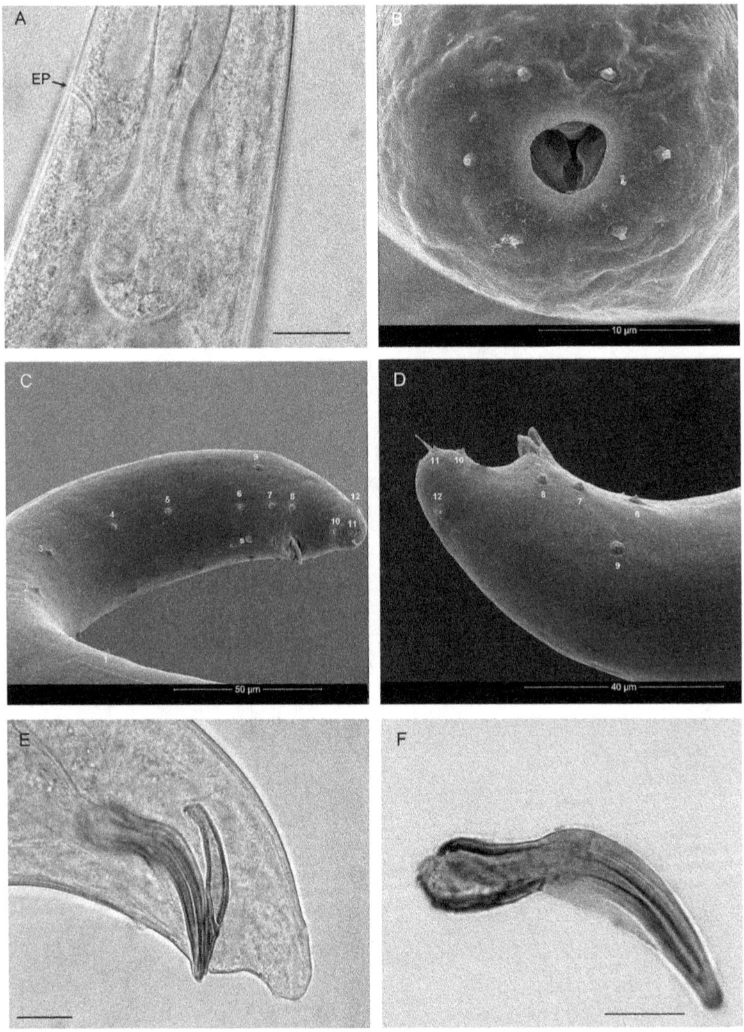

Fig. 4.69. *Male of* Steinernema sacchari. *A-C: First generation. A: Pharyngeal region showing excretory pore (EP); B:* En face *view; C: Tail region, showing 12 pairs of papillae and one midventral papilla (s). D: Lateral view of tail of second generation; E: Spicule and gubernaculum of tail of first generation; F: Spicule shape. (Scale bars: A, E, F = 20 μm; B = 10 μm; C = 50 μm; D = 40 μm.) After Nthenga* et al. *(2014), courtesy* Nematology.

pore anterior to nerve ring, well cuticularised, excretory gland not observed. Cardia prominent. Bacterial chamber obscure. Genital system monorchic, reflexed, comprising germinal zone, growth zone, *vas deferens*. Spicules paired, light brown in colour. Head region (manubrium) of spicules somewhat curved, shaft (calomus) short, blade (lamina) thick, tapering slightly posteriorly, with blunt terminus, velum present. Each spicule with two internal ribs. Gubernaculum boat-shaped in lateral view, anterior end curved. Copulatory papillae with single precloacal midventral papilla and 12 pairs of genital papillae arranged as follows: seven pairs precloacal subventral, one pair cloacal, one pair lateral, one pair subdorsal and two pairs subterminal. Tail conoid, hair-like mucron usually absent, but observed in 5% of first generation males.

Male (second generation)

Body length shorter and narrower than that of first generation. Tail shape same as first generation, but always with a mucron.

Female (first generation)

Anterior region typical for *Steinernema* female. Body C-shaped when heat-relaxed and fixed with TAF. Body cuticle smooth under a light microscope, but with annules under SEM. Lateral field not observed. Labial region rounded and continuous with body. Excretory cell obscure, but duct cuticularised. Deirid not observed. Excretory pore position variable, from near mid-pharynx to mid-basal bulb. Genital system amphidelphic, ovaries reflexed. Vulva a median transverse slit, not protruding from body surface, situated in mid-body region, with double-flapped epiptygmata. Postanal swelling absent. Tail bluntly conical to dome-shaped, shorter than anal body diam., peg-like mucron (8-10 μm long) present.

Female (second generation)

One-third shorter and narrower in body diam. than first generation. Two lateral lines with one ridge present. Vulva non-protruding from body surface, with double-flapped epiptygmata. No postanal swelling on posterior lip. Tail differing from first generation in being longer than anal body diam., tapering gently to a sharp point, without a mucron.

Infective juvenile

Anterior region typical for *Steinernema* IJ. Body straight when heat-relaxed, elongate and tapering both posteriorly and anteriorly. Cephalic

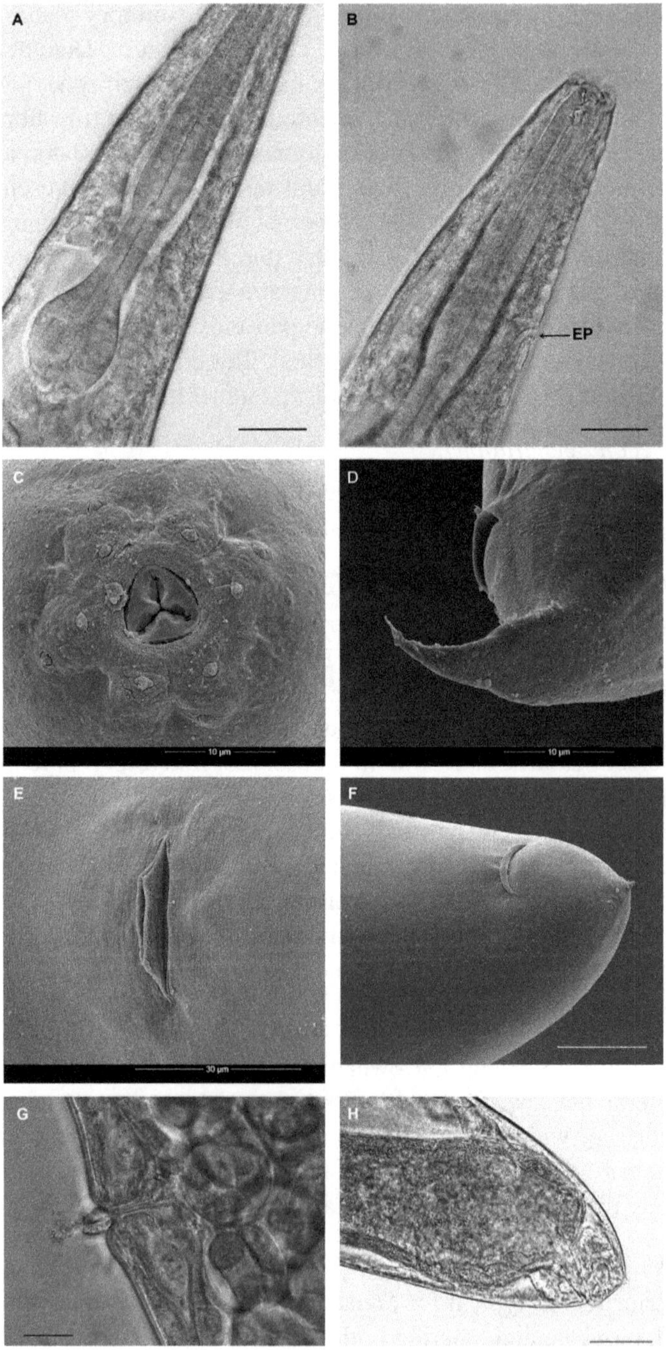
A
B
EP
C
10 μm
D
10 μm
E
30 μm
F
G
H

region gently rounded, continuous with body contour. Retained second-stage cuticle not usually observed after harvesting. Excretory pore at mid-pharynx level. Deirid not observed. Hemizonid distinct, located just anterior to basal bulb. Excretory pore prominently situated anterior to nerve ring in isthmus region. Lateral field starting with two ridges (three lines) at ninth annule from labial region, with five equal ridges (six lines) through mid-body and ending in two ridges (three lines) in vicinity of phasmid on mid-tail, almost disappearing at end of tail. Bacterial vesicle obscure. Phasmid prominent in mid-tail region just ventral to lateral field. Tail long conoid, evenly attenuated, *ca* four anal body diam. long. Hyaline region comprising 49% of tail length.

TYPE HABITAT AND LOCALITY

Steinernema sacchari, isolate SB10, was collected from a soil sample taken from a sugar cane field at coordinates 29°01′37″S, 31°35′37″E, in Gingindlovu, KwaZulu-Natal, South Africa. Nematodes were isolated by means of laboratory trapping with a sugar cane borer, *Eldana saccharina*. The natural host is unknown.

TYPE MATERIAL

Holotype (male first generation), isolated from haemocoel of *G. mellonella*, deposited in the Ghent University, Belgium. Paratype males (11) and females (19) of the first generation and third-stage IJ (42) in TAF deposited in the same collection. Permanent slides with each of the different stages are deposited in the National Collection of Nematodes, Biosystematics Division, Plant Protection Research Institute, Agricultural Research Council, Pretoria, South Africa (six IJ, four first generation males, three first generation females, five second generation males and six second generation females). Paratype males, females and IJ are also deposited in the United States Department of Agriculture Nematode Collection (USDANC), Beltsville, MD, USA (18 IJ, five first generation

Fig. 4.70. *Female* Steinernema sacchari. *A-C, E-H: First generation. A: Anterior region showing pharynx; B: Excretory pore position and stoma; C:* En face *view; E: Double-flapped epiptygmata; F: First generation tail; G: Vulva with epiptygmata; H: Tail of first generation. D: Second generation tail. (Scale bars: A, B, G, H = 20 μm; C, D = 10 μm; E = 30 μm; F = 40 μm.) After Nthenga* et al. *(2014), courtesy* Nematology.

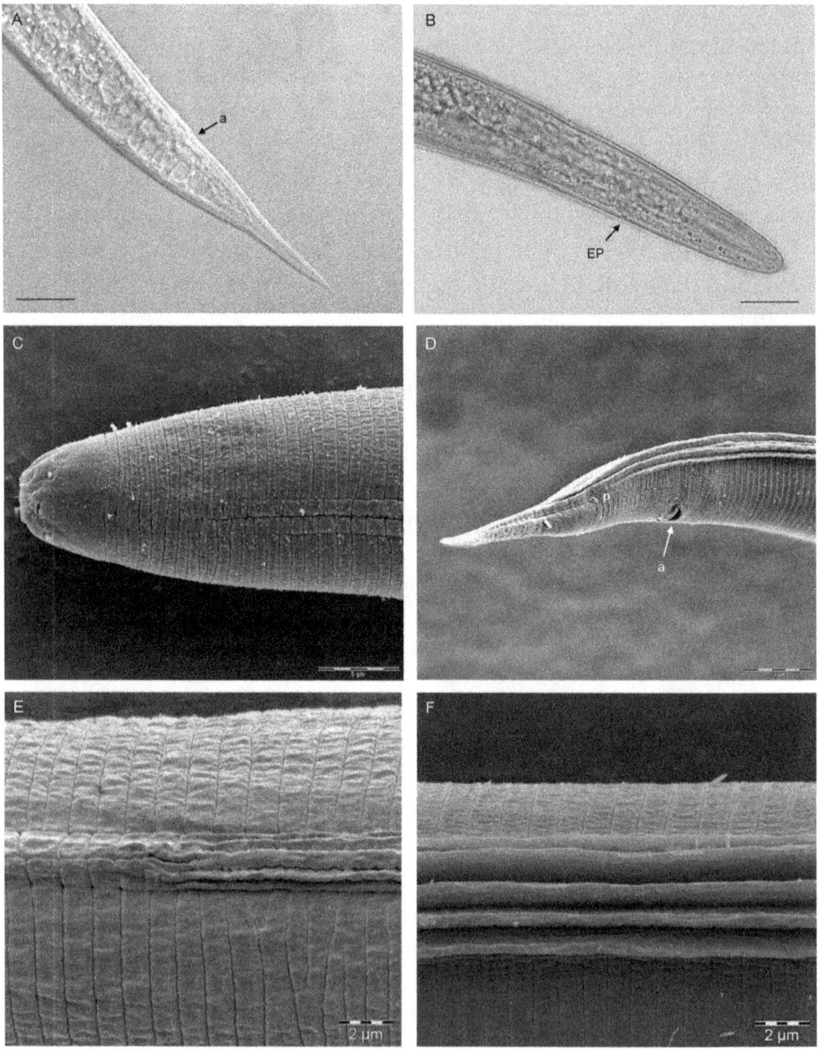

Fig. 4.71. *Infective juvenile of* Steinernema sacchari. *A: Tail region showing shape of tail and anus (a); B: Anterior end showing excretory pore (EP); C: Pharynx showing start of two lateral ridges; D: Tail with anus (a) and phasmid (p); E: Splitting of ridges in lateral field from two to five (from anterior); F: Five ridges in lateral field in mid-body. (Scale bars: A, B = 20 μm; C = 5 μm; D = 10 μm; E, F = 2 μm.) After Nthenga* et al. *(2014), courtesy* Nematology.

Table 4.27. *Morphometrics of* Steinernema sacchari. *All measurements are in μm and in the form: mean ± s.d. (range).*

Character	First generation			Second generation		Infective juvenile
	Male		Female	Male	Female	
	Holotype	Paratypes	Paratypes	Paratypes	Paratypes	
n	–	20	20	20	20	25
L	1867	1829 ± 228 (1406-2181)	4316 ± 623 (3163-5600)	1101 ± 134 (877-1441)	1394 ± 150 (1244-1849)	680 ± 27 (630-722)
a	12	14 ± 1.6 (9.3-16)	15 ± 2.6 (11-20)	15 ± 0.9 (14-18)	13 ± 0.7 (11-14)	19 ± 2.1 (14-23)
b	11	11 ± 1.0 (9.0-13)	20 ± 2.7 (15-26)	8.3 ± 0.6 (7.0-9.1)	8.0 ± 0.6 (7.4-9.7)	6.0 ± 0.2 (5.6-6.5)
c	58	54 ± 7.1 (38-68)	160 ± 35 (102-218)	45.2 ± 6.4 (34-56)	33 ± 7.1 (26-57)	10.5 ± 0.8 (9.6-12.3)
c′	–	0.7 ± 0.1 (0.6-0.9)	0.4 ± 0.1 (0.28-0.64)	–	–	–
V	–	–	54 ± 1.8 (50-57)	–	56 ± 1.1 (54-58)	–
Max. body diam.	153	145 ± 27 (86-205)	290 ± 25 (243-347)	72 ± 6.4 (61-82)	109 ± 7.4 (100-130)	37 ± 4.1 (30-47)
Excretory pore	107	110 ± 16 (84-158)	103 ± 21 (76-150)	92 ± 5.8 (81-102)	86 ± 5.4 (78-97)	53 ± 2.3 (49-58)
Nerve ring	113	116 ± 10 (97-135)	144 ± 14 (126-182)	96 ± 8.2 (82-119)	130 ± 6.4 (119-145)	84 ± 3.8 (78-97)
Pharynx length	168	164 ± 10 (149-183)	216 ± 10 (199-244)	133 ± 11 (119-169)	175 ± 7.6 (163-192)	113 ± 4.5 (104-127)
Hemizonion	–	–	–	–	–	99 ± 5.6 (86-101)
Testis reflex	435	419 ± 75 (262-541)	–	299 ± 58 (208-405)	–	–
Tail length	32	34 ± 3.1 (29-40)	28 ± 4.1 (21-37)	25 ± 2.4 (19-29)	44 ± 4.6 (33-51)	64 ± 5 (51-74)
Anal body diam.	47	49 ± 5.3 (40-57)	66 ± 9.6 (50-83)	35 ± 3.2 (28-40)	36 ± 1.8 (33-39)	19 ± 1.1 (17-22)
Hyaline region	–	–	–	–	–	32 ± 3.4 (25-39)
Spicule length	89	83 ± 4.6 (73-89)	–	70 ± 3.5 (64-78)	–	–
Spicule width	14	14 ± 1.6 (12-18)	–	12 ± 0.8 (11-14)	–	–
Gubernaculum length	65	61 ± 5.3 (50-68)	–	46 ± 3.9 (39-53)	–	–
Gubernaculum width	10	9 ± 0.9 (7.7-10.6)	–	7.6 ± 1.0 (6.0-9.7)	–	–
D%	63	67 ± 7.2 (54-88)	48 ± 10 (35-70)	69 ± 5.8 (51-78)	49 ± 3.3 (43-56)	47 ± 2.6 (41-54)
E%	228	225 ± 22 (185-285)	372 ± 90 (236-604)	376 ± 39 (318-442)	201 ± 24 (167-255)	82 ± 8.7 (70-109)
SW%	191	171 ± 18 (146-210)	–	205 ± 20 (172-244)	–	–
GS%	73	73 ± 5.0 (66-81)	–	65 ± 5.1 (58-77)	–	–
H%	–	–	–	–	–	49 ± 3.7 (43-57)

males, three first generation females, four second generation males and five second generation females).

RELATIONSHIPS

Steinernema sacchari is characterised by differences in the morphology and morphometrics of the IJ and adults. The IJ of the new species can be recognised by the pattern of the lateral field of 2, 5, 2 ridges (3, 6, 3 lines), the body length of 680 (630-722) μm, body diam. of 37 (30-47) μm, distance from anterior end to excretory pore of 53 (49-58) μm, distance from anterior to nerve ring of 88 (78-97) μm, distance from anterior to end of pharynx of 113 (104-127) μm, tail length of 64 (51-74) μm, anal body diam. of 19 (17-22) μm, D% = 47 (41-54), E% = 82 (70-109) and H% = 49 (43-57). The first generation male has a long spicule measuring 83 (73-89) μm and gubernaculum of 61 (50-68) μm, while other diagnotic characters include D% = 67 (54-88), E% = 255 (185-285), SW% = 171 (146-210) and GS% = 73 (66-81). The first generation male lacks a mucron, while the second generation male always has one. Both generations of males have 24 pairs of genital papillae and a single midventral papilla. Females have a non-protruding vulva with double-flapped epiptygmata, no postanal swelling and the first generation female tail is dome-shaped with a terminal peg.

Steinernema sacchari is similar to two African species from Cameroon, namely *S. cameroonense* Kanga, Trinh, Waeyenberge, Spiridonov, Hauser & Moens, 2012 and *S. nyetense* Kanga, Trinh, Waeyenberge, Spiridonov, Hauser & Moens, 2012 (Kanga *et al.*, 2012), and the *monticolum*-group, including *S. ashiuense* Phan, Takemoto & Futai, 2006 (Japan), *S. monticolum* Stock, Choo & Kaya, 1997 (Korea), *S. robustispiculum* Phan, Subbotin, Waeyenberge & Moens, 2005 (Vietnam), *S. schliemanni* Spiridonov, Waeyenberge & Moens, 2010 (Germany) and *S. rarum* (de Doucet, 1986) Mamiya, 1988 (Argentina) (Doucet *et al.*, 2003). Nthenga *et al.* (2014) discuss these relationships in more detail.

MOLECULAR CHARACTERISATION

Steinernema sacchari is characterised genetically by sequences of the ITS (KC633095) and the D2-D3 (KC633096) rDNA regions. The sequence of ITS regions of *S. sacchari*, including ITS1 + 5.8S + ITS2, can be recognised by its length of 764 bp (ITS1 = 311 bp; ITS2 =

296 bp) with a composition of A = 0.2251, C = 0.1963, G = 0.2383, T = 0.3403. *Steinernema sacchari* differs from the closest species, *S. cameroonense* and *S. nyetense*, in terms both of the ITS1 length (311 bp *vs* up to 291 bp and 282 bp, respectively) and ITS2 length (296 *vs* 284 bp for both species). Pairwise distances using the ITS regions show that the new species differs from *S. nyetense* by 22 bp, and from *S. cameroonense* by 29 bp, as well as from its most divergent species, *S. rarum*, by 152 bp.

The sequence of the D2-D3 region of *S. sacchari* is 877 bp long and its base composition is: A = 0.2406, C = 0.1950, G = 0.3079, T = 0.2566. Pairwise comparison using the D2-D3 regions revealed that the closest relatives to *S. sacchari*, with respect to mean character difference calculated by means of the Jukes-Cantor method, were *S. nyetense* and *S. cameroonense*, with 0.8% dissimilarity, and differing in terms of only four nucleotides, compared to *S. rarum*, with a dissimilarity of 14% and differing by 61 nucleotides in the sequence.

Morphological and molecular data confirmed *S. sacchari* as belonging to a new monophyletic group, the *Cameroonense*-clade, consisting of *S. cameroonense*, *S. nyetense* and, with low support, for *S. schliemanni* also belonging to this group. This group is closely related to the *feltiae/kraussei/oregonense* Clade III (Spiridonov *et al.*, 2004).

BACTERIAL SYMBIONT

No information.

BIONOMICS

Nthenga *et al.* (2014) comment that cross-breeding with closely related species such as *S. cameroonense* and *S. nyetense* would support the separation of species. However, no living specimens of the two Cameroonian species were available for comparison.

Steinernema schliemanni[59,60] Spiridonov, Waeyenberge & Moens, 2010
(Figs 4.72, 4.73)

MEASUREMENTS

See Table 4.28.

DIAGNOSTIC DESCRIPTION

Male (first generation)

Anterior region as for *Steinernema* male. Body C- or J-shaped after fixation. Anterior end hemispherical, 20-25 μm in diam., with broad funnel-shaped buccal cavity. Strongly cuticularised stoma walls 2-3 μm long, stegostom at least 5 μm long. Prominent deirids at cardia level. Excretory duct 1.0-1.5 μm wide, curving posteriad 5 μm from

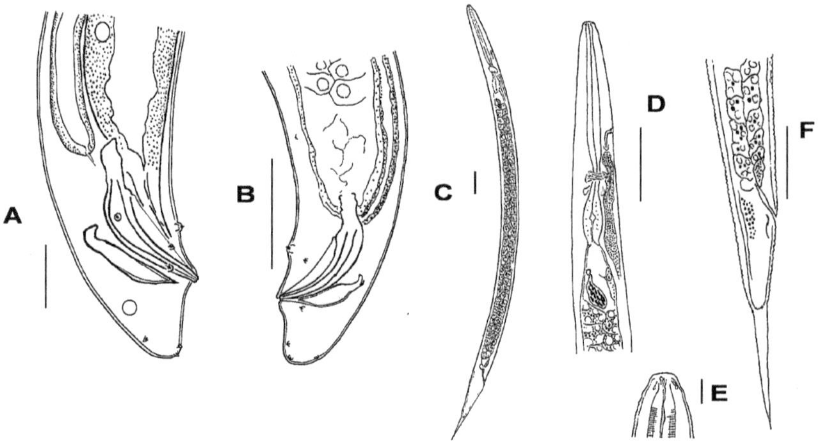

Fig. 4.72. Steinernema schliemanni. *A: First generation male, tail; B: Second generation male, tail; C: Infective juvenile, entire view; D: Infective juvenile, anterior end; E: Infective juvenile, labial region; F: Infective juvenile, tail. (Scale bars: A-C = 50 μm; D, F = 25 μm; E = 10 μm.) After Spiridonov* et al. *(2010), courtesy* Russian Journal of Nematology.

[59] Patronym in honour of entrepreneur and scientist Heinrich *Schliemann* (1822-1890) who was born in Neubukow (Mecklenburg-Vorpommern) and used his fortune gained in Russia to conduct the famous excavations of Troy.

[60] Description after Spiridonov *et al.* (2010). Revised by Sergei E. Spiridonov.

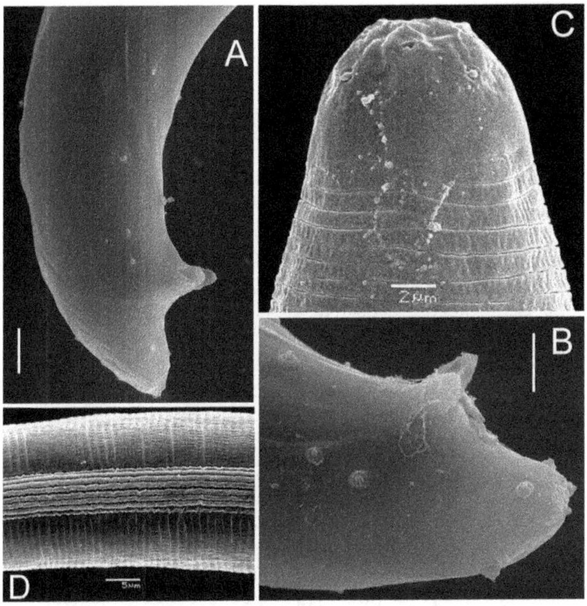

Fig. 4.73. Steinernema schliemanni. *A: First generation male, posterior end; B: First generation male, tail; C: Infective juvenile, labial region; D: Infective juvenile, lateral field. (Scale bars: A, B = 10 μm; C = 2 μm; D = 5 μm.) After Spiridonov* et al. *(2010), courtesy* Russian Journal of Nematology.

surface. Flexure of testis 50-60 μm posterior to basal bulb. Prominent pseudocoelomocytes (15 μm diam.) anterior to flexure. Testis filled with spermatocytes with round 4 μm diam. nuclei at level of flexure. Nuclei in spermatocytes disappearing at mid-body level, with broader testis lumen posteriorly filled with 7-10 μm spermatids with granular cytoplasm. Strongly vacuolated *vas deferens*, walls narrow to ejaculatory duct, starting at level of spicule manubrium. Narrow posterior part of intestine squeezed between *vas deferens* and body wall. Spicules moderately curved (approx. 120°), with 12 μm long and 10 μm wide manubrium and 7-8 μm wide shaft. Anteriorly directed projection on ventral edge of spicule proximal end. Proximal end of spicules covered with thin cuticular membrane. Velum 4 μm wide, ending 2-12 μm from rounded tip of spicules. Gubernaculum with protruding crurae, conspicuous median hole and proximal beak-like projection, directed toward ventral body surface. Cuneus absent. Total count of genital papillae: 22 + 1. Two pairs of postcloacal papillae on ventral side near tail terminus. Tiny (1 μm high) peg-like mucron. Pair of

Table 4.28. *Morphometrics of type population of* Steinernema schliemanni. *All measurements are in μm and in the form: mean ± s.d. (range).*

Character	First generation			Second generation		Infective juvenile
	Holotype (male)	Male	Female	Male	Female	
n	–	20	20	20	20	20
L	1140	1463 ± 368 (896-2213)	6408 ± 1269 (4431-8340)	1143 ± 71 (1059-1273)	2497 ± 278 (2112-2991)	934 ± 54 (842-1008)
a	14.8	15.5 ± 2.7 (10.3-20.5)	37.5 ± 9.4 (25.6-50.4)	19.2 ± 1.5 (17.3-21.4)	17.9 ± 1.5 (16.2-20.9)	26 ± 1.8 (23-30)
b	8.2	8.9 ± 1.2 (7.4-11.8)	35.8 ± 7.1 (24.3-44.4)	8.3 ± 0.7 (7.7-9.7)	14.1 ± 1.0 (13.1-16.1)	6.3 ± 0.3 (5.6-6.7)
c	43.8	49.3 ± 14.8 (35.9-87.2)	141.3 ± 39.9 (91.4-204.6)	42.5 ± 7.8 (24.9-53.3)	71.2 ± 11.3 (54.9-88.5)	10.6 ± 0.6 (9.6-12.2)
V	–	–	54 ± 2.5 (50.8-59.6)	–	56 ± 1.2 (54.1-58.1)	–
Max body diam.	77	87 ± 13 (76-120)	176 ± 38 (134-261)	59 ± 5.6 (50-70)	139 ± 12.4 (121-169)	35 ± 2.4 (30-38)
Excretory pore	–	85 ± 13 (68-106)	180 ± 23 (152-221)	137 ± 4.5 (125-142)	177 ± 17.5 (156-209)	148 ± 9.2 (127-162)
Pharynx	139	152 ± 13 (121-170)	114 ± 3.8 (106-117)	87 ± 5.8 (79-98)	106 ± 4.1 (102-112)	72 ± 5.3 (61-80)
Tail length	26	28 ± 3.7 (22-32)	46 ± 6.1 (36-55)	28 ± 7.6 (23-51)	35 ± 4.4 (31-46)	88 ± 5.4 (76-95)
Hyaline region	–	–	–	–	–	55 ± 7.3 (40-70)
Anal body diam.	–	36 ± 3.6 (31-42)	35.7 ± 3.7 (30-42)	60 ± 5.6 (54-69)	42 ± 2.7 (37-47)	17 ± 0.8 (16-18)
Spicule length along chord	65	72 ± 5.4 (61-81)	–	53 ± 3.8 (50-61)	–	–
Gubernaculum length	49	53 ± 5.4 (43-64)	–	38 ± 4.8 (31-48)	–	–
D%	–	54.4 ± 3.1 (50-58)	61.4 ± 4.9 (52-70)	63.3 ± 6.2 (58-74)	59.6 ± 5.2 (50-66)	47.6 ± 3.4 (42-55)
H%	–	–	–	–	–	66 ± 0.2 (51-74)

dorsolateral papillae near tail terminus. One pair of papillae at cloacal aperture level and one pair situated in 8-10 μm anterior to cloacal aperture. Unpaired median papilla as flattened elevation (4 μm diam.) *ca* 20 μm anterior to cloacal aperture. Five pairs of genital papillae discernible in precloacal area anterior to unpaired median papilla. A pair of post-deirids in dorsolateral position anterior to anteriormost genital papillae.

Male (second generation)

Body rarely C-shaped, more often J-shaped after fixation. Main morphological features similar to those of first generation male. Strongly cuticularised stoma walls 2-3 μm long.

Female (first generation)

Anterior region as for *Steinernema* female. Labial region rounded, *ca* 30 μm diam. Excretory duct 1-2 μm wide. Excretory channel lumen in lateral chords visible up to vulva, its walls collapsing posteriorly. Intestinal cells at proventriculus level large, more than 50 μm in size. Ovary flexures 150-200 μm posterior to basal bulb and 20-50 μm anterior to anal level. Ovary sometimes forming an additional loop. Spherical chamber at mid-oviduct, composed of radially arranged cells. Numerous spermatozoa (diam. of nucleus containing end 7-9 μm) forming chains of 3-8 cells scattered in lumen of spherical chamber and neighbouring parts of ovary and uterus. Anterior vulva lip 20-30 μm high, overhanging smaller (10 μm high) posterior lip. Tail end cupula-shaped with acute 20 μm long terminus.

Female (second generation)

No pronounced difference between first and second generation female in morphology and even in morphometrics of cephalic end, buccal cavity and pharynx. Basal bulb usually narrower (30-35 μm diam.), cardia lobes of same size. Intestinal cells with thinner walls. Ovary flexures situated at longer distance from basal bulb or anal opening (300-400 and 100-150 μm, respectively). No morphologically distinct chambers or widened parts of oviduct. Vulva lips equal, 10 μm high, usually covered with transparent substance. Tail end conical with rounded terminus.

Infective juvenile

Anterior region as for *Steinernema* IJ. Heat-relaxed body broadly C-shaped. Anterior end slightly set-off from body contour. Lateral field starting 60-70 μm from anterior end as four ridges, with central, fifth ridge, appearing in few micrometers from beginning. At 10 μm from beginning, lateral field containing five ridges. Eight equal ridges discernible at mid-body. This pattern changing into six ridges near anal opening level, and then into four ridges at mid-tail level. Lateral field formula = 2, 5, 8, 6, 4.

Type habitat and locality

The laboratory culture of the type isolate was established from juveniles collected from and around cadaver of a beetle larva (*Osmoderma eremita* Scopoli, 1763) found on the soil surface in woodland in the Mecklenburg-Vorpommern province of Germany in April, 2003.

Type material

Holotype male (N°1109) and paratype 1st generation female (N°1110) deposited in the collection of the Center of Parasitology, A.N. Severtsov Institute of Ecology and Evolution, Russian Academy of Sciences, Moscow, Russia. A paratype male is present in the collection of the Julius Kühn-Institut in Münster, Germany. Paratypes (male and female) are also deposited in the collection of Ghent University (Belgium).

Relationships

In all phylogenetic trees, *S. schliemanni* clusters with species of the *monticolum*-group. Apart from three described species (*S. monticolum*, *S. ashiuense* and *S. robustispiculum*), a strain identified as *Steinernema* sp. MY5 (Yoshida *et al.*, 1998) is a permanent member of this lineage. The lateral field of IJ of *S. schliemanni* consists of eight ridges at mid-body. In two species of the aforementioned group, the lateral field has a completely different appearance at mid-body, *viz.*, two well pronounced submarginal lines and two lower ridges in the centre of the field. Such a lateral field morphology was described for both *S. robustispiculum* (Phan *et al.*, 2005) and *S. monticolum* (see Fig. 4F in Spiridonov *et al.*, 2010). The lateral field of *S. ashiuense* was described as containing five ridges (Phan *et al.*, 2006). Inside the *monticolum* clade, *S. schliemanni* can be easily distinguished from the three other species by the morphometric characters of IJ. IJ of *S. schliemanni* are significantly longer (842-1008 μm) than those of all three other species. The ranges of the body length of IJ in *S. monticolum*, *S. ashiuense* and *S. robustispiculum* are 612-821 μm, 720-800 μm and 642-778 μm, respectively. The EP (distance from anterior end to excretory pore) and tail length are proportionally higher in the longer juveniles of *S. schliemanni*. The ratios D% and ratio c of IJ are virtually the same in all four species under comparison. The size of spicules of first generation males of *S. schliemanni* resembles that of the spicules of *S. monticolum* (64-81 *vs*

61-80 μm); spicules of first generation males of *S. ashiuense* and *S. robustispiculum* are shorter (not longer than 65 μm).

Molecular characterisation

The ITS rDNA 862 bp long sequence of *S. schliemanni* is deposited in NCBI GenBank as HM778112, and D2-D3 sequence of LSU rDNA (539 bp long) as HM778113. An analysis of ITS rDNA sequence of *S. schliemanni* demonstrates its close relationships with *S. monticolum*, *S. ashiuense* and *S. robustispiculum*. The nucleotide differences between the known species of the *monticolum*-group are on the level of 18-39 bp of ITS rDNA region.

Symbiotic bacterium

The IJ harbour cells of *Xenorhabdus* bacteria. According to the analysis of 16S rDNA, *X. hominickii* is the most similar bacterial species (Spiridonov *et al.*, 2010) differing by 28 bp (or 2% of total sequence).

Bionomics

Cross-hybridisation studies with *S. monticolum* produced no progeny. *Osmoderma eremita* grubs are the only known natural host of *S. schliemanni*. Under laboratory conditions this nematode can develop in *G. mellonella* larvae (Spiridonov *et al.*, 2010).

Steinernema surkhetense[61,62] Khatri-Chhetri, Waeyenberge, Spiridonov, Manadhar & Moens, 2011

(Figs 4.74, 4.75)

Measurements

See Table 4.29.

[61] Specific epithet named for *Surkhet* district of Nepal from where the species was described.

[62] Description after Khatri-Chhetri *et al.* (2011b). Revised by Sergei E. Spiridonov.

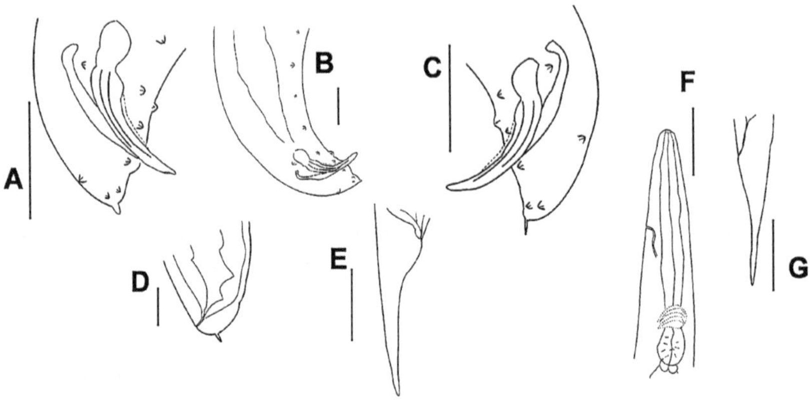

Fig. 4.74. Steinernema surkhetense. *A: First generation male, tail; B: First generation male, posterior end; C: Second generation male, tail; D: First generation female, tail; E: Second generation female, tail; F: Infective juvenile, anterior end; G: Infective juvenile, tail. (Scale bars: A-E = 50 μm; F, G = 25 μm.) After Khatri-Chhetri* et al. *(2011b), courtesy* Russian Journal of Nematology.

DIAGNOSTIC DESCRIPTION

Male (first generation)

Anterior region as for *Steinernema* male. Body curved ventrally posteriorly, tapering toward anterior end, mostly C-shaped, rarely J-shaped in heat-relaxed specimens. Cuticle with very faint transverse striae, lateral field absent. Labial region rounded, slightly offset from rest of body. Excretory pore always anterior to nerve ring, mostly located at metacorpus level. Well developed and reflexed testis. Testis reflexion ventral in position, extending for *ca* half of testis length. Paired, symmetrical, moderately curved, spicules, brownish in colour. Spicule manubrium elongated oval, tapering anteriorly, manubrium *ca* one-fourth of spicule length, calomus separating manubrium from lamina. Lamina well curved anteriorly and dorsally, with two internal ribs less prominent, slender and almost straight posteriorly. Rostrum not developed, velum thin, margin parallel to spicule edge, not covering distal part of lamina. Spicule terminus blunt. In lateral view, gubernaculum boat-shaped, anterior part curved up. In ventral view, cuneus Y-shaped, posteriorly forked, corpus with two opened wings. In dorsal view, gubernaculum with two marginal ridges. Tail bluntly conoid with a short (2.78 $\pm$ 0.66 μm) mucron. Genital papillae 23/25, comprising 11/12 pairs and a single midventral papilla. Specimens with 23 genital papillae bearing seven precloacal pairs (six pairs subventral and one pair lateral) and four post cloacal pairs (one pair subventral, two pairs subterminal and one pair subdorsal). Posteriormost two pairs consistently located near tail tip. Specimens with 25 genital papillae having an extra pair of precloacal papillae located close to second anterior pair. In addition to the genital papillae, one pair of sensory/papillae-like structures protruding from edges of cloacal aperture also present.

Male (second generation)

Second generation male similar to first generation except for length and diam. of body, spicule and gubernaculum.

Female (first generation)

Anterior region as for *Steinernema* female. Body variable in length, usually C-shaped, sometimes coiled on heat relaxation. Cuticle faintly striated. Lateral field not observed. Labial region truncated and slightly offset. Cheilorhabdions prominent. Cardia prominent. Excretory pore located anterior to nerve ring, duct well cuticularised. Genital tract amphidelphic with reflexed ovaries. Vulva transverse, slit-like, lips protruding, mostly equal, small epiptygmata present. Vulval opening located around mid-body level. Short vagina leading into paired uteri. Postanal swelling usually not present, if present usually not well developed. Tail shape variable, mostly dome-shaped. Tail terminus with short mucron. *Endotokia matricida* rarely observed.

Female (second generation)

Similar to first generation female in general morphology, but smaller. Body straight or slightly curved on heat relaxation. Vulval opening located at mid-body. Tail long, tapering to pointed end. Vulva slightly protruding, mostly with very small epiptygmata. Postanal swelling slightly or not developed.

Infective juvenile

Anterior region as for *Steinernema* IJ. Body of heat-relaxed specimens short, slender, slightly curved or straight, slightly tapering towards anterior and posterior ends. Cuticular annulation indistinct under light microscope, 1 μm wide in mid-body region. Lateral field beginning anteriorly with a single ridge at second annule. At annule 14 or 15,

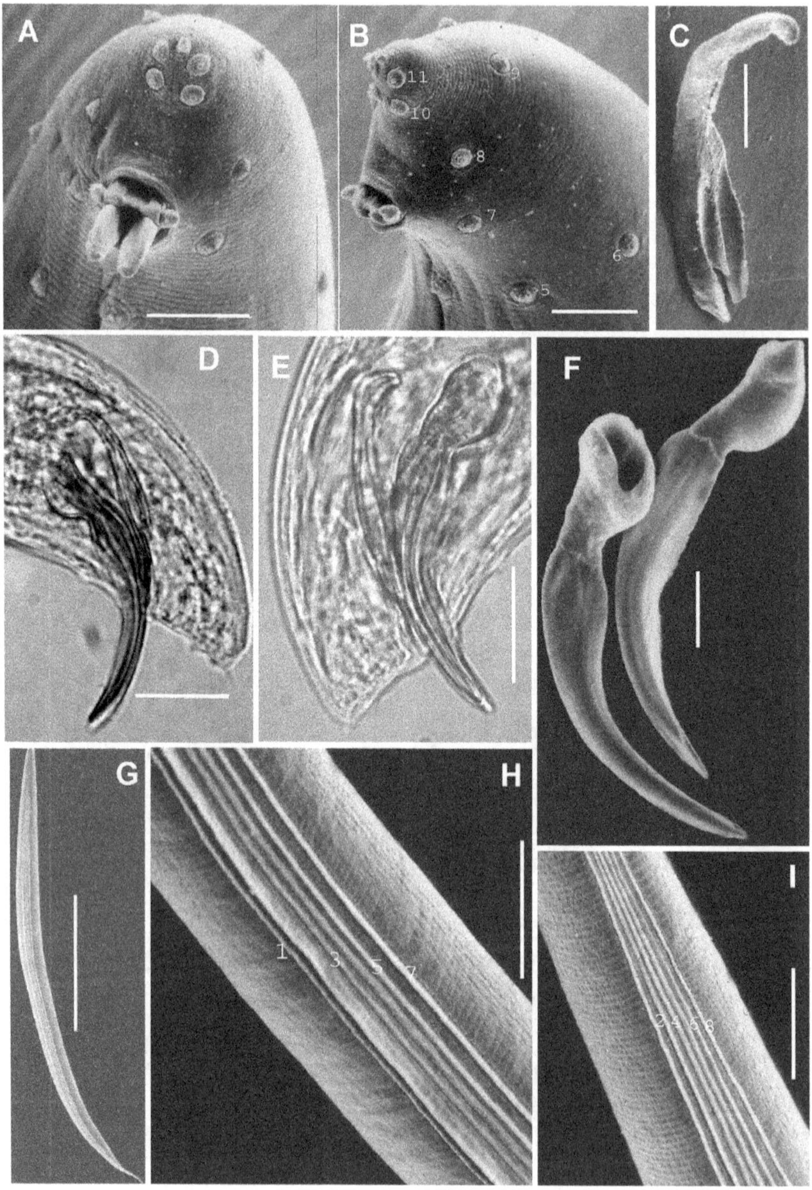

Fig. 4.75. Steinernema surkhetense. *A: First generation male, tail; B: First generation male, tail (papillae numbered); C: First generation male, gubernaculum; D: First generation male, tail; E: Second generation male, tail; F: First generation male, spicules; G: Infective juvenile, total view; H: infective juvenile, lateral field at mid-body (ridges numbered); I: Infective juvenile, lateral*

two additional lines appearing to form two ridges. At excretory pore level, each ridge dividing into four, ridge number increasing from two to eight, although two submarginal ridges not prominent. This arrangement remaining for anterior half of body. In posterior body, one submarginal ridge of one side disappearing, ridge number becoming seven, although another ridge of that side widening. One marginal and one submarginal ridge gradually disappearing to leave five ridges at phasmid level. Posterior to phasmid, central three ridges disappearing and number of ridges becoming two on tail. Lateral field formula = 2, 8, 7, 5, 2. Deirid not observed. Phasmid pore-like, posterior to anus, located at *ca* 30% distance of tail from anus. Labial region slightly truncate, usually slightly offset from body contour, not annulated. Stoma slightly open. Labial papillae not observed, amphids slit-like, four cephalic papillae distinct, posterior to amphids. Hemizonid distinct, lens-shaped, located at beginning of basal bulb. Excretory pore in anterior third of pharynx region, cuticularised, *ca* 1 μm wide. Cardia present. Bacterial vesicle oval or round, *ca* 8 × 5 μm in size. Intestine filled with numerous fat globules, lumen not visible. Genital primordia prominent. Rectum straight, anus sickle-shaped. Hyaline region occupying more than half of tail length. Tail attenuate with pointed terminus.

TYPE HABITAT AND LOCALITY

Steinernema surkhetense was recovered from a soil sample collected from Birendranagar (81°38′E, 28°35′N, 674 m a.s.l.), Surkhet district of western Nepal. Soil samples were collected from an agricultural field cultivated with black gram (*Vigna mungo* (L.) Hepper). The soil was loamy acidic (pH 5.2) with low organic matter content (1.14%).

TYPE MATERIAL

Holotype male (UGMD 104191), five male paratypes (three in UGMD 104191 and two in UGMD 104192) and 12 IJ paratypes (UGMD 104193) deposited in the nematode collection of the Institute of Zoology, Ghent University, K.L. Ledeganckstraat 35, Ghent, Belgium. Eight male

field on nerve ring level (ridges numbered). (Scale bars: A-C, F, H-I = 10 μm; D, E = 20 μm; G = 100 μm.) After Khatri-Chhetri et al. *(2011b), courtesy* Russian Journal of Nematology. *Figures D and E were kindly presented by Dr Khatri-Chhetri.*

Table 4.29. *Morphometrics of type population of* Steinernema surkhetense. *All measurements are in μm and in the form: mean ± s.d. (range).*

Character	First generation			Second generation		Infective juvenile
	Holotype (male)	Male	Female	Male	Female	
n	–	20	20	20	20	20
L	1194	1224 ± 194 (887-1576)	2970 ± 922 (1894-5692)	824 ± 89 (621-945)	1072 ± 101 (875-1290)	415 ± 15 (393-450)
a	10.4	12 ± 1.3 (10-14)	19 ± 2.7 (13-25)	12 ± 1.9 (8.8-15)	13 ± 1.4 (11-16)	19 ± 1.8 (16-24)
b	10.8	11 ± 1.6 (8.0-15)	24 ± 8.8 (14-49)	6.8 ± 0.8 (5.2-8.0)	5.6 ± 0.5 (4.6-6.7)	5.0 ± 0.2 (4.3-5.8)
c	66.3	65 ± 12 (43-85)	161 ± 54 (91-309)	44 ± 6.9 (31-59)	11 ± 1.6 (8.7-15)	9 ± 0.5 (8.3-10)
V	–	–	–	52 ± 3.2 (46-58)	58 ± 5.8 (41-70)	–
Max. body diam.	115	105 ± 19 (75-132)	154 ± 34 (102-236)	73 ± 13 (56-96)	80 ± 10 (64-97)	21 ± 1.8 (18-25)
Excretory pore	44	55 ± 10 (43-78)	53 ± 13 (33-86)	44 ± 4 (35-50)	137 ± 21 (96-195)	32 ± 1.5 (28-34)
Pharynx	110	115 ± 15 (86-146)	132 ± 31 (52-190)	122 ± 11 (106-141)	193 ± 16 (173-236)	92 ± 4.1 (84-101)
Tail length	18	19 ± 2.1 (16-23)	19 ± 5.0 (12-30)	19 ± 3.6 (15-27)	103 ± 15 (65-135)	45 ± 3.2 (38-53)
Hyaline region	–	–	–	–	–	26 ± 3.9 (20-32)
Anal body diam.	28	31 ± 3.8 (23-38)	47 ± 10 (25-68)	27 ± 2.3 (23-32)	28 ± 3.2 (21-34)	12 ± 0.8 (11-14)
Spicule length along chord	74	70 ± 5.2 (58-78)	–	64 ± 5.4 (55-76)	–	–
Gubernaculum length	55	52 ± 5.1 (42-63)	–	44 ± 5.6 (35-53)	–	–
D%	40	48 ± 8.7 (37-64)	44 ± 20 (28-117)	36 ± 3.3 (31-42)	71 ± 10 (50-87)	35 ± 2.2 (31-40)
E%	244	295 ± 66 (205-429)	292 ± 84 (156-453)	237 ± 45 (167-313)	137 ± 28 (85-189)	72 ± 6.4 (54-84)
SW%	264	230 ± 38 (168-330)	–	239 ± 31 (191-292)	–	–
GS%	74.32	75 ± 5.0 (66-84)	–	68 ± 6.8 (57-79)	–	–
H%	–	–	–	–	–	57 ± 7.4 (44-73)

paratypes and five female paratypes of *S. nepalense* and eight male paratypes and 14 IJ paratypes of *S. surkhetense* are also deposited in the nematode collection of Centre of Parasitology of A.N. Severtsov Institute of Ecology, and Evolution of Russian Academy of Sciences, Moscow, Russia, respectively.

RELATIONSHIPS

Steinernema surkhetense is a member of *carpocapsae*/*scapterisci*/*tami*-group *sensu* Spiridonov *et al.* (2004) and close to *S. nepalense*. The IJ of *S. surkhetense* are slightly shorter than those of *S. nepalense* (L = 415 (393-450) *vs* 479 (424-513) μm), shorter EP distance (32 (28-34) *vs* 38 (32-46) μm), slightly shorter tail length of 45 (38-53) *vs* 53 (48-61) μm and lateral line formula = 2, 8, 7, 5, 2 *vs* 2, 7, 8, 6, 5, 4, 2. First generation males of *S. surkhetense* can be distinguished from *S. nepalense* by the absence of a single flattened ridge in the lateral field, higher SW% value of 230 (168-330) *vs* 218 (165-260), and slightly shorter gubernaculum (52 (42-63) *vs* 56 (47-60) μm), respectively.

MOLECULAR CHARACTERISATION

The 817 bp long sequence of ITS rDNA region and 614 bp long sequence of D2-D3 expansion segment of LSU rDNA are deposited in GenBank as HQ190042 and HQ190043, respectively.

SYMBIOTIC BACTERIUM

No information.

BIONOMICS

Cross-hybridisation tests with *S. carpocapsae* and *S. siamkayai* produced progeny in 10% of cases, but these were non-fertile (Khatri-Chhetri *et al.*, 2011b).

Besides *G. mellonella*, the species reproduced well on rice moth, *Corcyra cephalonica* Stainton, lesser wax moth larvae, *Achroia grisella* Fabricius, and silk worm *Bombyx mori* L. (Khatri-Chhetri *et al.*, 2011b).

DISTRIBUTION

In addition to the type locality in Nepal, the species was also recovered from a sample taken nearby (81°37′E, 28°35′N, 690 m a.s.l.) in a sissoo

(*Dalbergia sissoo* Roxb.) forest with ground cover of various broad- and narrow-leaved weeds (Khatri-Chhetri *et al.*, 2011b).

Steinernema texanum[63,64] Nguyen, Stuart, Andalo, Gozel & Rogers, 2007
(Figs 4.76-4.80)

MEASUREMENTS

See Table 4.30.

DIAGNOSTIC DESCRIPTION

Male (first generation)

Anterior region typical for *Steinernema* male. Excretory pore anterior to nerve ring. Gonad monorchic, reflexed. Testis with ventral reflexion, germinal zone, growth zone and *vas deferens*. Distance from base of pharynx to anterior end of testis variable, longer than pharynx length. Spicules paired, dark brown in colour. Head region (manubrium) of spicules, somewhat elongate (spicule head region length/width = 1.2-1.4); shaft (calomus) very short or absent; rostrum present; blade (lamina) thick, tapering slightly posteriorly; blade terminus blunt; velum prominent. Each spicule with two internal ribs. Under SEM, spicule with three lobes: dorsal lobe prominent, well curved anteriorly; lateral lobe well defined on dorsal edge and divided into two posteriorly; ventral lobe short. Gubernaculum boat-shaped in lateral view, cuneus needle-shaped, anterior end curved ventrally. Usually a single precloacal papilla and 11 pairs of genital papillae arranged in normal position for *Steinernema* (*i.e.*, six or seven pairs precloacal subventral, one pair adcloacal, one pair lateral, two pairs subterminal and one pair subdorsal). Number of caudal papillae constant (three pairs), but number of precloacal subventral pairs variable. Tail conoid, without phasmid; tail terminus without mucron.

Male (second generation)

Similar to first generation male but smaller. Labial region usually somewhat swollen. Excretory pore more anterior, in posterior part of

[63] Specific epithet derived from the state of *Texas* where the species was collected.

[64] Description after Nguyen *et al.* (2007b). Revised by Khuong B. Nguyen.

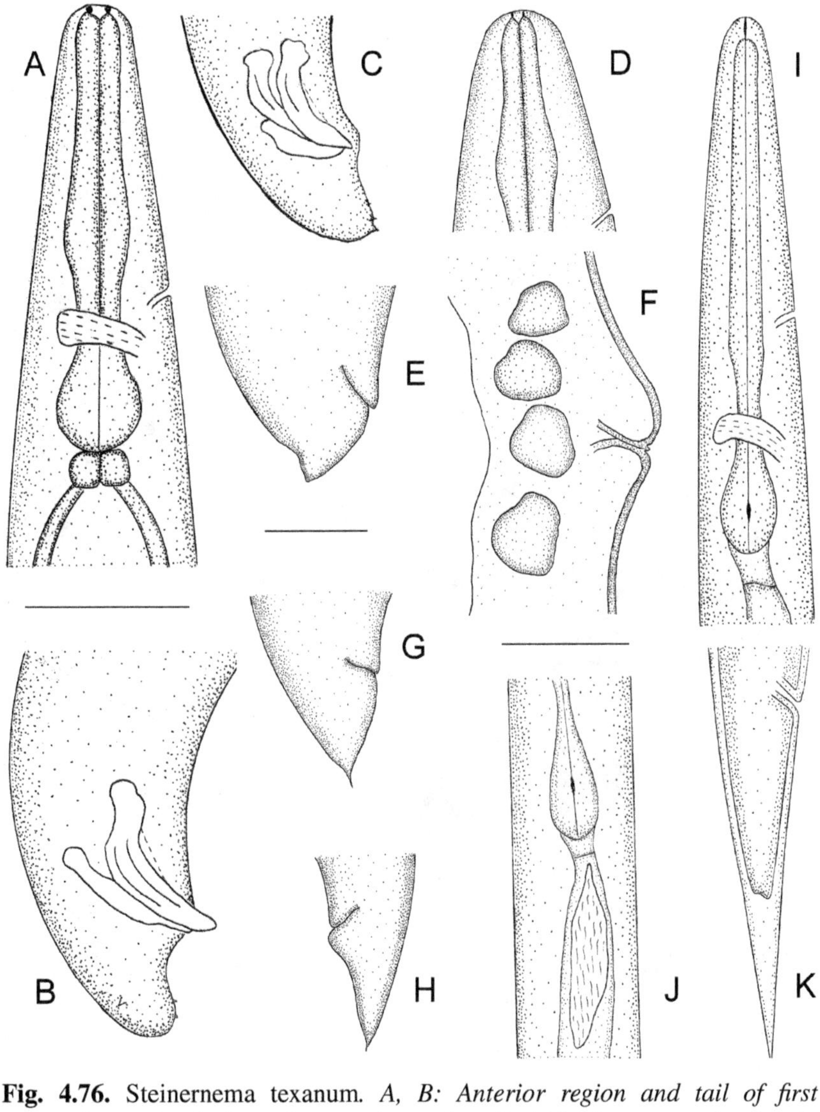

Fig. 4.76. Steinernema texanum. *A, B: Anterior region and tail of first generation male; C: Second generation male tail; D-F: First generation female labial region, tail and vulva; G, H: Second generation female tail shapes. I-K: Infective juvenile. I: Anterior region; J: Posterior part of pharynx showing basal bulb and bacterial chamber; K: Tail. (Scale bars: Left and central bars for males and females = 50 μm; right bar for infective juvenile = 25 μm.) After Nguyen* et al. *(2007b), courtesy* Nematology.

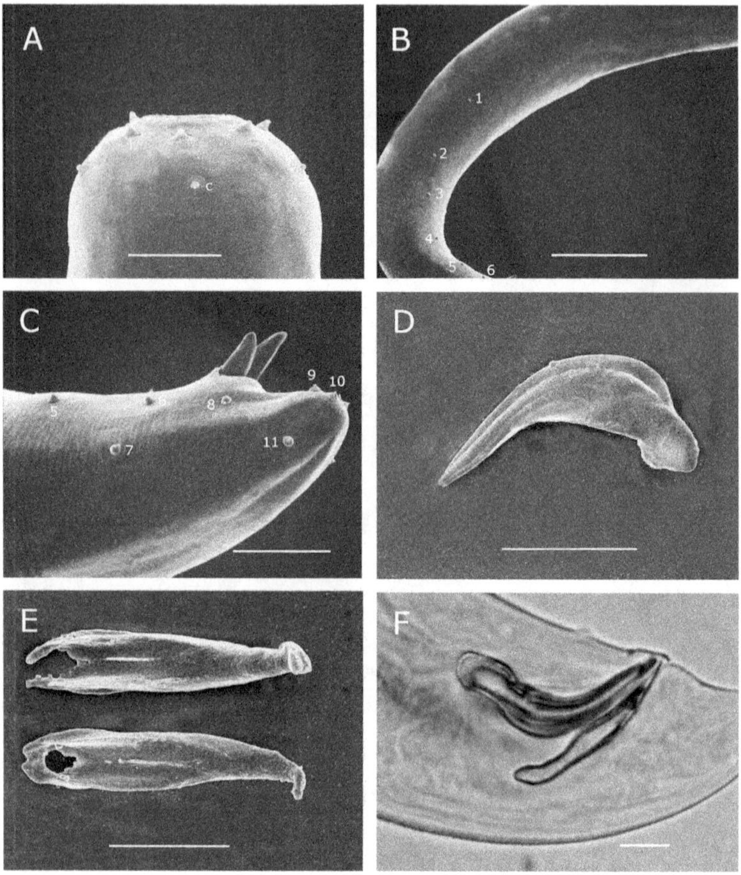

Fig. 4.77. *SEM of* Steinernema texanum *first generation males. A: Male labial region showing labial papillae (l), and cephalic papillae (c); B: Posterior region showing genital papillae from pair 1 to pair 6; C: Papillae pairs 5-11; D: Spicule; E: Gubernaculum ventral view showing needle-shaped cuneus; F: LM photograph of spicules and gubernaculum. (Scale bars: A = 12.0 μm; B = 66.7 μm; C = 20.0 μm; D = 27.3 μm; E = 16.7 μm; F = 12.4 μm.) After Nguyen* et al. *(2007b), courtesy* Nematology.

corpus. Posterior part of body with a single precloacal papilla and 13-14 pairs of genital papillae of which four pairs caudal compared to three in first generation. Lateral field prominent with single ridge. Tail terminus with a mucron, usually subterminal subventral in position.

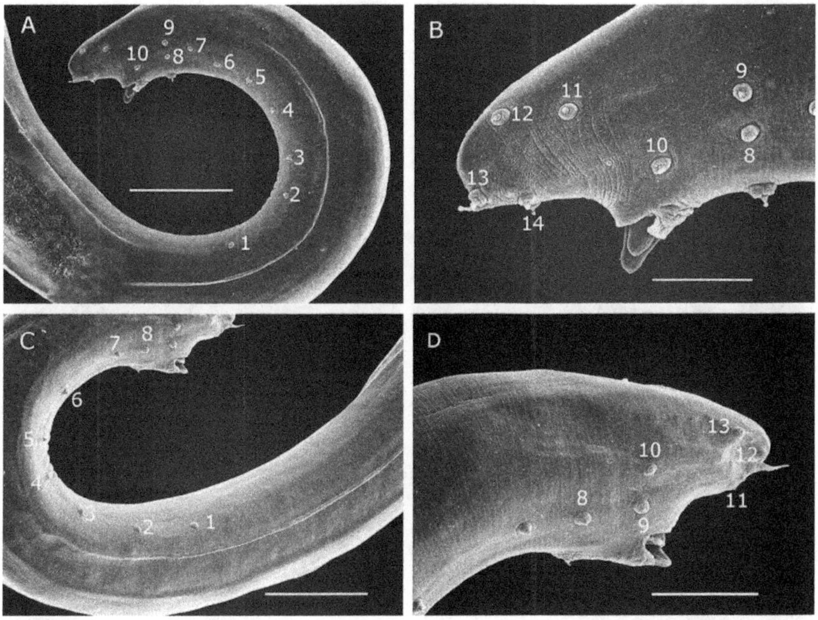

Fig. 4.78. *SEM of* Steinernema texanum. *A-D: Two second generation males showing 13 and 14 pairs of genital papillae, single-ridged lateral field and variation of mucron position. (Scale bars: A, C = 30 μm; B = 10 μm; D = 15 μm.) After Nguyen* et al. *(2007b), courtesy* Nematology.

Female (first generation)

Anterior region typical for *Steinernema* female. Lateral fields with one line present on all females observed. Excretory pore position variable, near mid-pharynx. Vulva a transverse slit on a protruding area, small epiptygmata present. Tail with bluntly pointed tip, one or two projections present. Postanal swelling present in most females, more pronounced in fully mature females. Tail shorter than anal body diam., tapering to a blunt terminus. Two projections usually present at tail end. Two wart-like structures anterior to tail tip always present on ventral region.

Female (second generation)

Similar to first generation female but smaller. Body diam. greater anterior to vulva than posterior. Vulva on asymmetrical protuberance and situated at mid-body, epiptygmata not observed. Postanal swelling

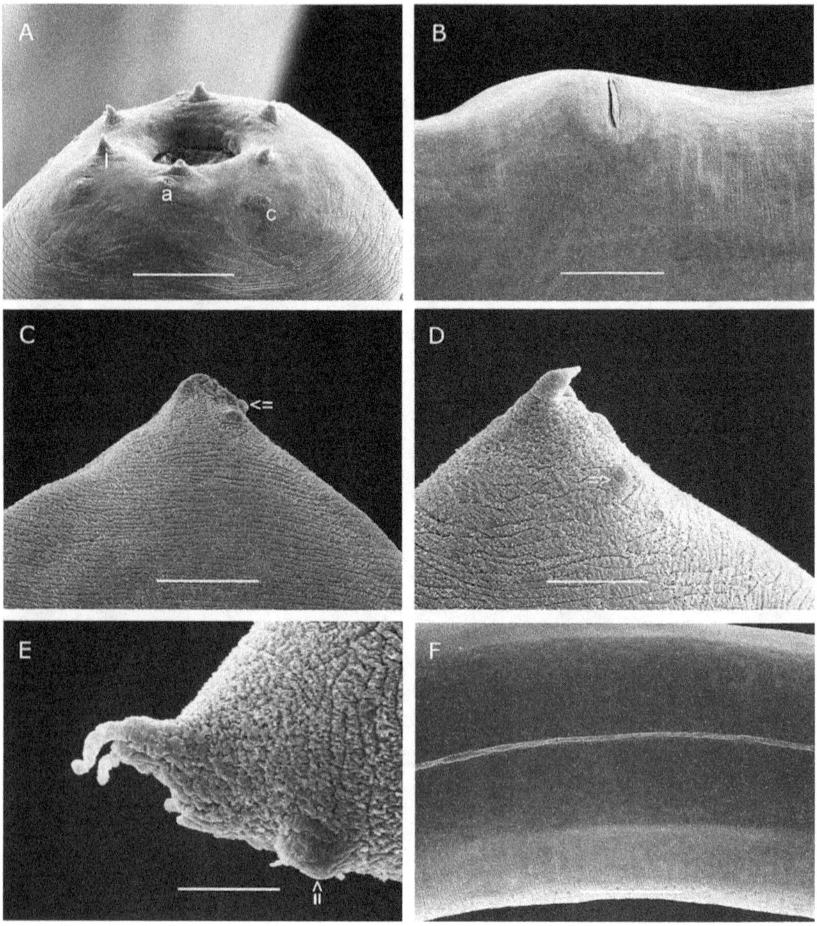

Fig. 4.79. *SEM of* Steinernema texanum *female. A: First generation female face view showing mouth, amphidial aperture (a), six labial papillae (l) and four cephalic papillae (c); B: First generation female vulva; C-E: Tail tip with posterior projections and wart-like structures (arrow) on tail; F: First generation female body showing lateral field with single line. (Scale bars: A = 8.57 μm; B = 37.5 μm; C = 6.0 μm; D = 3.7 μm; E = 2.0 μm; F = 15.0 μm.) After Nguyen* et al. *(2007b), courtesy* Nematology.

present. Tail mostly longer than anal body diam., tapering gently to a sharp point. For some large females (about 5%), tail shorter than anal body diam.

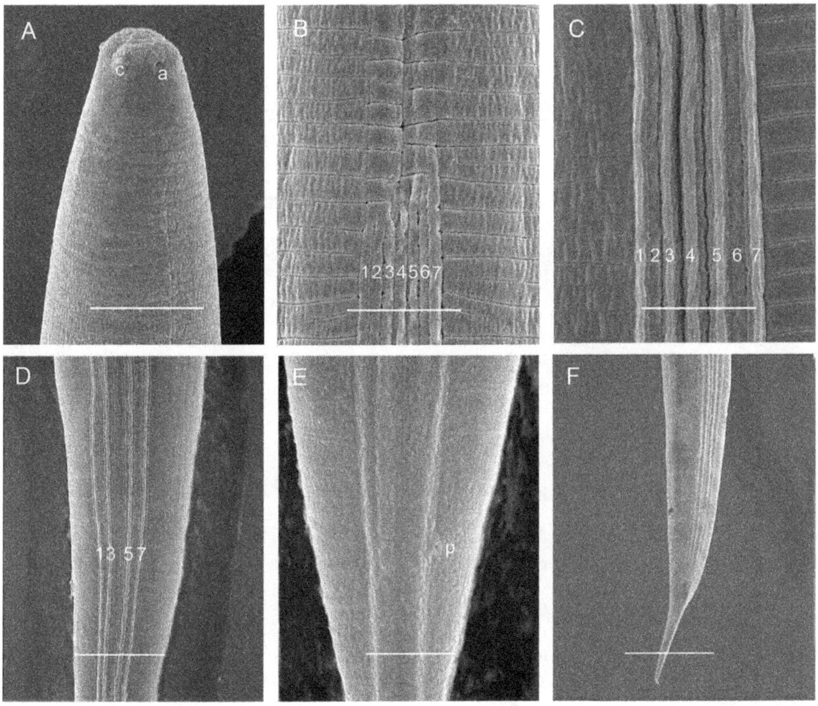

Fig. 4.80. *SEM of* Steinernema texanum *infective juvenile. A: Anterior region showing one of two amphidial apertures (a), three of four cephalic papillae (c) and single line in lateral field; B: Lateral field showing the change from two to seven ridges; C: Lateral field showing seven ridges; D: Lateral field showing lateral field narrowing posteriorly; E: Lateral field in posterior region showing change to two ridges and phasmid (p); F: Posterior region showing lateral field and anus. (Scale bars: A = 6.6 μm; B, C = 5.0 μm; D = 10.0 μm; E = 3.3 μm; F = 20.0 μm.) After Nguyen* et al. *(2007b), courtesy* Nematology.

Infective juvenile

Anterior region typical for *Steinernema* IJ. Excretory pore anterior to nerve ring. Lateral field beginning anteriorly with one line at fourth or fifth annule. A short distance posteriorly, two additional lines appearing to form two ridges. Near excretory pore level, number of ridges in lateral fields increasing from two to seven. Seven ridges remaining unchanged from excretory pore to anus except that two submarginal ridges not as raised as others in middle part of body. Near anus all ridges become smaller, only ridges 1, 3, 5, 7 raised and seen clearly, others gradually disappearing. Near phasmid, only two ridges (poorly

Table 4.30. *Morphometrics of* Steinernema texanum. *All measurements are in μm and in the form: mean ± s.d. (range).*

Character	First generation			Second generation		Infective juvenile
	Holotype (male)	Male	Female	Male	Female	
n	–	20	20	20	20	20
L	1384	1296 ± 66 (1197-1406)	3058 ± 268 (2720-3623)	828 ± 34 (773-890)	1717 ± 227 (1250-2105)	756 ± 13.4 (732-796)
a	–	–	–	–	–	25 ± 1.5 (22-27)
b	–	–	–	–	–	6.5 ± 0.2 (6.2-7.0)
c	–	–	–	–	–	10.4 ± 0.6 (9.6-12.5)
c′	–	–	–	–	–	4.1 ± 0.3 (3.3-4.6)
V	–	–	52 ± 1.4 (50-55)	–	55 ± 1.8 (52-59)	–
Body diam.	128	99 ± 10 (81-116)	163 ± 15 (130-202)	50 ± 3 (45-58)	95 ± 11 (75-114)	30 ± 2.2 (29-34)
Excretory pore	80	90 ± 4.7 (79-100)	88 ± 7.5 (78-107)	75 ± 5.4 (64-83)	94 ± 8 (68-104)	59 ± 2.6 (52-62)
Nerve ring	123	104 ± 4.8 (94-114)	122 ± 7.8 (111-135)	97 ± 8.7 (81-117)	127 ± 11 (113–159)	92 ± 4.8 (84-102)
Pharynx	155	135 ± 6.8 (123-147)	172 ± 9 (160-189)	129 ± 10.2 (109-151)	178 ± 17 (159-236)	115 ± 2.7 (111-120)
Testis reflexion	267	269 ± 15.8 (273-325)	–	130 ± 24.6 (90-165)	–	–
Tail length	28	23 ± 3.7 (19-30)	40 ± 5.7 (30-52)	28 ± 1.7 (25-30)	53 ± 6 (43-61)	73 ± 4.5 (60-79)
Hyaline region	–	–	–	–	–	43 ± 2.9 (37-47)
Anal body diam.	38	38 ± 4.1 (31-45)	60 ± 5.6 (50-71)	31 ± 1.7 (27-33)	43 ± 4 (38-52)	18 ± 0.9 (17-20)
Spicule length	58	60 ± 2.9 (55-66)	–	42 ± 1.9 (39-46)	–	–
Spicule width	10	12.6 ± 1.4 (10-14)	–	7.4 ± 1.0 (5.8-8.7)	–	–
Gubernaculum length	49	45 ± 3.9 (39-53)	–	24 ± 2.1* –	–	–
Gubernaculum width	7.2	6.7 ± 1.1 (5.8-10.0)	–	–	–	–
D%	52	67 ± 4.7 (58-73)	–	–	–	51 ± 2.2 (46-53)
E%	–	–	–	–	–	81 ± 3.2 (76-88)
SW%	152	157 ± 19.2 (127-203)	–	–	–	–
GS%	84	75 ± 5.5 (62-84)	–	–	–	–
H%	–	–	–			59 ± 2.7 (53-64)

* Range data missing in original paper.

separated) observed in lateral field. With above arrangement, lateral field formula = 2, 7, 2. Bacterial vesicle located just posterior to cardia, 7.0-8.5 μm in diam., 26-31 μm in length, containing bacterial cells, variable in shape, mostly fusiform or oval. Hemizonion and hemizonid not observed. Tail four times as long as anal body diam. and attenuate. Phasmid present near mid-tail, just ventral to lateral field. Hyaline region occupying 58 (54-61)% of tail length.

Type habitat and locality

Natural host unknown as *S. texanum* was isolated using the *Galleria*-baiting technique from soil samples taken at a site along state road Route 77 south of Kingsville between the towns of Sarita and Armstrong in Kennedy County, TX, USA, in an area with sandy soil, weedy vegetation, and shaded by mature live oak trees (*Quercus* sp.).

Type material

Holotype (male, first generation): isolated from haemocoel of *G. mellonella* deposited in the United States Department of Agriculture Nematode Collection (USDANC), Beltsville, MD, USA. Allotype (female, first generation): same data as holotype, deposited in the USDANC, Beltsville, MD, USA. Paratypes: same data as holotype. Many males and females of the first generation and several IJ in 4% formalin deposited in USDANC, Beltsville, MD, USA. Several males, females and IJ in formalin deposited in the Department of Entomology and Nematology, University of Florida, Gainesville, FL, USA.

Relationships

Steinernema texanum is characterised by morphometrics of the IJ with body length of 756 μm, distance from anterior end to the excretory pore = 59 μm, tail length = 73 μm, ratio a = 25, H% = 59 and E% = 81. Lateral field pattern of the new species is 2, 7, 2, and very typical for the taxon. The first generation male can be recognised by the length and shape of the spicule and the gubernaculum, position of the excretory pore, D% = 67 and GS% = 75. The female can be recognised by the vulva with very low epiptygmata, and two wart-like structures anterior to tail tip always present on ventral region. *Steinernema texanum* is closely related to species of the *feltiae*-group. More information on relationships is available in Nguyen *et al.* (2007b).

MOLECULAR CHARACTERISATION

Steinernema texanum is characterised genetically by the ITS (accession no. EF152568) and D2-D3 regions (accession no. EF152569). The sequence of ITS regions of *S. texanum* flanked by primers 18S and 26S is characterised by its length 956 bp, ITS1 = 263 bp, ITS2 = 286 bp. The sequence length of D2-D3 regions flanked by primers D2F and 536 is 860 bp. More information on molecular characterisation and phylogenetic analysis is available in Nguyen *et al.* (2007b).

SYMBIOTIC BACTERIUM

No information.

BIONOMICS

No information.

DISTRIBUTION

Steinernema texanum is known only from the type locality in Texas, USA.

Steinernema thesami[65,66] (Gorgadze, 1988) Gorgadze & Lortkipanidze, 2004

= *Neoaplectana thesami* Gorgadze, 1988
= *Steinernema tbilisiense*[67] Gorgadze, Lortkhipanidze, Ogier, Tailliez & Burjanadze, 2015
(Fig. 4.81)

MEASUREMENTS

See Table 4.31.

[65] Specific epithet derived from the River *Thesami*, the locality from where the species was originally found. The junior synonym name refers to *Tbilisi*, the city in Georgia near where the species was found.
[66] Description after Gorgadze *et al.* (2016). Revised by Sergei E. Spiridonov.
[67] Corrected from the original, and wrongly formed, *tbilisensis* (see Chapter 2).

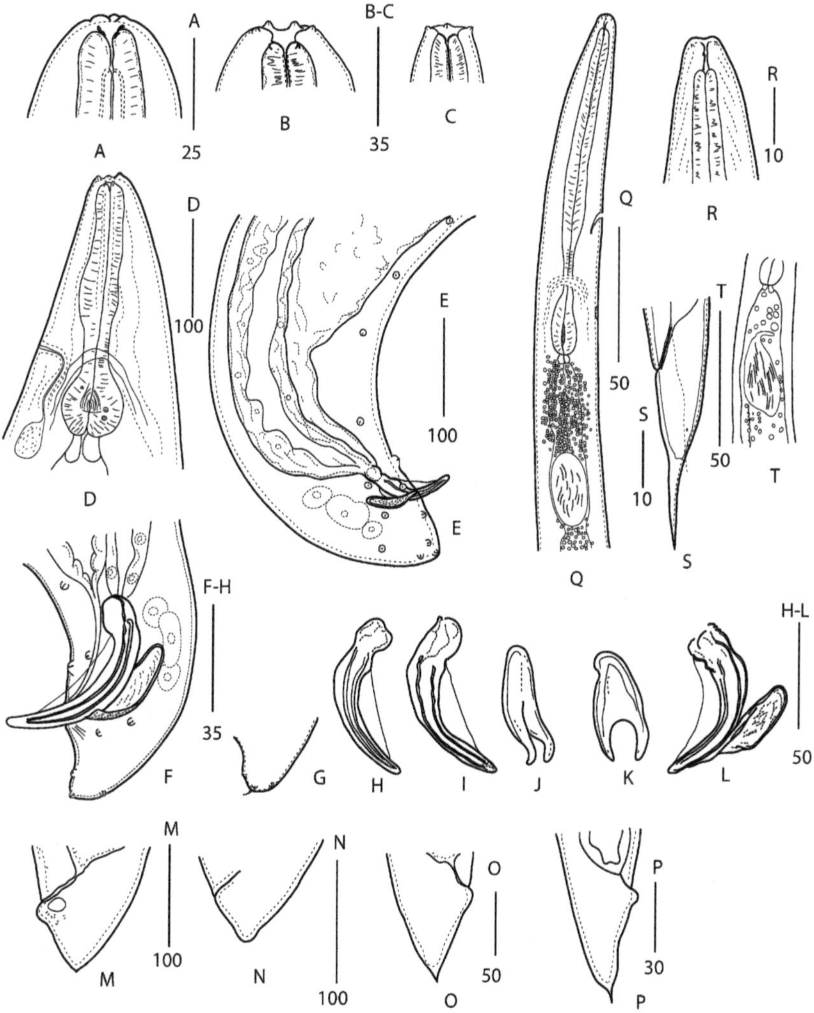

Fig. 4.81. Steinernema thesami *(Gorgadze, 1988). B, E, I-L: First generation male; C, F, G: Second generation male; A, D, M, N: First generation female; O, P: Second generation female; Q-T: Infective juvenile. A-C: Head end; D: Anterior region; E, F: Tail; G: Tail extremity; H, I: spicules; J, K: Gubernaculum; L: Spicules and gubernaculum; M-P: Tail shape; Q: Anterior region; R: Head; S: Tail; T: Bacterial vesicle. All in lateral position. (Scale bars in μm.) After Gorgadze* et al. *(2016), courtesy* Russian Journal of Nematology.

Table 4.31. *Morphometric characters of* Steinernema thesami *(Tbilisi material). All measurements are in μm and in the form: mean ± s.d. (range).*

Character	First generation			Second generation		Infective juvenile
	Holotype (male)	Male	Female	Male	Female	
n	–	20	20	20	20	50
L	2153	2030 ± 239 (1606-2332)	4763 ± 569 (3332-5795)	1393 ± 157 (1201-1630)	2683 ± 216 (2356-3117)	801 ± 31 (714-880)
a	10.5	10.4 ± 1.1 (9.0-13.2)	17.9 ± 2 (14.1-23.2)	15.5 ± 2.1 (13-23.6)	15.4 ± 1.3 (11.3-17.3)	23 ± 2 (20.4-27.4)
b	10.7	10.5 ± 1.1 (8.7-12.6)	18 ± 2 (13.8-21.5)	7.9 ± 0.7 (7.0-10.4)	12.6 ± 1.1 (11.2-15)	5.7 ± 0.2 (5.4-6.2)
c	34	38 ± 4.7 (30-47)	61 ± 10 (42.7-81.4)	33.6 ± 5.1 (23.5-39)	42.2 ± 4.7 (18-53.7)	11.4 ± 0.8 (10-13.3)
c′	0.9	0.9 ± 0.02 (0.7-1.0)	0.5 ± 0.03 (0.5-0.6)	0.6 ± 0.05 (0.6-0.7)	0.6 ± 0.01 (0.5-0.6)	4.2 ± 0.3 (4.0-4.4)
V	–	–	55 ± 3 (48.1-60.2)	–	56.5 ± 2 (52.3-64.8)	–
Body diam.	205	195 ± 25 (122-235)	266 ± 26 (214-321)	90 ± 9 (78-107)	170 ± 7 (151-181)	34 ± 3 (29-39)
Excretory pore	132	130 ± 10 (107-151)	143 ± 16 (117-181)	119 ± 9 (107-137)	128 ± 11 (112-151)	72 ± 3 (67-78)
Pharynx	200	191 ± 10 (171-205)	265 ± 12 (240-289)	174 ± 8 (161-186)	212 ± 7 (200-225)	140 ± 5 (127-151)
Tail length	63	53 ± 6 (44-68)	78 ± 10 (63-98)	42 ± 4 (34-53)	64 ± 6 (53-73)	70 ± 5 (63-78)
Hyaline region	–	–	–	–	–	31 ± 3 (26-39)
Anal body diam.	78	72 ± 8 (53-88)	94 ± 9 (78-107)	48 ± 4 (39-58)	58 ± 5 (49-68)	19 ± 2 (18-19)
Spicule length along chord	68	68 ± 6 (53-78)	–	55 ± 3 (51-61)	–	–
Gubernaculum length	53	49 ± 2 (44-53)	–	37 ± 3 (31-44)	–	–
D%	66	67 ± 0.4 (57-75)	53 ± 7 (42-72)	69 ± 0.8 (60-100)	60 ± 5 (53-73)	50 ± 2 (47-53)
E%	209	247 ± 32 (186-300)	183 ± 23 (153-253)	285 ± 34 (243-351)	207 ± 31 (153-269)	100 ± 0.7 (87-123)
SW%	87	94 ± 1.4 (60-118)	–	114 ± 1 (100-143)	–	–
GS%	77	72 ± 1.4 (62-92)	–	66 ± 0.6 (57-78)	–	–
H%	–	–	–	–	–	45 ± 0.4 (37-54)

DIAGNOSTIC DESCRIPTION

Male (first generation)

Anterior region typical of *Steinernema* male. Body plump, ventrally curved posteriorly, J-shaped when heat-relaxed, tapering to anterior end. Cuticle annulated. Lateral fields weakly developed, seen in SEM only as slight differentiation of cuticle. Deirids situated at level of pharyngeal bulb. Head continuous with body, truncate to rounded, six low, fused lips present, each bearing prominent labial papilla. Four slightly smaller cephalic papillae (two subdorsal + two subventral) located posterior to labial papillae and a pair of slit-like amphids in lateral position at level of cephalic papillae. Mouth aperture subtriangular. Perioral disk absent. Stoma short, shallow 8-13 μm long and 7-10 μm wide, cheilostom broad, with thick rounded rhabdia, gymnostom short, stegostom funnel-shaped. Pharynx clavate, muscular, comprising corpus reaching stoma base anteriorly and slightly swollen posteriorly, isthmus short, basal bulb valvate, slightly wider than metacorpal expansion, pyriform. Nerve ring surrounding isthmus. Excretory pore situated anterior to nerve ring, excretory duct weakly cuticularised in distal *ca* 10 μm, *ca* 60 μm long overall, excretory cell located at bulb level. Cardia prominent, projecting into intestine. Intestine with wide lumen. Reproductive system monorchic, reflexed. Testis reflexion *ca* 210-372 μm posterior to basal bulb, 590-640 μm long. Spermatids located in *vas deferens*, large, 18-21 $\times$ 20-24 μm, broadly elliptical in shape. Ejaculatory duct *ca* 170 μm long, with wide muscular walls. Immature sperm in ejaculatory duct *ca* 4-5 μm diam. Spicules paired, equal, colourless or yellowish, stout, with moderately defined, rounded manubria slightly longer than wide (*ca* 16-18 $\times$ 14-16 μm in size), shaft very short, *ca* 11-13 μm wide, blade moderately curved, 16-17 μm wide, rostrum absent, velum prominent, broad, not extending to spicule tip, distal tips bluntly rounded, 4 μm wide, sometimes appeared bifid. Blade with two lateral, prominent internal ribs. Gubernaculum *ca* 75% of spicule length and darker in colour than latter, broad (8-10 μm wide), proximal end spoon-like, usually wide, rounded but sometimes narrowed distally, gubernaculum corpus open, forming short wings curved upwards. Cuneus absent. Complement of 23 genital papillae arranged as typical for genus: single midventral precloacal papilla, five pairs of precloacal subventral papillae, two pairs adcloacal, one pair lateral postcloacal, two pairs subterminal and one pair subdorsal. Phasmids indistinct. Tail short, ventrally curved, conoid, tip rounded, mucron absent.

Male (second generation)

Similar to first generation but smaller (mean body length 1393 *vs* 2030 μm). Lips more pronounced. Stoma slightly longer. Spicules and gubernaculum shorter and thinner. Tail mucron *ca* 4 (2-5) μm long, present in 35% of specimens.

Female (first generation)

Anterior region typical of *Steinernema* female. Body variable in length, usually C-shaped on heat relaxation. Head bluntly rounded, slightly tapering anteriorly, not offset from body contour. Stoma *ca* 4-7 μm long and 10-14 μm broad. Genital tract amphidelphic with reflexed ovaries. Vulva distinct, mostly symmetrical, rarely slightly asymmetrical, in form of a transverse slit, not or only slightly (3-4 μm), protruding from body contour. Vulval opening always posterior to mid-body, on average 55%. Vulva slightly protruding (*ca* 12 μm), epiptygmata not developed. Vagina short, leading into paired uteri. Rectum narrow, anal opening distinct. Tail of mature female obese, bearing two minute protuberances. Postanal swelling slightly (2-8 μm) or not developed in mature female). Pygmy forms observed. *Endotokia matricida* rarely seen.

Female (second generation)

Smaller in size, more slender, and with fewer eggs. Postanal swelling stronger than in first generation. Short epiptygmata present or not. Tail longer than wide, conoid, pointed, rarely mucronate.

Infective juvenile

Anterior region typical of *Steinernema* IJ. Body C-shaped, tapering to both ends. Head end rounded, continuous with body contour, bearing four rather high cephalic papillae and small oval amphids. Cuticle annulated posterior to labial region, annuli less than 1 μm wide. Lateral fields extending from *ca* 15 μm from anterior to tail terminus. Lateral field 18-20 μm wide at mid-body. Initially, appearing as three incisures, then as a slightly elevated area mid-body where two prominent, marginal ridges positioned at short distance from field lateral margins and two lower ridges positioned in centre; in tail region, central ridges disappearing. Stoma *ca* 4-7 μm long and 2 μm wide. Hemizonid situated opposite basal bulb. Pharynx with slender corpus slightly expanded at posterior, narrow isthmus and small elongated bulb with valves. Excretory pore at level of metacorpal expansion, cuticularised part of excretory duct short, entire duct extending to basal bulb level. Cardia distinct. Intestine containing numerous fat globules. Bacterial vesicle located in anterior part of intestine, oval or drop-like, 28 (15-47) μm long and 15 (8-18) μm wide (measurements of Moscow culture) or 37-74 μm long $\times$ 21 μm wide (measurements of Tbilisi culture), usually with distinct, thin (*ca* 1 μm wide or less) wall, sometimes vesicle wall inconspicuous. Vesicle containing a number of rod-like bacterial cells (*ca* 5 μm long). Tail conical, slightly asymmetrical, with anterior part conical and posterior hyaline region somewhat cylindroid tapering to a point. Anus crescent-shaped, inverted. Rectum *ca* 14 μm long. Phasmids present at border of hyaline region. Hyaline region lacking spine but refractive droplet-like inclusions sometimes present.

TYPE HABITAT AND LOCALITY

Alder thickets along the River Thesami, Mtskheta-Mtianeti Region, Thesami village, Georgia. Altitude: 1105 m a.s.l.; coordinates: 44°52′52″N, 44°57′18″E.

OTHER HABITAT AND LOCALITY

Soil in a deciduous forest at the Kus Tba (Turtle Lake), Tbilisi, Georgia, 550 m a.s.l., coordinates: 41°42.645′N, 44°45.708′E. Mountains near Krasnaya Poliana (43°42′N, 40°15′E, 900 m a.s.l.), above the Black Sea city resort of Sotchi, Russian Federation.

TYPE MATERIAL

Original slides of *S. thesami* (isolated in 1987). Holotype first generation male (slide no. D-412), paratype first and second generation males (slide no. D-416-430), paratype first and second generation females (slide no. D-431-445) and infective juveniles (slide no. D-446-457) are deposited in the collection of the museum of the Institute of Zoology, Ilia State University Tbilisi, Georgia.

Other material

The types of *S. tbilisiense*, which was isolated in 2011 and is herein regarded as a junior synonym of *S. thesami*, are deposited as follows: holotype first generation male (slide no. D-337), paratype first and second generation males (slide no. D-338-358), paratype first and second generation females (slide no. D-359-374) and IJ (slide no. D-375-391) in the collection of the museum of the Institute of Zoology, Ilia State University, Tbilisi, Georgia.

Voucher materials: first generation male no. 14255; first generation female no. 14256 and IJ no. 14257 are deposited in the Museum of the Helminthological Collections, Centre of Parasitology, Severtsov Institute of Ecology and Evolution, Moscow, Russian Federation.

Relationships

Morphology and molecular data indicate that *S. thesami* belongs to the *affine-intermedium*-group. There are six other known species: *S. affine*, *S. intermedium*, *S. beddingi*, *S. sichuanense*, *S. arasbaranense* and *S. poinari*, all of which share a relatively uniform morphology. The members of the *affine-intermedium*-group are characterised by 500-900 μm long IJ which are C-shaped when heat-relaxed and, unlike all other steinernematids, possess lateral fields with two prominent central ridges. Infective juveniles of *S. thesami* differ from *S. affine*, *S. arasbaranense*, *S. intermedium*, *S. poinari* and *S. sichuanense* in the number of ridges in lateral field (four *vs* six to eight). From *S. beddingi*, which has similarly sized IJ with a similar structure to the lateral field, *S. thesami* can be differentiated by smaller manubria, presence *vs* absence of velum, reduced *vs* developed postanal swelling, and absence *vs* presence of a tail spine in the IJ.

Molecular characterisation

The ITS rDNA sequence of *S. thesami* (797 bp) was used for BLAST-search in NCBI GenBank and revealed two most similar deposited sequences: *Steinernema* sp. AY171291 from Russia and *Steinernema* sp. KC287223 from Georgia. The first sequence originates from *Steinernema* sp. isolated from spruce wood soil collected in 2000 near Krasnaya Poliana (43°42′N, 40°15′E, 900 m a.s.l.), in the mountains above the Black Sea city-resort of Sotchi and baited with *Galleria* larvae.

The invasive ability of the strain decreased rapidly and it was lost after sustained attempts to culture it on *Galleria*. The sequence of *S. thesami* is identical to that of *Steinernema* sp. AY171291 from Sotchi.

The sequence KC287223 was obtained for the type culture of the recently described species *S. tbilisiense* (Gorgadze *et al.*, 2015). A 1 bp difference was reported between the sequences of *S. thesami* and *S. tbilisiense*. Hence, this latter species is considered as junior synonym of *S. thesami*. The interspecific differences for this group account for 17-23 bp.

Bionomics

Originally, the nematode was found in the infected pupa of *Orerophthera brumata* L., 1758 (Lepidoptera: Geometridae). The infection of the laboratory insects *G. mellonella* and *Tenebrio molitor*, in addition to the Colorado beetle, *Leptinotarsa decemlineata* Say, by *S. thesami* was reported by Gorgadze & Lortkipanidze (2004).

Cross-breeding studies of the *S. tbilisiense* isolate with *S. affine* and *S. intermedium* produced no progeny. In the control, when all species were self-crossed, males and females produced offspring normally. No progeny was observed in the single female control plates. Gorgadze *et al.* (2015) studied the development of the species they named as *S. tbilisiense* and provided details of life cycle times, *etc.*

Symbiotic bacterium

The five protein coding sequences tested (*recA*, KF946000; *gyr*B, KF946004; *dna*N, KF946009; *glt*X, KF946014 and *inf*B, KF946023) and 16S rDNA sequence (KF945996) of the symbiotic bacterium isolated from the *S. tbilisiense* population were highly similar to those of *X. bovienii* type strain T228T (98.6% for the five concatenated protein coding sequences on 4363 nucleotide and 99.7% for the 16S rDNA sequence on 1319 nucleotide). This result is consistent with the classification of the nematode in the *affine*/*intermedium*-group (see Gorgadze *et al.*, 2015).

Distribution

In addition to the type locality in Georgia, this species was also isolated by baiting with *Galleria* larvae from a soil sample collected in

2000 (43°42′N, 40°15′E, 900 m a.s.l.) near Krasnaya Poliana, Russian Federation, in the mountains above the Black Sea city resort of Sotchi.

Steinernema tielingense[68,69] **Ma, Chen, Li, Han, Khatri-Chhetri, De Clercq & Moens, 2012**
(Figs 4.82-4.85)

MEASUREMENTS

See Table 4.32.

DIAGNOSTIC DESCRIPTION

Male (first generation)

Anterior region typical for *Steinernema* male. Body J-shaped when heat-relaxed. Cuticle smooth under light microscope. Lateral fields absent. Labial region smooth, rounded and offset from rest of body. Excretory pore located more or less opposite to nerve ring (at 64-78% of length from anterior body end to pharynx base). Genital system monorchic, well developed, ventrally reflexed. Spicules paired, symmetrical, well curved, yellowish in colour. Manubrium of spicules with rounded to ellipsoid anterior end, slightly offset from shaft, calomus very short. Lamina well curved anteriorly and dorsally, with two internal sclerotised ribs, tapering slightly posteriorly. Spicule terminus blunt. Rostrum not well developed, velum distinct. Gubernaculum boat-shaped in lateral view, tapering gradually posteriorly, corpus with two wings opening posteriorly. Twenty-three genital papillae comprising 11 pairs and single midventral papilla located just anterior to cloacal opening. Seven pairs located precloacal (six pairs subventral and one pair lateral), one pair adcloacal and three pairs postcloacal, consistently near tail terminus (one subdorsal, two at tail tip). Lateral papillae usually located 40-50 μm anterior to cloacal aperture. Tail dorsally convex, terminus bluntly rounded, terminal mucron absent.

[68] Named after *Tieling*, the city in the vicinity of which the type population was collected.

[69] Description after Ma *et al.* (2012b). Revised by David J. Hunt.

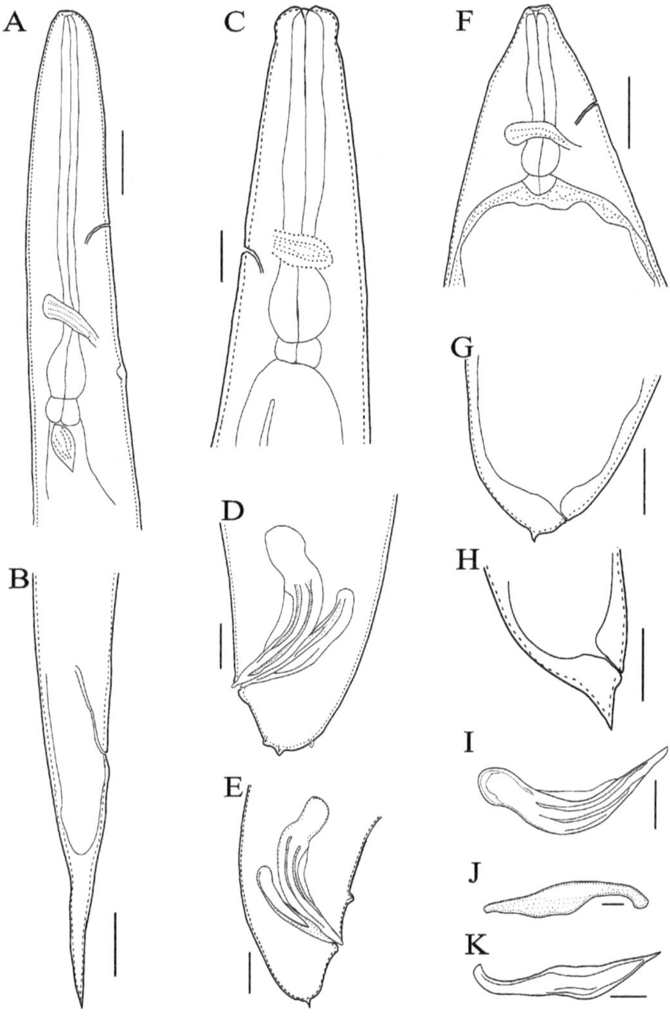

Fig. 4.82. Steinernema tielingense. *A, B: Infective juvenile. A: Pharyngeal region showing basal bulb and bacterial chamber; B: Tail region. C, D, I-K: First generation male. C: Labial and pharyngeal region; D: Tail region; I: Spicule; J, K: Gubernaculum. E: Second generation male, tail region. F, G: First generation female. F: Labial region; G: Tail region. H: Second generation female tail. (Scale bars: A-E, I = 20 μm; F, G = 100 μm; H = 50 μm; J, K = 10 μm.) After Ma* et al. *(2012b), courtesy* Nematology.

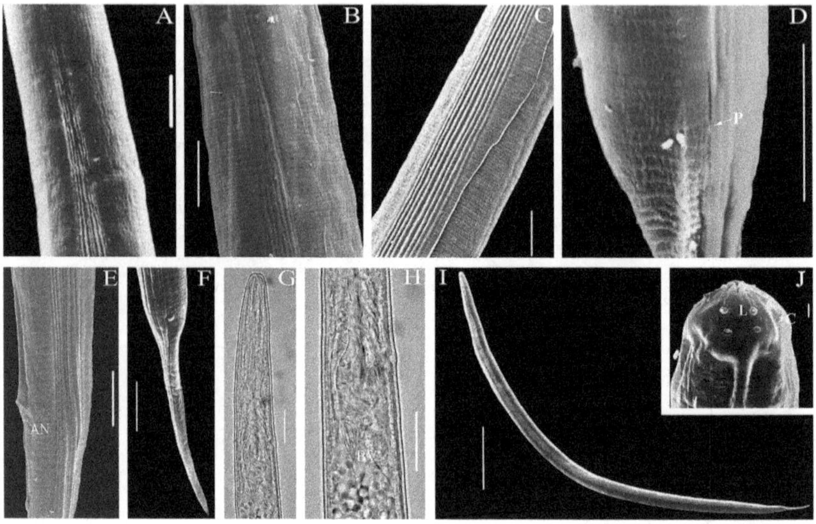

Fig. 4.83. Steinernema tielingense. *SEM and LM pictures of infective juvenile. A, B: Lateral field showing change in anterior part; C: Lateral field with eight ridges; D: Tail region showing phasmid (P); E: Posterior part showing lateral field and anus (AN); F: Tail region showing gradual reduction in number of ridges so as to become two after phasmid level; G: Labial and pharyngeal region; H: Pharyngeal region showing basal bulb and bacterial vesicle (BV); I: Entire view; J: Labial region showing labial papillae (L) and cephalic papillae (C). (Scale bars: A-F = 10 μm; G, H = 20 μm; I = 100 μm; J = 1 μm.) After Ma* et al. *(2012b), courtesy* Nematology.

Male (second generation)

Similar to first generation, but smaller in size. Tail terminus mostly with a mucron, usually subterminal subventral in position. Spicules and gubernaculum shorter and thinner.

Female (first generation)

Anterior region typical for *Steinernema* female. Body coiled when heat-relaxed, cuticle with longitudinal bands under SEM. Labial region truncate to slightly rounded. Amphidial apertures usually situated posterior to labial disc but anterior to cephalic papillae. Morphology of labial region and pharynx similar to male. Cheilorhabdions well sclerotised. Excretory pore located anterior to nerve ring. Gonads amphidelphic, reflexed dorsally. Vulva asymmetrical, slightly protruding from body contour, anterior lip larger than posterior. Vulval opening

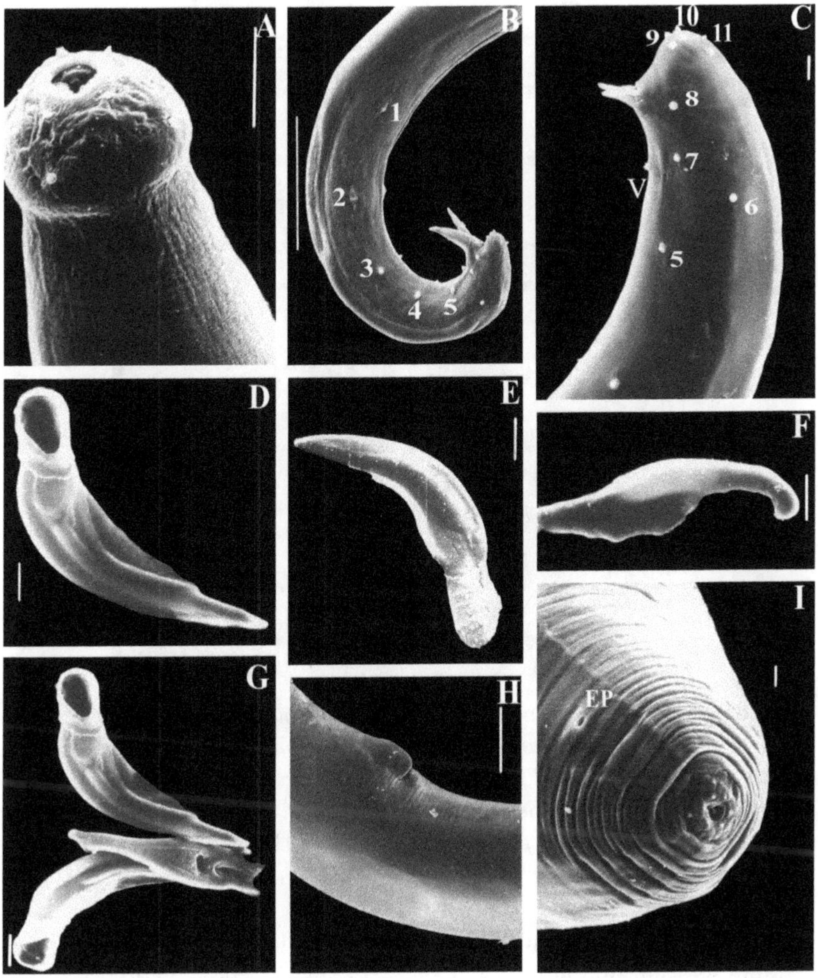

Fig. 4.84. Steinernema tielingense. *SEM pictures of first generation male and female. A-G: Male. A: Labial region; B: Posterior region showing arrangement of pairs 1-5 of genital papillae; C: Posterior part of tail region showing genital papillae pairs 5-11 and a single precloacal papilla (V); D, E: Spicules; F: Gubernaculum; G: Spicules and gubernaculum. H, I: Female. H: Vulval region; I: Labial region showing four cephalic, six labial papillae and excretory pore (EP). (Scale bars: A, C-G, I = 10 μm; B, H = 100 μm.) After Ma* et al. *(2012b), courtesy* Nematology.

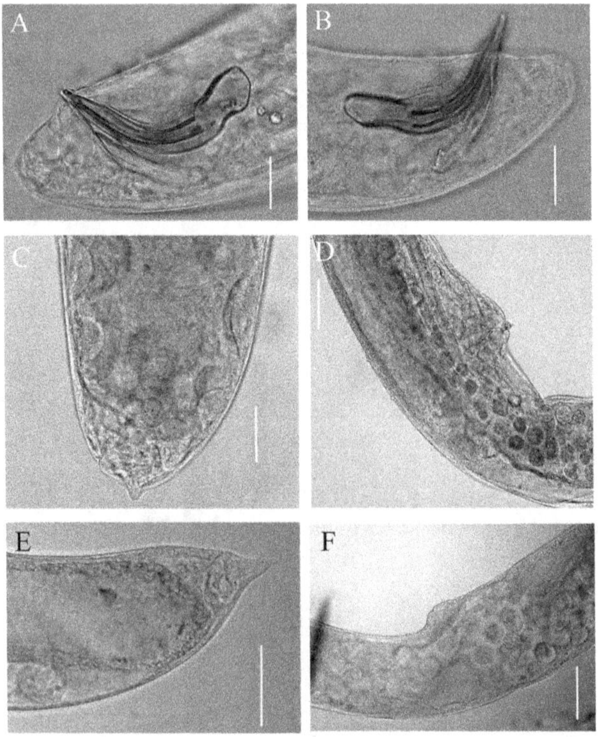

Fig. 4.85. Steinernema tielingense. *LM pictures of male and female. A, B: Male. A: First generation tail showing spicule; B: Second generation tail showing spicule. C-F: Female. C: First generation tail; D: First generation vulva; E: Second generation tail; F: Second generation vulva. (Scale bars: A, B = 20 μm; C, E = 50 μm; D, F = 100 μm.) After Ma* et al. *(2012b), courtesy* Nematology.

mostly located near mid-body. Vagina short, leading into paired uteri. Rectum narrow, anal opening distinct. Tail shorter than anal body diam. Tail shape variable, mostly blunt, with small papilla-like projection on tip. Postanal swelling not present.

Female (second generation)

Similar to first generation in general morphology, but shorter and narrower. Body arcuate when heat-relaxed. Body depressed after vulva. Tail conical and sharply pointed. Terminal mucron absent.

Table 4.32. *Morphometrics of* Steinernema tielingense. *All measurements are in μm and in the form: mean $\pm$ s.d. (range).*

Character	First generation			Second generation		Infective juvenile
	Male		Female	Male	Female	
	Holotype	Paratypes	Paratypes	Paratypes	Paratypes	Paratypes
n	–	20	20	20	20	20
L	1691	1778 ± 238 (1430-2064)	6190 ± 1260 (4028-8538)	1359 ± 74 (1302-1528)	2532 ± 301 (2094-3070)	915 ± 49 (824-979)
a	14	14 ± 2.2 (11-18)	23 ± 3.8 (17-32)	19 ± 1.7 (16-22)	15 ± 1.7 (11-18)	26 ± 1.8 (23-28)
b	10	11 ± 1.2 (9-13)	29 ± 6.3 (21-45)	9 ± 0.7 (8-11)	14 ± 2.2 (11-19)	7 ± 0.4 (6-8)
c	77	70 ± 11.7 (57-85)	117 ± 25.6 (72-158)	50 ± 5.9 (43-59)	65 ± 11 (47-79)	11 ± 1.2 (9-13)
c′	0.5	0.5 ± 0.09 (0.3-0.6)	0.6 ± 0.18 (0.5-0.9)	0.7 ± 0.07 (0.6-0.8)	1 ± 0.14 (0.9-1.3)	4 ± 0.3 (3.5-4.6)
V	–	–	51 ± 2.3 (49-54)	–	53 ± 2.4 (50-57)	–
Max. body diam.	123	129 ± 11.2 (111-159)	251 ± 26.8 (200-307)	74 ± 4.2 (63-84)	174 ± 28 (144-196)	35 ± 1.6 (32-38)
Excretory pore	114	114 ± 16.6 (94-133)	84 ± 10.4 (82-103)	101 ± 7.1 (91-115)	77 ± 10.1 (64-91)	69 ± 3 (64-73)
Nerve ring	113	112 ± 5.6 (96-132)	136 ± 10.6 (111-144)	125 ± 8.8 (108-140)	130 ± 8 (115-132)	98 ± 4.6 (90-105)
Pharynx length	164	160 ± 10.4 (145-173)	202 ± 15.7 (186-263)	160 ± 6.3 (147-169)	177 ± 8.8 (165-193)	128 ± 6.6 (120-135)
Tail length	22	26 ± 4.4 (22-33)	45 ± 8.8 (40-69)	28 ± 3.3 (21-33)	41 ± 4.5 (31-50)	81 ± 4 (74-85)
Anal body diam.	45	47 ± 4.5 (40-52)	69 ± 9.9 (56-92)	39 ± 3.5 (32-45)	42 ± 5.3 (32-53)	21 ± 1.1 (19-23)
Testis reflexion	456	492 ± 80.6 (416-661)	–	220 ± 60 (183-289)	–	–
Spicule length	85	88 ± 7.1 (79-98)	–	74 ± 3.3 (65-79)	–	–
Spicule width	14	13 ± 1.4 (12-15)	–	10 ± 0.9 (9-12)	–	–
Gubernaculum length	54	61 ± 5.8 (49-70)	–	43 ± 4 (41-48)	–	–
Gubernaculum width	6	6 ± 1.1 (6-8)	–	5 ± 0.6 (4-6)	–	–
Labial region to vulva	–	–	3095 ± 675 (2374-4270)	–	1455 ± 259 (1160-1680)	–
Hyaline region	–	–	–	–	–	41 ± 4 (37-47)
D%	69	71 ± 5.8 (64-78)	41 ± 5.2 (32-49)	63 ± 5.6 (53-72)	45 ± 9.2 (35-62)	55 ± 3.8 (47-61)
E%	513	441 ± 95 (309-642)	214 ± 35 (156-235)	370 ± 42 (317-435)	186 ± 53 (150-212)	88 ± 6 (85-94)
SW%	188	191 ± 15 (176-212)	–	188 ± 16 (168-209)	–	–
GS%	64	73 ± 6.6 (59-82)	–	59 ± 5.2 (54-66)	–	–
H%	–	–	–	–	–	58 ± 5.8 (53-64)

Infective juvenile

Anterior region typical for *Steinernema* IJ. Body of heat-relaxed specimens almost straight or slightly curved, slender, slightly tapering towards anterior and posterior ends. Body cuticle with faint annules. Sheath (second-stage cuticle) present after harvesting, but usually lost during storage. Labial region slightly truncate, continuous with body contour. Excretory pore anterior to nerve ring. Hemizonid distinct, located at level of isthmus and basal bulb junction. Pharynx long and narrow, cardia present. Nerve ring encircling narrow isthmus. Lateral field beginning with one cuticular depression or line at six to eight annules from anterior end. Two additional lines appearing posteriorly forming two thin ridges, further posteriorly each ridge dividing into three with number of ridges increasing from two to six. Near excretory pore a central ridge emerging, although central ridge dividing into two over a very short distance, making a total of eight, evenly spaced and developed ridges at mid-body. Ridges remaining unchanged until phasmid. A short distance posterior to phasmid only two prominent ridges observed. Lateral field formula = 2, 6, 7, 8, 2. Bacterial vesicle dorsally displaced, just posterior to cardia, well defined with visible walls, usually teardrop-shaped and tapering posteriorly, 6.3 (5.4-7.9) μm in diam., 12.2 (10.2-13.6) μm in length, containing rod-shaped bacteria. Tail long conoid, prominent hyaline region occupying half of tail length. Phasmid pore-like, located *ca* 35% of tail length posterior to anus.

TYPE HABITAT AND LOCALITY

Natural host unknown as it was isolated by *Galleria* baiting a soil sample taken in the vicinity of Tieling, Liaoning province, China (42.08°N, 124.28°E). The type locality is characterised by low hills with shrubs, weeds and pines. The soil was a humid, loamy-sand under dense grass with moderate shading by shrubs.

TYPE MATERIAL

Holotype male and 12 male paratypes, four female paratypes and ten IJ paratypes are deposited in the nematode collection of the Institute of Zoology of Ghent University, K.L. Ledeganckstraat 35, Ghent, Belgium. Slides with ten first generation male paratypes, eight first generation female paratypes and 15 IJ paratypes are also deposited in the Institute of

Plant Protection, Hebei Academy of Agricultural and Forestry Sciences, Baoding, Hebei, China.

RELATIONSHIPS

This species can be recognised by the labial region of the first generation male having a depression visible at the junction with the body wall, and well curved yellowish spicules with a prominent and short velum. The gubernaculum is boat-shaped in lateral view; the corpus has two wings opening posteriorly. The length of the spicules and gubernaculum of first generation males is 88 (79-98) μm and 61 (49-70) μm, respectively. Genital papillae total 23 comprising 11 pairs and a single midventral papilla. Of these, seven pairs are located precloacal (six pairs subventral and one pair lateral), one pair adcloacal and three pairs postcloacal. The females have a protruding vulva with the anterior lip larger than the posterior. IJ have a body length of 915 (824-979) μm, the distance from anterior end to the excretory pore averages 69 (64-73) μm, the lateral field has eight ridges which are evenly spaced at the mid-body, the tail is 81 (74-85) μm long and the hyaline portion occupies more than one half the tail length.

Steinernema tielingense is a member of the *krausseilfeltiae/oregonense*-group (Spiridonov *et al.*, 2004). Ma *et al.* (2012b) discuss the relationships in greater detail.

MOLECULAR CHARACTERISTICS

Steinernema tielingense is characterised genetically by sequences of ITS (GU994201) and D2-D3 regions (GU994202) of rDNA. The sequence length of the ITS region, including primers TW81 and AB28, is 806 bp. The length of the ITS1 + 5.8S + ITS2 sequence is 742 bp, ITS1 = 265 bp and ITS2 = 320 bp, and its composition is: A = 0.24528, C = 0.18059, G = 0.22237, T = 0.35175. The length of the total sequence as well as of the ITS2 is the greatest amongst the 16 closely related species (Ma *et al.*, 2012b). *Steinernema tielingense* differs from *S. kraussei*, its closest taxon, by 39 bp. The sequence of the D2-D3 region of *S. tielingense* is 874 bp long and its composition is: A = 0.2494, C = 0.1968, G = 0.2998, T = 0.2540. Pairwise distances show that *S. tielingense* differs from its sister taxa *S. oregonense*, *S. cholashanense* and *S. kraussei*, by 11, 12 and 13 bp, respectively (Ma *et al.*, 2012b).

In a consensus tree, three species, *S. tielingense*, *S. kraussei* and *S. silvaticum*, formed a clade with bootstrap support of 78%. For the D2-D3 region, maximum parsimony analysis shows that the alignment resulted in 1034 characters of which 443 are constant, 243 variable characters are parsimony-uninformative and 348 characters are parsimony-informative. In a tree with 46 *Steinernema* species, *S. tielingense* was placed as a sister taxon of *S. cholashanense*.

SYMBIOTIC BACTERIUM

No information.

BIONOMICS

Cross-hybridisation assays between *S. tielingense*, *S. kraussei*, *S. feltiae* and *S. hebeiense* yielded no progeny. In the self-cross controls, offspring were produced normally (Ma *et al.*, 2012b).

The life cycle is comparable with that of other species of the *kraussei/feltiae/oregonense*-group. *Galleria mellonella* larvae were killed after 2 days and first generation adults developed after 4-5 days at 25°C. Second generation adults developed after 8 days. It usually took more than 10 days for IJ to emerge from the insect cadavers.

Steinernema tophus[70,71] Çimen, Lee, Hatting, Hazir & Stock, 2014
(Figs 4.86, 4.87)

MEASUREMENTS

See Table 4.33.

DIAGNOSTIC DESCRIPTION

Male (first generation)

Anterior region typical for *Steinernema* male. Body slender, ventrally curved posteriorly, J-shaped when heat-killed. Cuticle smooth under

[70] The name *tophus* is the Latin word for sandstone, sandstone formations being characteristic in the Western State of South Africa and from where the species was isolated.

[71] Description after Çimen *et al.* (2014). Revised by David J. Hunt.

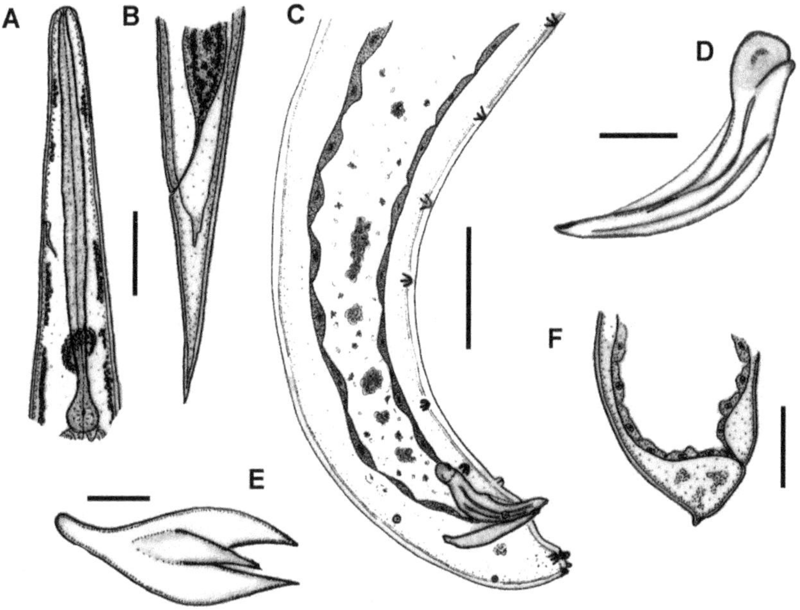

Fig. 4.86. Steinernema tophus. *A: Infective juvenile anterior region; B: Infective juvenile tail region; C: Tail of first generation male, lateral view showing genital papillae; D: First generation male spicule, lateral view; E: First generation male gubernaculum, ventral view; F: Tail of first generation female, lateral view. (Scale bars: A, B = not given in original; C = 75 μm; D = 21 μm; E = 16 μm; F = 60 μm.) Modified after Çimen* et al. *(2014), with acknowledgement to* Zootaxa.

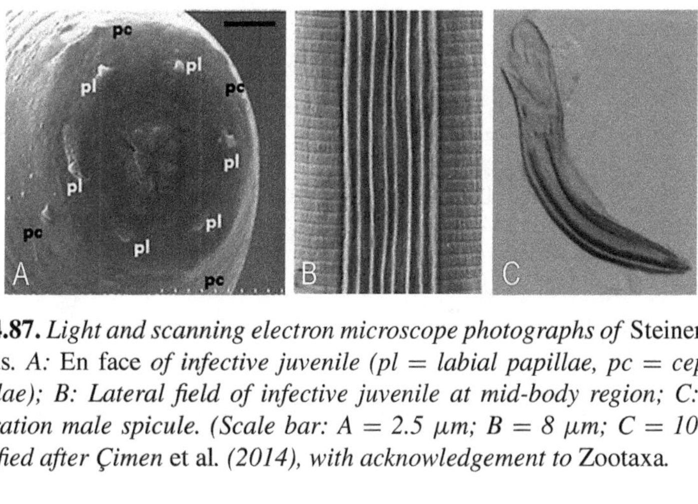

Fig. 4.87. *Light and scanning electron microscope photographs of* Steinernema tophus. *A: En face of infective juvenile (pl = labial papillae, pc = cephalic papillae); B: Lateral field of infective juvenile at mid-body region; C: First generation male spicule. (Scale bar: A = 2.5 μm; B = 8 μm; C = 10 μm.) Modified after Çimen* et al. *(2014), with acknowledgement to* Zootaxa.

Table 4.33. *Morphometrics of* Steinernema tophus. *All measurements are in μm and in the form: mean ± s.d. (range).*

Character	First generation			Second generation		Infective juvenile
	Male		Female	Male	Female	
	Holotype	Paratypes	Paratypes	Paratypes	Paratypes	Paratypes
n	–	24	25	25	25	25
L	1912	1832 ± 177 (1503-2096)	4658 ± 810 (3303-5994)	1360 ± 86 (1029-1448)	2498 ± 156 (2142-2778)	1046 ± 40 (966-1126)
a	–	–	–	–	–	28 ± 4 (21-34)
b	–	–	–	–	–	7.2 ± 0.3 (6.5-8.1)
Labial region to vulva	–	–	2326 ± 431 (1473-2958)	–	1296 ± 76 (1156-1430)	–
Max. body diam.	169	162 ± 18 (112-185)	17? ±28 (162-287)	81 ± 7 (68-93)	129 ± 7 (118-144)	38 ± 6 (30-50)
Pharynx	169	168 ± 7 (151-180)	196 ± 11 (171-218)	148 ± 4 (136-155)	192 ± 7 (173-202)	146 ± 5 (134-156)
Excretory pore	172	154 ± 15 (119-177)	172 ± 23 (116-208)	123 ± 6 (110-132)	159 ± 11 (131-180)	92 ± 3 (86-97)
Nerve ring	126	118 ± 7 (107-132)	146 ± 13 (122-166)	115 ± 6 (102-127)	141 ± 11 (118-160)	105 ± 4 (98-111)
Tail	45	45 ± 6 (36-67)	61 ± 8 (43-73)	39 ± 2 (36-43)	72 ± 7 (60-92)	70 ± 5 (58-80)
Anal body diam.	68	66 ± 6 (53-77)	81 ± 11 (61-112)	44 ± 2 (41-47)	55 ± 4 (45-64)	24 ± 4 (19-32)
Spicule	78	79 ± 6 (70-91)	–	60 ± 2 (56-64)	–	–
Gubernaculum	59	59 ± 4 (47-65)	–	45 ± 2 (42-49)	–	–
D%	101	92 ± 8 (71-103)	–	83 ± 4 (74-89)	–	63 ± 3 (58-71)
E%	381	344 ± 42 (230-425)	–	316 ± 20 (259-354)	–	132 ± 10 (120-156)
SW%	110	120 ± 20 (100-160)	–	140 ± 10 (130-160)	–	–
GS%	80	70 ± 10 (60-90)	–	80 ± 0 (70-80)	–	–
H	–	–	–	–	–	32 ± 3 (26-38)
H%	–	–	–	–	–	46 ± 3 (38-51)

light microscopy. Lateral field and phasmids inconspicuous under light microscopy. Labial region truncate to slightly round, continuous with body. Deirids inconspicuous with DIC microscopy but distinct under SEM. Nerve ring usually surrounding isthmus or anterior part of basal bulb. Excretory pore opening circular, located anterior to nerve ring at anterior one-third of metacorpus. Single reflexed testis, consisting of germinal growth zone leading to seminal vesicle. *Vas deferens* with inconspicuous walls. Spicules paired, symmetrical, curved, with ochre-brown colouration. Manubrium elongated. Calomus (shaft) distinct. Lamina with rostrum or retinaculum and two internal ribs. Velum present, narrow. Blade terminus blunt. Gubernaculum arcuate, *ca* 75% of spicule length. Manubrium of gubernaculum curved ventrally. Tail conoid, no mucron. There are 23 genital papillae (11 pairs and one single) arranged as follows: six subventral precloacal pairs, one single midventral precloacal papilla (located between precloacal pairs 5 and 6), one pair lateral subdorsal precloacal papillae, one pair of adcloacal subventral papillae, three pairs postcloacal (one pair subdorsal, two pairs subventral terminal).

Male (second generation)

General morphology similar to that of first generation but smaller. Tail with or without mucron. Spicules with manubrium morphology similar to those of first generation. Gubernaculum more slender and longer than that of first generation.

Female (first generation)

Anterior region typical for *Steinernema* female. Body C-shape when heat-killed. Excretory pore located about mid-procorpus level. Ovaries opposed, reflexed in dorsal position, oviduct well developed; glandular spermatheca and uterus in ventral position. Vagina short, with muscular walls. Vulva located near middle of body with slightly protruding lips and mostly symmetric. Tail blunt, conoid, with a mucron. Postanal lips slightly protruding and asymmetric, anterior lip smaller than posterior.

Female (second generation)

Body open C-shape when heat-killed. Similar to first generation female but smaller. Vulva with protruding and slightly asymmetric lips. Tail conoid, without post-anal swelling.

Infective juvenile

Anterior region typical for *Steinernema* IJ. Body of heat-relaxed specimens almost straight, slender, gradually tapering posteriorly. Cuticle with fine transverse striae. Labial region continuous with body, slightly truncate. Lip region smooth, continuous; stoma closed. Lateral field distinct with nine distinct lines (*i.e.*, eight ridges) evenly spaced and developed in mid-body region. Pharynx long, narrow, with slightly expanded procorpus, narrower isthmus and pyriform basal bulb with valve. Cardia present. Nerve ring located at isthmus level. Excretory pore located about mid-corpus. Anterior portion of intestine with small bacterial vesicle. Rectum long, straight, anus distinct. Tail conoid with pointed terminus. Hyaline region comprising *ca* 32% of tail length.

TYPE HABITAT AND LOCALITY

Natural host unknown. The type isolate, ROOI-352, was collected from a vineyard in Clanwilliam, Western State, South Africa.

TYPE MATERIAL

Holotype male, first generation, five first generation paratype males, five first generation paratype females, and five paratype IJ deposited in the USDA Nematode Collection, MD, USA. Five first generation paratype males, five first generation paratype females, and five paratype IJ deposited at the National Collection of Nematodes, Biosystematics Division, Plant Protection Research Institute, Agricultural Research Council, Pretoria, South Africa.

RELATIONSHIPS

Based on its morphological and morphometric traits, the new species is considered a member of the *glaseri*-group. Nematodes in this group are characterised by having the largest IJ (L = >1000 μm) and by the presence of conspicuous deirids which are generally situated in the region of the pharyngeal bulb or, in some members, slightly posterior. The lateral field of *glaseri*-group infective juveniles is characterised by nine equally developed lines (= eight ridges) arranged evenly in the mid-body region. *Steinernema tophus* has IJ with long bodies (L = 1046 (966-1126) μm) and nine lines in the mid-body region of the lateral field. Most of the morphometric traits of the IJ overlap with other species in the *glaseri*-group. Presence of deirids in IJ was not observed in the new species. Additionally, first generation males can be distinguished from other related species by the morphology of the spicules and gubernaculum, the arrangement of the genital papillae and the values of ratios SW = 120 (100-160) and GS = 70 (60-90). Phylogenetic analyses placed *S. tophus* in Clade V (as depicted by Spiridonov *et al.*, 2004). Within this clade, the new species was found most closely related to a newly described South African species, *S. innovationi*, followed by *S. khoisanae* (Nguyen *et al.*, 2006) in the 28S topology. In particular, for the more variable ITS dataset, there were 59 bp differences between *S. tophus* and its sister species, *S. innovationi*. Further details are provided by Çimen *et al.* (2014).

Molecular characterisation

Sequence length for the 28S fragment was 856 bp (GenBank accession no. KJ701240). Bootstrap MP analysis of this three-gene dataset placed *S. tophus* as a member of Clade V, and sister to *S. innovationi* (Çimen *et al.*, 2015), another species described from South Africa. This clade comprises *Steinernema* spp. with IJ having an exceptionally large body size (average > 1000 μm). The sequenced length of the ITS fragment for the new species was 797 bp (GenBank accession no. KJ701241). The MP bootstrap analysis of ITS data also placed *S. tophus* as a member of Clade V. Again, the closest sister taxon in this topology was *S. innovationi*. The close sister taxon, *S. innovationi*, differed by 59 bp. A minimum evolution distance model provided 100% bootstrap support for the branch containing *S. tophus* and *S. innovationi*. Further details are provided by Çimen *et al.* (2014).

Bacterial symbiont

The bacterial symbiont isolated from *S. tophus* is closely related to a *Xenorhabdus* sp. from Japan, isolate MY8.

Bionomics

Cross-hybridisation assays between males and females of *S. tophus* with *S. khoisanae*, *S. innovationi* and *S. glaseri* yielded no progeny. In the controls, offspring were produced in all self-crossed species. No progeny were observed in the single female control plates.

The life cycle of *S. tophus* is one of the shortest observed for the genus; at 25°C the IJ killed *G. mellonella* larvae 20-28 h after initial exposure and moulted to the J4 at 23-36 h after inoculation. Mature first generation males and females developed 28-40 h after initial exposure. Gravid first generation females were observed from 48 h after exposure and their progeny developed from 50-72 h after exposure. A massive emergence of J3 was observed 10 days from the initial exposure.

Steinernema unicornum[72,73] Edgington, Buddie, Tymo, France, Merino & Hunt, 2009

(Figs 4.88-4.91)

MEASUREMENTS

See Table 4.34.

DIAGNOSTIC DESCRIPTION

Male (first generation)

Anterior region typical for *Steinernema* male. Body J-, S- or C-shaped when heat-relaxed (55, 20 and 25% of 20 individuals observed, respectively) with labial region directed in opposite direction to curve in 40% of individuals. Cuticle smooth under light microscope. Cheilorhabdions visible, sclerotised, other rhabdions not distinguishable. Pharynx with cylindrical procorpus, swollen non-valvate metacorpus, distinct isthmus and valvate basal bulb. Nerve ring surrounding isthmus between posterior margin of metacorpus and anterior margin of basal bulb. Excretory pore located anterior to nerve ring, anterior portion of excretory duct well cuticularised, posterior section slightly sinuate, less easily seen. Non-overlapping junction between pharynx and intestine, cardia present. In 80% of specimens observed intestine pushing anteriad beyond pharyngo-intestinal junction to form a sleeve, sometimes extending as far as anterior margin of basal bulb. Prominent papilliform deirid located laterally at level of pharyngo-intestinal junction. Genital tract monorchic, ven-

[72] Specific epithet derived from the Latin *unus* (= one) and *cornu* (= a horn), and referring to the presence of a single, horn-like tooth on the labial region of the IJ.

[73] Description after Edgington *et al.* (2009b). Revised by David J. Hunt.

trally reflexed. Spherical, unpolarised structures observed in *vas deferens*, diam. 16 (14.9-17.3) μm, full of a granular cytoplasm, probably corresponding to mature pre-spermatozoa. Spicules paired, brownish-amber in colour. Manubrium shape distinctive, moderately elongate (average length:width ratio = 1.5:1), although occasionally slightly wider than long, slightly offset from shaft; shaft short; blade strongly curved, almost hemispherical from base of shaft to tip; terminus of blade smooth, slightly rounded; high velum present, not extending to spicule tip; two internal ribs visible on each spicule. Gubernaculum boat-shaped in lateral view, corpus enlarged and closed posteriorly, cuneus Y-shaped, large hole or lacuna present dorsally. Copulatory papillae numbering 23 in 85% of specimens observed, comprising a single precloacal, verrucose, midventral papilla, transversely ovate (average width 5 μm) in shape and with a raised central region, and 11 pairs, five pairs being precloacal subventral, very slightly posteriorly staggered on right side, and with anteriormost pair slightly more widely spaced, two pairs of adcloacal subventral papillae, ranging in position from just anterior to slightly posterior to cloacal aperture, one lateral pair located close to cloacal aperture level, two postcloacal subventral pairs arranged more or less in tandem and located subterminally and one postcloacal subdorsal pair, also subterminal. A sixth (15% of specimens) and seventh (5% of specimens) precloacal subventral papilla, lacking opposite papilla, sometimes present at anterior end of sequence, not side-dependent. Tail more or less conoid, dorsally convex to a broad, usually bluntly rounded, terminus, small terminal mucron present in 10% of specimens observed.

Male (second generation)

Body shorter and much narrower than that of first generation. Pharyngeal region as for first generation, except excretory pore more posterior to cephalic extremity and closer to location of nerve ring. Deirid located slightly more posterior, just posterior to pharyngo-intestinal junction. Testis reflexion much shorter. Spicules shorter than first generation, manubrium slightly more elongate (average length:width ratio = 1.6:1). Gubernaculum shorter than first generation. Copulatory papillae arranged similarly to first generation, although first pair of precloacal papillae anterior to cloacal aperture was slightly more posteriorly located relative to midventral papilla and in a number of specimens the posterior staggering of precloacal papillae was more exaggerated, particularly the first pair, also one pair at the tail tip was more lateral than

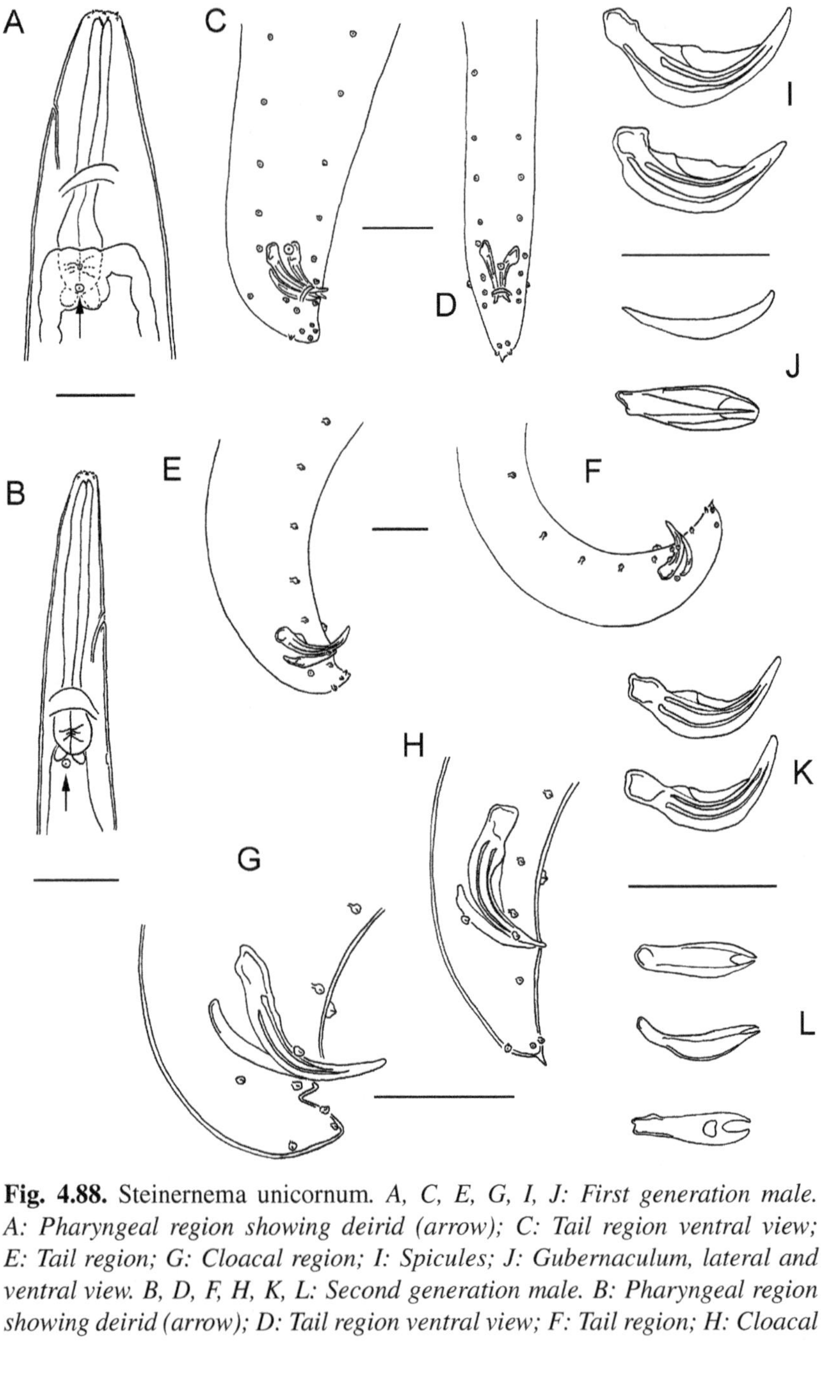

Fig. 4.88. Steinernema unicornum. *A, C, E, G, I, J: First generation male. A: Pharyngeal region showing deirid (arrow); C: Tail region ventral view; E: Tail region; G: Cloacal region; I: Spicules; J: Gubernaculum, lateral and ventral view. B, D, F, H, K, L: Second generation male. B: Pharyngeal region showing deirid (arrow); D: Tail region ventral view; F: Tail region; H: Cloacal*

subventral and an adcloacal papilla in one specimen was duplex (*i.e.*, two nerve endings situated in a single, presumably fused, papilla). In 50% of specimens observed, an additional precloacal subventral papilla present at anterior end of sequence but lacking its opposite papilla, this lack not being side-dependent, hence papillae numbered 24. In 10% of specimens observed there was a sixth pair of precloacal subventral papillae at anterior end of sequence. Tail longer and narrower than first generation, terminal mucron present.

Female (first generation)

Anterior region typical for *Steinernema* female. Body arcuate when heat-relaxed, cuticle smooth. Labial region low, rounded, continuous with body contour, although with an occasional slight depression present in some specimens. No lateral fields observed. Excretory pore usually located anterior to both nerve ring and base of metacorpus, duct heavily cuticularised but excretory cell obscure. Non-overlapping junction between pharynx and intestine, cardia present. In 40% of specimens observed intestine pushing anteriad beyond pharyngo-intestinal junction, although never enveloping basal bulb beyond its anterior extremity. Genital system amphidelphic, ovaries reflexed. Vulva a transverse slit, located approximately 55% from labial region end. Vulval lips hypertrophied in 90% of individuals, protruding 6 (3-8) μm above body contour. Cuticle anterior and posterior to vulva thickened, forming a number of transverse ridges separated by deep grooves. Ridges more obvious on mature adults and often higher anterior to vulva. Epiptygmata absent. Dentate appearance to vagina wall observed in a number of specimens. With SEM, minute, lateral pore-like structure observed at either end of vulval slit and at convergence point of several transverse somatic striae. Prominent post-anal swelling absent in all individuals, occasional slight swelling observed. Tail more or less conoid, much shorter than anal body diam., tapering strongly both dorsally and ventrally to a subcylindroid terminal section with a bluntly rounded terminus. Small mucron-like projection present in dorsal region of tail terminus, present on all specimens observed. Mucron occasionally more or less regularly duplex.

---

region; K: Spicules; L: Gubernaculum, ventral, lateral and dorsal view. (Scale bars = 50 μm.) After Edgington et al. *(2009b), courtesy* Journal of Nematode Morphology and Systematics.

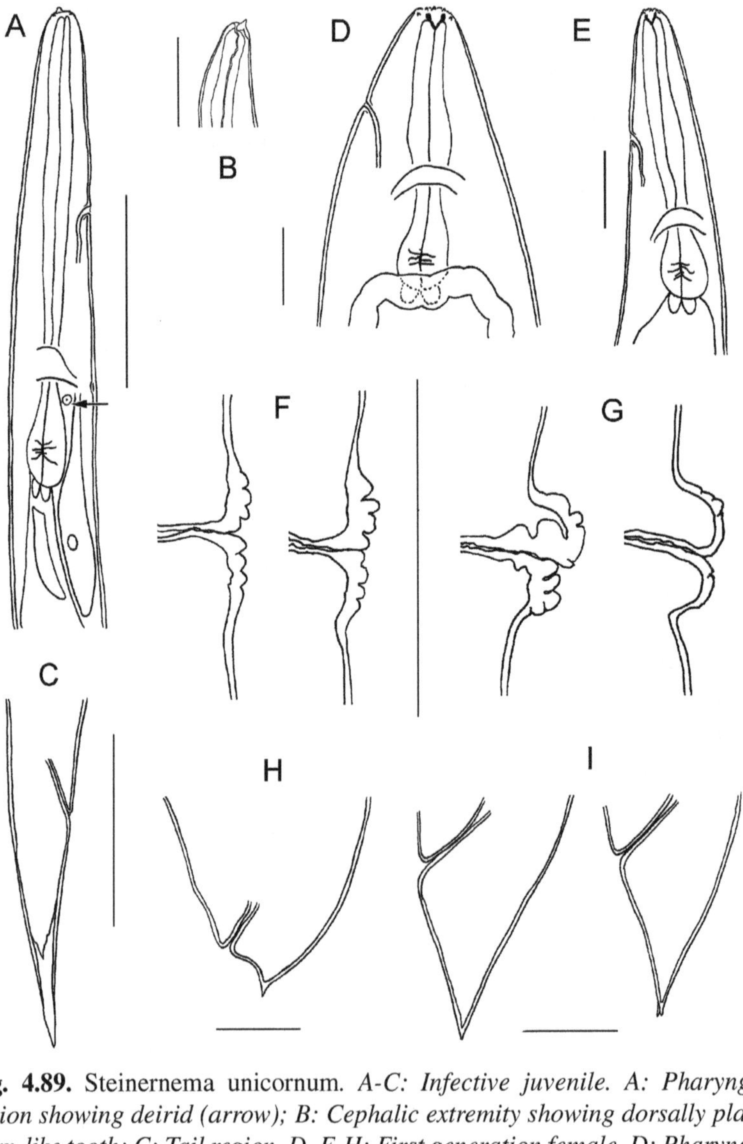

Fig. 4.89. Steinernema unicornum. *A-C: Infective juvenile. A: Pharyngeal region showing deirid (arrow); B: Cephalic extremity showing dorsally placed horn-like tooth; C: Tail region. D, F, H: First generation female. D: Pharyngeal region; F: Vulva showing cuticular folds; H: Tail region. E, G, I: Second generation female. E: Pharyngeal region; G: Vulva showing cuticular folds; I: Tail region showing variation in mucron development. (Scale bars = 50 μm.) After Edgington* et al. *(2009b), courtesy* Journal of Nematode Morphology and Systematics.

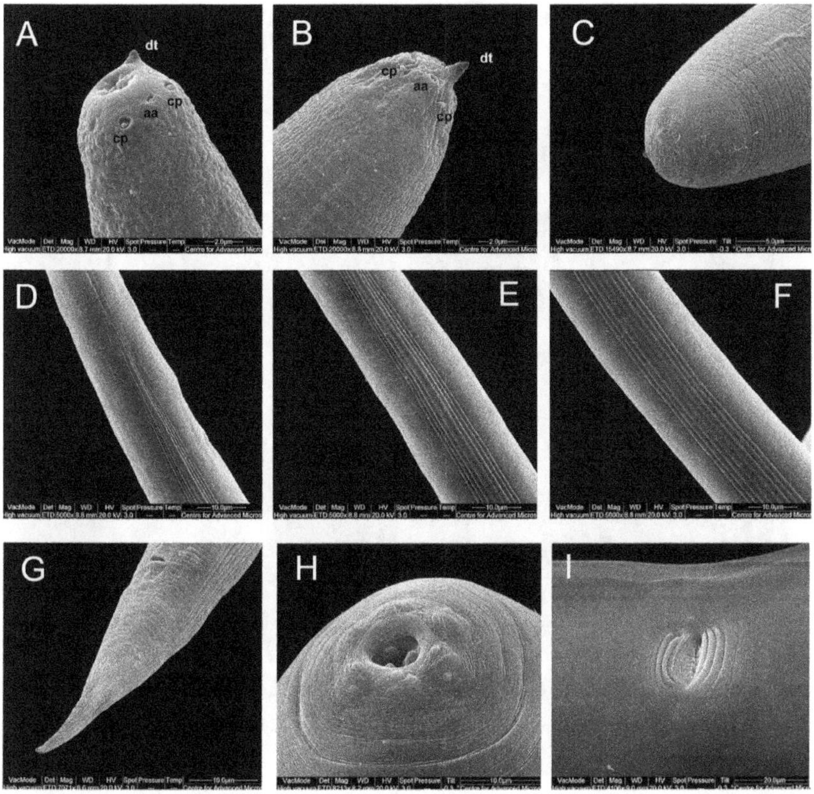

Fig. 4.90. Steinernema unicornum *(SEM). A-G: Infective juvenile. A-C: Cephalic extremity showing dorsally placed horn-like tooth (dt), cephalic papillae (cp) and amphidial aperture (aa); D-F: Lateral field showing a maximum of eight ridges; G: Tail region. H, I: First generation female. H: Cephalic extremity; I: Vulva showing cuticular ridges and lateral pore-like structures. After Edgington* et al. *(2009b), courtesy* Journal of Nematode Morphology and Systematics.

Female (second generation)

Similar to first generation female in general morphology, but shorter and much narrower. Body slightly arcuate when heat-relaxed. Metacorpus and isthmus poorly demarcated. Vulva protruding 7 (3-9) μm on 65% of specimens observed, epiptygmata absent. Thickened cuticle around anterior and posterior side of vulva, bearing pronounced transverse ridges. Ridges more obvious on mature females and generally higher than on first generation female. Epiptygmata absent. Post-anal

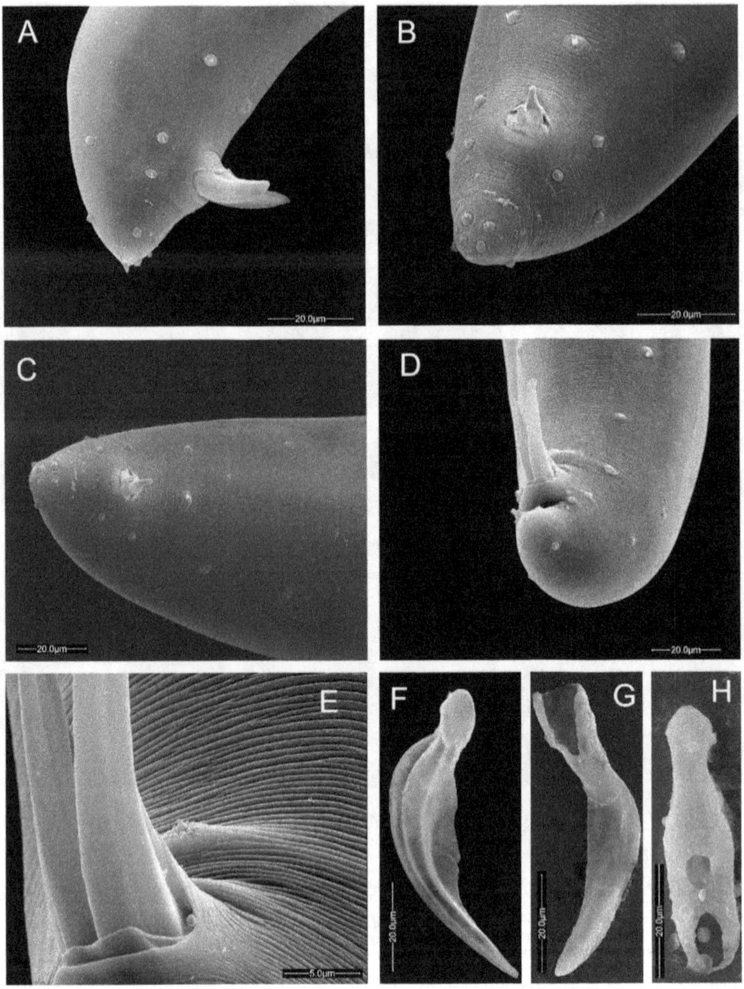

Fig. 4.91. Steinernema unicornum. *First generation male (SEM). A-E: Posterior region. A-D: Position of copulatory papillae; E: Posterior region, ventral view, showing verrucose midventral papilla. F, G: Spicules; H: Gubernaculum, dorsal view. After Edgington* et al. *(2009b), courtesy* Journal of Nematode Morphology and Systematics.

swelling absent. Tail medium conoid, longer than anal body diam., tapering to a sharp, mucron-like terminus, occasional regular duplex termination observed.

Table 4.34. *Morphometrics of* Steinernema unicornum. *All measurements are in μm and in the form: mean ± s.d. (range).*

Character	First generation			Second generation		Infective juvenile
	Male		Female	Male	Female	
	Holotype	Paratypes	Paratypes	Paratypes	Paratypes	Paratypes
n	–	20	20	20	20	20
L	1276	1224 ± 28.4 (918-1458)	2295 ± 67.9 (1858-3024)	1107 ± 28.3 (855-1326)	1506 ± 35.4 (1251-1754)	732 ± 9.8 (654-786)
a	–	–	–	–	–	25.4 ± 0.5 (21.9-29.4)
b	–	–	–	–	–	5.9 ± 0.1 (5.3-7.3)
c	–	–	–	–	–	12.2 ± 0.2 (10.7-14.5)
c′	0.7	0.6 ± 0.03 (0.4-0.8)	0.3 ± 0.03 (0.1-0.6)	0.9 ± 0.02 (0.7-1.1)	1.4 ± 0.06 (1.0-2.1)	3.1 ± 0.1 (2.5-4.1)
V	–	–	54 ± 0.8 (48-60)	–	53 ± 0.4 (51-57)	–
Max. body diam.	143	130 ± 2.6 (94-151)	179 ± 6.1 (138-233)	57 ± 1.4 (44-69)	90 ± 2.4 (66-107)	29 ± 0.5 (25-33)
Excretory pore	55	57 ± 1.9 (41-75)	60 ± 3.6 (22-79)	71 ± 2.0 (53-86)	78 ± 1.8 (69-101)	55 ± 0.8 (48-62)
Nerve ring	96	103 ± 1.9 (85-119)	111 ± 2.3 (90-126)	109 ± 1.6 (99-123)	113 ± 1.8 (94-126)	88 ± 1.3 (77-99)
Pharynx	159	161 ± 2.3 (138-174)	191 ± 1.9 (168-204)	153 ± 1.6 (137-171)	174 ± 2.2 (154-189)	124 ± 1.9 (106-143)
Testis reflexion	385	373 ± 12.7 (283-440)	–	255 ± 10.2 (187-308)	–	–
Tail length	22	22 ± 1.3 (13-36)	21 ± 1.3 (8-35)	30 ± 1.0 (22-38)	55 ± 2.4 (42-88)	60 ± 1.1 (50-71)
Hyaline region	–	–	–	–	–	23 ± 1.0 (13-35)
Anal body diam.	33	37 ± 1.4 (24-45)	67 ± 2.9 (39-86)	33 ± 0.5 (28-38)	40 ± 0.8 (36-47)	19 ± 0.5 (17-25)
Spicule length (arc)	63	73 ± 1.9 (60-82)	–	59 ± 1.0 (52-66)	–	–
Spicule width	13	16 ± 0.3 (14-19)	–	12 ± 0.3 (9-14)	–	–
Gubernaculum length	47	48 ± 0.9 (44-52)	–	39 ± 0.7 (35-42)	–	–
Gubernaculum width	6	9 ± 0.5 (6-13)	–	9 ± 0.3 (6-9)	–	–
D%	35	36 ± 1.4 (24-49)	31 ± 1.9 (13-41)	46 ± 1.3 (35-59)	45 ± 1.0 (39-58)	45 ± 0.9 (38-55)
E%	–	–	–	–	–	91 ± 1.9 (77-116)
SW%	190	198 ± 3.4 (184-217)	–	188 ± 4.8 (165-228)	–	–
GS%	75	68 ± 1.4 (60-74)	–	65 ± 1.2 (58-71)	–	–
H%	–	–	–	–	–	37 ± 1.2 (24-49)

Infective juvenile

Anterior region typical for *Steinernema* IJ apart from presence of dorsal tooth (see below). Body slender, almost straight when heat-relaxed and tapering gently towards anterior and posterior ends. Cuticle finely annulated with delicate transverse striations. Cephalic region rounded not annulated, continuous with body contour. Prominent, dorsally placed, horn-like tooth located near oral opening, observed in both ensheathed and exsheathed live and fixed specimens. Horn-like tooth dorsally directed at approximately 45° to body axis, with thickened cuticular root. Retained second-stage cuticle commonly observed after harvesting. Excretory pore prominent, situated about mid-pharynx, anterior to nerve ring, excretory duct cuticularised. Deirid prominent under LM, located in middle of lateral field at level of anterior end of basal bulb. Hemizonid present, lenticular in shape, located at anterior margin of basal bulb. Hemizonion not seen. Lateral field beginning anteriorly with one line at annule 4. Two additional lines appearing between annules 12-19 to form two ridges. Just posterior to excretory pore number of ridges increasing from two to six. A further 15-20 annules posterior, ridges increasing from six to eight, *i.e.*, maximum number of ridges in lateral field. Portion with eight ridges forming longest part of lateral field. At anus level two central ridges and two most outer ridges disappearing, leaving four ridges. Therefore, lateral field formula = 2, 6, 8, 4. Bacterial vesicle well defined with visible walls, teardrop-shaped, sometimes slightly rectangular in form, 8 (7-9) μm in diam., 26 (20-31) μm in length, containing rod-shaped bacteria. Tail long conoid, evenly attenuated, *ca* three times as long as anal body diam., prominent hyaline portion occupying 37 (24-49)% of tail length. Phasmid distinct, pore-like, located at 48 (44-50)% of tail length posterior to anus, *i.e.*, anterior to beginning of hyaline region.

TYPE HABITAT AND LOCALITY

Steinernema unicornum (isolate D68) was isolated by *Galleria* baiting a soil sample taken in Tierra del Fuego, in the far south of Chile (53°28′2.7″S, 70°11′26″W). The soil was a relatively dry, loamy-sand soil within a native forest of *Nothofagus antarctica* (Forst.) Oerst. (Antarctic beech, or ñirre in Spanish). The pH of the sample, as evaluated in the laboratory, was 5.

Other habitats and localities

Isolates of *S. unicornum*, all of which carried the horn-like tooth on the cephalic extremity and were confirmed as conspecific by ITS sequencing, were collected from 43 soil samples from the south of Chile. Isolate D36 used for sequencing of the D2-D3 region was obtained by waxmoth baiting a soil sample from the Bio Bio Region (37°23′26.7″S, 71°24′14″W). The soil was moist and organic, within 5 m of a small waterfall and of pH 4.9.

Type material

Holotype (first generation male), isolated from haemocoel of *G. mellonella*, deposited in the CABI Nematode Collection, Europe-UK Centre, Egham, Surrey, UK. Five paratype males and females (first and second generations) and ten paratype IJ, all as permanent slide mounts, together with 20 paratypes of each stage in TAF, deposited in the nematode collections of CABI, Europe-UK Centre, Egham, Surrey, UK, INIA – Quilamapu, Chillán, Chile, and the United States Department of Agriculture Nematode Collection (USDANC), Beltsville, MD, USA. In addition, numerous paratype IJ and other stages preserved in 70% ethyl alcohol deposited in the CABI Nematode Collection, Europe-UK Centre, Egham, Surrey, UK.

Relationships

Steinernema unicornum is unique amongst the known species of the genus to date in the presence of a prominent, dorsally-placed, horn-like tooth on the cephalic extremity of the IJ. *Steinernema unicornum* can also be recognised by morphometrics of the IJ, which has a body length of 732 (654-786) μm, excretory pore located about mid-pharynx at 55 (48-62) μm from anterior end, tail 60 (50-71) μm long, H% = 37 (24-49), E% = 91 (77-116), a ratio = 25.4 (21.9-29.4). The first generation male is characterised by a body diam. of 130 (94-151) μm, spicule and gubernaculum lengths of 73 (60-82) μm and 48 (44-52) μm, respectively, shape of spicule and gubernaculum, the position of the excretory pore and SW% of 198 (184-217). The first generation female possesses transverse cuticular ridges immediately anterior and posterior to the vulva, lacks a prominent post-anal swelling and has a mucron (occasionally regularly duplex) on the tail tip. More information is avaiable in Edgington *et al.* (2009b).

MOLECULAR CHARACTERISATION

Sequences were deposited in GenBank with accession numbers: GQ497167, GU191461 and GU191463 for ITS (D68), ITS (D36) and D2-D3 (D36), respectively. The sequence of the ITS region of *S. unicornum*, including primers TW81 and AB28, is characterised by its length of 1000 base pairs (bp) and its composition A = 0.25700, C = 0.19300, G = 0.23700, T = 0.31300. The sequence length of ITS of the new species, 1000 bp, is shorter than that of *S. bicornutum* (1017 bp), *S. intermedium*, *S. beddingi* (1070 bp) and *S. riobrave* (1005 bp) but longer than that of all other selected species. The distance between *S. unicornum* and *S. aciari* is the shortest (167 bp). The least divergent species are *S. beddingi* and *S. intermedium* with 72 bp difference and the most divergent species are *S. beddingi* and *S. pakistanense*, with 345 bp difference. For D2-D3 regions, the sequence of *S. unicornum* is 673 bp, composition, A = 0.24010, C = 0.19157, G = 0.30907, T = 0.25926. Pairwise distances showed that the new species differed from its closest taxon, *S. australe*, by 62 bp and from the most divergent species, *S. colombiense*, by 249 bp. *Steinernema unicornum*, *S. aciari* and *S. rarum* formed a monophyletic group although one that was only weakly supported by bootstrap proportion (59%).

SYMBIOTIC BACTERIUM

No information.

BIONOMICS

No information.

DISTRIBUTION

Recorded from Chile by Edgington *et al.* (2009b).

Steinernema vulcanicum[74,75] Clausi, Longo, Rappazzo, Tarasco & Vinciguerra, 2011
(Figs 4.92-4.94)

MEASUREMENTS

See Table 4.35.

DIAGNOSTIC DESCRIPTION

Male (first generation)

Anterior region typical of *Steinernema* male. Body straight in anterior part but strongly ventrally curved posteriorly in specimens relaxed by gentle heat, J-shaped, tapering toward lip region. Cuticle slightly transversely striated, with striae more evident under SEM. Lateral field with a single ridge (two lines). Lip region rounded, continuous with the adjoining body. Excretory pore position variable, generally anterior to nerve ring, sometimes at same level, rarely posterior. Excretory duct wide, well sclerotised, excretory cell not always discernible, located at level of cardia. Cardia prominent, generally longitudinally split into two parts. Deirids small, usually at level of pharyngeal bulb valve, 150-165 μm from anterior end. Hemizonid usually at level of bulb base, 170-200 μm from anterior end. Gonad monorchic, ventrally reflexed. Distance between pharynx base and anterior end of testis very variable, ranging from as long as isthmus to as long as entire pharynx. Spicules paired, brown in colour. Manubrium almost rectangular, length:width ratio = 0.7:0.8. Calomus (shaft) short but well defined, rostrum absent. Blade (lamina) thick, moderately curved. Velum prominent, variable in length, starting 3-4.5 μm after calomus and extending to tip, posterior end not reaching tail tip, ventral lobe ending before tip. Spicule tip rounded, with ventral concavity. Each spicule with two internal ribs. Spicule length:width ratio = 4.5:5.2. Gubernaculum of same colour as spicules, boat-shaped in lateral view, proximal end, in ventral view, with ventrally, gradually tapering, curved manubrium, larger than neck, gubernaculum swollen at middle and distal end forked in ventral view, cuneus pointed, V-shaped. Eleven pairs of genital papillae plus single

[74] The specific epithet refers to the location of the type locality, on the slopes of the Mt Etna *volcano*.

[75] Description after Clausi *et al.* (2011). Revised by David J. Hunt.

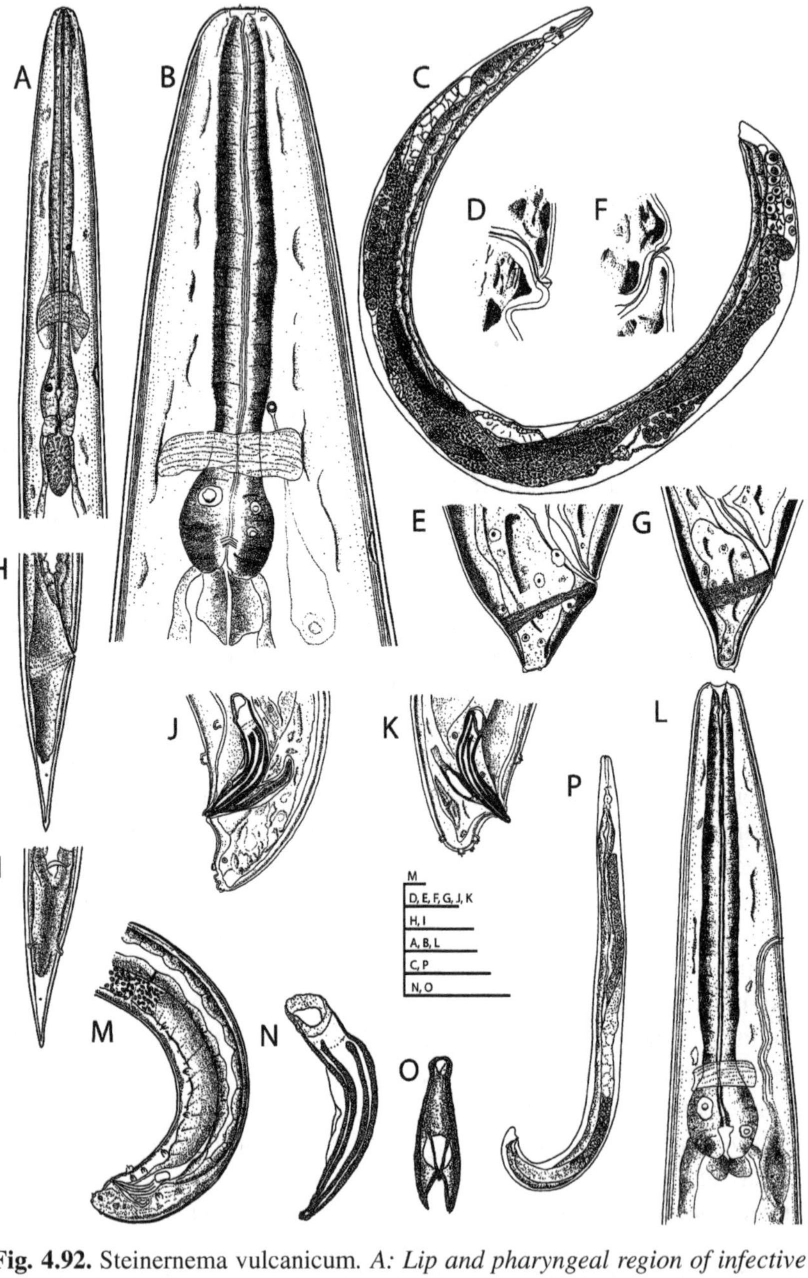

Fig. 4.92. Steinernema vulcanicum. *A: Lip and pharyngeal region of infective juvenile; B-E: First generation female. B: Lip and pharyngeal region; C: Entire body; D: Vulva region showing epiptygmata; E: Tail showing small*

midventral papilla, arranged as follows: six precloacal pairs, one pair lateral, one pair adcloacal and three pairs caudal (two pairs subterminal and one pair subdorsal). Tail conoid, terminus rounded, phasmids visible, mucron absent.

Male (second generation)

Similar to first generation male but shorter and generally smaller and less variable in morphometric characters. Excretory pore always anterior to nerve ring. Tail conoid, phasmids visible, no mucron.

Female (first generation)

Anterior region typical of *Steinernema* female. Body generally C-shaped in heat-relaxed specimens, largest females may be variously bent or curled in a spiral. Cuticle with very fine transverse striae visible only with SEM. Lateral field with a single ridge (two lines). Anterior body contour (up to 20-32 μm from anterior end) slightly narrower than adjoining body. Lip region rounded, stoma structure as in male, cheilorhabdions and prorhabdions prominent, well sclerotised, posterior part funnel-shaped, short. Excretory pore position variable, as in male. Deirids as in male, at 184-208 μm from anterior end. Hemizonid at 190-208 μm from anterior end. Cardia prominent. Vulva with very prominent lips and very evident double epiptygmata. Vagina curved, generally directed anteriad, then bent posteriad to open into uterus. Tail generally shorter than anal body diam. (but in the smallest specimens can be slightly longer), conoid, subdigitate, with posterior third offset by a constriction, tapering to a blunt terminus. One or two very tiny projections generally present near tip. Postanal swelling and phasmids present.

mucronate projection. F, G: Second generation female. F: Vulval region showing epiptygmata; G: Tail showing small mucronate projection. H, I: Tails of infective juvenile. J-O: First generation male. J: Tail; K: Lip and pharyngeal region; L: Tail showing location of genital papillae; M: Spicules; N: Ventral view of gubernaculum; O: Entire body. P: Tail of second generation male. (Scale bars: A, B, D-O = 30 μm; C, P = 300 μm.) After Clausi et al. *(2011), courtesy* Nematology.

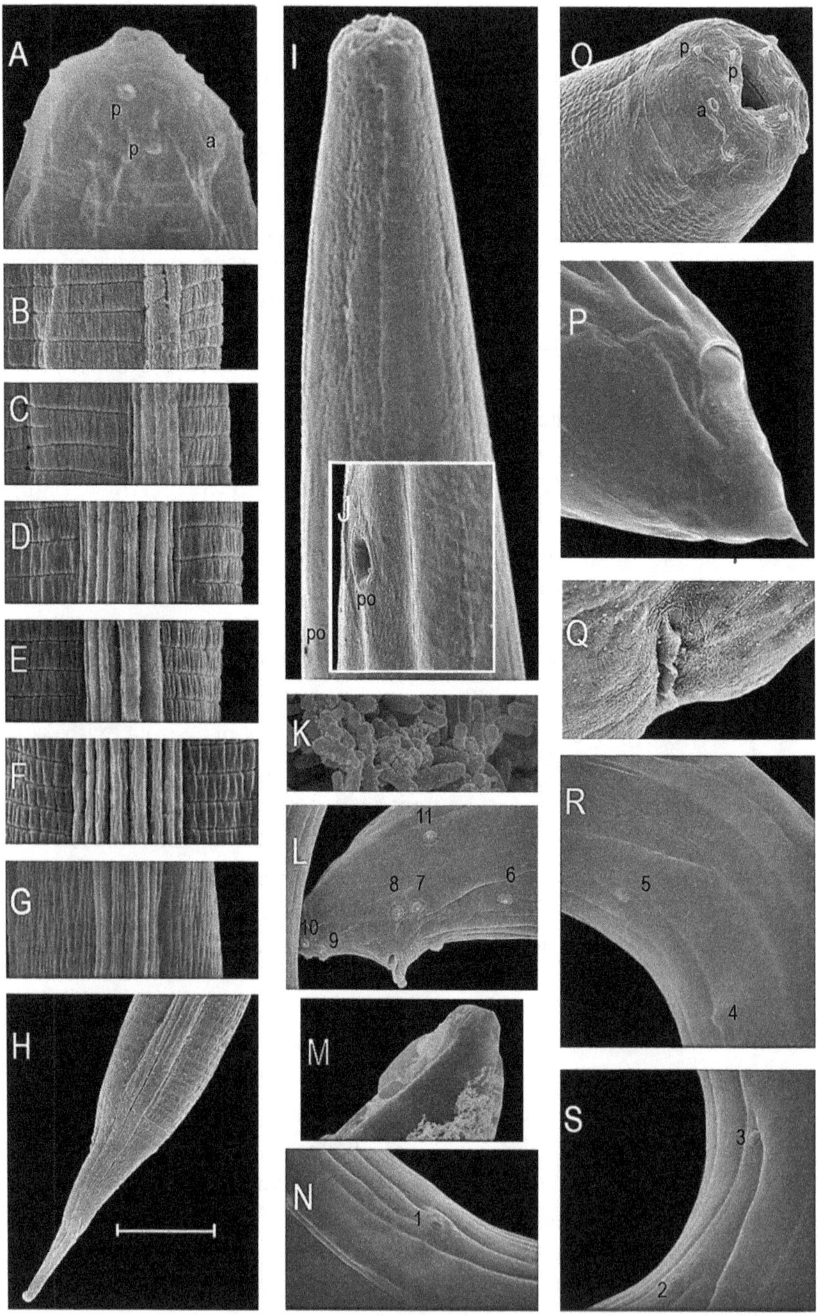
A
p
p
a
B
C
D
E
F
G
H
I
J
po
po
K
L
11
8 7
6
10
9
M
N
1
O
p
p
a
P
Q
R
5
4
S
3
2

Female (second generation)

Similar to first generation female, but smaller. Body C- or spiral-shaped. Labial papillae distinct. Pharynx stouter than in first generation female, with shorter isthmus, Vulval lips protruding, with double epiptygmata; vagina straight or directed posteriad. Tail, on average, slightly longer and thinner than in first generation, with one or two projections at tip.

Infective juvenile

Anterior region typical of *Steinernema* IJ. Body elongate, slender, straight in live specimens, only slightly bent in heat-relaxed worms, gradually tapering towards both extremities. Rarely with retained second-stage cuticle. Lip region smooth, rounded, anterior end slightly narrower than adjoining body due to shallow depression at 5.5-7.0 μm from anterior end. Amphidial openings pore-like, at level of cephalic papillae, fusus at *ca* 18-20 μm from amphidial opening, duct usually visible. Cuticle marked with clear transverse striations starting posterior to cephalic papillae. Lateral fields with maximum of eight, evenly spaced, ridges (nine incisures) in mid-body region, starting with a double ridge and gradually reaching eight ridges *ca* 75 μm from anterior end, subsequently decreasing to five but then increasing again to eight, then decreasing gradually, becoming two ridges at level of phasmids, lateral field formula = 2, 5, 8, 5, 8, 6, 4, 2. This peculiar and unusual arrangement was observed in all specimens studied with SEM and therefore seems not to be due to individual aberration. Excretory pore strongly sclerotised, generally located at level of posterior part of corpus, always anterior to nerve ring. Deirids at 136-159 μm from ante-

Fig. 4.93. *SEM photographs of* Steinernema vulcanicum. *A-H: Infective juvenile. A: Lip region showing papillae (p) and amphidial aperture (a); B-G: Lateral field from anterior end to tail; H: Tail. K: Symbiotic bacteria. I, J, O-Q: First generation female. I: Anterior end (po = excretory pore); J: Excretory pore (po); O: Lip region showing papillae (p) and amphidial aperture (a); P: Tail; Q: Epiptygmata. L-N, R, S: First generation male. L: Tail; M: Detail of spicule tip; N, R, S: Ventral surface of posterior region with numbered papillae. (Scale bar: A = 4.42 μm; B = 5.7 μm; C-G = 4.8 μm; H = 14.03 μm; I = 14.11 μm; J = 3.33 μm; K = 2.58 μm; L = 25.8 μm; M = 5.1 μm; N, R, S = 20 μm; O = 7.5 μm; P = 12.5 μm; Q = 13.3 μm.) After Clausi* et al. *(2011), courtesy* Nematology.

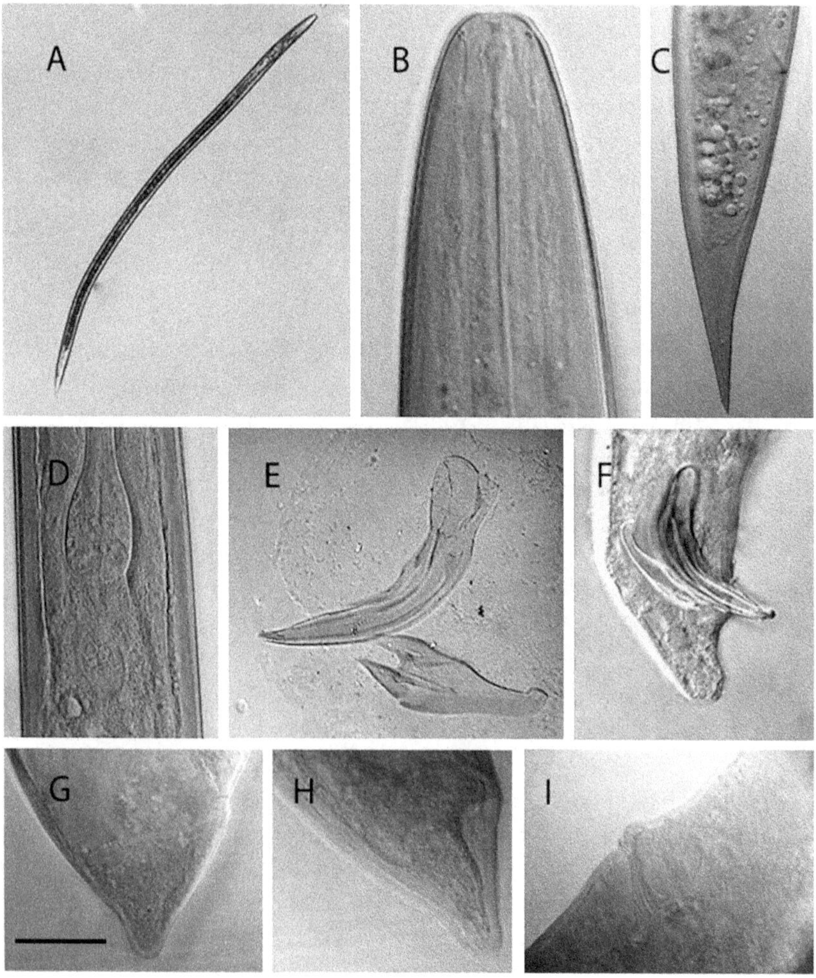

Fig. 4.94. Steinernema vulcanicum. *A-D: Infective juvenile. A: Entire body; B: Lip region; C: Tail; D: Basal bulb and bacterial sac. E, F: First generation male. E: Spicules and gubernaculum; F: Tail with spicules and gubernaculum. G-I: Female. G: Tail of first generation; H: Tail of second generation; I: Second generation, vulval region with epiptygmata. (Scale bar: A = 250 μm; B = 13 μm; C = 24 μm; D = 14.5 μm; E = 20 μm; F = 23 μm; G-I = 35 μm.) After Clausi* et al. *(2011), courtesy* Nematology.

rior end, at level of anterior half of pharyngeal bulb. Hemizonid at 135-146 μm from anterior end, level with anterior margin of pharyngeal bulb. Numerous large glands surrounding pharynx, gland duct opening at car-

Table 4.35. *Morphometrics of* Steinernema vulcanicum. *All measurements are in μm and in the form: mean ± s.d. (range).*

Character	First generation			Second generation		Infective juvenile
	Male		Female	Male	Female	
	Holotype	Paratypes	Paratypes	Paratypes	Paratypes	Paratypes
n	–	20	20	20	20	20
L	1918	1674 ± 267 (1318-2236)	5338 ± 1855 (2454-9222)	1494 ± 147 (1227-1714)	3688 ± 1121 (1667-5752)	1204 ± 72.5 (1081-1323)
a	17.3	14.8 ± 3.7 (11-18)	19.5 ± 6.2 (10.4-31.5)	19.4 ± 4.9 (16-24)	17.6 ± 6.3 (10-25)	31.5 ± 5.3 (21.6-39.0)
b	11.3	9.4 ± 1.5 (8-12)	21.5 ± 6.2 (11.6-33.3)	7.7 ± 0.9 (6-10)	15.8 ± 5.8 (8-23.2)	7.7 ± 0.6 (6.8-9.2)
c	38.4	47.4 ± 12.5 (25-69)	90.1 ± 37.4 (31-168)	43.1 ± 8.2 (32-64)	60.8 ± 24.0 (28-101)	14.1 ± 1.2 (11.3-16.8)
c′	0.9	0.68 ± 0.11 (0.5-0.7)	0.8 ± 0.2 (0.49-1.45)	0.8 ± 0.09 (0.6-0.9)	1.0 ± 0.3 (0.65-1.75)	3.2 ± 0.4 (2.6-4)
V	–	–	52.4 ± 19.2 (46-66)	–	52.4 ± 11.5 (45-57)	–
Max. body diam.	111	110 ± 47.3 (48-159)	271 ± 96.1 (123-400)	78 ± 22.3 (60-110)	212 ± 80.7 (131-357)	39 ± 7.7 (29-55)
Pharynx	169	181 ± 31.4 (122-227)	247 ± 31.6 (148-285)	195 ± 18.0 (157-233)	235 ± 58.0 (152-288)	157 ± 9.3 (134-173)
Excretory pore	132	128 ± 36.3 (54-170)	169 ± 37.0 (94-234)	127 ± 11.0 (107-148)	165 ± 59.2 (91-226)	93 ± 8 (77-107)
Nerve ring	140	134 ± 38.7 (93-170)	183 ± 22.5 (118-211)	152 ± 48.4 (126-179)	187 ± 77.8 (121-233)	117 ± 9.6 (98-136)
Tail length	50	38 ± 8.2 (22-59)	64.5 ± 17.8 (41-102)	36 ± 6.3 (24-48)	64 ± 20.1 (43-98)	86 ± 7.2 (74-100)
Anal body diam.	54	55 ± 13.8 (30-84)	84 ± 29.9 (34-136)	43 ± 5.7 (36-55)	66 ± 18.6 (45-90)	27 ± 4.2 (23-38)
Spicule length	82	71 ± 10.2 (46-88)	–	71.5 ± 4.7 (62-82)	–	–
Gubernaculum length	61	50 ± 8.4 (32-65)	–	47.2 ± 4.8 (31-53)	–	–
SW	1.5	1.4 ± 0.3 (0.9-2.1)	–	1.6 ± 0.3 (1.0-2.1)	–	–
GS	1.3	0.7 ± 0.1 (0.6-0.9)	–	0.7 ± 0.1 (0.5-0.8)	–	–
D%	78	69.7 ± 12.6 (42-90)	68.4 ± 11.2 (44-96)	65.6 ± 8.0 (50-78)	72.0 ± 25.5 (31.6-92)	59.6 ± 6.7 (47-68)
E%	264	354 ± 127.1 (121-630)	290 ± 83.7 (155-513)	369 ± 73.9 (249-531)	272 ± 59.8 (159-375)	109 ± 11.9 (83-135)
Hyaline region	–	–	–	–	–	40.5 ± 4.9 (28-47.6)
H%	–	–	–	–	–	48.8 ± 4.9 (38-56)

dia level. Bacterial vesicle, barely visible because of its particularly thin wall, located at beginning of intestinal lumen, 23-27 × 14-15 μm in size. Intestine with wide lumen and thin wall, containing granular and more massive material. Rectum long and straight, anus visible. Tail conoid with acute terminus, narrowing abruptly just after beginning of hyaline region which forms *ca* 50% of tail length. Phasmids at 18-40 μm from anus or at 43-65% of tail length.

TYPE HABITAT AND LOCALITY

Natural host unknown. Collected by baiting soil from a chestnut grove at Salto del Cane, Mount Etna, Sicily, Italy.

TYPE MATERIAL

Holotype male, first generation; five paratype males, five paratype females (first generation) five paratype IJ and five paratype males and five paratype females (second generation), deposited in the collection of the Department of Animal Biology 'Marcello La Greca', University of Catania, Italy. Three paratype males, three paratype females (first generation) and three paratype IJ deposited in the collection of the Department of Agricultural and Environmental Biology and Chemistry, Section of Entomology and Zoology (DIBCA), University of Bari, Italy. Two paratype males, two paratype females (first generation) and two paratype IJ deposited in the USDA Nematode Collection, Beltsville, MD, USA.

RELATIONSHIPS

Steinernema vulcanicum is mainly characterised by the male spicules which are long and slender, with the manubrium wider than long, velum prominent and variable in length, and with a rounded tip with a ventral concavity. Other characters include the male tail lacking a mucron, female vulva with double epiptygmata in both generations, IJ with slender body, lateral fields with eight ridges (nine incisures) at mid-body, decreasing to five, then increasing to eight again before decreasing to two, lateral field formula = 2, 5, 8, 5, 8, 6, 4, 2, phasmids always evident, hyaline region of tail comprising *ca* 50% of its length.

The species, because of its morphological characteristics and the length of the IJ (L = 1081-1323 μm), can be placed in the *glaseri*-group.

The most similar species are *S. apuliae*, *S. arenarium* and *S. boemarei* Lee, Sicard, Skeie & Stock, 2009, all of which have the same symbiotic bacterium, *Xenorhabdus kozodoii* Talliez, Pagès, Ginibre, Boemare, 2006. For further details, see Clausi *et al.* (2011).

MOLECULAR CHARACTERISATION

This species is molecularly characterised by ITS1 (accession no. GU929442) 12S (accession no. GU929443) and COXI (accession no. GU599902). In the trees obtained using the different reconstruction methods, *S. vulcanicum* was found to be closer to *S. arenarium* and *S. boemarei* within a monophyletic branch that included all species of the *glaseri*-group. More detailed discussion is provided by Clausi *et al.* (2011).

SYMBIOTIC BACTERIUM

Steinernema vulcanicum, together with other species of the *glaseri*-group: *S. arenarium*, *S. apuliae*, *S. boemarei*, and a yet to be identified species found in Spain (Emelianoff *et al.*, 2008), all host the same symbiotic bacterium – *X. kozodoii*. BLASTn analysis demonstrated a very high homology (99% over a >1500 bp interval) of the bacterial isolate from *S. vulcanicum* with *X. kozodoii* Talliez, Pagès, Ginibre, Boemare, 2006, with some interspersed base substitution (Clausi *et al.*, 2011).

BIONOMICS

The life cycle of *S. vulcanicum* is quite variable in many aspects. The infection rate in *G. mellonella* ranged from 30 to 70% at 25°C. From the cadaver, sometimes only dauer juveniles emerge, otherwise only third-stage, non-dauer, juveniles or even juveniles of various stages and adults. The J3 non-dauer larvae are able to become IJ outside of the insect body and subsequently to enter and kill a new host. Such behaviour is clearly not occasional and occurs in different seasons of the year and at different temperatures between 15° and 30°C, although more frequently at the higher temperatures. Moreover, at 25°C, the duration of the life cycle can vary as well, IJ sometimes emerging very quickly, even on the 3rd day after insect death, whilst in other cases the cycle shows a more regular trend and first generation adults can be found 2 days after insect death and second generation adults 6 days after insect death.

During individual cross-breeding experiments, mating attempts were observed between the Sicilian species and *S. apuliae* and *S. arenarium*, but no fertile offspring occurred in any of the crosses. Cross hybridisation tests with *S. apuliae*, *S. glaseri*, *S. arenarium* and It-C31 showed that *S. vulcanicum* was reproductively isolated (Clausi *et al.*, 2011).

DISTRIBUTION

Known only from the type locality in Sicily, Italy (Clausi *et al.*, 2011).

Steinernema xinbinense[76,77] Ma, Chen, De Clercq, Waeyenberge, Han & Moens, 2012
(Figs 4.95-4.98)

MEASUREMENTS

See Table 4.36.

DIAGNOSTIC DESCRIPTION

Male (first generation)

Anterior region typical for *Steinernema* male. Body J-shaped when heat-relaxed, tapering toward labial region. Cuticle with faint transverse striae, lateral fields mostly present, with one or two ridges. Labial region smooth, slightly flattened, not offset from rest of body. Papilliform deirids conspicuous. Excretory pore located opposite to nerve ring, cuticularised, *ca* 2 μm wide. Small overlapping junction between pharynx and intestine. Intestine generally with a wide lumen. Genital system monorchic, well developed, testis ventrally reflexed. Paired symmetrical and well curved spicules, slightly ochre-brown or yellowish in colour. Spicule manubrium usually with rounded end, slightly offset from shaft, manubrium less than one-fifth of spicule length (*ca* 16-20%), spicule head region length/width = 0.6-1.1. Calomus very short. Lamina well curved anteriorly and dorsally, with two internal ribs, tapering slightly posteriorly. Spicule terminus blunt. Rostrum not well developed, velum bow-like, strong, short, prominent, not extending

[76] Specific epithet derived from *Xinbin* county, the locality in China from where it was described.
[77] Description after Ma *et al.* (2012c). Revised by David J. Hunt.

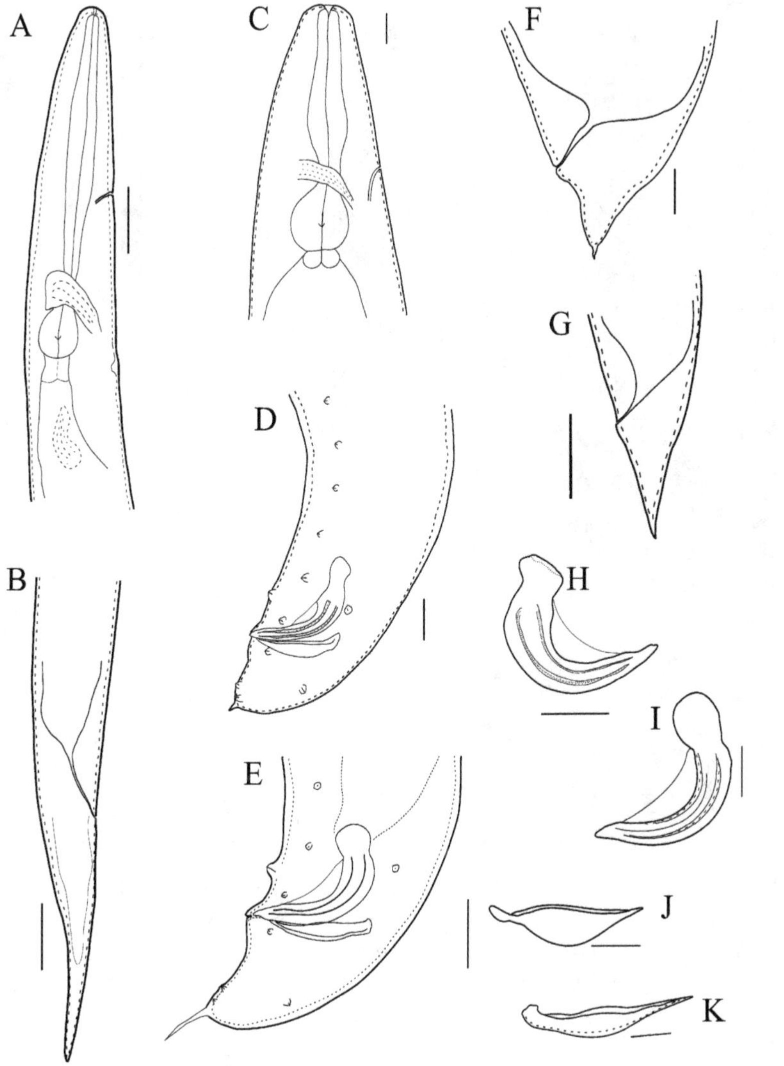

Fig. 4.95. Steinernema xinbinense. *A, B: Infective juvenile. A: Pharyngeal region showing basal bulb and bacterial chamber; B: Tail region. C, D, H, J, K: First generation male. C: Labial and pharyngeal region; D: Tail region; H: Spicule; J, K: Gubernaculum. E, I: Second generation male. E: Tail region; I: Spicule. F: First generation female, tail region; G: Second generation female, tail region. (Scale bars: A-F, H = 20 μm; G = 50 μm; I-K = 10 μm.) After Ma* et al. *(2012c), courtesy* Nematology.

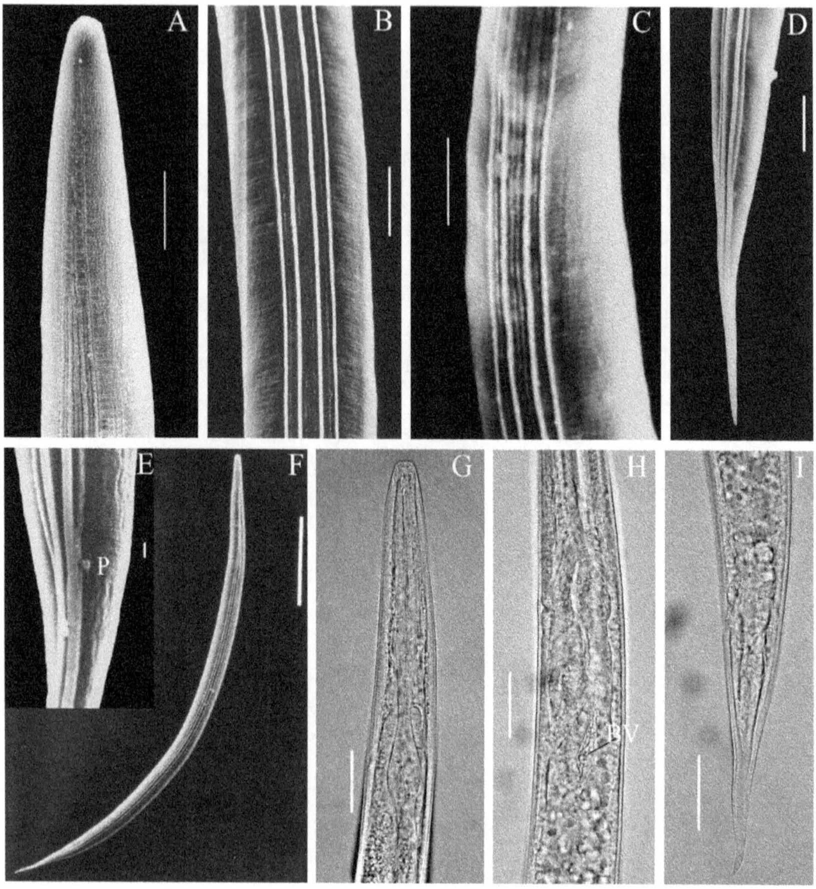

Fig. 4.96. Steinernema xinbinense. *SEM and LM pictures of infective juvenile. A: Lateral field showing change in anterior part; B, C: Lateral field with eight ridges; D: Posterior part showing lateral field; E: Tail region showing phasmid (P) and gradual reduction in number of ridges so as to become two after phasmid level; F: Entire view; G: Labial and pharyngeal region; H: Pharyngeal region showing basal bulb and bacterial vesicle (BV); I: Tail region. (Scale bars: A-D = 10 μm; G-I = 20 μm; E = 1 μm; F = 100 μm.) After Ma* et al. *(2012c), courtesy* Nematology.

to spicule tip. Gubernaculum boat-shaped in lateral view, tapering gradually posteriorly, corpus with two wings opening posteriorly. Tail bluntly conoid, usually concave on ventral surface, small mucron mostly present (3-5 μm). Twenty-three genital papillae comprising 11 pairs and single midventral papilla located just anterior to cloacal opening. Seven pairs located precloacal (six pairs subventral and one pair lateral), one pair adcloacal and three pairs postcloacal, consistently located near tail terminus (one subdorsal, two at the tail tip).

Male (second generation)

Similar to first generation male, but smaller. Tail mucron more distinct, developed as a medium-long filamentous projection (9-15 μm), usually subterminal subventral in position. Spicules and gubernacula shorter and thinner.

Female (first generation)

Anterior region typical for *Steinernema* female. Body variable in length, usually C-shaped, sometimes coiled on heat relaxation. Cuticle faintly striated. Labial region truncate to slightly round, tapering anteriorly, continuous with body contour. Morphology of labial region and pharynx similar to male. Excretory pore located anterior to nerve ring, always before mid-pharynx. Genital tract amphidelphic with reflexed ovaries. Vulva a symmetrical, transverse slit slightly protruding from body contour. Vulval opening mostly located slightly posterior to mid-body. Vagina short, leading into paired uteri. Narrow rectum, distinct anal opening. Tail shorter than anal body diam. Tail shape variable, mostly conical with pointed terminus, one or two small mucrons present. Postanal swelling not present.

Female (second generation)

Similar to first generation in general morphology, but shorter and narrower. Body slightly curved when heat-relaxed. Excretory pore position more anterior. Vulva located near mid-body with protruding (7-10 μm) and slightly asymmetrical lips (anterior lip larger than posterior one). Tail conical and sharply pointed, longer than anal body diam., tapering to a mucron-like terminus. No subsidiary mucron visible.

Infective juvenile

Anterior region typical for *Steinernema* IJ. Body of heat-relaxed specimens almost straight or slightly curved, slender; slightly tapering towards anterior and posterior ends. Body cuticle finely annulated with delicate transverse striations. Retained second-stage cuticle not usually observed after harvesting. Labial region continuous with body, rounded to slightly truncate, not annulated. Cephalic region gently rounded,

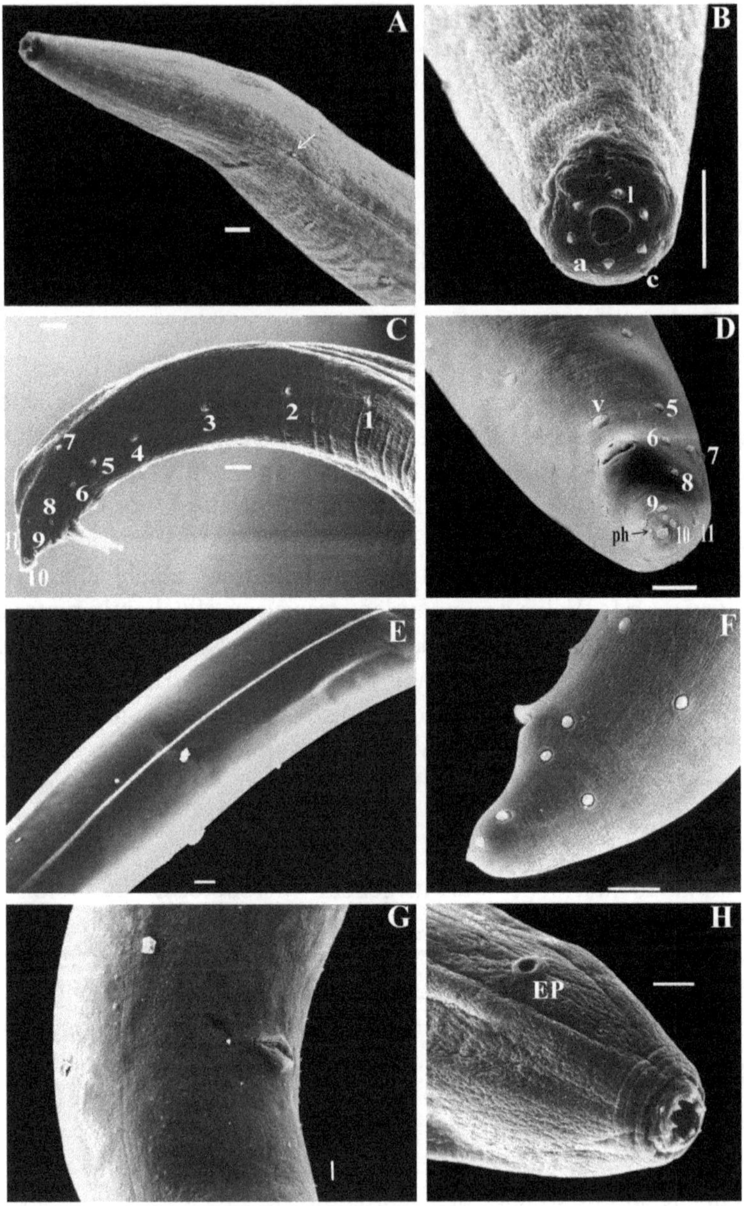

Fig. 4.97. Steinernema xinbinense. *SEM pictures of first generation male and female. A-F: Male. A: Anterior region showing deirid (arrow); B: Labial region showing amphidial aperture (a), cephalic papilla (c) and labial papilla (l); C: Posterior region of male showing arrangement of 11 pairs of genital*

continuous with body contour. Excretory pore anterior to nerve ring, located at mid-pharynx level, cuticularised, *ca* 1 μm wide. Hemizonid distinct, located near middle of basal bulb, posterior to nerve ring. Lateral field beginning with one cuticular depression or line at 5-6 annules from anterior end. Two additional lines appearing posteriorly forming two thin ridges, further posteriorly each ridge dividing into three and a central ridge emerging with number of ridges increasing from two to seven. Central ridge dividing into two over a very short distance, making a total of eight, the maximum number of ridges in lateral field (*i.e.*, nine lines). Ridges unevenly spaced and developed, with four less distinct submarginal ridges (2, 4, 5, 7). Ridges remaining unchanged until phasmid. Only two prominent ridges observed posterior to phasmid. Lateral field formula = 2, 7, 8, 2. Bacterial vesicle elongate, dorsally displaced, just posterior to cardia, variable in shape, approximately 21 $\times$ 6 μm in size, containing bacterial cells. Rectum long, straight; anus distinct. Tail conoid, hyaline portion occupying 30-42% of tail length. Phasmid pore-like, located 34-37% of tail length posterior to anus.

TYPE HABITAT AND LOCALITY

The natural host of *S. xinbinense* is unknown since the species was recovered by *Galleria* baiting from a soil sample collected in the vicinity of Xinbin county, Liaoning province, China (41.88°N, 124.47°E) during a survey in April 2008. The type locality is characterised by a mountain slope with pines, grasses and shrubs. The soil is a humid, loamy soil under pines.

TYPE MATERIAL

Holotype male, ten paratype males, six paratype females and ten paratype IJ deposited in the nematode collection of the Institute of Zoology, Ghent University, Ghent, Belgium. Slides with paratype first (14) and second generation males (16), first (11) and second (15)

---

papillae (1-11); D: Posterior part of tail region showing genital papillae pairs 5-11, a single precloacal papilla (V) and phasmid (ph); E: Lateral field; F: Tail region. G, H: Female. G: Vulval region; H: Labial region showing excretory pore (EP). (Scale bars: A-H = 10 µm.) After Ma et al. *(2012c), courtesy* Nematology.

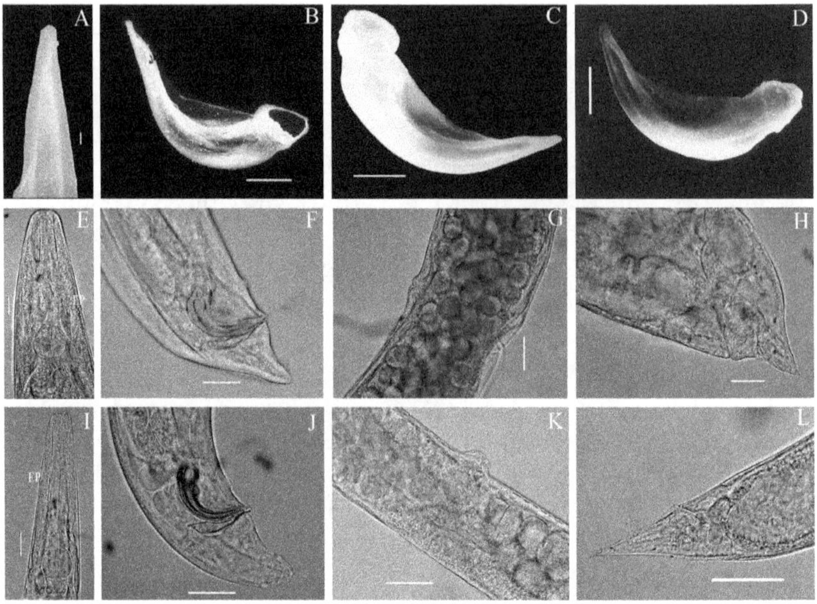

Fig. 4.98. Steinernema xinbinense. *SEM and LM pictures of fixed male and female. A-F: First generation male. A: Spicule tip; B-D: Spicules; E: Labial and pharyngeal region showing excretory pore (EP); F: Tail region. G, H: First generation female. G: Vulva region; H: Tail region. I, J: Second generation male. I: Pharynx region (EP = excretory pore); J: Tail showing spicule and gubernaculum. K, L: Second generation female. K: Vulva region; L: Tail region. (Scale bars: A = 1 μm; B-D = 10 μm; E, F, H-J = 20 μm; G, K, L = 50 μm.) After Ma* et al. *(2012c), courtesy* Nematology.

generation females and IJ (12) deposited at the Institute of Plant Protection, Hebei Academy of Agricultural and Forestry Sciences, Baoding, Hebei, China.

RELATIONSHIPS

Steinernema xinbinense is characterised by the morphology and morphometrics of adults and IJ. The male of the first generation can be recognised by the lateral fields and a well curved, yellowish spicule with a prominent velum. Males of both first and second generation have a tail mucron. The length of the spicule and gubernaculum of first generation males is 56 (49-62) μm and 35 (30-41) μm, respectively. Twenty-three genital papillae comprising 11 pairs and a single midventral papilla

Table 4.36. *Morphometrics of* Steinernema xinbinense. *All measurements are in μm and in the form: mean ± s.d. (range).*

Character	First generation			Second generation		Infective juvenile
	Male		Female	Male	Female	
	Holotype	Paratypes	Paratypes	Paratypes	Paratypes	Paratypes
n	–	20	20	20	20	20
L	1318	1265 ± 93 (1133-1440)	4037 ± 680 (3025-5121)	942 ± 36 (874-1020)	1739 ± 189 (1433-2060)	694 ± 29 (635-744)
a	13	12 ± 0.7 (11-13)	22 ± 2.3 (19-25)	17 ± 1.0 (15-18)	18 ± 1.6 (16-21)	24 ± 1.4 (21-25)
b	8.9	8.5 ± 0.6 (7-9)	22 ± 3.3 (17-26)	7.0 ± 0.3 (6-8)	11 ± 1.2 (9-13)	6.1 ± 0.3 (5-7)
c	35	34 ± 3 (31-39)	106 ± 24.6 (79-123)	28 ± 2.2 (25-32)	28 ± 2.6 (24-32)	9.7 ± 0.7 (8-11)
c′	0.9	0.9 ± 0.08 (0.7-1.0)	0.6 ± 0.1 (0.5-0.8)	1.0 ± 0.06 (0.9-1.2)	1.6 ± 0.17 (1.2-1.9)	4.2 ± 0.3 (3-5)
V	–	–	52 ± 2.9 (46-57)	–	54 ± 2.4 (51-57)	–
Max. body diam.	105	103 ± 8.7 (90-126)	176 ± 12 (159-200)	56 ± 3.3 (48-62)	98 ± 8.5 (88-115)	30 ± 0.9 (28-31)
Nerve ring	101	106 ± 9.3 (91-120)	126 ± 9.9 (106-141)	105 ± 2.5 (100-109)	122 ± 3.7 (117-130)	86 ± 3.8 (75-90)
Pharynx length	153	149 ± 5.5 (138-159)	186 ± 6.2 (167-192)	136 ± 3.3 (127-140)	163 ± 3.8 (156-168)	116 ± 6.6 (109-125)
Excretory pore	68	68 ± 6.8 (57-75)	80 ± 8.2 (70-87)	59 ± 1.4 (57-62)	65 ± 3 (58-69)	51 ± 2.6 (46-53)
Tail length	37	37 ± 3.1 (30-41)	40 ± 7.1 (30-53)	34 ± 1.9 (30-37)	61 ± 3 (54-63)	73 ± 5.2 (61-81)
Anal body diam.	43	41 ± 3.5 (36-46)	61 ± 7.2 (50-67)	34 ± 1.1 (31-36)	38 ± 3.4 (31-45)	17 ± 1 (16-19)
Testis reflexion	279	295 ± 46 (240-345)	–	135 ± 12 (117-165)	–	–
Spicule length	56	56 ± 5 (49-62)	–	46 ± 3.5 (36-52)	–	–
Spicule width	8	9 ± 1.5 (7-12)	–	7.9 ± 0.9 (6-10)	–	–
Gubernaculum length	40	35 ± 6.7 (30-41)	–	25 ± 2.7 (21-32)	–	–
Gubernaculum width	8	6.7 ± 1.1 (5-8)	–	4.7 ± 0.8 (3-6)	–	–
Labial region to vulva	–	–	2333 ± 151 (1732-2860)	–	953 ± 104 (840-1088)	–
Hyaline region	–	–	–	–	–	24 ± 3.3 (18-28)
E%	179	181 ± 12 (152-198)	166 ± 26 (134-174)	176 ± 12 (153-198)	107 ± 4.7 (98-112)	71 ± 4 (65-78)
D%	45	45 ± 3.9 (41-50)	40 ± 3.7 (38-45)	44 ± 1.6 (40-46)	41 ± 2.7 (36-46)	44 ± 1.9 (40-47)
SW%	130	137 ± 10 (114-156)	–	136 ± 9.4 (118-153)	–	–
GS%	67	63 ± 5.8 (54-72)	–	53 ± 4.2 (44-58)	–	–
H%	–	–	–	–	–	35 ± 3.8 (30-42)

are located just anterior to the cloacal opening. The females have a protruding symmetrical vulva and a tail that is mostly conical with a pointed terminus. IJ are characterised by a short body length of 694 (635-744) μm, distance from anterior end to the excretory pore of 51 (46-53) μm, lateral field with eight unevenly spaced ridges (*i.e.*, nine lines) in mid-body, tail length of 73 (61-81) μm, and hyaline portion occupying only 30-42% of tail length. Using the grouping of EPN proposed by Spiridonov *et al.* (2004), *S. xinbinense* belongs to group III, the *kraussei*/*feltiae*/*oregonense*-group, based on features of ITS rDNA sequence. Ma *et al.* (2012c) provide further details of the relationships.

Molecular characteristics

Steinernema xinbinense, a member of the *feltiae*/*kraussei*/*oregonense*-group, is characterised genetically by sequences of the ITS (JN171593) and D2-D3 regions (GU994204) of rDNA. The length of the ITS1 + 5.8S + ITS2 sequence is 729 bp, ITS1 = 265 bp and ITS2 = 307 bp, and its composition is: A = 0.2483, C = 0.1660, G = 0.2222, T = 0.3635. *Steinernema xinbinense* differs from *S. tielingense*, its closest taxon, by 33 bp. The sequence of the D2-D3 region of *S. xinbinense* is 868 bp long and its composition is: A = 0.2488, C = 0.1982, G = 0.2972, T = 0.2558. Pairwise distances show that *S. xinbinense* differs from its sister taxon, *S. cholashanense*, by 10 bp; it differs from *S. tielingense*, *S. oregonense* and *S. kraussei* by 12, 12 and 15 bp, respectively.

Symbiotic bacterium

No information.

Bionomics

Cross-hybridisation assays between the new species and *S. tielingense*, *S. kraussei*, *S. feltiae* and *S. hebeiense* yielded no progeny. In the controls, offspring were produced normally.

Distribution

Recorded only from the type locality in China by Ma *et al.* (2012c).

Steinernema xueshanense[78,79] **Mráček, Liu & Nguyen, 2009**
(Figs 4.99, 4.100)

MEASUREMENTS

See Table 4.37.

DIAGNOSTIC DESCRIPTION

Male (first generation)

Anterior region typical for *Steinernema* male. Excretory pore mostly near nerve ring. Cardia rather long, intestine generally with a wide lumen. Distinct, papilliform deirid situated at the pharyngeal bulb to nerve ring. Testis well developed, reflexed. Paired symmetrical, finely curved and slender spicules, slightly yellowish, distally sharply pointed. Spicule manubrium rounded to ellipsoid, with approximate length:width ratio of 1:1, manubrium mostly less than one-fourth of spicule length, calomus distinct. Lamina with two ribs, slender and finely curved. Rostrum rarely developed and if developed only slightly sclerotised, velum bow-like and distinct, extending from calomus part to distal part of lamina. In lateral view, gubernaculum boat-shaped, short proximal part varying from hook- to head-like, posterior part usually tapering abruptly to distal tip. In ventral view, cuneus small, needle-shaped, or absent, corpus with two wings. Tail bluntly conoid, usually concave at ventral side, mucron mostly absent, if developed enlarged papilla-like (never filamentous). There are 23 genital papillae comprising 11 pairs and a single midventral papilla located just anterior to cloacal opening. Of these, eight pairs are located precloacal (five pairs subventral, one pair lateral and two pairs adcloacal). Three postcloacal pairs consistently located near tail terminus (one subdorsal, two at tail tip).

Male (second generation)

Similar to first generation male, but slightly smaller. Tail mucron developed as a medium long filamentous projection.

---

[78] Named after the *Xue Shan* Mountains, China, where the species was collected.

[79] Description after Mráček *et al.* (2009). Revised by Khuong B. Nguyen.

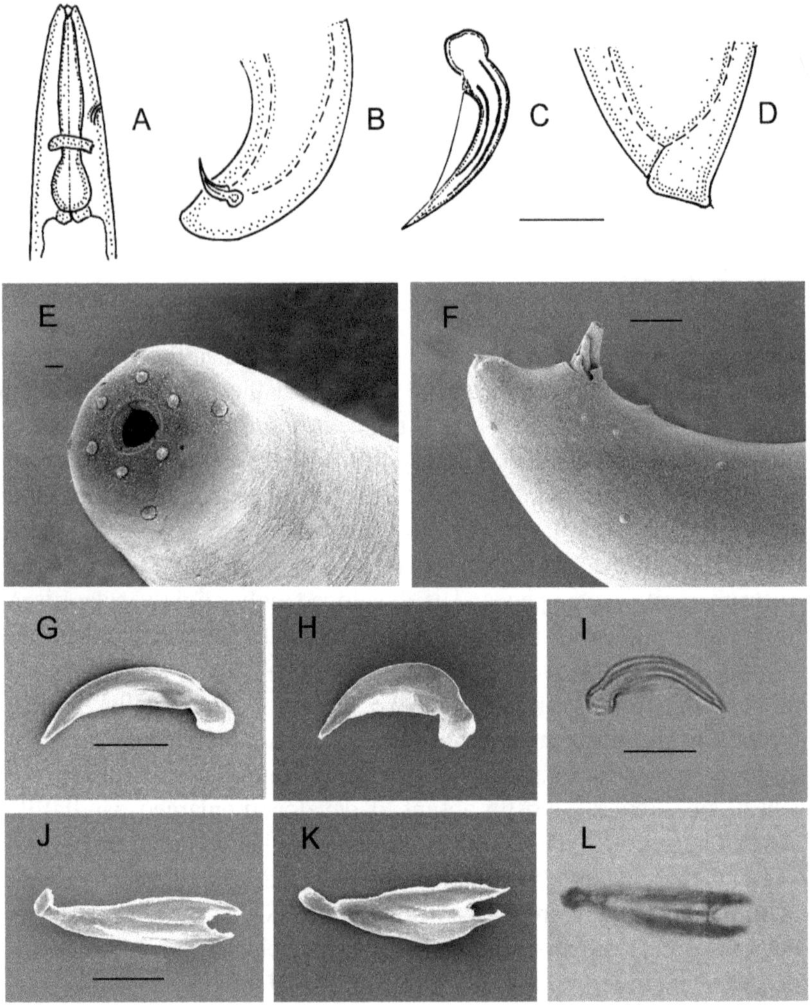

Fig. 4.99. Steinernema xueshanense. *A-C: Anterior and posterior region and spicule of first generation male; D: First generation female tail; E: Anterior region of male; F: Posterior region of male; G-I: Spicules by SEM and LM; J-L: Gubernaculum by SEM and LM. (Scale bars: (A-D: using top bar): A, B = 75 μm; C = 30 μm; D = 55 μm; E = 1 μm; F = 10 μm; G, H = 15 μm (scale in G); J, K = 12 μm (scale in J); I = 32.2 μm; L = 18 μm (scale in I).) Modified after Mráček* et al. *(2009), courtesy* Journal of Invertebrate Pathology.

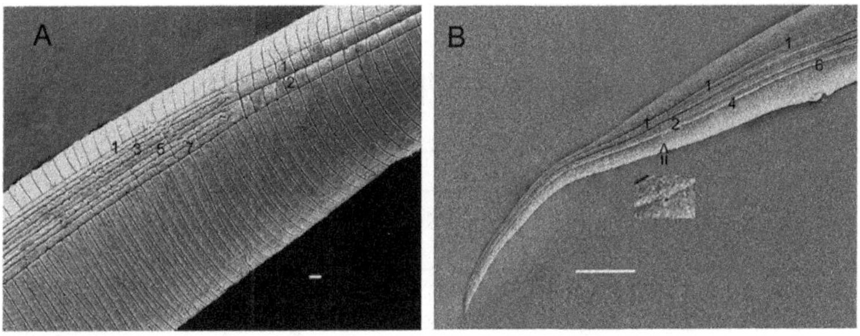

Fig. 4.100. Steinernema xueshanense. *SEM of infective juvenile showing pattern of lateral field. (Scale bars: A = 1 μm; B = 10 μm.) Modified after Mráček* et al. *(2009), courtesy* Journal of Invertebrate Pathology.

Female (first generation)

Anterior region typical for *Steinernema* female. Excretory pore located anterior to nerve ring. Genital tract amphidelphic with reflexed ovaries. Vulva a transverse slit, protruding from body contour. Vulval opening slightly posterior to mid-body. Vagina short, slanted, leading into paired uteri. Tail of mature females with a characteristic cone, never obtuse as in most *Steinernema* species. Mucron-like projection on cone tip, in younger specimens sometimes enlarged to filiform projection. Shorter bifurcation can be observed in older specimens replacing filiform projection. Postanal swelling not, or very slightly, developed. Pygmy forms not observed.

Female (second generation)

Similar to first generation female in general morphology, but smaller. Vulva symmetrical. Excretory pore located more anteriorly. Conical and sharply pointed tail, postanal swelling not developed, tail longer than anal body diam.

Infective juvenile

Anterior region typical for *Steinernema* IJ. Excretory pore in mid-pharynx region, cuticularised, *ca* one μm wide. Lateral field beginning anteriorly with one line close to annule four from anterior end. A short distance posteriorly, two additional lines appearing to form two ridges. At about excretory pore level, lateral field changing from two ridges to seven ridges, more posteriorly, central ridge dividing into two, making a total of eight ridges. Lateral field pattern characteristic with two, less distinct, submarginal ridges. In posterior part of body, number of ridges

Table 4.37. *Morphometrics of* Steinernema xueshanense *(holotype is the infective juvenile). All measurements are in μm and in the form: mean ± s.d. (range).*

Character	First generation			Second generation		Infective juvenile
	Holotype (IJ)	Male	Female	Male	Female	
n	–	20	20	20	20	20
L	869	1589 ± 197 (1313-2040)	5952 ± 1196 (4181-8181)	1024 ± 83 (929-1232)	2892 ± 340 (2081-3333)	860 ± 45 (768-929)
a	30	–	–	–	–	28 ± 1.7 (26-32)
b	6.6	–	–	–	–	6.4 ± 0.3 (5.8-7.0)
c	11	–	–	–	–	9.9 ± 0.7 (9.0-11.0)
c′	–	–	–	–	–	4.6 ± 0.3 (3.8-5.1)
V	–	–	54 ± 2.6 (52-62)	–	55 ± 2.0 (52-60)	–
Body diam.	29	144 ± 13 (97-159)	257 ± 46* (182-343)	67 ± 7.3 (56-81)	153 ± 7.3 (131-182)	30 ± 1.4 (29-33)
Excretory pore	68	128 ± 5.6 (113-137)	148 ± 15* (117-148)	79 ± 7.6 (64-92)	88 ± 6.3 (78-102)	67 ± 2.8 (60-72)
Nerve ring	94	–	–	–	–	91 ± 4.8 (81-96)
Pharynx	132	160 ± 6.8 (151-175)	230 ± 23 (196-274)	149 ± 10 (129-164)	195 ± 7.8 (180-207)	135 ± 3.8 (130-143)
Tail length	78	38 ± 5.0 (29-48)	55 ± 5.4 (43-66)	30 ± 2.4 (27-37)	72 ± 6.0 (59-82)	87 ± 4.6 (80-92)
Hyaline region	38	–	–	–	–	44 ± 2.8 (40-48)
Anal body diam.	19	50 ± 6.4 (37-67)	84 ± 12* (38-72)	36 ± 2.0 (33-41)	54 ± 7.6 (47-74)	19 ± 0.7 (17-21)
Spicule length	–	76 ± 5.9 (66-91)	–	59 ± 27 (54-66)	–	–
Gubernaculum length	–	49 ± 4.8 (41-60)	–	41 ± 1.4 (38-43)	–	–
Gubernaculum width	–	7.2 ± 0.8 (5.6-8.7)	–	6.8 ± 0.5 (6.4-8.0)	–	–
E%	–	338 ± 34 (280-400)	272 ± 44 (190-380)	265 ± 34 (210-322)	120 ± 20 (100-170)	78 ± 5.2 (70-90)
SW%	–	152 ± 19 (93-172)	–	164 ± 11* (119-159)	–	–
GS%	–	76 ± 9.1 (58-95)	–	82 ± 3.3 (77-87)	–	–
H%	49	–	–	–	–	51 ± 2.7 (46-55)

* Mean and range data as cited in original paper, but clearly in error.

of lateral field reducing gradually. About two anal body diam. anterior to anus, two middle ridges merging forming seven ridges in lateral field. Close to anus, lateral field reducing to six ridges which merge between anus and phasmids to make four. Near phasmid two marginal ridges disappearing to form two-ridge pattern in lateral field which extends almost to tail tip. With above arrangement, lateral field formula = 2, 7, 8, 7, 6, 4, 2. Phasmids pore-like, posterior to anus, approximately at 33-35% of tail length from anus, located at ventral edge of lateral fields. Cardia present. Bacterial vesicle oval, *ca* 40-50 × 15-20 μm in size. Rectum straight, anus sickle-shaped. Tail conoid, sharply tapering with slight ventral constriction situated slightly before proximal portion of hyaline region which comprises about half of tail length. Tail terminus pointed.

TYPE HABITAT AND LOCALITY

Type host unknown. The type locality (altitude *ca* 3200 m a.s.l.) was characterised by dense bush vegetation in a deep valley, northeast of Dequen city, Yunnan, China.

TYPE MATERIAL

Holotype IJ, paratype IJ, males and females of first and second generation mounted on slides and deposited in the EPN collection at the Laboratory of Insect Pathology, Institute of Entomology, Biology Centre, CAS, Česke Budějovice, Czech Republic. Another batch of paratype IJ and adults deposited in the nematode collection at the Department of Entomology and Nematology, University of Florida, Gainesville, FL, USA.

RELATIONSHIPS

Steinernema xueshanense belongs to the *feltiae*-group and can be distinguished from other species by characters of the male, female and IJ. In the male, the spicules are moderately curved with a pointed tip and mostly with a developed rostrum, the gubernaculum has a blunt anterior end, and a needle-shaped cuneus. The second generation male has a distinct mucron. The female has a characteristic cone on the tail tip bearing one to two mucron-like projections, and postanal swelling slightly developed or absent. For the IJ, the body length of *ca*

860 μm, slightly flattened labial region, lateral field with eight ridges, the submarginal pair being less distinct, and lateral field formula = 2, 7, 8, 7, 6, 4, 2, tail lacking a depression and hyaline region occupying *ca* half of tail length. More information on relationship is available in Mráček *et al.* (2009).

MOLECULAR CHARACTERISATION

Steinernema xueshanense is characterised genetically by sequences of ITS (accession no. FJ666052) and D2-D3 (FJ666053) regions of rDNA. The sequence of ITS regions including ITS1 + 5.8S + ITS2 can be recognised by its length of 714 bp. The sequence length D2-D3 fragment is 862 bp. More information about molecular characterisation and phylogenetic analysis is available in Mráček *et al.* (2009).

SYMBIOTIC BACTERIUM

No information.

BIONOMICS

Cross-breeding tests with *S. feltiae*, *S. kraussei* and *S. cholashanense* produced no progeny (Mráček *et al.*, 2009).

DISTRIBUTION

Steinernema xueshanense was collected in Dequen city, Yunan, China.

Addendum

The following species descriptions were published during 2016 yet too late for full inclusion in the book. They are represented below by their modified original abstracts plus molecular sequence accession numbers.

Steinernema fabii[80] Abate, Malan, Tiedt, Wingfield, Slippers & Hurley, 2016

Steinernema fabii was isolated by trapping with wax moth (*Galleria mellonella*) larvae from soil in an *Acacia mearnsii* plantation in the Mpumalanga province of South Africa. The species is morphologically characterised by the length of the infective juvenile (IJ) of 641 (590-697) μm, by a tail length of 58 (52-64) μm, ratio a = 24 (21-41), H% = 53 (37-61) and E% = 93 (83-105). The pattern of the IJ lateral field = 2, 5, 2 ridges (3, 6, 3 incisures). The first generation male can be recognised by the long spicule of 90 (79-106) μm and gubernaculum of 66 (56-77) μm; D% = 64 (52-75) and GS% = 73 (63-86). The first generation female can be recognised by a protruding vulva with short, double-flapped epiptygmata, and the lack of a postanal swelling, while the second generation differs in having a postanal swelling and a conical, sharply pointed tail. Analysis of DNA sequences for the ITS (KR527216) and D2-D3 (KR527217) gene regions showed *S. fabii* differed from all other *Steinernema* species and belonged to a new monophyletic group, the *Cameroonense*-clade, consisting of *S. cameroonense*, *S. nyetense*, *S. sacchari* and *S. fabii*, all from the African continent. This group is closely related to species in the *feltiae/kraussei/oregonense* Clade III.

Steinernema goweni[81] San-Blas, Morales-Montero, Portillo, Nermut' & Půža, 2016

Steinernema goweni, a member of the *bicornutum*-group, was isolated in Zulia state north-western Venezuela. *Steinernema goweni* is charac-

[80] Specific epithet derived from the Forestry and Agricultural Biotechnology Institute (*FABI*), University of Pretoria, Pretoria, South Africa.
[81] Patronym in honour of Dr Simon *Gowen*.

terised by morphometrics of its infective juveniles, with body length 640 (531-719) μm, pharynx length of 119 (109-126) μm, tail of 67 (59-89) μm, c = 9 (6-11) and E% = 77 (48-94). The first generation male can be recognised by the spicule shape and length of 55 (50-57) μm, gubernaculum length of 35 (30-40) μm, and D% value 42 (28-59), which is at the lower limit within the *bicornutum*-group. The number of male genital papillae is also distinctive, presenting up to 27 (13 pairs + one) papillae in 15% of specimens. The phasmids can be seen in scanning electron microscopy in all stages, characters not previously reported within the *bicornutum*-group. Phylogenetic analyses of the *bicornutum*-group based on both ITS (KR781038) and D2-D3 (KR781039) regions showed a clear separation of *S. goweni* from the other species and formed a strongly supported group of American species.

Steinernema beitlechemi[82] Çimen, Půža, Nermut', Hatting, Ramakuwela, Faktorová & Hazir, 2016

Steinernema beitlechemi was isolated from soil samples collected in Bethlehem, Free State province, South Africa. The infective juveniles of *S. beitlechemi* are characterised by a body length of 696 (606-768) μm, nerve ring position of 96 (86-105) μm from the labial region and lateral fields with six ridges (*i.e.*, seven lines) at the mid-body. The first generation male can be distinguished by a spicule length of 93 (88-100) μm and presence of 12 pairs of genital papillae. The first generation female has a moderately protruding vulva and a rounded tail tip bearing one spine-like projection. The species is further characterised by sequences of the ITS and partial 28S regions of the ribosomal DNA. Phylogenetic analyses showed *S. beitlechemi* to be a member of a strongly supported monophyletic clade with three other African species: *S. sacchari*, *S. cameroonense* and *S. nyetense*. The sequences were deposited in GenBank under accession numbers KT373856 (ITS sequence, strain SGI-197), KT373855 (ITS sequence, strain NWGA-33), KT580949 (28S sequence, strain SGI-197) and KT878316 (18S sequence, strain SGI-197)

[82] Specific epithet referring to *Bethlehem*, the town in the wheat-growing area from which the species was isolated and derived from the Jewish word *Beit Lechem*, meaning 'house of bread'.

Steinernema nguyeni[83] Malan, Knoetze & Tiedt, 2016

Steinernema nguyeni was recovered by baiting from beneath an *Olea africana* tree in South Africa. The combination of morphological and molecular features suggests that it is a member of the *feltiae/kraussei/oregonense*-group, clustering with members of this group in Clade III. The species is morphologically characterised by the infective juvenile body length of 737 (673-796) μm and the infective juvenile lateral field pattern = 2, 8, 2. The first generation male can be recognised by the spicule length of 66 (58-75) μm and a gubernaculum length of 43 (30-55) μm. The first generation female can be recognised by the slightly protruding vulval lips and the presence of low, double-flapped epiptygmata. Analysis of the ITS (KP325084) and D2-D3 (KR815816) regions of the ribosomal DNA confirmed that *S. nguyeni* differed from all other known *Steinernema* species.

Steinernema pwaniense[84] Půža, Nermut', Mráček, Gengler & Haukeland, 2016

Steinernema pwaniense, belonging to the *glaseri*-group, is described from Tanzania. The infective juveniles are characterised by a body length of 978 (808-1131) μm, distance from anterior end to excretory pore of 86 (80-95) μm and a tail length of 87 (75-95) μm. The ratios a, c and E% are substantially lower than those of all other African *glaseri*-group members. The first generation male can be distinguished by a large spicule length of 92 (80-97) μm and by the absence of the caudal mucron, while the second generation male possesses a short spine-like mucron. First generation females have a peg-like tail tip bearing three spine-like projections. Second generation females can be recognised by a slightly protruding vulva and well developed postanal swelling. The species is further characterised by sequences of the ITS (KT358811) and partial 28S regions of the ribosomal DNA (KT358813). Phylogenetic analyses show that *S. pwaniense* forms a

[83] Patronym in honour of Dr Khuong B. *Nguyen* for his work on entomopathogenic nematodes.

[84] Emended ending from the incorrect *-ensis*. The specific epithet is derived from the regional name *Pwani*, a Swahili word meaning coast or coastal.

strongly supported monophyletic clade with two other East African species, *S. ethiopiense* and *S. karii*.

References

ABATE, B.A., MALAN, A.P., TIEDT, L.R., WINGFIELD, M.J., SLIPPERS, B. & HURLEY, B.P. (2016). *Steinernema fabii* n. sp. (Rhabditida: Steinernematidae), a new entomopathogenic nematode from South Africa. *Nematology* 18, 235-255.

ÇIMEN, H., LEE, M.-M., HATTING, J., HAZIR, S. & STOCK, S.P. (2014). *Steinernema tophus* sp. n. (Nematoda: Steinernematidae), a new entomopathogenic nematode from South Africa. *Zootaxa* 3821, 337-353.

ÇIMEN, H., LEE, M.M., HATTING, J., HAZIR, S. & STOCK, S.P. (2015). *Steinernema innovationi* n. sp. (Panagrolaimomorpha: Steinernematidae), a new entomopathogenic nematode species from South Africa. *Journal of Helminthology* 89, 415-427.

ÇIMEN, H., PŮŽA, V., NERMUT', J., HATTING, J., RAMAKUWELA, T., FAKTOROVÁ, L. & HAZIR, S. (2016). *Steinernema beitlechemi* n. sp., a new entomopathogenic nematode (Nematoda: Steinernematidae) from South Africa. *Nematology* 18, 439-453.

CLAUSI, M., LONGO, A., RAPPAZZO, G., TARASCO, E. & VINCIGUERRA, M.T. (2011). *Steinernema vulcanicum* n. sp. (Rhabditida: Steinernematidae), a new entomopathogenic nematode species from Sicily (Italy). *Nematology* 13, 409-423.

DOUCET, M., BERTOLOTTIE, M.A. & DOUCET, M.E. (2003). Morphometric and molecular studies of isolates of *Steinernema rarum* (de Doucet, 1986) Mamiya, 1988 (Nematoda: Steinernematidae) from the province of Córdoba, Argentina. *Journal of Nematode Morphology and Systematics* 6, 27-36.

EDGINGTON, S., BUDDIE, A.G., TYMO, L., HUNT, D.J., NGUYEN, K.B., FRANCE, A.I., MERINO, L.M. & MOORE, D. (2009a). *Steinernema australe* n. sp. (Panagrolaimomorpha: Steinernematidae), a new entomopathogenic nematode from Isla Magdalena, Chile. *Nematology* 11, 699-717.

EDGINGTON, S., BUDDIE, A.G., TYMO, L., FRANCE, A., MERINO, L. & HUNT, D.J. (2009b). *Steinernema unicornum* sp. n. (Panagrolaimomorpha: Steinernematidae), a new entomopathogenic nematode from Chile. *Journal of Nematode Morphology and Systematics* 12, 113-131.

EMELIANOFF, V., LE BRUN, N., PAGÈS, S., STOCK, S.P., TAILLIEZ, P., MOULIA, C. & SICARD, M. (2008). Isolation and identification of entomopathogenic nematodes and their symbiotic bacteria from Hérault and Gard (southern France). *Journal of Invertebrate Pathology* 98, 211-217.

FAYYAZ, S., YAN, X., QIU, L., HAN, R., GULSHER, M., KHANUM, T.A. & JAVED, S. (2014). A new entomopathogenic nematode, *Steinernema bifurcatum* n. sp. (Rhabditida: Steinernematidae) from Punjab, Pakistan. *Nematology* 16, 821-836.

FAYYAZ, S., KHANUM, T.A., ALI, S., SOLANGI, G.S., GULSHER, M. & JAVED, S. (2015). *Steinernema balochiense* n. sp. (Rhabditida: Steinernematidae) a new entomopathogenic nematode from Pakistan. *Zootaxa* 3904, 387-402.

GORGADZE, O.A. (1988). [A new species of nematode *Neoaplectana thesami* sp. n. found in *Operophthera brumata* L.] *Sakartvelos SSR Mecnierebata Akademiis Moambe* 130, 405-408.

GORGADZE, O.A. & LORTKIPANIDZE, M.G. (2004). [The invasion of Colorado beetle (*Leptinotarsa decemlineata* Say) by using the nematode *Steinernema thesami* (Steinernematidae).] *Sacartvelos Mecnierebata Akademiis Moambe* 169, 592-594.

GORGADZE, O., LORTKHIPANIDZE, M., OGIER, J.-C., TAILLIEZ, P. & BURJANADZE, M. (2015). *Steinernema tbilisiensis* sp. n. (Nematoda: Steinernematidae) – a new species of entomopathogenic nematode from Georgia. *Journal of Agricultural Science and Technology A* 5, 264-276.[85]

GORGADZE O.A., IVANOVA, E.S., LORTKHIPANIDZE, M.G. & SPIRIDONOV, S.E. (2016). Redescription of *Steinernema thesami* Gorgadze, 1988 (Rhabditida: Steinernematidae) from Georgia. *Russian Journal of Nematology* 24, 17-28.

IVANOVA, E.S., SHEPELEVA, N.S. & SPIRIDONOV, S.E. (2013). Morphological and molecular characterisation of *Steinernema costaricense* Uribe-Lorío, Mora & Stock, 2007 (Panagrolaimorpha: Steinernematidae) isolate from Bush Augusta State Park, Missouri, USA. *Systematic Parasitology* 85, 219-234.

KANGA, F.N., TRINH, P.Q., WAEYENBERGE, L., SPIRIDONOV, S.E., HAUSER, S. & MOENS, M. (2012). Two new species of *Steinernema* Travassos, 1927 from the humid forest of southern Cameroon. *Russian Journal of Nematology* 20, 15-36.

KANGA, F.N., IVANOVA, E.S., SHEPELEVA, N.S. & SPIRIDONOV, S.E. (2014). Additional data on *Steinernema cameroonense* Ngo Kanga, Phap Quang Trinh, Wayenberge, Spiridonov, Hauser & Moens, 2012. *Russian Journal of Nematology* 22, 67-76.

KHATRI-CHHETRI, H.B., WAEYENBERGE, L., SPIRIDONOV, S.E., MANANDHAR, H.K. & MOENS, M. (2011a). *Steinernema lamjungense* n. sp.

[85] The species epithet is wrongly formed and is corrected to '*tbilisiense*' herein – see Chapter 2.

(Rhabditida: Steinernematidae), a new species of entomopathogenic nematode from Lamjung district, Nepal. *Nematology* 13, 589-605.

KHATRI-CHHETRI, H.B., WAEYENBERGE, L., SPIRIDONOV, S.E., MANADHAR, H.M. & MOENS, M. (2011b). Two new species of *Steinernema* Travassos, 1927 with short infective juveniles from Nepal. *Russian Journal of Nematology* 19, 53-74.

LEE, M.M., SICARD, M., SKEIE, M. & STOCK, S.P. (2009). *Steinernema boemarei* n. sp. (Nematoda: Steinernematidae), a new entomopathogenic nematode from southern France. *Systematic Parasitology* 72, 127-141.

LÓPEZ-NÚÑEZ, J.C., PLICHTA, K., GÓNGORA-BOTERO, C.E. & STOCK, S.P. (2008). A new entomopathogenic nematode, *Steinernema colombiense* n. sp. (Nematoda: Steinernematidae) from Colombia. *Nematology* 10, 561-574.

LUCERO, V.V. (2011). *Symbiont specificity of* Steinernema puntauvense *(Nematoda: Steinernematidae).* B.A. Thesis, University of Arizona, AZ, USA.

MA, J., CHEN, S., DE CLERCQ, P., HAN, R. & MOENS, M. (2012a). *Steinernema changbaiense* n. sp. (Rhabditida: Steinernematidae), a new species of entomopathogenic nematode from Northeast China. *Russian Journal of Nematology* 20, 97-112.

MA, J., CHEN, S., LI, X., HAN, R., KHATRI-CHHETRI, H.B., DE CLERCQ, P. & MOENS, M. (2012b). A new entomopathogenic nematode, *Steinernema tielingense* n. sp. (Rhabditida: Steinernematidae), from north China. *Nematology* 14, 321-338.

MA, J., CHEN, S., DE CLERCQ, P., WAEYENBERGE, L., HAN, R. & MOENS, M. (2012c). A new entomopathogenic nematode, *Steinernema xinbinense* n. sp. (Nematoda: Steinernematidae), from north China. *Nematology* 14, 723-739.

MALAN, A.P., KNOETZE, R. & TIEDT, L.R. (2015). *Steinernema jeffreyense* n. sp. (Rhabditida: Steinernematidae), a new entomopathogenic nematode from South Africa. *Journal of Helminthology* 90, 262-278.

MALAN, A.P., KNOETZE, R. & TIEDT, L.R. (2016). *Steinernema nguyeni* n. sp. (Rhabditida: Steinernematidae), a new entomopathogenic nematode from South Africa. *Nematology* 18, 571-590.

MANEESAKORN, P., GREWAL, P.S. & CHANDRAPATYA, A. (2010). *Steinernema minutum* sp. nov. (Rhabditida: Steinernematidae): a new entomopathogenic nematode from Thailand. *International Journal of Nematology* 20, 27-42.

MRÁČEK, Z., LIU, Q. & NGUYEN, K.B. (2009). *Steinernema xueshanense* n. sp. (Rhabditida, Steinernematidae), a new species of entomopathogenic nematode from the province of Yunnan, southeast Tibetan Mts., China. *Journal of Invertebrate Pathology* 102, 69-78.

MRÁČEK, Z., PŮŽA, V. & NERMUT', J. (2014). *Steinernema poinari* sp. n. (Nematoda: Steinernematidae) a new entomopathogenic nematode from the Czech Republic. *Zootaxa* 3760, 336-350.

MURFIN, K.E., LEE, M.-M., KLASSEN, J.L., MCDONALD, B.R., LARGET, B., FORST, S., STOCK, S.P., CURRIE, C.R. & GOODRICH-BLAIR, H. (2015). *Xenorhabdus bovienii* strain diversity impacts coevolution and symbiotic maintenance with *Steinernema* spp. nematode hosts. *mBio* 6(3), e00076-15. DOI:10.1128/mBio.00076-15.

NGUYEN, K.B. & BUSS, E.A. (2011). *Steinernema phyllophagae* n. sp. (Rhabditida: Steinernematidae), a new entomopathogenic nematode from Florida, USA. *Nematology* 13, 425-442.

NGUYEN, K.B. & HUNT, D.J. (Eds) (2007). *Entomopathogenic nematodes: taxonomy, phylogeny and bacterial symbionts. Nematology Monographs and Perspectives 5* (Series editors: Hunt, D.J. & Perry, R.N.). Leiden, The Netherlands, Brill.

NGUYEN, K.B., MALAN, A.P. & GOZEL, U. (2006). *Steinernema khoisanae* n. sp. (Rhabditida: Steinernematidae), a new entomopathogenic nematode from South Africa. *Nematology* 8, 157-175.

NGUYEN, K.B., HUNT, D.J. & MRÁČEK, Z. (2007a). Steinernematidae: species descriptions. In: Nguyen, K.B. & Hunt, D.J. (Eds). *Entomopathogenic nematodes: taxonomy, phylogeny and bacterial symbionts. Nematology Monographs and Perspectives 5* (Series editors: Hunt, D.J. & Perry, R.N.). Leiden, The Netherlands, Brill, pp. 121-609.

NGUYEN, K.B., STUART, R.J, ANDALO, V., GOZEL, U. & ROGER, M.E. (2007b). *Steinernema texanum* n. sp. (Rhabditida: Steinernematidae), a new entomopathogenic nematode from Texas, USA. *Nematology* 9, 379-396.

NGUYEN, K.B., PŮŽA, V. & MRÁČEK, Z. (2008). *Steinernema cholashanense* n. sp. (Rhabditida, Steinernematidae) a new species of entomopathogenic nematode from the province of Sichuan, Chola Shan Mountains, China. *Journal of Invertebrate Pathology* 97, 251-264.

NGUYEN, K.B., GINARTE, C.M.A., LEITE, L.G., SANTOS, J.M. & HARAKAVA, R. (2010). *Steinernema brazilense* n. sp. (Rhabditida: Steinernematidae), a new entomopathogenic nematode from Mato Grosso, Brazil. *Journal of Invertebrate Pathology* 103, 8-20.

NIKDEL, M., NIKNAM, G. & YE, W. (2011). *Steinernema arasbaranense* n. sp. (Nematoda: Steinernematidae), a new entomopathogenic nematode from Arasbaran forests, Iran. *Nematologia Mediterranea* 39, 17-28.

NTHENGA, I., KNOETZE, R., BERRY, S., TIEDT, L.R. & MALAN, A.P. (2014). *Steinernema sacchari* n. sp. (Rhabditida: Steinernematidae), a new entomopathogenic nematode from South Africa. *Nematology* 16, 475-494.

PHAN, K.P., MRÁČEK, Z., PŮŽA, V., NERMUT', J. & JAROŠOVÁ, A. (2014). *Steinernema huense* sp. n., a new entomopathogenic nematode (Nematoda: Steinernematidae) from Vietnam. *Nematology* 16, 761-775.

PHAN, L.K., SUBBOTIN, S.A., WAEYENBERGE, L. & MOENS, M. (2005). A new entomopathogenic nematode, *Steinernema robustispiculum* n. sp. (Rhabditida: Steinernematidae), from Chumomray National Park in Vietnam. *Systematic Parasitology* 60, 23-32.

PHAN, L.K., TAKEMOTO, S. & FUTAI, K. (2006). *Steinernema ashiuense* sp. n. (Nematoda: Steinernematidae), a new entomopathogenic nematode from Japan. *Nematology* 8, 681-690.

PŮŽA, V., NERMUT', J., MRÁČEK, Z., GENGLER, S. & HAUKELAND, S. (2016). *Steinernema pwaniensis* n. sp., a new entomopathogenic nematode (Nematoda: Steinernematidae) from Tanzania. *Journal of Helminthology*. DOI:10.1017/S0022149X15001157.

QIU, L., ZHAO, J., WU, Z., LV, Z. & PANG, Y. (2011). *Steinernema pui* sp. n. (Rhabditida, Steinernematidae), a new entomopathogenic nematode from Yunnan, China. *Zootaxa* 2767, 1-13.

SAN-BLAS, E., PORTILLO, E., NERMUT', J., PŮŽA, V. & MORALES-MONTERO, P. (2015). *Steinernema papillatum* n. sp. (Rhabditida: Steinernematidae), a new entomopathogenic nematode from Venezuela. *Nematology* 17, 1081-1097.

SAN-BLAS, E., MORALES-MONTERO, P., PORTILLO, E., NERMUT', J. & PŮŽA, V. (2016). *Steinernema goweni* n. sp. (Rhabditida: Steinernematidae), a new entomopathogenic nematode from Zulia State, Venezuela. *Zootaxa* 4067, 200-214.

SPIRIDONOV, S.E., REID, A.P., PODRUCKA, K., SUBBOTIN, S.A. & MOENS, M. (2004). Phylogenetic relationships within the genus *Steinernema* (Nematoda: Rhabditida) as inferred from analyses of sequences of the ITSI-5.8S-ITS2 region of rDNA and morphological features. *Nematology* 6, 547-566.

SPIRIDONOV, S.E., WAEYENBERGE, L. & MOENS, M. (2010). *Steinernema schliemanni* sp. n. (Rhabditida, Steinernematidae) – a new species of steinernematid of the '*monticolum*' group from Europe. *Russian Journal of Nematology* 18, 175-190.

STOKWE, N.F., MALAN, A.P., NGUYEN, K.B., KNOETZE, R. & TIEDT, L. (2011). *Steinernema citrae* n. sp. (Rhabditida: Steinernematidae), a new entomopathogenic nematode from South Africa. *Nematology* 13, 569-587.

TAILLIEZ, P., PAGES, S., GINIBRE, N. & BOEMARE, N. (2006). New insight into diversity in the genus *Xenorhabdus*, including the description of ten novel species. *International Journal of Systematic and Evolutionary Microbiology* 56, 2805-2818.

TAILLIEZ, P., PAGÈS, S., EDGINGTON, S., TYMO, L.M. & BUDDIE, A.G. (2012). Description of *Xenorhabdus magdalenensis* sp. nov., the symbiotic bacterium associated with *Steinernema australe*. *International Journal of Systematic and Evolutionary Microbiology* 62, 1761-1765. DOI:10.1099/ijs.0.034322-0.

TAMIRU, T., WAEYENBERGE, L., HAILU, T., EHLERS, R.-U., PŮŽA, V. & MRÁČEK, Z. (2012). *Steinernema ethiopiense* sp. n. (Rhabditida: Steinernematidae), a new entomopathogenic nematode from Ethiopia. *Nematology* 14, 741-757.

TARASCO, E., MRÁČEK, Z., NGUYEN, K.B. & TRIGGIANI, O. (2008). *Steinernema ichnusae* sp. n. (Nematoda: Steinernematidae), a new entomopathogenic nematode from Sardinia Island (Italy). *Journal of Invertebrate Pathology* 99, 173-185.

URIBE-LORÍO, L., MORA, M. & STOCK, S.P. (2007). *Steinernema costaricense* n. sp. and *S. puntauvense* n. sp. (Rhabditida: Steinernematidae), two new entomopathogenic nematodes from Costa Rica. *Systematic Parasitology* 68, 167-182.

YOSHIDA, M., REID, A.P., BRISCOE, B.R. & HOMINICK, W.M. (1998). Survey of entomopathogenic nematodes (Rhabditida: Steinernematidae and Heterorhabditidae) in Japan. *Fundamental and Applied Nematology* 21, 185-198.

Nematology Monographs & Perspectives, 2016, Vol. 12, 377-411

Chapter 5

Heterorhabditidae: species descriptions

Khuong B. NGUYEN [1] and David J. HUNT [2]

[1] *Entomology and Nematology Department, Institute of Food and Agricultural Sciences, University of Florida, Gainesville, FL 32611-0620, USA*

[2] *CABI Europe-UK, Bakeham Lane, Egham, Surrey, TW20 9TY, UK*

This chapter is an update of Chapter 5 by Nguyen & Hunt (2007a) in Nguyen & Hunt (2007b). Since then, several new species of *Heterorhabditis* have been proposed although two of these, *H. gerrardi* and *H. sonorensis*, are herein synonymised under *H. indica* and *H. taysearae*, respectively. In this chapter, descriptions and illustrations of valid species described since Nguyen & Hunt (2007a) to the end of 2015 are presented in a condensed format as stated in Chapter 4. Everything related to the order of description, the collection of morphological, morphometric data and illustrations mentioned in Nguyen & Hunt (2007a) is followed in this chapter.

As suggested by Nguyen *et al.* (2008) and Spiridonov & Subbotin (see Chapter 6), the valid species of *Heterorhabditis* are divided into three phylogenetic clades: *i*) the *bacteriophora*-clade, which includes *H. bacteriophora*, *H. georgiana* and *H. beicherriana*; *ii*) the *megidis*-clade, which includes *H. atacamensis*, *H. downesi*, *H. megidis*, *H. marelatus*, *H. safricana* and *H. zealandica*; and *iii*) the *indica*-clade, which includes *H. amazonensis*, *H. baujardi*, *H. floridensis*, *H. indica*, *H. mexicana*, *H. noenieputensis* and *H. taysearae*.

Notes concerning the species descriptions

For the reasons stated in Chapter 4, when presenting the species descriptions we omit, except for diagnostic characters, detailed descriptions of the morphological characteristics of the anterior regions of males, hermaphroditic females, amphimictic females and infective juveniles of *Heterorhabditis* species. However, for the sake of completeness, descriptions of the anterior region in *Heterorhabditis* are given below.

Male

Body curved ventrally when killed by gentle heat. Head truncate, sometimes slightly swollen. Face view with six labial papillae, cephalic papillae not observed. Amphidial aperture present or inconspicuous. Stoma with sclerotised rod- or barrel-shaped cheilorhabdions, other rhabdions not distinguishable, forming a funnel-shaped structure posterior to cheilorhabdions. Pharynx with cylindrical corpus, metacorpus not observed but posterior part of corpus sometimes slightly swollen. Nerve ring located surrounding isthmus just anterior to basal bulb. Basal bulb with reduced valve. Cardia present, protruding into intestine (Fig. 5.1A, B).

Hermaphroditic female

C-shaped after killing with gentle heat. Body robust, always with many eggs in young females and many eggs and juveniles in mature females. Cuticle smooth under light microscope but finely annulated with SEM. Lateral field and phasmid not observed. Head region tapering anteriorly. Labial region with six prominent lips, each with a terminal labial papilla. Cephalic papillae present or not clearly seen when observed with SEM. In face view, mouth hexagonal in shape. Amphidial aperture pore-like. Stoma with refractile cheilorhabdions. Posterior part of stoma funnel-shaped, enclosed by anterior part of pharynx. Pharynx with cylindrical corpus. Isthmus present. Nerve ring located at middle of isthmus. Basal bulb prominent with inconspicuous valve, but lumen of pharynx in basal bulb well sclerotised. Cardia present (Fig. 5.1C, D).

Note on appearance of cephalic papillae

In the above paragraph, the cephalic papillae of the hermaphroditic female are described as 'present' or 'not clearly seen'. The cephalic papillae are described as present or prominent when all ten of them are seen clearly (see Fig. 5.2A). Sometimes, a number of anterior projections are seen in face view, but they are not papilliform so we cannot say that they are papillae. In this case we use the term 'not clearly seen' (see Fig. 5.2B-D).

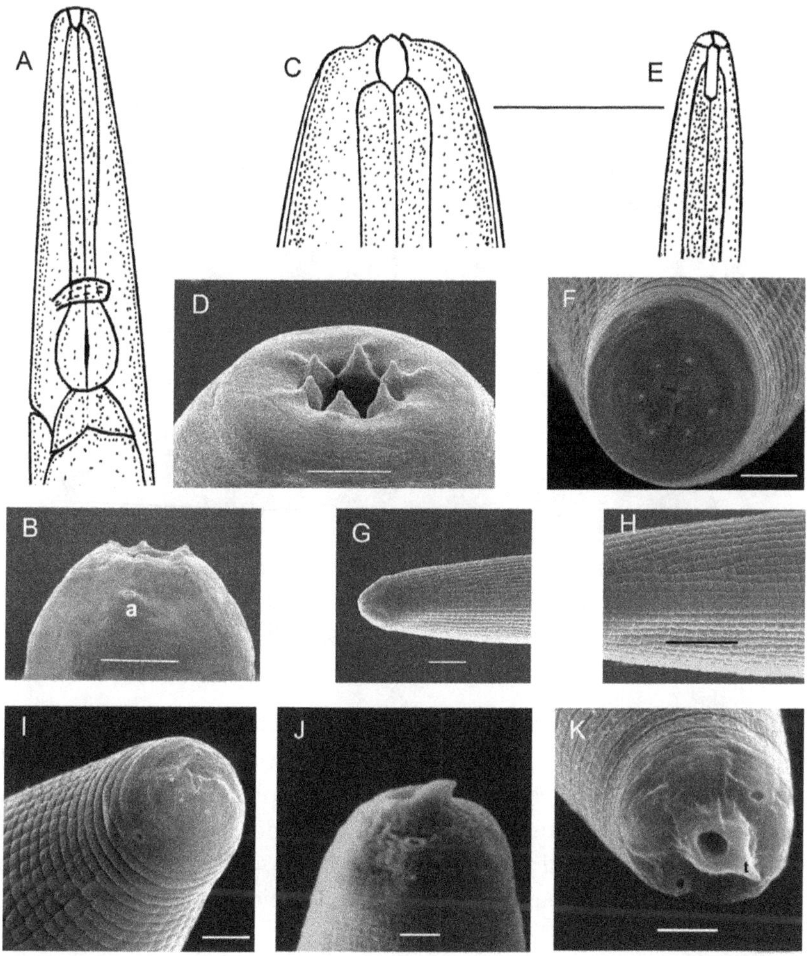

Fig. 5.1. *Anterior region of* Heterorhabditis. *A, B: Male; C, D: Female; E-K: Infective juvenile showing anterior part of body with head, labial papillae (F, I), tessellate pattern (G, H), dorsal tooth and amphids (J, K). (Scale bars: A, C (top bar) = 50 μm; B = 3.8 μm; D = 6.7 μm; E (top bar) = 43 μm; F = 2 μm; G = 8.6 μm; H = 4.3 μm; I = 2 μm; J = 3.8 μm; K = 2 μm.) Compiled by Khuong B. Nguyen.*

INFECTIVE JUVENILE (IJ)

Body elongate. Sheath (second-stage cuticle) present immediately after harvesting but many IJ losing sheath in storage. Anterior part (*ca* 3-5 body diam. from anterior end) of ensheathed IJ with tessellate

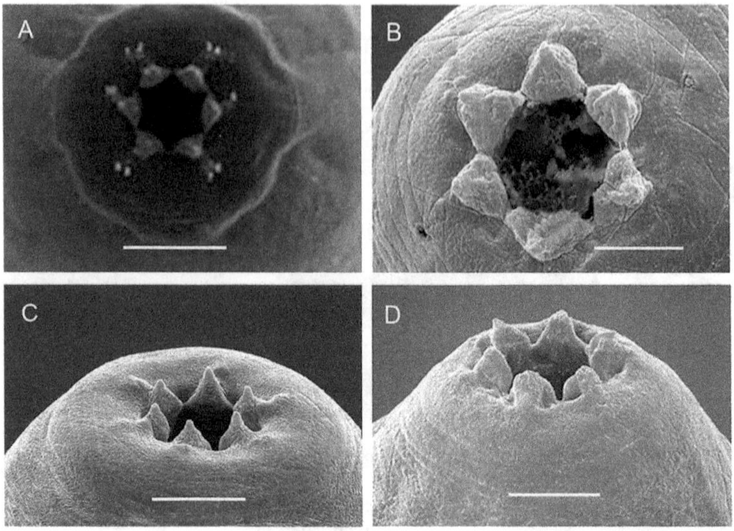

Fig. 5.2. *Cephalic papillae of* Heterorhabditis. *A: Prominent cephalic papillae; B-D: Cephalic papillae not clearly seen. (Scale bars: A = 10 μm; B = 3 μm; C = 6.7 μm; D = 6 μm.) Compiled by Khuong B. Nguyen.*

pattern, six labial papillae, mouth opening and amphids, cephalic papillae not observed. Posterior part with longitudinal ridges. Tail long, pointed in ensheathed IJ and bluntly pointed in exsheathed IJ. Exsheathed IJ body annulated, without longitudinal ridges. Labial region with dorsal tooth, labial and cephalic papillae absent. Amphidial aperture present or inconspicuous. Pharynx reduced, typical for *Heterorhabditis*. Cardia present (Fig. 5.1E-K).

ABBREVIATIONS

In addition to the standard de Man ratios, the following abbreviations are used throughout the text and tables:

D% = excretory pore/pharynx length × 100,
E% = excretory pore/tail length × 100,
SW% = spicule length/cloacal body diam. × 100,
GS% = gubernaculum length/spicule length × 100.

Species descriptions

Heterorhabditis atacamensis[3,4] Edgington, Buddie, Moore, France, Merino & Hunt, 2011

= *H.* cf. *safricana apud* Edgington *et al.* (2010)
(Figs 5.3, 5.4)

MEASUREMENTS

See Table 5.1.

DIAGNOSTIC DESCRIPTION

Male

Body J-shaped when heat-relaxed. Cuticle smooth under light microscope. Head low, truncate, sometimes slightly offset from body contour. No lateral fields observed. Excretory pore located posterior to basal bulb, anterior portion of excretory duct cuticularised, posterior section short, leading to excretory cell. Non-overlapping junction between pharynx and intestine, cardia present, extending into intestine. Testis monorchic, ventrally reflexed. Spicules paired, colourless. Manubrium shape somewhat square, offset from shaft, shaft short, blade almost straight, terminus of blade pointed, velum absent. Gubernaculum linear. Bursa peloderan, papillae totalling nine pairs in all specimens observed, comprising (anterior to posterior): pair 1 well anterior to cloacal aperture; pairs 2 and 3 forming a group anterior to cloacal aperture; pairs 4, 5 and 6 forming a group posterior to cloacal aperture; and pairs 7, 8 and 9 forming another group around the tail tip, although pair 7 somewhat separated from pairs 8 and 9 on 60% of specimens observed. All pairs curved ventrally apart from pairs 4 and 7 which are curved laterally. Pairs 8 and 9 shorter than pair 7, particularly pair 8 which is substantially reduced on 90% of specimens observed. No obvious post-cloacal swelling. Tail conoid, curved ventrally, slightly longer than cloacal body diam., with pointed tip.

[3] Specific epithet derived from the *Atacama* Desert from where the species was collected.

[4] Description after Edgington *et al.* (2011). Revised by David J. Hunt.

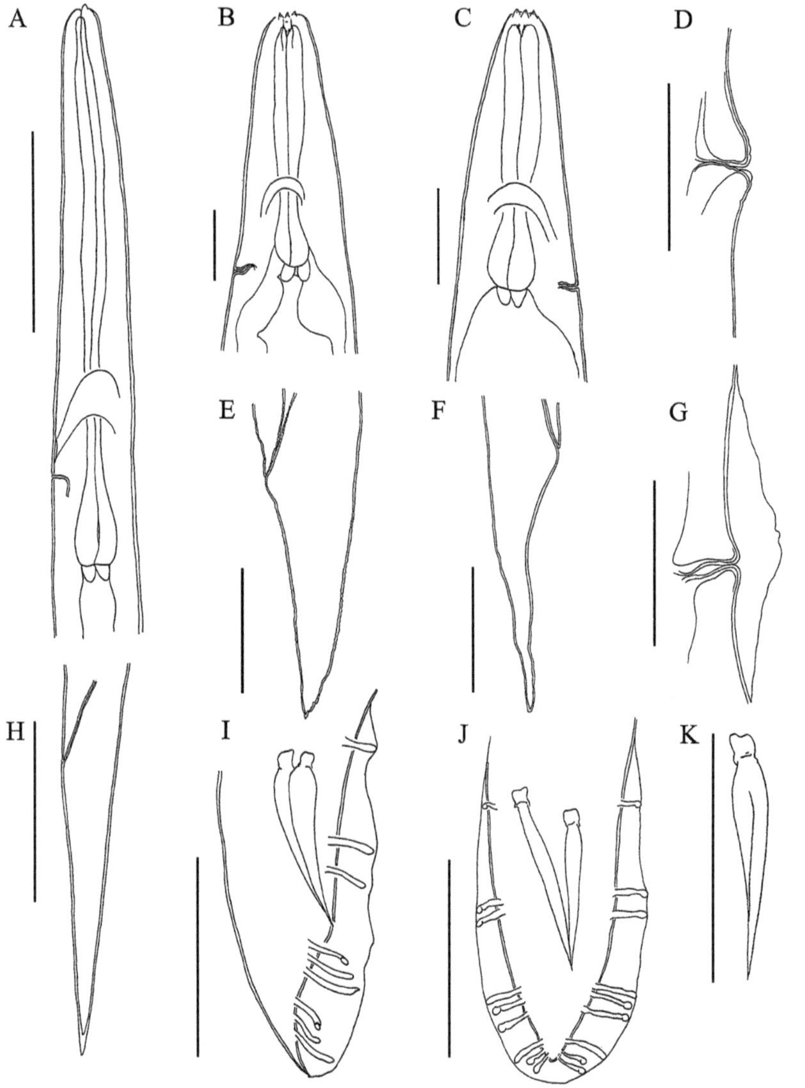

Fig. 5.3. Heterorhabditis atacamensis, *line drawings. A, H: Infective juvenile (exsheathed). A: Pharyngeal region; H: Tail region. B, D, E: Hermaphrodite. B: Pharyngeal region; D: Vulva; E: Tail region. C, F, G: Adult female. C: Pharyngeal region; F: Tail region; G: Vulva showing exudates. I-K: Adult male. I: Tail in lateral view showing bursa and arrangement of genital papillae; J: Tail in ventral view showing bursa and arrangement of genital papillae; K: Spicule. (Scale bars = 50 μm.) After Edgington et al. (2011), courtesy* Journal of Helminthology.

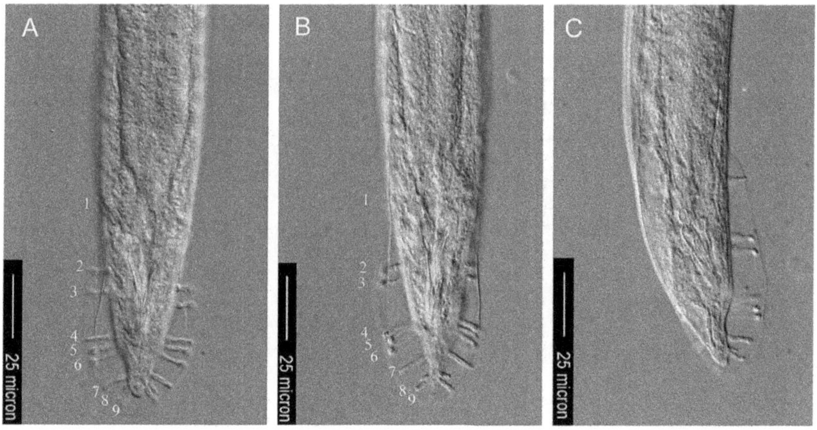

Fig. 5.4. Heterorhabditis atacamensis, *light micrographs of adult male. A-C: Tail region, showing bursa and position and orientation of nine pairs of genital papillae (numbered 1-9). A: Pair 7 grouped with pairs 8 and 9; B: Pair 7 somewhat separated from pairs 8 and 9; C: Lateral view. After Edgington et al. (2011), courtesy* Journal of Helminthology.

Hermaphroditic female

Body C-shaped when heat-relaxed, always with 15 eggs. Cuticle with slight annulation, more prominent at tail. Head low, relatively truncate, continuous with body contour. No lateral field observed. Excretory pore usually located close to posterior level of basal bulb. Non-overlapping junction between pharynx and intestine, cardia present. Genital system didelphic, amphidelphic. Vulva a transverse slit, located approximately mid-body. Vulval lips protruding 5 (3-8) μm above body contour, epiptygmata absent. Shallow folding of cuticle around vulva visible under light microscope. Vagina extending 16 (11-19) μm into body. Slight post-anal swelling present. Tail conoid, *ca* two to three anal body diam. long, with finely rounded terminus.

Amphimictic female

Similar to hermaphrodite in general morphology, but slightly shorter. Body C-shaped when heat-relaxed. Vulva protruding on 60% of specimens observed, 2 (2-6) μm above body contour, epiptygmata absent. Exudates (copulatory plug?) around vulva on 70% of specimens observed. No cuticular folds around vulva when viewed under light microscope. Slight post-anal swelling present. Tail conoid, *ca* three anal body diam.

Table 5.1. *Morphometrics of* Heterorhabditis atacamensis. *All measurements are in μm and in the form: mean ± s.d. (range).*

Character	Male		Hermaphrodite (first generation)	Female (second generation)	Infective juvenile
	Holotype	Paratypes	Paratypes	Paratypes	Paratypes
n	–	20	20	20	20
L	1025	918 ± 13.0 (842-1025)	2277 ± 54.5 (1791-2904)	2071 ± 58.6 (1754-2628)	611 ± 4.9 (578-666)
a	–	–	–	–	27.9 ± 0.4 (24.6-30.7)
b	–	–	–	–	5.3 ± 0.1 (4.8-5.7)
c	–	–	–	–	6.1 ± 0.1 (5.7-7.1)
V	–	–	45 ± 0.4 (39-48)	46 ± 0.4 (43-49)	–
Max. body diam.	52	48 ± 0.7 (42-55)	105 ± 2.8 (88-122)	105 ± 2.9 (86-129)	22 ± 0.4 (19-26)
Stoma length	7.1	8.1 ± 0.5 (4.5-12.2)	6.8 ± 0.2 (5.1-7.7)	12.1 ± 0.3 (9.6-13.5)	–
Stoma diam.	5.1	4.4 ± 0.1 (3.8-5.1)	4.8 ± 0.3 (3.2-7.9)	5.9 ± 0.3 (4.5-7.7)	–
Head to excretory pore	143	134 ± 2.0 (116-149)	184 ± 2.8 (165-206)	161 ± 1.8 (154-182)	115 ± 1.2 (101-126)
Head to nerve ring	80	82 ± 1.2 (69-93)	120 ± 1.5 (101-132)	103 ± 2.4 (79-119)	92 ± 1.3 (79-101)
Head to pharynx	107	112 ± 1.2 (99-119)	184 ± 1.5 (174-200)	150 ± 2.4 (129-167)	135 ± 1.1 (124-144)
Testis reflexion	138	120 ± 2.5 (109-135)	–	–	–
Tail length with sheath	–	–	–	–	101 ± 1.0 (94-107)
Tail length without sheath	35	29 ± 0.9 (24-36)	88 ± 2.2 (72-112)	94 ± 1.9 (80-108)	69 ± 0.9 (62-79)
Anal body diam.	23	22 ± 0.4 (19-27)	36 ± 0.9 (30-46)	29 ± 0.6 (24-33)	14 ± 0.2 (12-16)
Spicule length	44	45 ± 0.6 (40-49)	–	–	–
Spicule width	6.4	6.8 ± 0.2 (4.7-7.9)	–	–	–
Gubernaculum length	23	20 ± 0.5 (17-22)	–	–	–
D%	134	117 ± 1.2 (108-126)	100 ± 1.4 (90-114)	108 ± 0.9 (100-113)	85 ± 1.0 (79-94)
E%	–	–	–	–	165 ± 2.2 (149-182)
SW%	191	205 ± 4.7 (179-249)	–	–	–
GS%	52	44 ± 0.9 (38-51)	–	–	–

long, tapering to a finely rounded terminus, occasional regular duplex termination observed.

Infective juvenile

Body slender, almost straight when heat-relaxed, tapering gently towards anterior and posterior ends. Cuticle finely annulated with delicate transverse striations. Excretory pore prominent, situated between nerve ring and anterior end of basal bulb, excretory duct cuticularised. Deirid sometimes visible under light microscope, located between nerve ring and anterior end of basal bulb. Nerve ring surrounding isthmus about mid-way between posterior of metacorpus and anterior of basal bulb. Junction between pharynx and intestine non-overlapping, cardia prominent extending into intestine. Hemizonid present, lenticular in shape, located at anterior end of basal bulb, usually three to four annules anterior to excretory pore. Hemizonion not seen. Lateral field starting as two lines around nerve ring level, forming four lines (two ridges) at about basal bulb level, with both ridges separated by 2-3 μm, continuing as such until approximately mid-tail level. Rod-shaped bacteria seen throughout anterior portion of intestine, observations further posterior obscured by body contents. Tail conoid, *ca* five anal body diam. long, with finely rounded tip. Length of exsheathed tail *ca* 70% of ensheathed tail.

TYPE HABITAT AND LOCALITY

The natural host is unknown as the species was isolated by *Galleria* baiting a soil sample from the Atacama Desert, Chile (23811016.700S, 67859031.600W). The sample was taken from a sandy bank, shaded by trees, at 2499 m a.s.l. (Edgington *et al.*, 2011).

TYPE MATERIAL

Holotype male isolated from haemocoel of *Galleria mellonella*, deposited in the CABI Nematode Collection, Europe-UK Centre, Egham, Surrey, UK. Five paratype males and hermaphroditic adults and ten paratype IJ (permanent slide mounts), together with 20 paratypes of each stage in TAF, deposited in the nematode collections of CABI, UK and INIA – Quilamapu, Chillán, Chile. Twenty paratypes of each stage in TAF, deposited in the United States Department of Agriculture Nematode Collection (USDANC), Beltsville, MD, USA. In addition, numerous paratype IJ and other stages preserved in 70% ethyl alcohol deposited in the CABI Nematode Collection, UK.

Relationships

Heterorhabditis atacamensis can be recognised by morphometrics of the IJ, with body length of 611 (578-666) μm, EP = 115 (101-126) μm, tail without sheath = 69 (62-79) μm long, and ratio a = 28 (25-31). The male is characterised by the length of spicules and gubernaculum, at 45 (40-49) μm and 20 (17-22) μm, respectively, and the number and position of the genital papillae. The hermaphroditic female has a slight postanal swelling and a conoid tail 88 (72-112) μm long with a slightly rounded terminus. According to Edgington *et al.* (2011), and based on the morphology of IJ, males and hermaphroditic females, *H. atacamensis* is most similar to *H. marelatus*, *H. downesi*, *H. amazonensis*, *H. bacteriophora*, *H. georgiana* and *H. safricana*. However, *H. atacamensis* can be separated from these species by a number of morphometric features and by the following features: the genital papillae of males of *H. atacamensis* consist of nine pairs with pairs 4 and 7 curved laterally, *vs* 7-8 pairs in *H. amazonensis* with only pair 4 curved laterally; the excretory pore of males of *H. atacamensis* is located posterior to the basal bulb, *vs* anterior in *H. downesi*. There is equal protrusion of anterior and posterior vulval lips of hermaphroditic females of *H. atacamensis*, while the anterior lip protrudes more than the posterior lip in *H. downesi*. More information on relationships is available in Edgington *et al.* (2011).

Molecular characterisation

The ITS rDNA regions (accession no. HM230723) are characterised by sequence length (730 bp). The D2-D3 expansion regions (accession no. HM230724) flanked by the primer D2A and D3B are characterised by sequence length (592 bp) with composition A = 25%, C = 19%, G = 31% and T = 25%. More information on relationships and phylogenetic analysis is available in Edgington *et al.* (2011).

Symbiotic bacterium

No information.

BIONOMICS

No progeny were produced when *H. atacamensis* and *H. safricana* were crossed. No progeny were produced from virgin females in the absence of males. When both species were self-crossed, progeny were produced (Edgington *et al.*, 2011).

DISTRIBUTION

Known only from the Atacama Desert, Chile.

Heterorhabditis beicherriana[5,6] Li, Liu, Nermut', Půža & Mráček, 2012

(Fig. 5.5)

MEASUREMENTS

See Table 5.2.

DIAGNOSTIC DESCRIPTION

Male

Anterior region typical for *Heterorhabditis* male. Excretory pore usually near pharynx base or slightly posterior. Testis single, reflexed anteriorly. *Vas deferens* well developed. Spicules paired, separate, without rostrum, manubrium slightly longer than wide, distinctly separated from lamina. Lamina of spicule slightly curved ventrally. Gubernaculum narrow, slightly curved ventrally and anteriorly, about half size of spicules in length. Tail conoid, with peloderan bursa, slightly curved ventrally. Nine pairs of genital papillae (bursal rays) with typical *Heterorhabditis* formula = 1, 2, 3, 3 (counting from anterior end of bursa). Pairs 1, 2 and 3 anterior to cloaca; pairs 4 and 7 curved outward; pairs 7-9 forming a terminal group at end of bursa. No variation in bursal formula observed.

Hermaphroditic female

Anterior region typical for *Heterorhabditis* hermaphroditic female. Excretory pore located posterior to nerve ring, mostly at level of basal

[5] The name *beicherriana* is derived from *Beijing* and the *cherry* orchard from where the nematode species was collected.

[6] Description after Li *et al.* (2012). Revised by Khuong B. Nguyen.

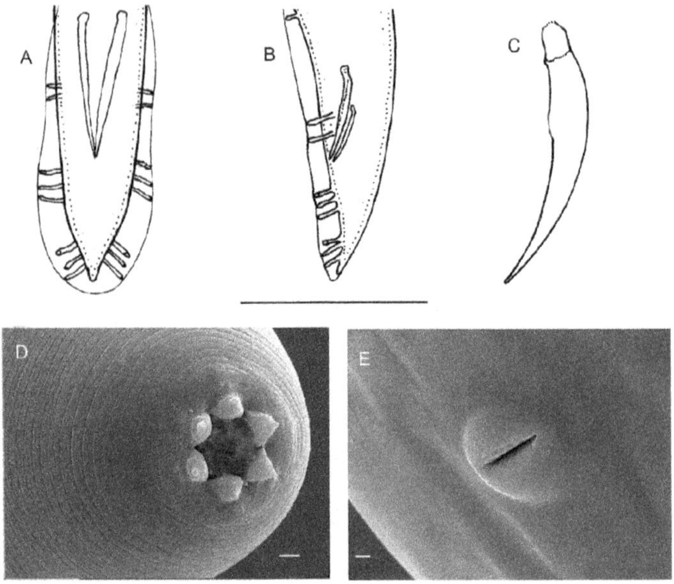

Fig. 5.5. Heterorhabditis beicherriana. *A, B: Male tail, ventral and lateral views showing terminal group with three pairs of papillae; C: Spicule; D: Face view of hermaphroditic female showing six labial papillae and inconspicuous cephalic papillae; E: Vulvar pattern showing its almost smooth character. (Scale bars: A, B = 90 μm; C = 30 μm; D, E = 2 μm.) Modified after Li* et al. *(2012), with acknowledgement to* Zootaxa.

bulb. Gonad didelphic, amphidelphic. Vulva in form of a transverse slit, located slightly anterior to mid-body region. Vulval lips smooth, slightly to moderately protruding (averaging 27 μm), symmetric. Vulvar pattern smooth, rounded or slightly elliptical. Mucus plug usually absent. Vagina short. Rectum *ca* 80-98 μm long. Tail conoid with a moderately prominent post-anal swelling, slightly longer than anal body diam. Phasmids inconspicuous.

Amphimictic female

Similar to hermaphroditic female, but smaller. Excretory pore located near pharynx base or slightly posterior. Indistinct vulva in form of a transverse slit, located slightly anterior to mid-body. Copulatory mucus plug sometimes observed. Vulval lips non-protruding. Vagina short. Rectum 59-78 μm long. Tail conical without or with a slightly developed post-anal swelling, twice as long as anal body diam. Phasmids inconspicuous.

Table 5.2. *Morphometrics of* Heterorhabditis beicherriana. *All measurements are in μm and in the form: mean ± s.d. (range).*

Character	Male		Hermaphrodite	Female	Infective juvenile
	Holotype	Paratypes	Paratypes	Paratypes	Paratypes
n	–	20	20	20	20
L	1040	1028 ± 79 (889-1192)	4617 ± 543 (3671-5543)	2250 ± 444 (1581-3026)	639 ± 38 (566-687)
a	14	18 ± 1.8 (15-23)	17 ± 2.0 (13-20)	14 ± 2.0 (10.3-18.1)	27 ± 1.2 (24-29)
b	79	8.2 ± 6 (7.2-10.0)	19 ± 3.3 (13.2-25.0)	15 ± 2.9 (10-23)	4.7 ± 017 (4.2-4.9)
c	33	27 ± 2.6 (22-34)	46 ± 7.4 (34-63)	27 ± 4.7 (19-34)	6.4 ± 0.26 (5.9-6.8)
V	–	–	45 ± 2.1 (41-49)	47 ± 2.4 (41-49)	–
Max. body diam.	73	58 ± 4.9 (51-73)	279 ± 47 (198-374)	165 ± 25 (125-218)	24 ± 1 (21-25)
Excretory pore	157	140 ± 7.1 (130-157)	206 ± 35 (165-297)	130 ± 18 (95-165)	113 ± 6.3 (100-122)
Nerve ring	92	95 ± 5.8 (81-108)	169 ± 26 (135-243)	106 ± 21 (59-138)	95 ± 5.4 (85-106)
Pharynx	132	126 ± 6.4 (116-143)	243 ± 43 (192-343)	142 ± 20 (105-186)	137 ± 8.0 (118-146)
Testis reflexion	111	107 ± 7 (90-117)	–	–	–
Tail length	32	38 ± 3.1 (32-45)	103 ± 16 (68-130)	85 ± 10 (68-105)	101 ± 6.2 (86-111)
Tail length without sheath	–	–	–	–	67 ± 5.1 (52-75)
Anal body diam.	24	25 ± 2.9 (19-39)	65 ± 12 (51-92)	44 ± 9.9 (35-81)	15 ± 1.0 (13-17)
Spicule length	51	44 ± 2.2 (40-49)	–	–	–
Gubernaculum length	24	24 ± 1.5 (22-27)	–	–	–
D%	118	112 ± 4.2 (102-120)	85 ± 6 (76-94)	96 ± 3 (88-98)	83 ± 3.0 (80-93)
E%	–	–	–	–	112 ± 6 (103-121)
SW%	211	180 ± 18 (153-208)	–	–	–
GS%	47	54 ± 3.1 (48-59)	–	–	–

Infective juvenile

Anterior region typical for *Heterorhabditis* IJ. Hemizonid located near posterior part of isthmus. Excretory pore located close to base of pharynx. Anterior part of intestine swollen, bacterial cells present throughout intestine. Phasmids inconspicuous. Tail long, conoid, with pointed terminus. Hyaline region absent, hypodermis extending to tail tip.

TYPE HABITAT AND LOCALITY

Natural host unknown. This species was collected from a cherry orchard near Beijing, China.

TYPE MATERIAL

Holotype male, paratype third-stage juveniles, males, females and hermaphrodites deposited in the EPN collection at the Laboratory of Entomopathogenic Nematodes, Institute of Entomology, Biology Centre, CAS, České Budejovice, Czech Republic; paratype third-stage juveniles and adults also in the nematode collection at the China Agricultural University, Beijing, China.

RELATIONSHIPS

Heterorhabditis beicherriana is characterised by a combination of morphological and morphometric characters of the male, female and IJ. This species is similar to *H. bacteriophora* and *H. georgiana*. For the male, spicule length, gubernaculum length, EP and the ratios E% and SW% can be used to differentiate between these three related species. For the hermaphroditic female, the vulvar pattern is the only character that can be used to distinguish between the three species. For the IJ, body length, EP, ES and tail can be used to recognise the three species. More information on relationships is available in Li *et al.* (2012).

MOLECULAR CHARACTERISATION

The ITS rDNA regions (accession no. HQ896630), flanked by primers 18S and 26S, of *H. beicherriana* are characterised by the sequence lengths (994 base pairs, ITS1 = 389 bp and ITS2 = 230 bp). The D2-D3 expansion regions (accession no. HQ896631), flanked by the primer

D2F and 536, are characterised by the sequence length (890 bp). More information on molecular characterisation and phylogenetic analysis is available in Li *et al.* (2012).

SYMBIOTIC BACTERIUM

No information.

BIONOMICS

Heterorhabditis beicherriana reproduced well on *G. mellonella* larvae in an incubator at 20 and 25°C. At 20°C nematode-infected *G. mellonella* larvae died 4 days after exposure to nematodes. Hermaphrodites appeared 2 days later, second generation adults filled the cadavers after 2 more days and the first IJ emerged 3 days later (10 days after exposure). At 25°C the nematodes developed as stated above but more rapidly. The IJ emerged from the cadavers after 8 days.

DISTRIBUTION

Only known from the type locality in China.

Heterorhabditis georgiana[7,8] Nguyen, Shapiro-Ilan & Mbata, 2008 (Figs 5.6-5.9)

MEASUREMENTS

See Table 5.3.

DIAGNOSTIC DESCRIPTION

Male

Anterior region typical for *Heterorhabditis* male. Excretory pore usually posterior to basal bulb. Testis monorchic, reflexed, distance from end of basal bulb to testis flexure variable. *Vas deferens* well developed. Spicules paired, separate, rostrum absent, ventral side flattened. Gubernaculum straight or slightly curved ventrally. Bursa peloderan. Nine

[7] The specific epithet is derived from the state of *Georgia*, USA, from where the nematode was collected.

[8] Description after Nguyen *et al.* (2008). Revised by Khuong B. Nguyen.

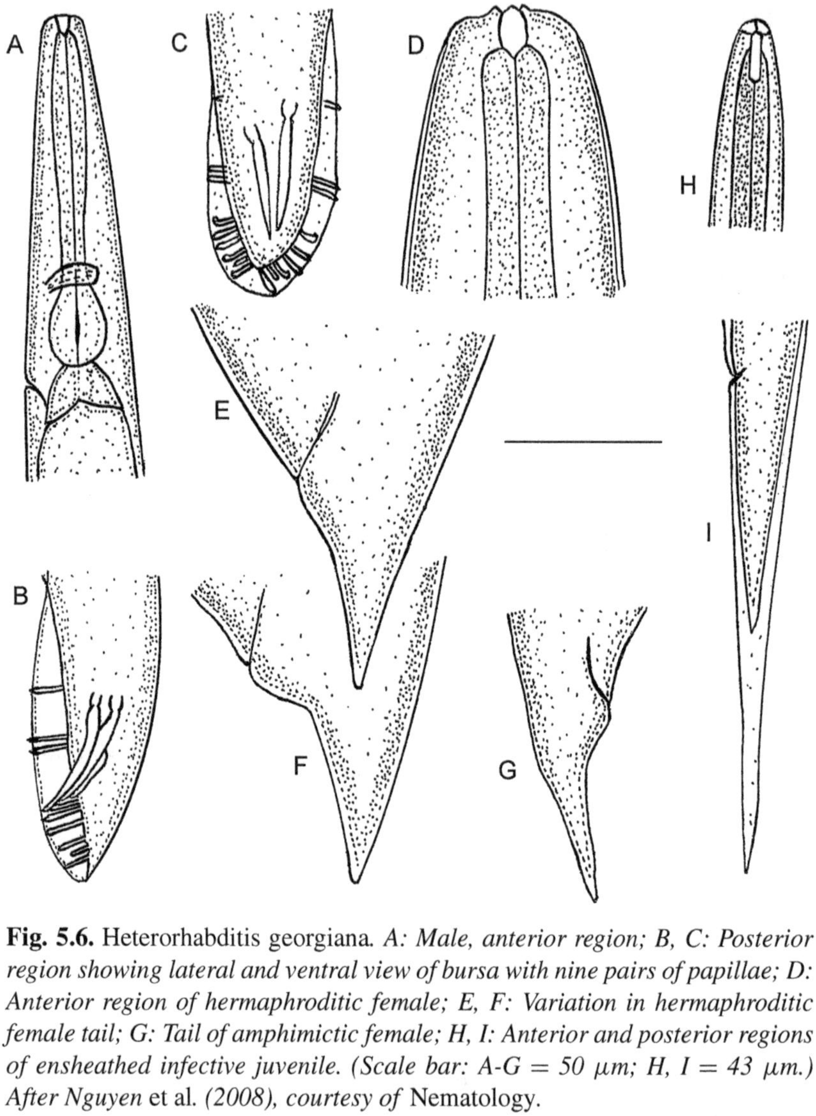

Fig. 5.6. Heterorhabditis georgiana. *A: Male, anterior region; B, C: Posterior region showing lateral and ventral view of bursa with nine pairs of papillae; D: Anterior region of hermaphroditic female; E, F: Variation in hermaphroditic female tail; G: Tail of amphimictic female; H, I: Anterior and posterior regions of ensheathed infective juvenile. (Scale bar: A-G = 50 μm; H, I = 43 μm.) After Nguyen* et al. *(2008), courtesy of* Nematology.

pairs of bursal papillae present or nine papillae on one side and eight on the other. Moving from anterior to posterior: pair 1 well anterior with papillal tips extending beyond bursal rim; pairs 2 and 3 in a group anterior to cloacal aperture, also extending beyond bursal rim; pairs 4, 5 and 6 forming a group posterior to cloacal aperture with pair 4 usually curved outward (laterally). Genital papillae 1-6 (from anterior end) typical for

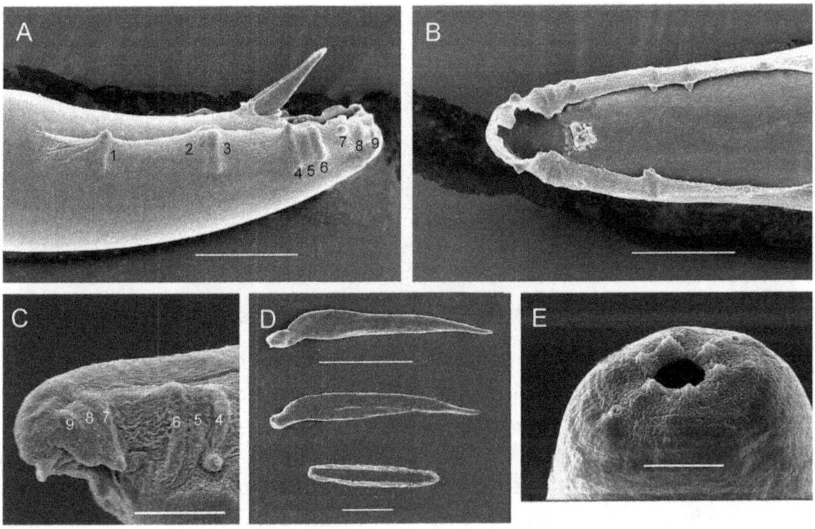

Fig. 5.7. Heterorhabditis georgiana. *A-C: SEM photographs of posterior region of male (numbers refer to bursal papillae). A: Bursa, lateral view showing nine pairs of genital papillae; B: Bursa, ventral view showing abnormal distribution of papillae 2 and 3: on one side, they are well separated whereas on the other side they are almost fused; C: Enlargement of posterior region showing papillae from 4 to 9. D: Two spicules and gubernaculum (below); E: Male head showing six labial papillae, mouth and two amphidial apertures. (Scale bars: A, B = 15 μm; C = 6 μm; D: spicules = 17 μm, gubernaculum = 8.6 μm; E = 3.8 μm.) After Nguyen* et al. *(2008), courtesy of* Nematology.

Heterorhabditis spp.; pairs 7, 8 and 9 forming a terminal group of papillae at end of bursa; pair 7 curved outward, sometimes with two branches (10%), one curved outward and other straight but never reaching bursa rim. Number and distribution of papillae in terminal group variable. Observation of papillae in terminal group of bursa (n = 10) gave the following results: 60% with three pairs of papillae and 40% with three papillae on one side and two on other. In one male, papilla number 4 was not observed on one side of bursa. Tail conoid, slightly curved ventrally.

Hermaphroditic female

Anterior region typical for *Heterorhabditis* hermaphroditic female. Gonads didelphic, amphidelphic. Vulva in form of a transverse slit, located on a slightly protruding area, anterior to mid-body (V = 42-47). In ventral view, vulva top smooth, elliptical, forming a pattern

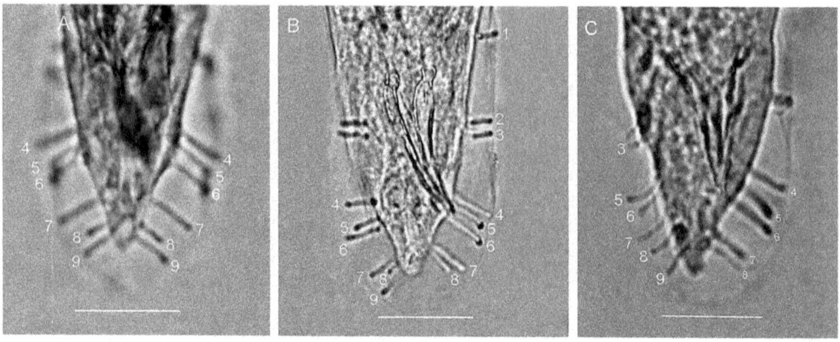

Fig. 5.8. Heterorhabditis georgiana. *Male tail with three type of bursal papillae distribution (numbers refer to bursal papillae). A: Bursa with three pairs (7, 8 and 9) of bursal papillae in terminal group; B: Bursa with three bursal papillae on one side and two papillae on other side in terminal group; C: Bursa with three bursal papillae on one side and two papillae on other side in terminal group (as in B, but papilla 4 is missing). (Scale bars: A-C = 26 μm.) After Nguyen* et al. *(2008), courtesy of* Nematology.

with prominent annules around. Vulval opening dividing pattern into two parts, anterior part smaller than posterior part. Pattern present on all hermaphroditic females and different from those of similar species. Vagina short. Tail longer than anal body diam., conoid with pointed terminus. Tail sometimes widening near tip then narrowing to a pointed terminus. Post-anal swelling present or absent. Phasmid inconspicuous.

Amphimictic female

Amphimictic female similar to hermaphroditic female but smaller. Vulva rarely protruding. Exudates or copulation plug after mating not observed. Post-anal swelling present.

Infective juvenile

Anterior region typical for *Heterorhabditis* infective juvenile. Excretory pore posterior to nerve ring but anterior to base of pharynx. Hemizonid, when observed, located two or three annules anterior to excretory pore. Pharynx reduced, typical for *Heterorhabditis*. Cardia present. Bacterial pouch not seen but bacterial cells visible posterior to cardia. Lateral field with two ridges. Phasmid inconspicuous. Tail elongate conoid with pointed terminus. Exsheathed tail length *ca* 61% of tail length with sheath.

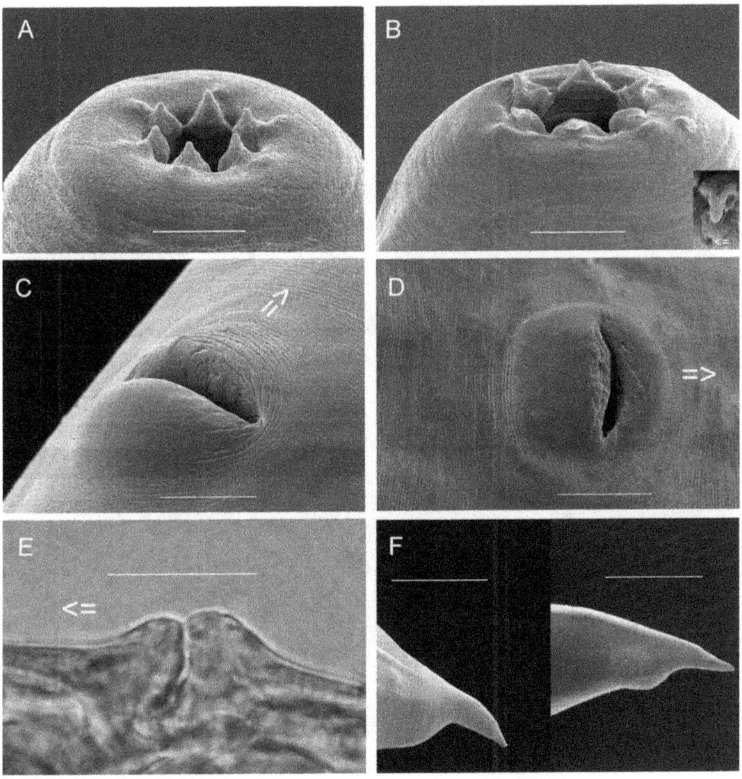

Fig. 5.9. Heterorhabditis georgiana. *Hermaphroditic female. A: Anterior region of young female showing anteriorly directed lips and papillae; B: Anterior region of adult female showing outwardly curved lips and an amphidial aperture (inset at higher magnification); C, D: Different views of vulval pattern (arrows indicate anterior); E: Lateral view of vulva (LM) (arrow indicates anterior); F: Tail of a hermaphroditic female (left) and an amphimictic female (right). (Scale bars: A-C = 6.7 μm; D = 12 μm; E = 30 μm; F = 42 μm (left), 28 μm (right).) After Nguyen* et al. *(2008), courtesy of* Nematology.

Type habitat and locality

Natural host unknown. The nematodes were collected by *Galleria* baiting from soil collected around a pecan tree in Byron, GA, USA.

Type material

Holotype male, paratypes (two hermaphrodites, two females, two males and five IJ) on slides and other paratypes in a vial (hermaphrodites,

Table 5.3. *Morphometrics of* Heterorhabditis georgiana. *All measurements are in* μm *and in the form: mean* $\pm$ *s.d. (range).*

Character	Male		Hermaphrodite	Female	Infective juvenile
	Holotype	Paratypes	Paratypes	Paratypes	Paratypes
n	–	20	20	20	25
L	860	838 ± 48 (721-913)	4090 ± 484 (3232-4928)	(2098 ± 294) (1640-2779)	598 ± 27 (547-651)
a	–	–	–	–	27 ± 2.6 (23-34)
b	–	–	–	–	4.7 ± 0.3 (4.1-5.3)
c	–	–	–	–	6.1 ± 0.4 (5.5-6.9)
V	–	–	47 ± 2.4 (44-55)	50 ± 2.2 (46-53)	–
Max. body diam.	52	49 ± 3.3 (43-55)	225 ± 31 (157-267)	133 ± 25 (101-188)	22 ± 2 (17-26)
Excretory pore	129	120 ± 12.0 (101-145)	228 ± 24 (200-277)	142 ± 14 (111-177)	104 ± 4 (97-113)
Nerve ring	91	82 ± 5.6 (72-93)	183 ± 25 (143-217)	105 ± 27 (96-152)	85 ± 5 (74-94)
Pharynx	112	109 ± 6.0 (100-122)	225 ± 32 (132-271)	150 ± 18 (136-219)	127 ± 7 (110-139)
Testis reflexion	100	101 ± 17.0 (75-120)	–	–	–
Tail length	33	33 ± 3.6 (29-41)	83 ± 9.0 (65-96)	75 ± 6.0 (62-88)	98 ± 5.0 (86-108)
Tail length without sheath	–	–	–	–	61 ± 3.4 (57-72)
Anal body diam.	23	26 ± 1.7 (23-28)	–	33 ± 4.5* –	15 ± 1.5 (13-17)
Spicule length	43	44 ± 2.0 (41-49)	–	–	–
Spicule width	–	–	–	–	–
Gubernaculum length	22	25 ± 2.7 (20-28)	–	–	–
D%	115	110 ± 6.0 (100-122)	–	–	82 ± 6.0 (70-93)
E%	–	–	–	–	106** –
SW%	186	172 ± 14.0 (150-200)	–	–	–
GS%	51	56 ± 5.6 (51-64)	–	–	–

* No range cited in original description.
** Calculated value as E% not cited in original publication.

females, male and infective juveniles in 3% formalin) isolated from haemocoel of *G. mellonella* deposited in the United States Department of Agriculture Nematode Collection (USDANC), Beltsville, MD, USA. One male and one female paratype of the second generation and several paratype IJ deposited in the California Collection of Nematodes, University of California Davis Nematode Collection, Davis, CA, USA. Several hermaphrodites, females, male and all stages in 3% formalin deposited in the Entomology and Nematology Department, University of Florida, Gainesville, FL, USA.

RELATIONSHIPS

Heterorhabditis georgiana is characterised by a combination of morphological and morphometric characters of the male, female and IJ. For males, the spicule length of 44 μm, the length of the spicule relative to anal body diam. (SW% = 172) and the percentage of the gubernaculum to spicule length (GS% = 56) are differentiating characteristics. Also, the number of papillae in the terminal group of the bursa typically has 60% with three pairs of papillae and 40% with three papillae on one side and two on the other. The female possesses an elliptical vulva pattern in ventral view with one lip smooth and larger than the other. The IJ has a body length of 598 (547-651) μm (similar to that of *H. bacteriophora* and *H. mexicana*), excretory pore = 104 (97-113) μm from anterior end, pharynx length = 127 (110-139) μm, tail length = 98 (86-108) μm, and a = 27 (23-34). More information on the relationships is available in Nguyen *et al.* (2008).

MOLECULAR CHARACTERISATION

The ITS rDNA regions (accession no. EU099032), flanked by primers 18S and 26S, of *H. georgiana* are characterised by the sequence lengths (1021 base pairs (bp), ITS1 = 389 bp, ITS2 = 228 bp). The D2-D3 expansion regions (accession no. EU099033), flanked by the primer D2F and 536, are characterised by the sequence length (890 bp). More information on molecular characterisation and phylogenetic analysis is available in Nguyen *et al.* (2008).

SYMBIOTIC BACTERIUM

No information.

BIONOMICS

No information.

DISTRIBUTION

Only known from the type locality of Byron, GA, USA.

Heterorhabditis noenieputensis[9,10] Malan, Knoetze & Tiedt, 2014
(Fig. 5.10)

MEASUREMENTS

See Table 5.4.

DIAGNOSTIC DESCRIPTION

Male

Anterior region typical for *Heterorhabditis* male. Excretory pore usually posterior to basal bulb. Testis monorchic, reflexed anteriorly. *Vas deferens* well developed. Spicules paired, rostrum absent, lamina of spicules slightly curved ventrally. Gubernaculum slightly curved ventrally, *ca* half spicule length in size. Tail conoid slightly curved ventrally, with peloderan bursa. Seven or eight pairs of bursal papillae (bursal rays) present on bursa. From anterior to posterior, pair 1 well anterior from cloacal aperture, with papilla tips extending to bursal rim; pairs 2 and 3 in a group anterior to cloacal aperture, also extending beyond bursal rim; pairs 4, 5 and 6 forming a group posterior to cloacal aperture, extending beyond bursal rim, with pair 4 directed outwards (laterally). Number of genital papillae from pairs 1-7 (from anterior end) typical for *Heterorhabditis* species. Observation of papillae of terminal group in bursa (n = 10) gave the following results: eight with both pairs 8 and 9 absent; two with pair 8 present but one with one papilla of pair 8 missing.

Hermaphroditic female

Anterior region typical for *Heterorhabditis* hermaphroditic female except *en face* view with an outer ring of ten cephalic papillae

[9] The species was named after the settlement of *Noenieput*, South Africa.
[10] Description after Malan *et al.* (2014). Revised by Khuong B. Nguyen.

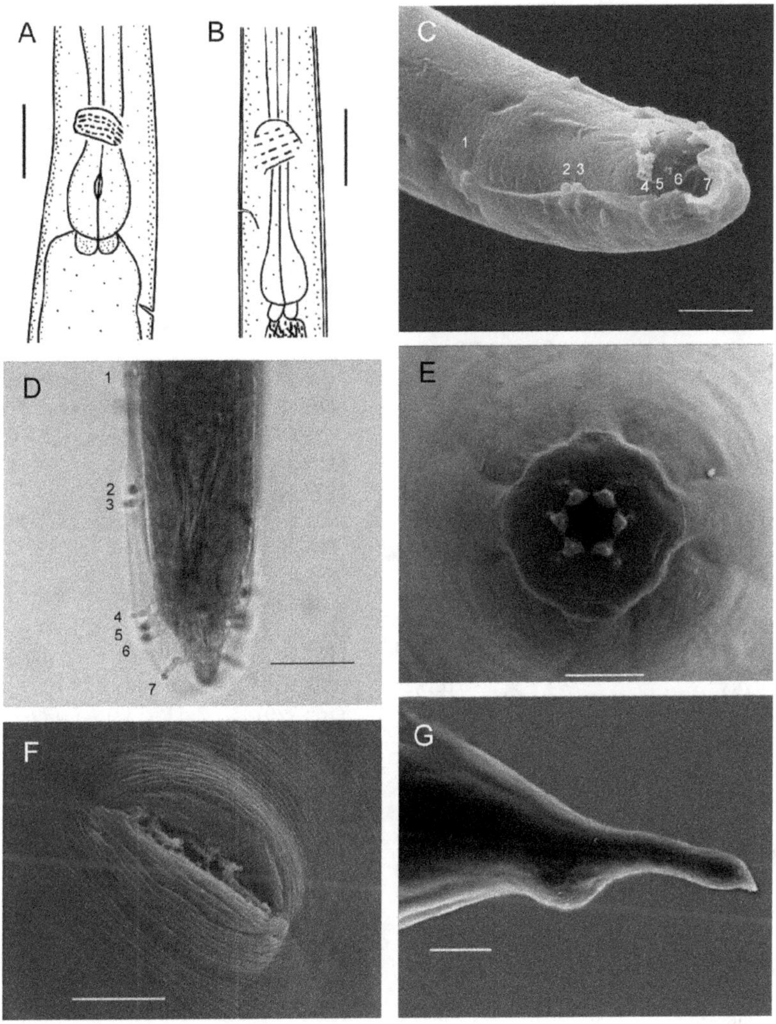

Fig. 5.10. Heterorhabditis noenieputensis. *A: Male anterior region showing position of excretory pore; B: Infective juvenile showing position of excretory pore; C: SEM of male tail, subventral view, showing bursa and genital papillae (numbered 1-7); D: Male tail showing bursa and terminal group of genital papillae comprising only a single pair (papillae numbered 1-7); E: Hermaphroditic female, en face view; F: Vulval region of amphimictic female showing elliptical pattern; G: Hermaphroditic female tail showing post-anal swelling and bulbous structure anterior to tail tip. (Scale bars: A, B, D = 20 μm; C, E, F = 10 μm; G = 20 μm.) Modified after Malan* et al. *(2014), with acknowledgement to* Journal of Helminthology.

Table 5.4. *Morphometrics of* Heterorhabditis noenieputensis. *All measurements are in μm and in the form: mean ± s.d. (range).*

Character	Male		Hermaphrodite	Female	Infective juvenile
	Holotype	Paratypes	Paratypes	Paratypes	Paratypes
n	–	20	20	20	25
L	644	649 ± 72 (530-775)	4323 ± 555 (2987-5498)	1409 ± 151 (1075-1697)	536 ± 21 (484-578)
a	17	16 ± 1.2 (14-18)	18 ± 2.3 (14-23)	15 ± 1.0 (13-17)	24 ± 1.4 (21-27)
b	6.8	6.8 ± 0.6 (5.6-7.9)	22 ± 2.2 (18-28)	12 ± 1.3 (9-14)	4.9 ± 0.2 (4.3-5.2)
c	24	26 ± 4 (21-33)	47 ± 6.3 (37-58)	21 ± 1.9 (17-24)	6.2 ± 0.3 (5.5-6.8)
V	–	–	43 ± 6.4 (39-47)	47 ± 2.6 (40-53)	–
Max. body diam.	39	41 ± 3 (34-46)	239 ± 37 (168-289)	96 ± 13 (76-129)	23 ± 1 (21-25)
Excretory pore	102	86 ± 7.1 (75-102)	183 ± 17 (152-209)	113 ± 7 (102-125)	97 ± 3.4 (88-105)
Nerve ring	67	67 ± 3 (64-75)	133 ± 12 (112-152)	84 ± 4.7 (73-90)	81 ± 63 (69-96)
Pharynx	94	95 ± 5.2 (88-106)	199 ± 13 (166-220)	123 ± 5 (115-132)	106 ± 9.1 (79-115)
Testis reflexion	97	82 ± 11 (67-104)	–	–	–
Tail length	27	25 ± 2.4 (21-32)	94 ± 12 (79-120)	69 ± 4 (63-75)	86 ± 3 (78-95)
Tail length without sheath	–	–	–	–	60 ± 2.8 (55-66)
Anal body diam.	19	19 ± 1.9 (15-22)	45 ± 7.7 (26-56)	26 ± 3 (22-32)	14 ± 1.0 (12-16)
Spicule length	47	43 ± 3.5 (37-49)	–	–	–
Spicule width	5	5.2 ± 0.6 (3.9-6.6)	–	–	–
Gubernaculum length	21	20 ± 2 (17-24)	–	–	–
D%	108	90 ± 6.2 (81-108)	93 ± 8.7 (77-112)	92 ± 6 (83-104)	89 ± 3 (81-95)
E%	–	–	–	–	113 ± 6.1 (99-125)
SW%	247	231 ± 24 (202-301)	–	–	–
GS%	45	47 ± 3.9 (38-56)	–	–	–

(two lateral, four dorso-lateral and four ventro-lateral). Opening of excretory pore located posterior to end of pharynx. Cardia present and well developed. Gonads didelphic, amphidelphic. Vulva in form of a transverse slit, located anterior to mid-body on a slightly protruding area. In ventral view, vulva elliptical, encircled by arch-shaped annules anteriorly and posteriorly. Vagina short. Tail longer than anal body diam., in some cases forming a thickening before ending in a pointed terminus. Slight post-anal swelling present. Phasmid inconspicuous.

Amphimictic female

Similar to hermaphroditic female, but smaller. Vulva non-protuberant, usually covered with exudates or copulation plug after mating. Post-anal swelling less pronounced. Tail not forming bulbous structure as found in hermaphrodite, but always conoid with a pointed terminus.

Infective juvenile

Anterior region typical for *Heterorhabditis* IJ. Excretory pore located anterior to basal bulb. Tail long, pointed in ensheathed IJ and bluntly pointed in exsheathed IJ. Excretory duct not pronounced or cuticularised. Exsheathed IJ with two-ridged lateral field.

TYPE HABITAT AND LOCALITY

Natural host unknown. This species was collected from soil samples near the settlement of Noenieput (27°27.149′S/20°05.716′E) and in an orchard near Nelspruit area (25°28.639′S/31°02.549′E), South Africa.

TYPE MATERIAL

Holotype male (mounted) and 20 paratypes each of hermaphrodites, amphimictic females, males, and several IJ in TAF, isolated from the haemocoel of *G. mellonella*, deposited in the United States Department of Agriculture Nematode Collection (USDANC), Beltsville, MD, USA. In addition, 20 paratypes each of hermaphrodites, males, amphimictic females and several IJ in TAF deposited in the National Collection of Nematodes, Biosystematics Division, Plant Protection Research Institute, Agricultural Research Council, Pretoria, South Africa. Slides of one male and one female of the second generation, and of several IJ deposited in the California Collection of Nematodes, University of California-Davis Nematode Collection, Davis, CA, USA. Several slides of hermaphrodites, males, females and all stages in TAF are located in the Department of Entomology at Stellenbosch University.

RELATIONSHIPS

Heterorhabditis noenieputensis can be distinguished from other *Heterorhabditis* species by a combination of morphological and morphometric traits of males, hermaphrodites and IJ. Males of the new species can be differentiated from other species by the terminal group of bursal papillae: pair 7 with tip curved outward; pair 8 may be present in 20% of the population, of which 10% have only one papilla on one side; pair 9 absent. The hermaphrodite can be distinguished from other species by the outwardly protruding vulval lips, and vulvar pattern. The tail is unique in forming a thickening before ending in a pointed terminus. The IJ is characterised by body and EP lengths and the ratios D% and SW%. More information on relationships is available in Malan *et al.* (2014).

MOLECULAR CHARACTERISATION

The ITS rDNA regions (accession no. JN620538), flanked by primers 18S and 26S, of *H. noenieputensis* are characterised by the sequence lengths of 1032 base pairs (ITS1 = 371 bp and ITS2 = 216 bp). The D2-D3 regions (accession no. JX624110) of the 28S rDNA gene, flanked by primers D2F and 536, is characterised by the sequence length of 898 bp. More information on molecular characterisation and phylogenetic analysis is available in Malan *et al.* (2014).

SYMBIOTIC BACTERIUM

The associated symbiotic bacterium of *H. noenieputensis* differs from that of *H. indica* and has been described as *Photorhabdus luminescens* ssp. *noenieputensis* (Ferreira *et al.*, 2013), whilst the bacterium associated with *H. indica* is *P. luminescens* ssp. *akhurstii* (Fischer-Le Saux *et al.*, 1999).

BIONOMICS

No information.

DISTRIBUTION

Known from two locations: Northern Cape Province, northwest South Africa, and Mpumalanga near Nelspruit, northeast South Africa, approximately 1130 km away from the first location (Malan, A.P., pers. comm.).

Heterorhabditis safricana[11,12] Malan, Nguyen, De Waal & Tiedt, 2008

(Figs 5.11-5.14)

MEASUREMENTS

See Table 5.5.

DIAGNOSTIC DESCRIPTION

Male

Anterior region typical for *Heterorhabditis* male. Excretory pore usually posterior to basal bulb. Testis monorchic, reflexed. *Vas deferens* well developed. Spicules paired. Bursa peloderan. Genital papillae normally comprising nine pairs and typical for genus. From anterior to posterior: pair 1 well anterior to cloacal aperture; pairs 2 and 3 forming a group; pairs 4, 5 and 6 forming a group with anterior pair in group curved laterally and slightly separated from posterior two, terminal group formed by pairs 7, 8 and 9, with pair 7 curved laterally. In general, anterior two pairs of papillae in terminal group weaker and shorter. Tail conoid, slightly curved ventrally.

Hermaphroditic female

Anterior region typical for *Heterorhabditis* hermaphroditic female. Opening of excretory pore located posterior to end of pharynx. Cardia present, well developed. Gonads didelphic, amphidelphic. Vulva in form of a transverse slit, located anterior to mid-body on slightly protruding area. In ventral view, vulva elliptical, encircled by arch-shaped annules anteriorly and posteriorly. Vagina short. Tail longer than anal body

[11] The species was derived from *South Africa*, the country from where it was described.

[12] Description after Malan *et al.* (2008). Revised by Khuong B. Nguyen.

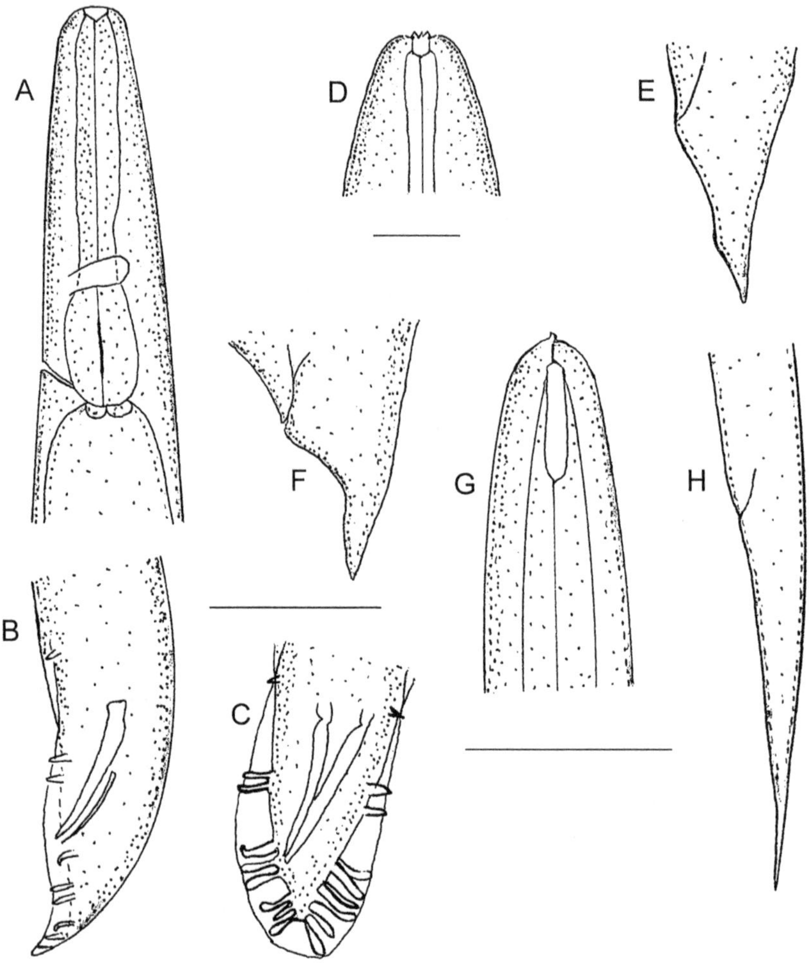

Fig. 5.11. Heterorhabditis safricana. *A-C: Male. A: Anterior region; B, C: Lateral and ventral view of tail showing spicules, gubernaculum and genital papillae. D-F: Female head and variation of tail. G, H: Infective juvenile. G: Anterior region; H: Posterior region. (Scale bars: A-C, E, F, H (left bar) = 50 μm; D (top bar) = 50 μm; G (right bar) = 25 μm.) After Malan* et al. *(2008), courtesy of* Nematology.

diam., conoid with pointed terminus. Tail widening near terminus, then narrowing to a pointed tip. Post-anal swelling present. Phasmid inconspicuous.

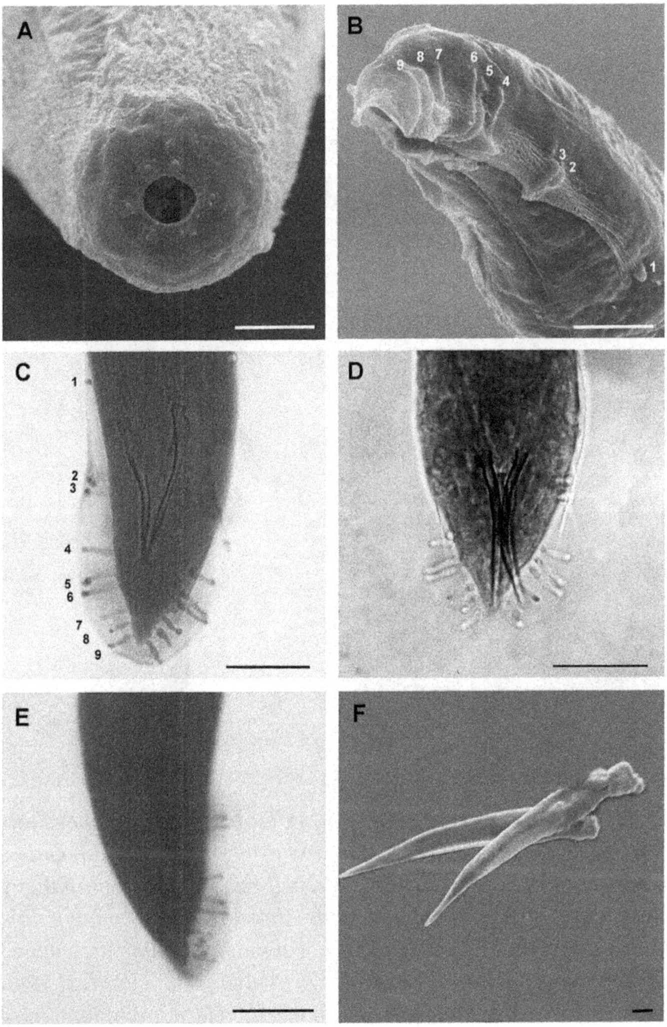

Fig. 5.12. Heterorhabditis safricana. *A, B, F: SEM photographs of male; C-E: Light microscope photographs. A: Head region showing mouth opening and six labial papillae; B: Posterior region showing cloacal aperture and bursa with full complement of papillae (numbered 1-9); C, D: Male tail with bursal papillae (numbered 1-9); E: Lateral view of male tail with bursa and papillae; F: Spicules. (Scale bars: A = 5 μm; B = 20 μm; C-E = 24 μm; F = 3 μm.) After Malan* et al. *(2008), courtesy of* Nematology.

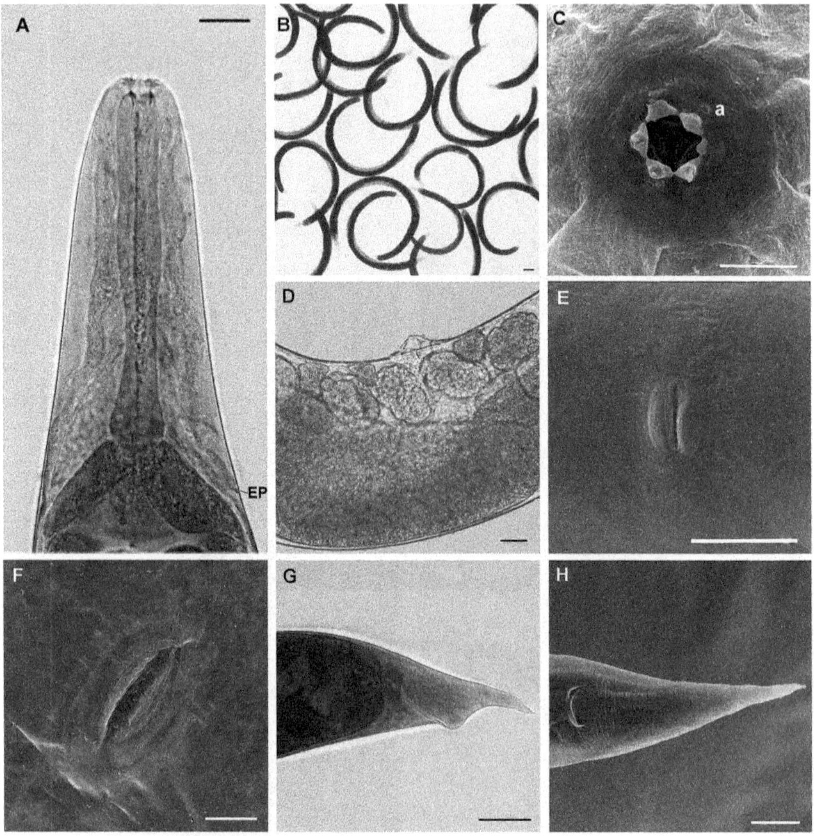

Fig. 5.13. Heterorhabditis safricana. *A, B, D, G: Light microscope photographs; C, E, F, H: SEM photographs. A: Pharynx of hermaphroditic female showing position of excretory pore (EP); B: C-shaped body of hermaphroditic females; C:* En face *view of amphimictic female showing six prominent lips, labial papillae and amphidial aperture (a); D: Vulva of amphimictic female covered with exudates or copulation plug; E, F: Vulva with elliptical pattern; G: Hermaphroditic female tail, lateral view; H: Hermaphroditic female tail, ventral view. (Scale bars: A, D, E, G = 20 μm; B = 150 μm; C, F, H = 10 μm.) After Malan* et al. *(2008), courtesy of* Nematology.

Amphimictic female

Similar to hermaphroditic female but smaller. Vulva not protruding, usually (*i.e.*, in 60% of specimens) covered with exudates or copulation plug after mating. Post-anal swelling much smaller.

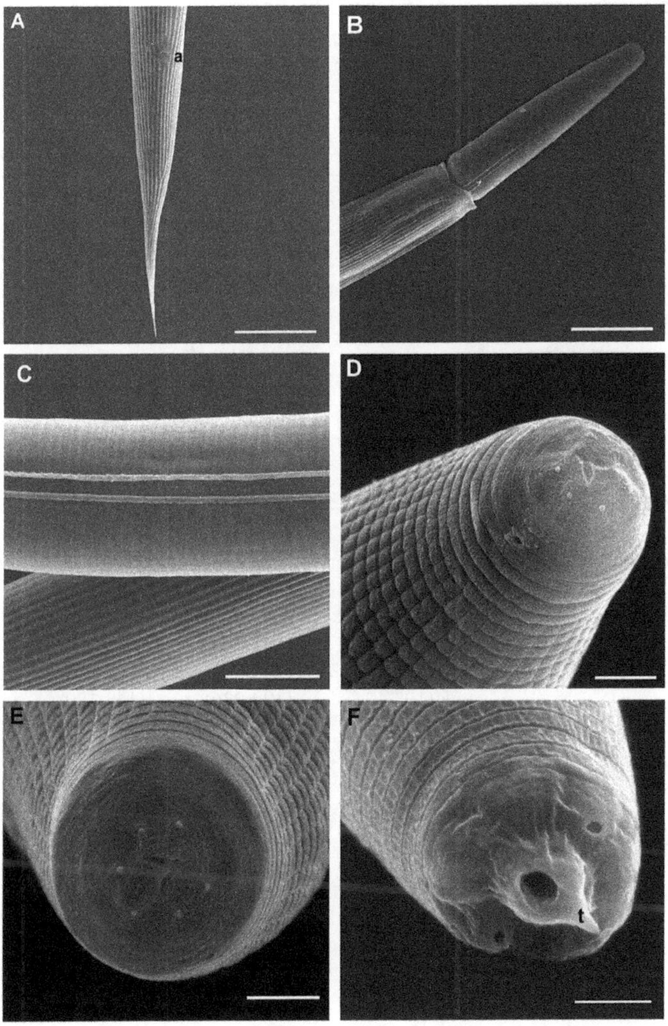

Fig. 5.14. Heterorhabditis safricana. *A-F: SEM photographs of infective juvenile (IJ). A: Tail of ensheathed IJ showing anus (a) and longitudinal ridges; B: Anterior end of a partially sheathed IJ; C: Longitudinal ridges of ensheathed IJ (bottom nematode) and lateral field at mid-body with two ridges of exsheathed IJ (top nematode); D: Anterior end of ensheathed IJ showing tessellate pattern, labial papillae, mouth opening and amphidial aperture; E: Anterior end of ensheathed IJ showing six labial papillae; F: Exsheathed IJ showing dorsal tooth (t). (Scale bars: A, B = 20 μm; C-F = 2 μm.) After Malan* et al. *(2008), courtesy of* Nematology.

Table 5.5. *Morphometrics of* Heterorhabditis safricana. *All measurements are in μm and in the form: mean $\pm$ s.d. (range).*

Character	Male		Hermaphrodite	Female	Infective juvenile
	Holotype	Paratypes	Paratypes	Paratypes	Paratypes
n	–	20	20	20	25
L	946	892 ± 66 (777-1009)	3713 ± 203 (3373-4073)	2292 ± 302 (1679-2937)	600 ± 27 (550-676)
a	–	–	–	–	28.6 ± 1.7 (24.8-31.8)
b	–	–	–	–	4.5 ± 0.2 (3.9-4.9)
c	–	–	–	–	6.4 ± 0.6 (5.4-7.5)
V	–	–	44 ± 0.8 (43-46)	48 ± 1.2 (45-50)	–
Max. body diam.	47	49 ± 4.5 (40-58)	155 ± 16 (127-188)	171 ± 28.8 (102-229)	21 ± 1.2 (19-23)
Excretory pore	145	135 ± 11 (104-147)	230 ± 14 (210-267)	171 ± 11 (151-196)	110 ± 4.2 (103-122)
Nerve ring	66	64 ± 8 (52-61)	139 ± 12 (121-163)	111 ± 11 (87-139)	93 ± 4.0 (86-101)
Pharynx	119	115 ± 5 (105-126)	215 ± 9 (199-236)	168 ± 8 (148-180)	131 ± 3.7 (125-141)
Testis reflexion	88	93 ± 22 (52-129)	–	–	–
Tail length	33	35 ± 5 (27-49)	77 ± 8 (64-91)	69 ± 14 (55-111)	93 ± 5.5 (86-108)
Tail length without sheath	–	–	–	–	57 ± 3.0 (49-62)
Anal body diam.	22	24 ± 3 (18-27)	46 ± 5 (40-54)	35 ± 11 (25-72)	13 ± 0.6 (12-14)
Spicule length	47	45 ± 4 (35-54)	–	–	–
Spicule width	7.1	5.7 ± 1.1 (3.6-7.1)	–	–	–
Gubernaculum length	26	24 ± 2 (19-27)	–	–	–
D%	122	117 ± 10 (92-133)	107 ± 6 (98-119)	108 ± 6 (97-120)	84 ± 2.6 (80-90)
E%	–	–	–	–	119 ± 9 (99-133)
	–	–	–	–	
SW%	212	196 ± 32 (130-259)	–	–	–
GS%	55	53.9 ± 4.7 (43.3-61.7)	–	–	–

Infective juvenile

Anterior region typical for *Heterorhabditis* infective juvenile. Excretory duct pronounced, well cuticularised (except in anterior part) 3-5 μm from excretory pore. Pharynx typical of genus. Lateral field at mid-body with two ridges. Phasmid not observed under light microscope. Tail elongate conoid with pointed terminus. Tail length without sheath *ca* 61% of tail with sheath.

TYPE HABITAT AND LOCALITY

The natural host of *H. safricana* is unknown as the nematode was trapped by baiting with *G. mellonella*. Soil samples were obtained from a peach orchard on Hochland farm in the Western Cape Province (32°48.10′S/18°42.34′E), South Africa.

TYPE MATERIAL

Holotype male (permanent mount in glycerin) and 20 paratypes (all in TAF) of each of hermaphrodite, female and male, together with many IJ, isolated from the haemocoel of *G. mellonella*, deposited in the United States Department of Agriculture Nematode Collection (USDANC), Beltsville, MD, USA. In addition, 20 paratypes (all in TAF) of each of hermaphrodite, male, female, together with many IJ, deposited in the National Collection of Nematodes, Biosystematics Division, Plant Protection Research Institute, Agricultural Research Council, Pretoria, South Africa. Slides of one paratype male and one paratype female of the second generation, together with several IJ, deposited in the California Collection of Nematodes, University of California Davis Nematode Collection, Davis, CA, USA. Several slides of paratype hermaphrodites, males, females, together with all stages preserved in TAF, deposited in the Entomology and Nematology Department, University of Florida, Gainesville, FL, USA.

RELATIONSHIPS

Heterorhabditis safricana is characterised by a combination of morphological characters of the hermaphroditic female, male and IJ. The hermaphroditic female can be differentiated from other species by the unique vulvar pattern. In the male of *H. safricana*, the gubernaculum is longer than all other species at 24 (19-27) μm. The IJ has a body length of 600 (550-676) μm, EP = 110 (103-122) μm, tail length (with sheath) = 93 (86-108) μm and ratio a = 28.6 (24.8-31.8).

Molecular characterisation

The ITS rDNA regions (accession no. EF488006), flanked by primers 18S and 26S, of *H. safricana*, are characterised by the sequence lengths (995 bp, ITS1 = 379 bp, ITS2 = 211 bp). More information on molecular characterisation and phylogenetic analysis is available in Malan *et al.* (2008).

Symbiotic bacterium

No information.

Bionomics

No information.

Distribution

Heterorhabditis safricana was collected from a sample taken from a peach orchard in the Western Cape Province, South Africa.

References

Edgington, S., Buddie, A.G., Moore, D., France, A.I., Merino, L.M., Tymo, L. & Hunt, D.J. (2010). Diversity and distribution of entomopathogenic nematodes in Chile. *Nematology* 12, 915-928.

Edgington, S., Buddie, A.G., Moore, D., France, A., Merino, L. & Hunt, D.J. (2011). *Heterorhabditis atacamensis* n. sp. (Nematoda: Heterorhabditidae), a new entomopathogenic nematode from the Atacama Desert, Chile. *Journal of Helminthology* 85, 381-394.

Ferreira, T., Van Reenen, C., Pagès, S., Tailliez, P., Malan, A.P. & Dicks, L.M.T. (2013). Description of *Photorhabdus luminescens* subsp. *noenieputensis* subsp. nov., a symbiotic bacterium associated with a new *Heterorhabditis* species related to *Heterorhabditis indica. International Journal of Systematic and Evolutionary Microbiology* 63, 1853-1858.

Fischer-Le Saux, M., Viallard, V., Brunel, B., Normand, P. & Boermare, N.E. (1999). Polyphasic classification of the genus *Photorhabdus* and proposal of new taxa: *P. luminescens* subsp. *luminescens* subsp. nov.,

P. luminescens subsp. *akhurstii* subsp. nov., *P. luminescens* subsp. *laumondii* subsp. nov., *P. temperata* sp. nov., *P. temperata* subsp. *temperata* subsp. nov. and *P. asymbiotica* sp. nov. *International Journal of Systematic Bacteriology* 49, 1645-1656.

LI, X.Y., LIU, Q.Z., NERMUT', J., PŮŽA, V. & MRÁČEK, Z. (2012). *Heterorhabditis beicherriana* n. sp. (Nematoda: Heterorhabditidae), a new entomopathogenic nematode from the Shunyi district of Beijing, China. *Zootaxa* 3569, 25-40.

MALAN, A.P., NGUYEN, K.B., DE WAAL, J.Y. & TIEDT, L. (2008). *Heterorhabditis safricana* n. sp. (Nematoda: Heterorhabditidae), a new entomopathogenic nematode from South Africa. *Nematology* 10, 381-396.

MALAN, A.P., KNOETZE, R. & TIEDT, L. (2014). *Heterorhabditis noenieputensis* n. sp. (Rhabditida: Heterorhabditidae), a new entomopathogenic nematode from South Africa. *Journal of Helminthology* 88, 139-151.

NGUYEN, K.B. & HUNT, D.J. (2007a). Heterorhabditidae: species descriptions. In: Nguyen, K.B. & Hunt, D.J. (Eds). *Entomopathogenic nematodes: taxonomy, phylogeny and bacterial symbionts. Nematology Monographs and Perspectives 5* (Series editors: Hunt, D.J. & Perry, R.N.). Leiden, The Netherlands, Brill, pp. 611-692.

NGUYEN, K.B. & HUNT, D.J. (Eds) (2007b). *Entomopathogenic nematodes: taxonomy, phylogeny and bacterial symbionts. Nematology Monographs and Perspectives 5* (Series editors: Hunt, D.J. & Perry, R.N.). Leiden, The Netherlands, Brill.

NGUYEN, K.B., SHAPIRO-ILAN, D.I. & MBATA, G.N. (2008). *Heterorhabditis georgiana* n. sp. (Rhabditida: Heterorhabditidae) from Georgia, USA. *Nematology* 10, 433-448.

Nematology Monographs & Perspectives, 2016, Vol. 12, 413-427

Chapter 6

Phylogeny and phylogeography of *Heterorhabditis* and *Steinernema*

Sergei E. SPIRIDONOV [1] and Sergei A. SUBBOTIN [1,2]

[1] *Center of Parasitology, A.N. Severtsov Institute of Ecology and Evolution of the Russian Academy of Sciences, Leninskii prospect 33, Moscow 119071, Russia*

[2] *Plant Pest Diagnostic Center, California Department of Food and Agriculture, 3294 Meadowview Road, Sacramento, CA 95832, USA*

Origin of *Steinernema* and *Heterorhabditis*

The evolutionary relationships of EPN were outlined for the first time by Poinar (1981), when he speculated that Steinernematidae and Heterorhabditidae arose as two separate, independent monophyletic lineages. Poinar (1993) suggested that heterorhabditids evolved from a free-living *Pellioditis*-like ancestor in an arenicolous marine environment, whereas *Steinernema* probably evolved from a proto-*Rhabditonema* ancestor in a terrestrial environment. In his later publication, Poinar (2011) proposed a Permian origin (230-252 million years ago) for heterorhabditids and steinernematids.

The similarity of *Steinernema* to *Strongyloides* and affinity to free-living Panagrolaimoidea, as well as that of *Heterorhabditis* with *Pellioditis* and Rhabditidae, was proved by several analyses of 18S rRNA gene sequences (Blaxter *et al.*, 1998; Holterman *et al.*, 2006; van Megen *et al.*, 2009; Zhang *et al.*, 2012). Holterman *et al.* (2006) and van Megen *et al.* (2009) placed *Steinernema* in Clade 10 and *Heterorhabditis* in Clade 9, where bacterial-feeding nematode families dominated and only a few animal-parasitic nematodes are present. However, other analyses of 18S rRNA gene sequences and mitochondrial gene sequences studies have found the position of *Steinernema* to be unresolved (Smythe *et al.*, 2006; Park *et al.*, 2011).

Phylogeny and phylogeography of the genus *Heterorhabditis*

The molecular phylogeny of the genus *Heterorhabditis* has been reconstructed based on the analyses of partial sequences from the 28S rRNA gene (Curran & Driver, 1994; Nguyen *et al.*, 2004, 2008; Edgington *et al.*, 2011; Li *et al.*, 2012; Malan *et al.*, 2014), partial 18S of rRNA gene (Liu *et al.*, 1997), ITS1 of rRNA gene (Adams *et al.*, 1998; Phan *et al.*, 2003), ITS1-5.8S-ITS2 of rRNA gene (Malan *et al.*, 2008, 2014; Nguyen *et al.*, 2008; Edgington *et al.*, 2011; Li *et al.*, 2012), partial *nd*4 gene (Liu *et al.*, 1999) and partial *cox*I mtDNA gene (Kuwata *et al.*, 2007). Whereas the analysis of the partial 18S rRNA gene showed that this fragment is too conservative and not useful for phylogenetic study of this genus, the analyses of other regions, especially ITS and D2-D3 of 28S rRNA gene, revealed resolved relationships between a number of species. The phylogenetic estimations based on the *nd*4 gene (Liu *et al.*, 1999; Saeb & Grewal, 2014), however, differed in positions of some species, but it cannot be excluded that these differences occurred due to some misidentification of nematode samples.

The first comprehensive phylogenetic analysis of the genus *Heterorhabditis* was conducted by Adams *et al.* (1998). Nguyen *et al.* (2008) was the first proposed division of the genus into three clades (groups): '*Indica*', '*Bacteriophora*' and '*Megidis*', which are presently recognised by the majority of authors. The species of the *Indica*-clade formed a highly supported clade in the phylogenetic tree (Fig. 6.1). This clade might be subdivided into two subclades: *i*) '*Indica*' with *H. indica* (= *H. brevicaudis*, *H. gerrardi*, *H. hawaiiensis*) and *H. noenieputensis*; and *ii*) '*Baujardi*' with *H. amazonensis*, *H. baujardi* (= *H. somsookae*), *H. floridensis*, *H. taysearae* (= *H. sonorensis*) and *H. mexicana*. The proposed synonymisation of *H. gerrardi* with *H. indica* and *H. sonorensis* with *H. taysearae* (see Chapter 2) is based on almost identical ITS rRNA gene sequences in these two pairs of heterorhabditids. All species of the *Indica*-clade are found in the tropics and subtropics or in the warm temperature zone. *Heterorhabditis indica* is the only species of the group, widely distributed in Asia, Africa, Australia, North and South America and the Caribbean region. *Heterorhabditis baujardi* is found in southeast Asia and Africa, whilst *H. taysearae* is known from western Asia, Africa and Mexico. Four other species are each from a certain area only: *H. noenieputensis* –

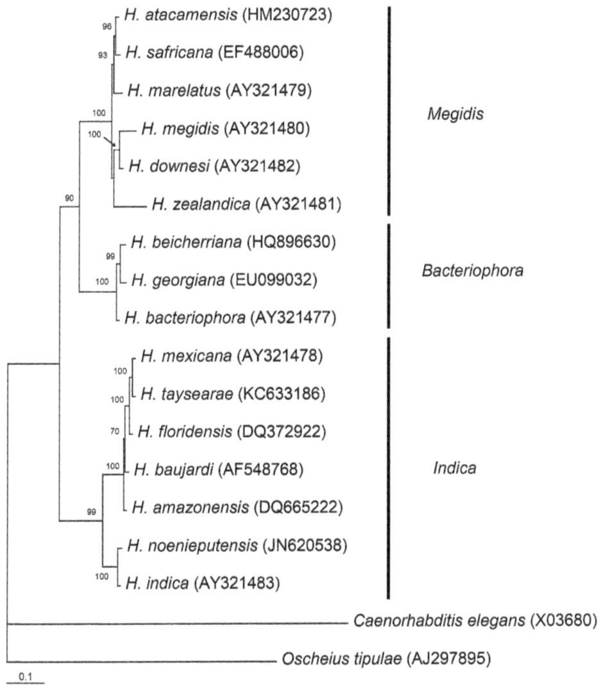

Fig. 6.1. *Phylogenetic relationships within* Heterorhabditis *species as inferred from Bayesian analyses of sequences of the ITS rRNA gene using the GTR + I + G model of DNA evolution. Posterior probability values more than 70% are given on appropriate clades.*

South Africa, *H. amazonensis* – Brazil, *H. floridensis* – southeastern USA, and *H. mexicana* – Mexico.

The *Bacteriophora*-clade includes the following species: *H. bacteriophora* (= *H. argentinensis*), *H. georgiana* and *H. beicherriana*. *Heterorhabditis bacteriophora* is currently known as a species with the widest cosmopolitan geographical distribution among heterorhabditids, whereas *H. georgiana* is only known from the USA and *H. beicherriana* only from China.

The *Megidis*-clade contains *H. zealandica*, *H. downesi*, *H. megidis*, *H. marelatus* (= *H. hepialius*), *H. safricana* and *H. atacamensis*. *Heterorhabditis zealandica* is widely distributed in New Zealand, USA, Australia, South Africa and China. Relationships between some species of this clade are not well resolved, except for two pairs of species, *H. downesi* and *H. megidis*, *H. safricana* and *H. atacamensis*, which

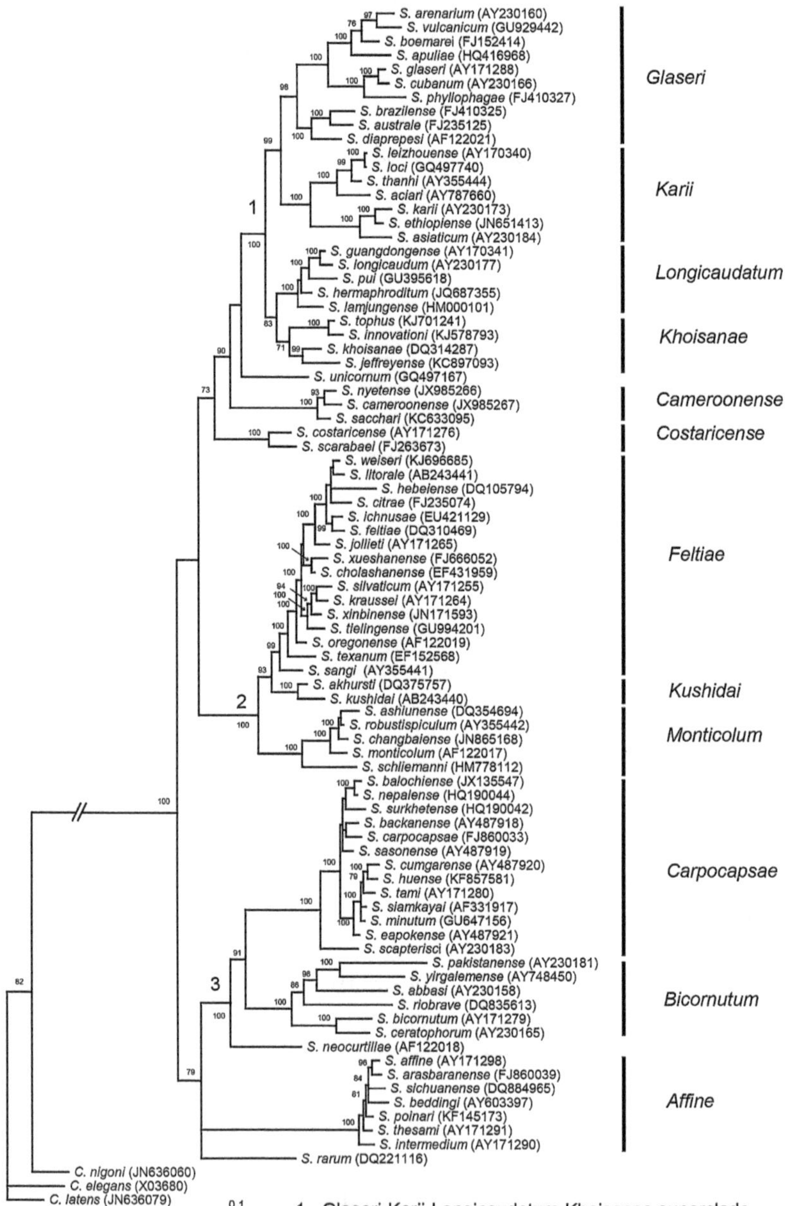

1 - Glaseri-Karii-Longicaudatum-Khoisanae superclade
2 - Feltiae-Kushidai-Monticolum superclade
3 - Carpocapsae-Bicornutum superclade

clustered together in most phylogenetic trees with moderate or high statistical supports: *i*) *H. downesi* found only in western Europe and *H. megidis* distributed across the entire Holarctic; and *ii*) *H. safricana* from South Africa and *H. atacamensis* from Chile. *Heterorhabditis marelatus*, which often clustered with *H. safricana* and *H. atacamensis*, has restricted geographic distribution and is currently found only in Oregon and California, USA.

Phylogeny and phylogeography of the genus *Steinernema*

Several attempts have been made to compare morphological and molecular approaches to reconstruct the phylogeny of the genus *Steinernema*. Liu & Berry (1996) presented phylogenetic relationships derived from the analysis of both RAPD fragments and 11 morphological features for 14 *Steinernema* species. Reid (1994) and Reid *et al.* (1997) analysed the restriction fragments of ITS1-5.8S-ITS2 PCR products generated by 17 restriction enzymes and proposed a phylogram in which the clades correspond to conditional groups of species such as 'steinernematids with long juveniles' or 'steinernematids with short juveniles'.

The first attempt to use sequencing data for *Steinernema* phylogenetic analysis was based on analyses of the partial 18S rRNA gene (Liu *et al.*, 1997). This region, however, was found to be too conservative to resolve phylogeny of this genus. Sequences of the ITS1 region of rRNA and mitochondrial *cox*II and 16S rRNA genes for seven species of *Steinernema* were analysed by Szalanski *et al.* (2000). The authors obtained trees that were in good concordance with the morphological features of the species and considered these genes to be ideal candidates for the phylogenetic reconstruction of the group. Analyses of sequences of ITS1-5.8S-ITS2 regions obtained from ten steinernematid species had revealed several groups with significant bootstrap supports (Nguyen *et al.*, 2001; Nguyen & Duncan, 2002). Comprehensive phylogenetic analyses based on combined morphological and D2-D3 expansion segments of 28S rRNA gene sequences of 21 species of *Steinernema* were made by Stock

Fig. 6.2. *Phylogenetic relationships within* Steinernema *species as inferred from Bayesian analyses of sequences of the ITS rRNA gene using the GTR + I + G model of DNA evolution. Posterior probability values more than 70% are given on appropriate clades.*

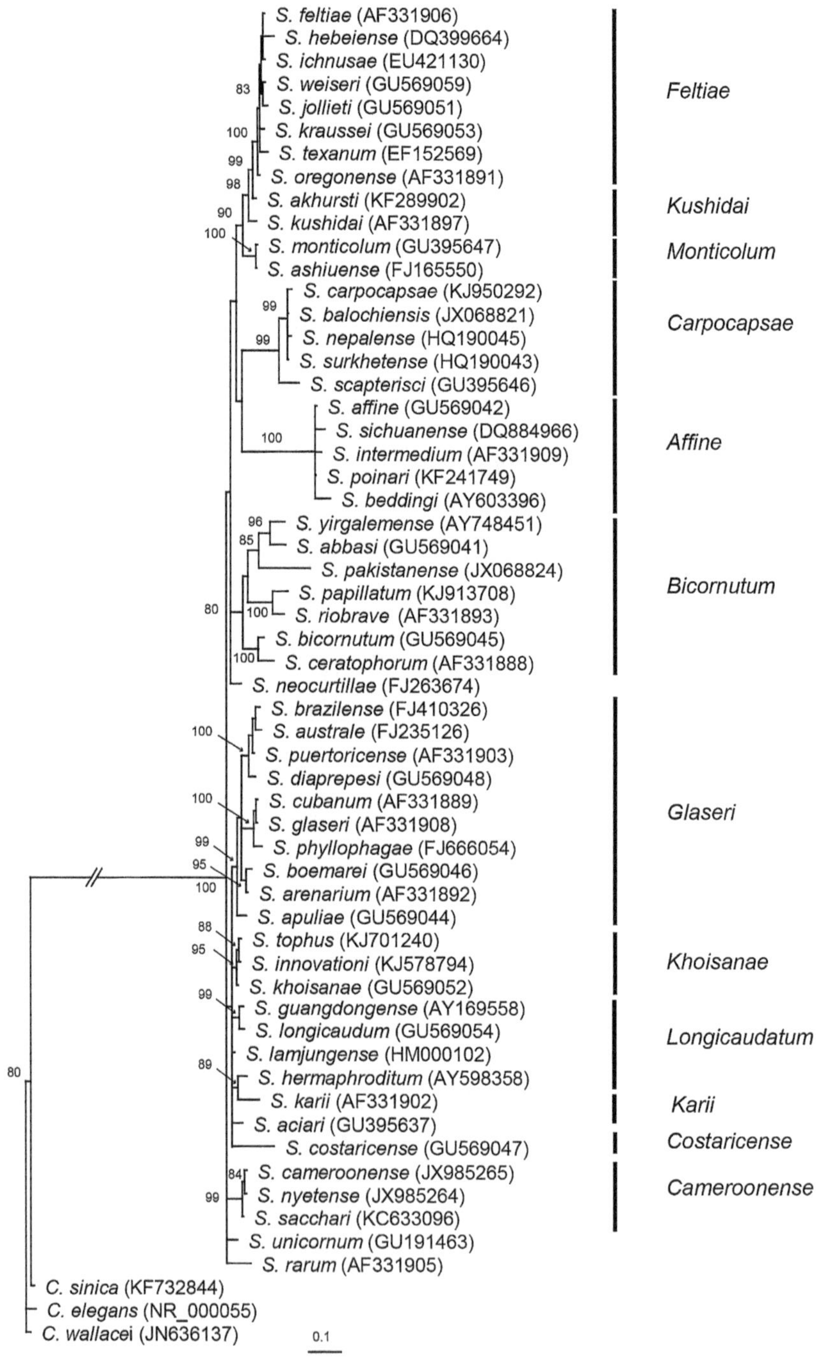
S. feltiae (AF331906)
S. hebeiense (DQ399664)
S. ichnusae (EU421130)
S. weiseri (GU569059)
S. jollieti (GU569051)
S. kraussei (GU569053)
S. texanum (EF152569)
S. oregonense (AF331891)
S. akhursti (KF289902)
S. kushidai (AF331897)
S. monticolum (GU395647)
S. ashiuense (FJ165550)
S. carpocapsae (KJ950292)
S. balochiensis (JX068821)
S. nepalense (HQ190045)
S. surkhetense (HQ190043)
S. scapterisci (GU395646)
S. affine (GU569042)
S. sichuanense (DQ884966)
S. intermedium (AF331909)
S. poinari (KF241749)
S. beddingi (AY603396)
S. yirgalemense (AY748451)
S. abbasi (GU569041)
S. pakistanense (JX068824)
S. papillatum (KJ913708)
S. riobrave (AF331893)
S. bicornutum (GU569045)
S. ceratophorum (AF331888)
S. neocurtillae (FJ263674)
S. brazilense (FJ410326)
S. australe (FJ235126)
S. puertoricense (AF331903)
S. diaprepesi (GU569048)
S. cubanum (AF331889)
S. glaseri (AF331908)
S. phyllophagae (FJ666054)
S. boemarei (GU569046)
S. arenarium (AF331892)
S. apuliae (GU569044)
S. tophus (KJ701240)
S. innovationi (KJ578794)
S. khoisanae (GU569052)
S. guangdongense (AY169558)
S. longicaudum (GU569054)
S. lamjungense (HM000102)
S. hermaphroditum (AY598358)
S. karii (AF331902)
S. aciari (GU395637)
S. costaricense (GU569047)
S. cameroonense (JX985265)
S. nyetense (JX985264)
S. sacchari (KC633096)
S. unicornum (GU191463)
S. rarum (AF331905)
C. sinica (KF732844)
C. elegans (NR_000055)
C. wallacei (JN636137)
0.1
Feltiae
Kushidai
Monticolum
Carpocapsae
Affine
Bicornutum
Glaseri
Khoisanae
Longicaudatum
Karii
Costaricense
Cameroonense

et al. (2001). Stock *et al.* (2001) found that these species were represented by five main clades. Later, Spiridonov *et al.* (2004) conducted detailed analyses of the ITS rRNA gene sequences for 24 nominal species and 16 unidentified *Steinernema* species. In the majority of the phylogenetic trees obtained from different ITS alignments and phylogenetic procedures, the 84 studied isolates formed five main, highly or moderately supported, clades, *viz.*, Clade I: '*affine/intermedium*'; Clade II: '*carpocapsae/scapterisci/tami*'; Clade III: '*feltiae/kraussei/oregonense*'; Clade IV: '*bicornutum/ceratophorum/riobrave*'; and Clade V: '*arenarium/glaseri/karii/longicaudum*'. The ITS rRNA gene sequence data were found to be of little utility in resolving relationships between these clades, but were useful in studying relationships between species within certain clades. The level of intra-specific variability was different between clades with a sequence divergence of 2.4-2.8% of ITS rRNA gene for some species.

Nadler *et al.* (2006) reconstructed the phylogeny of 25 *Steinernema* species analysing the sequences of three gene fragments: D2-D3 of 28S rRNA, 12S rRNA gene and *cox*I gene of mtDNA. The topology of the tree inferred from the total evidence dataset was generally congruent with that presented by Spiridonov *et al.* (2004), except for the positions of *S. siamkayai*, *S. scapterisci*, *S. carpocapsae* and *S. monticolum*, which formed a basal clade within *Steinernema*. Our analysis has revealed, however, that the D2-D3 of 28S rRNA gene sequences of these four *Steinernema* species provided by Stock *et al.* (2001) and Nadler *et al.* (2006) actually belong to *H. georgiana*. These four sequences have been incorrectly used in many further publications dealing with the phylogenetic analysis of *Steinernema* (Stock & Gress, 2006; Lopez-Nunez *et al.*, 2008; Lee *et al.*, 2009; Mráček *et al.*, 2009; Lee & Stock, 2010; Nguyen & Buss, 2011; Çimen *et al.*, 2015; Shahina *et al.*, 2015).

In our present analyses we used two separate datasets: *i*) the ITS rRNA sequences of 85 *Steinernema* species and three outgroup species; and *ii*) the D2-D3 of 28S rRNA sequences of 55 *Steinernema* species and three outgroup species. The phylogenetic trees inferred from the Bayesian analyses of these datasets are given in Figures 6.2

Fig. 6.3. *Phylogenetic relationships within* Steinernema *species as inferred from Bayesian analyses of sequences of the D2-D3 of 28S rRNA gene using the GTR + I + G model of DNA evolution. Posterior probability values more than 70% are given on appropriate clades.*

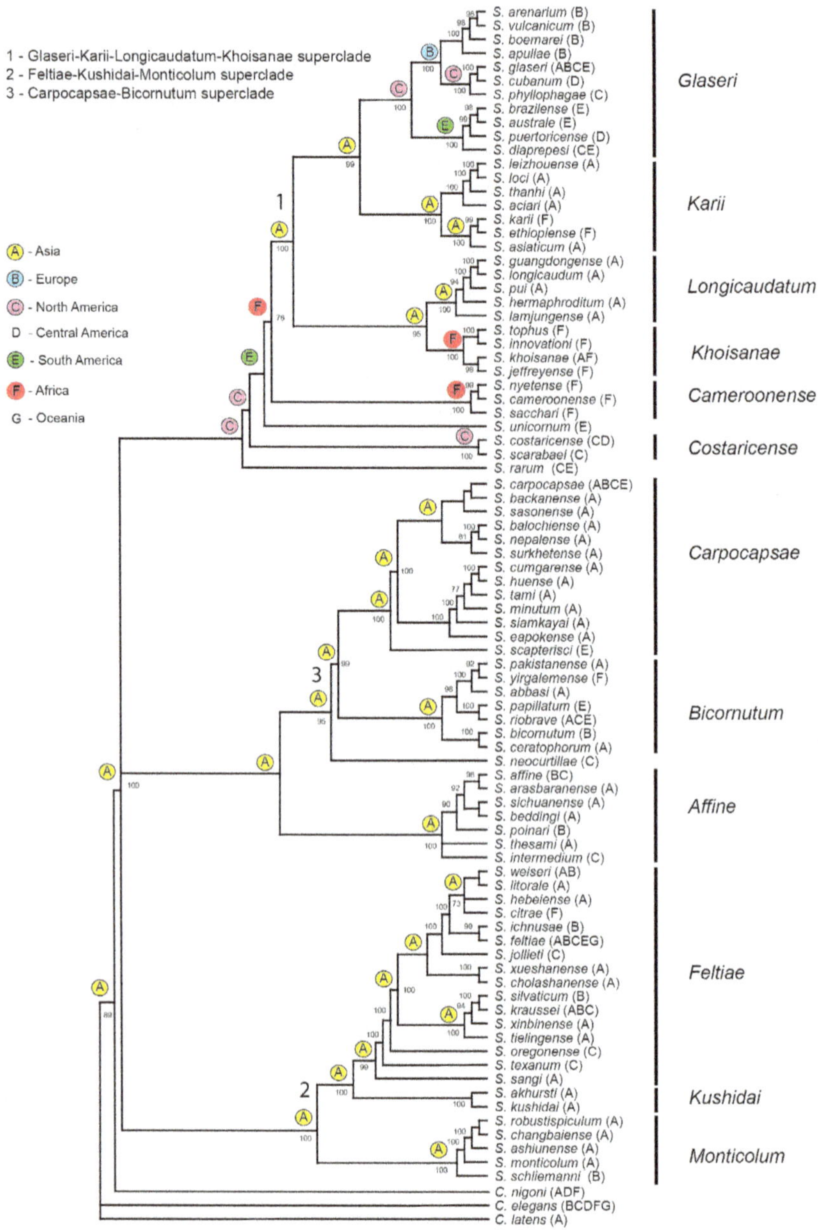
1 - Glaseri-Karii-Longicaudatum-Khoisanae superclade
2 - Feltiae-Kushidai-Monticolum superclade
3 - Carpocapsae-Bicornutum superclade
A - Asia
B - Europe
C - North America
D - Central America
E - South America
F - Africa
G - Oceania
S. arenarium (B)
S. vulcanicum (B)
S. boemarei (B)
S. apuliae (B)
S. glaseri (ABCE)
S. cubanum (D)
S. phyllophagae (C)
S. brazilense (E)
S. australe (E)
S. puertoricense (D)
S. diaprepesi (CE)
S. leizhouense (A)
S. loci (A)
S. thanhi (A)
S. aciari (A)
S. karii (F)
S. ethiopiense (F)
S. asiaticum (A)
S. guangdongense (A)
S. longicaudum (A)
S. pui (A)
S. hermaphroditum (A)
S. lamjungense (A)
S. tophus (F)
S. innovationi (F)
S. khoisanae (AF)
S. jeffreyense (F)
S. nyetense (F)
S. cameroonense (F)
S. sacchari (F)
S. unicornum (E)
S. costaricense (CD)
S. scarabaei (C)
S. rarum (CE)
S. carpocapsae (ABCE)
S. backanense (A)
S. sasonense (A)
S. balochiense (A)
S. nepalense (A)
S. surkhetense (A)
S. cumgarense (A)
S. huense (A)
S. tami (A)
S. minutum (A)
S. siamkayai (A)
S. eapokense (A)
S. scapterisci (E)
S. pakistanense (A)
S. yirgalemense (F)
S. abbasi (A)
S. papillatum (E)
S. riobrave (ACE)
S. bicornutum (B)
S. ceratophorum (A)
S. neocurtillae (C)
S. affine (BC)
S. arasbaranense (A)
S. sichuanense (A)
S. beddingi (A)
S. poinari (B)
S. thesami (A)
S. intermedium (C)
S. weiseri (AB)
S. litorale (A)
S. hebeiense (A)
S. citrae (F)
S. ichnusae (B)
S. feltiae (ABCEG)
S. jollieti (C)
S. xueshanense (A)
S. cholashanense (A)
S. silvaticum (B)
S. kraussei (ABC)
S. xinbinense (A)
S. tielingense (A)
S. oregonense (C)
S. texanum (C)
S. sangi (A)
S. akhursti (A)
S. kushidai (A)
S. robustispiculum (A)
S. changbaiense (A)
S. ashiuense (A)
S. monticolum (A)
S. schliemanni (B)
C. nigoni (ADF)
C. elegans (BCDFG)
C. latens (A)
Glaseri
Karii
Longicaudatum
Khoisanae
Cameroonense
Costaricense
Carpocapsae
Bicornutum
Affine
Feltiae
Kushidai
Monticolum

and 6.3. The following 12 multispecies clades are distinguished: '*Affine*' (7 species), '*Bicornutum*' (7), '*Cameroonense*' (3), '*Carpocapsae*' (13), '*Costaricense*' (2), '*Feltiae*' (16), '*Glaseri*' (11), '*Karii*' (7), '*Khoisanae*' (4), '*Kushidai*' (2), '*Longicaudum*' (5) and '*Monticolum*' (5) and three monospecies clades: *S. neocurtillae*, *S. unicornum* and *S. rarum*. Ten of these clades can be combined into three superclades.

Morphological features of steinernematids play a minor role in the phylogenetic reconstruction of this group. Some of these features provide an additional support for the separation of clades of steinernematids, but, in general, the relationships between the steinernematid species inferred from molecular data are not affected significantly by their morphological and morphometric characteristics (Spiridonov *et al.*, 2004). Still, some of these characters can be considered as a valuable addition to the clade diagnoses. To make morphology of steinernematids more useful for phylogenetic purposes, it is important to use consistent terminology. Presently, the different interpretations of the structure of lateral fields (bands/ridges) or other structures seen in SEM by various authors are misleading. Morphological studies have shown that some clades with numerous species are characterised by a uniform structure of the lateral field. For example, a lateral field with eight ridges of equal height and width is characteristic for the *Bicornutum*-clade, *Glaseri*-clade, *Khoisanae*-clade and *Longicaudum*-clade. All the species of the *Carpocapsae*-clade are characterised by a more or less uniform structure of the lateral field, *i.e.*, with four ridges in the centre of the field, poorly visible submarginal and strongly protruding marginal ridges (*i.e.*, eight ridges in total). In the *Affine*-clade, the structure of the lateral field is also quite uniform with two slightly protruding but well discernible central ridges, prominent submarginal ridges and less distinct marginal ones. In groups such as the *Feltiae*-clade or *Monticolum*-clade, the structure of the lateral field is variable. For example, closely related 'sister' species such as *S. kraussei* and *S. silvaticum* are characterised by seven unequal and eight equal ridges, respectively. In the *Monticolum*-clade,

Fig. 6.4. *Phylogenetic relationships within* Steinernema *species as inferred from Bayesian analyses of sequences of the D2-D3 of 28S and ITS rRNA genes using the GTR + I + G model of DNA evolution. Reconstruction of ancestral areas was obtained with Bayesian Binary MCMC method using the software RASP3.2 (Yu* et al., *2015). The distribution areas are marked from A to G.*

eight ridges of unequal width are characteristic for *S. monticolum* and eight equal ridges for *S. schliemanni*.

The number and distribution of genital papillae on the male posterior end are variable even between specimens of the same species. For example, in *S. diaprepesi* and *S. pui*, specimens with 25 (24 + 1) and 23 (22 + 1) papillae have been described. For the majority of steinernematid species, the constant number of 11 pairs of papillae and a single midventral precloacal papilla (22 + 1) is characteristic. Thus, this feature is of no help in distinguishing between clades.

There are two less labile traits of steinernematid morphology that we consider useful for phylogenetic analysis: the spermatozoon structure and the organisation of the bacterial vesicle. Both features reflect the basic biological adaptations of these nematodes. Two main types of spermatozoa have been reported for steinernematids. Representatives of the *Affine*-clade and *Feltiae*-clade produce spermatozoa forming chain-like groups of 2-25 cells (Spiridonov *et al.*, 1999). The size of cells in the chains differ between clades, being *ca* 4-7 μm in diam. in the *Feltiae*-clade and larger (usually 7-11 μm diam.) in the *Affine*-clade. Chain-forming spermatozoa were also observed in representatives of the *Costaricense*-, *Cameroonense*-, *Monticolum*- and *Kushidai*-clades, and in *S. rarum*. Apart from the chain-forming characteristic, such spermatozoa are similar to those of *Caenorhabditis elegans* and probably represent a basic ancestral state of spermatozoon evolution within Rhabditida. Up to now, the formation of spermatozeugmata (huge amoeboid cells with microscopic spermatozoa on the surface) (Yushin *et al.*, 2007) has been described for *Steinernema* species of the clades *Glaseri*, *Karii*, *Longicaudum*, *Carpocapsae* and *Bicornutum*. The monophyletic status of steinernematids with spermatozeugmata is not substantiated. It can be presumed that the formation of either spermatozeugmata or chains is an adaptation to the permanent and massive flow of oocytes from ovaries into uteri – smaller and separate cells may be flushed down with this flow of zygotes from the distal part of the uterus into the part of its lumen with the developing eggs.

The retention of bacterial cells in the anterior part of the mid-gut is a characteristic feature of steinernematids, unlike the situation in heterorhabditids where the bacteria are scattered throughout the mid-gut lumen of the infective juvenile (IJ). The shape of the bacterial vesicle in some clades is quite uniform. For example, in the *Feltiae*-clade the bacterial vesicle is usually more or less spindle-shaped with well discernible walls (*ca* 1 μm thick). Such a vesicle shape is formed during the first week of IJ existence in the soil and does not subsequently change. In the *Carpocapsae*-clade the vesicle is ellipsoid, also with well defined walls. The same structure for the bacterial vesicle was described for the *Bicornutum*-clade. For species of the *Glaseri*-clade, bacterial vesicle walls were either not described or were not discernible. In the *Affine*-clade, the bacterial vesicle increases in size for several weeks after migration from the host cadaver and, after several months of storage, looks like a vast space in the gut lumen filled with numerous bacterial cells. Vesicle walls in the *Affine*-clade are indiscernible or very thin.

Biogeographical analyses using Bayesian and parsimony reconstruction methods reveal an ancestral diversification of the genus *Steinernema* in Asia with repeated colonisation of other regions. Fifty out of 85 molecularly studied species currently inhabit various parts of Asia. Eight (*Monticolum*, *Kushidai*, *Feltiae*, *Affine*, *Bicornutum*, *Carpocapsae*, *Longicaudum* and *Karii*) out of 12 steinernematid clades have a greater probability of being Asian in origin (Fig. 6.4). The *Costaricense*-clade and *Glaseri*-clade appear to have originated in North America, some species of the latter clade subsequently colonising Central and South America and Europe. Steinernematids colonised Africa five times, leaving behind moderate species diversity. In the four ancestral colonisation events: *i*) *S. citrae*; *ii*) *S. yirgalemense*; *iii*) *S. tophus*, *S. innovationi*, *S. khoisanae* and *S. jeffreyense*; and *iv*) *S. karii* and *S. ethiopiense*, ancestors were surmised to have originated from Asia, whereas an ancestor of the *Cameroonense*-clade likely transitioned from the Americas. Colonisation of Europe occurred ten times with at least eight events from Asia and, probably, twice from North America.

References

ADAMS, B.J., BURNELL, A.M. & POWERS, T.O. (1998). A phylogenetic analysis of *Heterorhabditis* (Nematoda, Rhabditida) based on internal transcribed spacer DNA sequence data. *Journal of Nematology* 30, 22-39.

BLAXTER, M.L., DE LEY, P., GAREY, J.R., LIU, L.X., SCHELDEMAN, P., VIERSTRAETE, A., VANFLETEREN, J.R., MACKEY, L.Y., DORRIS, M., FRISSE, L.M. *ET AL.* (1998). A molecular evolutionary framework for the phylum Nematoda. *Nature* 392, 71-75.

ÇIMEN, H., LEE, M.M., HATTING, J., HAZIR, S. & STOCK, S.P. (2015). *Steinernema innovationi* n. sp. (Panagrolaimomorpha: Steinernematidae), a new entomopathogenic nematode species from South Africa. *Journal of Helminthology* 89, 415-427.

CURRAN, J. & DRIVER, F. (1994). Molecular taxonomy of *Heterorhabditis*. In: Burnell, A.M., Ehlers, R.-U. & Masson, J.-P. (Eds). *COST 812 Biotechnology: genetics of entomopathogenic nematode-bacterium complexes. Proceedings of symposium and workshop, St Patrick's College, Maynooth, Kildare County, Ireland.* Luxembourg, European Commission, DG EUR 15681, pp. 41-48.

EDGINGTON, S., BUDDIE, A.G., MOORE, D., FRANCE, A., MERINO, L. & HUNT, D.J. (2011). *Heterorhabditis atacamensis* n. sp. (Nematoda: Heterorhabditidae), a new entomopathogenic nematode from the Atacama Desert, Chile. *Journal of Helminthology* 85, 381-394.

HOLTERMAN, M., VAN DER WURFF, A., VAN DEN ELSEN, S., VAN MEGEN, H., BONGERS, T., HOLOVACHOV, O., BAKKER, J. & HELDER, J. (2006). Phylum-wide analysis of SSU rDNA reveals deep phylogenetic relationships among nematodes and accelerated evolution toward crown clades. *Molecular Biology and Evolution* 23, 1792-1800.

KUWATA, R., YOSHIGA, T., YOSHIDA, M. & KONDO, E. (2007). Phylogenetic relationships of Japanese *Heterorhabditis* nematodes and their symbiotic *Photorhabdus* bacteria. *Japanese Journal of Nematology* 37, 39-50.

LEE, M.-M. & STOCK S.P. (2010). A multilocus approach to assessing co-evolutionary relationships between *Steinernema* spp. (Nematoda: Steinernematidae) and their bacterial symbionts *Xenorhabdus* spp. (gamma-Proteobacteria: Enterobacteriaceae). *Systematic Parasitology* 77, 1-12.

LEE, M.-M., SICARD, M., SKEIE, M. & STOCK, S.P. (2009). *Steinernema boemarei* n. sp. (Nematoda: Steinernematidae), a new entomopathogenic nematode from southern France. *Systematic Parasitology* 72, 127-141.

LI, X.-Y., LIU, Q.-Z., NERMUT', J., PŮŽA, V. & MRÁČEK, Z. (2012). *Heterorhabditis beicherriana* n. sp. (Nematoda: Heterorhabditidae), a new entomopathogenic nematode from the Shunyi district of Beijing, China. *Zootaxa* 3569, 25-40

LIU, J. & BERRY, R.E. (1996). Phylogenetic analysis of the genus *Steinernema* by morphological characters and randomly amplified polymorphic DNA fragments. *Fundamental and Applied Nematology* 19, 463-469.

LIU, J., BERRY, R.E. & MOLDENKE, A.F. (1997). Phylogenetic relationships of entomopathogenic nematodes (Heterorhabditidae and Steinernematidae) inferred from partial 18S rDNA gene sequences. *Journal of Invertebrate Pathology* 69, 246-252.

LIU, J., BERRY, R.E. & BLOUIN, M.S. (1999). Molecular differentiation and phylogeny of entomopathogenic nematodes (Rhabditida, Heterorhabditidae) based on ND4 gene sequences of mitochondrial DNA. *Journal of Parasitology* 85, 709-715.

LOPEZ-NUNEZ, J.C., PLICHTA, K., GONGORA-BOTERO, C.E. & STOCK, S.P. (2008). A new entomopathogenic nematode, *Steinernema colombiense* n. sp. (Nematoda: Steinernematidae), from Colombia. *Nematology* 10, 561-574.

MALAN, A.P., NGUYEN, K.B., DE WAAL, J.Y. & TIEDT, L. (2008). *Heterorhabditis safricana* n. sp. (Rhabditida: Heterorhabditidae) a new entomopathogenic nematode from South Africa). *Nematology* 10, 381-396.

MALAN, A.P., KNOETZE, R. & TIEDT, L. (2014). *Heterorhabditis noenieputensis* n. sp. (Rhabditida: Heterorhabditidae), a new entomopathogenic nematode from South Africa. *Journal of Helminthology* 88, 139-151.

MRÁČEK, Z., LIU, Q.Z. & NGUYEN, K.B. (2009). *Steinernema xueshanense* n. sp. (Rhabditida, Steinernematidae), a new species of entomopathogenic nematode from the province of Yunnan, southeast Tibetan Mts., China. *Journal of Invertebrate Pathology* 102, 69-78.

NADLER, S.A., BOLOTIN, E. & STOCK, S.P. (2006). Phylogenetic relationships of *Steinernema* (Cephalobina: Steinernematidae) based on nuclear, mitochondrial, and morphological data. *Systematic Parasitology* 63, 159-179.

NGUYEN, K.B. & BUSS, E.A. (2011). *Steinernema phyllophagae* n. sp. (Rhabditida: Steinernematidae), a new entomopathogenic nematode from Florida, USA. *Nematology* 13, 425-442.

NGUYEN, K.B. & DUNCAN, L.W. (2002). *Steinernema diaprepesi* n. sp. (Rhabditida: Steinernematidae), a parasite of the citrus root weevil *Diaprepes abbreviatus* (L.) (Coleoptera: Curculionidae). *Journal of Nematology* 34, 159-170.

NGUYEN, K.B., MARUNIAK, J. & ADAMS, B.J. (2001). Diagnostic and phylogenetic utility of the rDNA internal transcribed spacer sequences of *Steinernema*. *Journal of Nematology* 33, 73-82.

NGUYEN, K.B., SHAPIRO-ILAN, D.I., STUART, R.J., MCCOY, C.W., JAMES, R.R. & ADAMS, B.J. (2004). *Heterorhabditis mexicana* n. sp. (Rhabditida: Heterorhabditidae) from Tamaulipas, Mexico, and morphological studies of the bursa of *Heterorhabditis* spp. *Nematology* 6, 231-244.

NGUYEN, K.B., SHAPIRO-ILAN, D.I. & MBATA, G. (2008). *Heterorhabditis georgiana* n. sp. (Rhabditida: Heterorhabditidae) from Georgia, USA. *Nematology* 10, 433-448.

PARK, J.-K., SULTANA, T., LEE, S.-H., KANG, S., KIM, H.K., MIN, G.-S., EOM, K.S. & NADLER, S.A. (2011). Monophyly of clade III nematodes is not supported by phylogenetic analysis of complete mitochondrial genome sequences. *BMC Genomics* 392, 1-16.

PHAN, K.L., SUBBOTIN, S.A., NGUYEN, N.C. & MOENS, M. (2003). *Heterorhabditis baujardi* sp. n. (Rhabditida: Heterorhabditidae) from Vietnam and morphometric data for *H. indica* populations. *Nematology* 5, 367-382.

POINAR JR, G.O. (1981). *The natural history of nematodes*. Englewood Cliffs, NJ, USA, Prentice Hall.

POINAR JR, G.O. (1993). Origins and phylogenetic relationships of the entomophilic rhabditids, *Heterorhabditis* and *Steinernema*. *Fundamental and Applied Nematology* 16, 333-338.

POINAR JR, G.O. (2011). *The evolutionary history of nematodes. Nematology Monographs and Perspectives 9* (Series editors: Hunt, D.J. & Perry, R.N.). Leiden, The Netherlands, Brill.

REID, A.P. (1994). Molecular taxonomy of *Steinernema*. In: Burnell, A.M., Ehlers, R.-U. & Masson, J.-P. (Eds). *COST 812 Biotechnology: genetics of entomopathogenic nematodes-bacterium complexes. Proceedings of symposium and workshop, St Patrick's College, Maynooth, Kildare County, Ireland.* Luxembourg, European Commission, DG EUR 15681, pp. 49-58.

REID, A.P., HOMINICK, W.M. & BRISCOE, B.R. (1997). Molecular taxonomy and phylogeny of entomopathogenic nematode species (Rhabditida: Steinernematidae) by RFLP analysis of the ITS region of the ribosomal DNA repeat unit. *Systematic Parasitology* 37, 187-193.

SAEB, A.T.M. & GREWAL, P.S. (2014). NADH dehydrogenase subunit 4 is not an ideal target for phylogeny reconstruction or genetic differentiation among entomopathogenic nematodes (Rhabditida: Heterorhabditidae). *SOP Transactions on Inheritance and Genetic Engineering*. http://www.scipublish.com/journals/IGE/papers/1146 (published online 2014-12-11).

SHAHINA, F. & MEHREEN, G. (2015). Phylogenetic analysis and biological studies of entomopathogenic nematode *Steinernema abbasi* isolates. *Pakistan Journal of Nematology* 33, 1-16.

SMYTHE, A.B., SANDERSON, M.J. & NADLER, S.A. (2006). Nematode small subunit phylogeny correlates with alignment parameters. *Systematic Biology* 55, 972-992.

SPIRIDONOV, S.E., HOMINICK, W.M. & BRISCOE, B.R. (1999). Morphology of amoeboid cells in the uterus of *Steinernema* species (Rhabditida; Steinernematidae). *Russian Journal of Nematology* 7, 49-56.

SPIRIDONOV, S.E., REID, A.P., PODRUCKA, K., SUBBOTIN, S.A. & MOENS, M. (2004). Phylogenetic relationships within the genus *Steinernema* (Nematoda: Rhabditida) as inferred from analyses of sequences of the ITS1-5.8S-ITS2 region of rDNA and morphological features. *Nematology* 6, 547-566.

STOCK, S.P. & GRESS, J.C. (2006). Diversity and phylogenetic relationships of entomopathogenic nematodes (Steinernematidae and Heterorhabditidae) from the Sky Islands of Southern Arizona. *Journal of Invertebrate Pathology* 92, 66-72.

STOCK, S.P., CAMPBELL, J.F. & NADLER, S.A. (2001). Phylogeny of *Steinernema* Travassos, 1927 (Cephalobina: Steinernematidae) inferred from ribosomal DNA sequences and morphological characters. *Journal of Parasitology* 87, 877-889.

SZALANSKI, A.P., TAYLOR, D.B. & MULLIN, P.G. (2000). Assessing nuclear and mitochondrial DNA sequence variation within *Steinernema* (Rhabditida: Steinernematidae). *Journal of Nematology* 32, 229-233.

VAN MEGEN, H., VAN DER ELSEN, S., HOLTHUIS, B.V., KARSSEN, G., MOOYMAN, P., BONGERS, T., HOLOVACHOV, O., BAKKER, J. & HELDER, J. (2009). A phylogenetic tree of nematodes based on about 1200 full-length small subunit ribosomal DNA sequences. *Nematology* 11, 927-950.

YU, Y., HARRIS, A.J., BLAIR, C. & HE, X.J. (2015). RASP (Reconstruct Ancestral State in Phylogenies): a tool for historical biogeography. *Molecular Phylogenetics and Evolution* 87, 46-49.

YUSHIN, V.V., YOSHIDA, M. & SPIRIDONOV, S.E. (2007). Riders on the sperm: sperm dimorphism and spermatozeugmata in nematodes from the genus *Steinernema* (Rhabditida: Steinernematidae). *Nematology* 9, 61-75

ZHANG, K.Y., LIU, X.H., TAN, J., WANG, Y., QIAO, L., YEDID, G., DAI, C.S., QIU, R.L., YAN, X.W., TAN, H.W. ET AL. (2012). *Heterorhabditidoides rugaoensis* n. sp. (Rhabditida: Rhabditidae), a novel highly pathogenic entomopathogenic nematode member of Rhabditidae. *Journal of Nematology* 44, 348-360.

Index